COMPOUND SEMICONDUCTOR INTEGRATED CIRCUITS

SELECTED TOPICS IN ELECTRONICS AND SYSTEMS

Editor-in-Chief: **M. S. Shur**

Published

Vol. 10: Emerging Optoelectronic Technologies and Applications
ed. *Y.-H. Lo*

Vol. 11: High Speed Semiconductor Lasers
ed. *S. A. Gurevich*

Vol. 12: Current Research on Optical Materials, Devices and Systems in Taiwan
eds. *S. Chi and T. P. Lee*

Vol. 13: High Speed Circuits for Lightwave Communications
ed. *K.-C. Wang*

Vol. 14: Quantum-Based Electronics and Devices
eds. *M. Dutta and M. A. Stroscio*

Vol. 15: Silicon and Beyond
eds. *M. S. Shur and T. A. Fjeldly*

Vol. 16: Advances in Semiconductor Lasers and Applications to Optoelectronics
eds. *M. Dutta and M. A. Stroscio*

Vol. 17: Frontiers in Electronics: From Materials to Systems
eds. *Y. S. Park, S. Luryi, M. S. Shur, J. M. Xu and A. Zaslavsky*

Vol. 18: Sensitive Skin
eds. *V. Lumelsky, M. S. Shur and S. Wagner*

Vol. 19: Advances in Surface Acoustic Wave Technology, Systems and Applications (Two volumes), volume 1
eds. *C. C. W. Ruppel and T. A. Fjeldly*

Vol. 20: Advances in Surface Acoustic Wave Technology, Systems and Applications (Two volumes), volume 2
eds. *C. C. W. Ruppel and T. A. Fjeldly*

Vol. 21: High Speed Integrated Circuit Technology, Towards 100 GHz Logic
ed. *M. Rodwell*

Vol. 22: Topics in High Field Transport in Semiconductors
eds. *K. F. Brennan and P. P. Ruden*

Vol. 23: Oxide Reliability: A Summary of Silicon Oxide Wearout, Breakdown, and Reliability
ed. *D. J. Dumin*

Vol. 24: CMOS RF Modeling, Characterization and Applications
eds. *M. J. Deen and T. A. Fjeldly*

Vol. 25: Quantum Dots
eds. *E. Borovitskaya and M. S. Shur*

Vol. 26: Frontiers in Electronics: Future Chips
eds. *Y. S. Park, M. S. Shur and W. Tang*

Vol. 27: Intersubband Infrared Photodetectors
ed. *V. Ryzhii*

Vol. 28: Advanced Semiconductor Heterostructures: Novel Devices, Potential Device Applications and Basic Properties
eds. *M. Dutta and M. A. Stroscio*

Selected Topics in Electronics and Systems – Vol. 29

COMPOUND SEMICONDUCTOR INTEGRATED CIRCUITS

Editor

Tho T. Vu

Top-Vu Technology, Inc.

World Scientific

New Jersey • London • Singapore • Hong Kong

Published by

World Scientific Publishing Co. Pte. Ltd.
5 Toh Tuck Link, Singapore 596224
USA office: Suite 202, 1060 Main Street, River Edge, NJ 07661
UK office: 57 Shelton Street, Covent Garden, London WC2H 9HE

British Library Cataloguing-in-Publication Data
A catalogue record for this book is available from the British Library.

COMPOUND SEMICONDUCTOR INTEGRATED CIRCUITS

ISBN 981-238-311-5

Printed in Singapore by Mainland Press

PREFACE

This special issue of the *International Journal of High Speed Electronics and Systems* reviews recent work in *Compound Semiconductor Integrated Circuits*. The special issue will also be reprinted in book form, providing a volume for the selected topics in electronics and systems. There are fourteen invited papers covering a wide range of applications, frequencies and materials. These papers deal with digital, analog, microwave and millimeter wave technologies, devices and integrated circuits (IC's) for wireline fiber-optic lightwave transmissions, and wireless radio-frequency microwave millimeter wave communications. In each case, the market is young and experiencing rapid growth for both commercial and military applications. Many new semiconductor technologies compete for these new markets, leading to a virtual alphabet soup of semiconductor materials described in these papers. This issue also includes three papers focused on radiation effects and reliability in III-V semiconductor electronics, which are useful for references and their future directions. Reliability is covered as well in several papers separately for certain process technology. After the first three general papers, the rest are roughly organized by IC technologies and briefly summarized below.

The first paper provides a tutorial review. Explosion of the Internet multimedia communications has rapidly penetrated the world, which demands the expansion of transmission network capacity. The transmission throughput of the backbone network should grow to a level of tens of terabits per second in the near future. With the emergence of WDM (Wavelength-Division Multiplexing) and OTDM (Optical Time-Division Multiplexing) technologies, transmission throughput has, at the experimental level, exceeded 10 Tbit/s with electrical time-division multiplexing (ETDM) base rates of up to 42.7 Gbit/s. In terms of the system reliability, ease of administration, and cost, advancing the electronic IC's to achieve base rates of 40 Gbit/s and beyond is the most promising way to build practical terabit network systems. The paper by Taiichi Otsuji of Kyushu Institute of Technology, "*Present and Future of High-Speed Compound Semiconductor IC's*" describes state-of-the-art of high-speed electronic device and IC technologies for very high-speed lightwave communication systems. The technology of interest is for over 40-Gbit/s transmitter and receiver operations. SiGe, GaAs and InP based heterostructure transistors, and circuit design for analog/digital/mixed-signal and optoelectronic IC's are reviewed. The speed limiting factors are discussed to address the future trends toward 100 Gbit/s and beyond.

The next paper presents the various process technologies being applied from a fabless design company. Scalable models for both active and passive components are essential for the design of highly integrated fiber-optic physical layer IC's. The paper by Sorin

Voinigescu et al. of Quake Technologies, "*A Comparison of Silicon and III-V Technology Performance and Building Block Implementations for 10Gb/s and 40Gb/s Optical Networking ICs*" focuses on the various technology options available for 10 Gb/s and 40 Gb/s applications, and on how their constituent components are modeled along with the characteristics and requirements of the basic building blocks.

The third paper discusses two new initiatives in microwave and millimeter-wave component technologies from the Defense Advanced Research Projects Agency (DARPA) that the editor and some reviewers found interesting, even though it is not a regular technical paper but the editor would like to make an exception and include it here. To meet the demands of future military and commercial sensors and mobile communication networks, a new paradigm shift in monolithic microwave IC (MMIC) and radio frequency (RF) IC technologies is anticipated in which component performance will be enhanced by new materials, device, and architectural concepts. The paper by Edgar Martinez of DARPA, "*Transforming MMIC*" describes two new DARPA initiatives namely the Wide Bandgap Compound Semiconductor materials and devices to improve the analog performance of future MMIC's by orders of magnitude over current GaAs MMIC technology, and the Intelligent RF Front-end aiming at a new class of highly integrated RF and mixed-signal components that through self-assessment and optimization can adapt to rapid changes in the environment or operational demands.

Papers 4 - 7 describe gallium arsenide (GaAs) metamorphic high electron mobility transistors (MHEMT's), pseudomorphic HEMT's (PHEMT's), and indium phosphide (InP) HEMT's. GaAs MHEMT technology has emerged as an attractive, low cost alternative to InP HEMT's. The strain-induced imperfections caused by high indium content layers on GaAs are eliminated in metamorphic devices by providing a properly grown lattice-matching buffer between the substrate and active device layers. With this limitation overcome, it is now possible to provide a superior performance of InP-based devices with the cost advantages of highly manufacturable 4- and 6-inch GaAs wafers that can easily be integrated on existing GaAs fabrication lines. The paper by Colin Whelan et al. of Raytheon RF Components, "*Metamorphic Low Noise Amplifiers and Optical Components*" reviews device performance, amplifiers, photodiodes and reliability data fabricated with this technology operating from 1 to 100 GHz.

Continuous improvement of high bit rate fiber optic networking up to 40 Gb/s, results in increased demand for III/V IC's offering ever lower current input noise, higher output voltage swing, lower jitter and lower DC power consumption. In their paper, "*GaAs PHEMT Chip Sets and IC Processes for High–End Fiber Optic Applications*," Rémy Leblanc et al. of OMMIC Philips discuss transimpedance amplifiers and laser and modulator drivers using production enhancement and depletion (E/D) 0.2 μm GaAs PHEMT processes. Future generations of products will be implemented using sub 0.1 μm processes based on E/D GaAs MHEMT's and InP HEMT's to maintain a long term performance edge over the silicon counterparts.

The paper by Hisao Shigematsu et al. of Fujitsu, "*Distributed Amplifier for Fiber-Optic Communication Systems*" describes 40 Gb/s $LiNbO_3$ modulator driver and preamplifiers for both the non-return-to-zero and return-to-zero applications using 0.15-μm double-doped AlGaAs/InGaAs/AlGaAs PHEMT's and InP-based HEMT's. Distributed amplifier has a wider bandwidth than a lumped amplifier and is designed with stabilization, negative resistance control, and broadband impedance matching.

40-Gbit/s per channel wavelength division multiplexing system experiments, based on an electrical time division multiplexing, were recently demonstrated. As for 1-Tbit/s transmission, the 40-Gbit/s based system has several advantages such as simplicity of the network management and the reduced node cost, compared to the system based on 2.5-Gbit/s or 10-Gbit/s. The paper by Yasuro Yamane and Koichi Murata of NTT, "*The InP-HEMT IC Technology for 40-Gbit/s Optical Communications*" describes the InP-HEMT circuit design, fabrication and reliability. This technology realizes successfully 40-Gbit/s small scale and medium scale integration circuits with practical performance and yield.

Papers 8 - 10 present silicon germanium (SiGe) and silicon (Si). Silicon-based devices, including the increasingly available SiGe-based devices, are now demonstrating f_T and f_{MAX} values over 200 GHz. These recent advances open the door to a wide range of silicon-based very high frequency, low power and highly integrated solutions. In the paper by Greg Freeman et al. of IBM, "*Integrated SiGe and Si Device Capabilities and Trends for Multi-GigaHertz Applications*," trends in silicon MOS, SiGe HBT, SiGe MODFET and SiGe strained silicon FET's are reported.

Under the auspices of DARPA's Microsystems Technology Office Low Power Electronics Program, the Mayo Foundation Special Purpose Processor Development Group is exploring ways to reduce circuit power consumption for existing military systems, while maintaining or increasing functionality. Applications presently being studied include all-digital radar receivers, electronic warfare receivers, and other types of digital signal processors. One of the IC technologies currently under investigation to support such military systems is the IBM SiGe BiCMOS process. The paper by Karl Fritz et al. of Mayo and IBM, entitled "*High-Speed, Low-Power Digital and Analog Circuits Implemented in IBM SiGe BiCMOS Technology*" deals with design methodology, simulation and test results from demonstration circuits developed for these applications and implemented in the IBM SiGe BiCMOS 5HP (50 GHz f_T HBT's with 0.5 μm CMOS) and 7HP (120 GHz f_T HBT's with 0.18 μm CMOS) technologies.

The paper by Toru Masuda et al. of Hitachi, "*A SiGe HBT IC Chipset for 40-Gb/s Optical Transmission Systems*" describes seven analog and digital IC's using a 0.2-μm self-aligned epitaxial-growth SiGe HBT technology. A common-base input stage is introduced in the transimpedance amplifier to increase the bandwidth. Active load circuits were used to have high gain and wide bandwidth, simultaneously.

Paper 11 discusses gallium nitride (GaN) on silicon. GaN-on-silicon provides a low-cost manufacturable platform that could lead to the commercialization of GaN-based power devices for wireless applications. The paper by S. Manohar et al. of University of California at Davis and Nitronex Corporation, "*Microwave GaN-based Power*

Transistors on Large-Scale Silicon Wafers" presents the development of microwave GaN heterostructure FET's on silicon.

Papers 12 - 14 review radiation effects and reliability. The paper by Todd Weatherford of Naval Postgraduate School, "*Radiation Effects in High Speed III-V Integrated Circuits*" provides a summary of the most common radiation effects, and outlines the past history and future directions of radiation effects in high-speed compound semiconductors. For most applications, III-V IC's have provided sufficient radiation hardness.

In the paper by Brad Weaver et al. of Naval Research Laboratory and Defense Threat Reduction Agency, "*Radiation-Effects in III-V Semiconductor Electronics*," particle irradiation-effects in III-V semiconductor devices and selected circuits are reviewed. Single-event effects are a serious concern for high-speed III-V semiconductor technologies operating in radiation-intense environments.

A thorough understanding of reliability and radiation hardness is required in order to use compound semiconductors in space, or in other environments involving radiation and/or extended temperature operation. The paper by Sammy Kayali and Allan Johnston of Jet Propulsion Laboratory, "*Reliability and Radiation Hardness of Compound Semiconductors*" discusses those issues for several types of compound semiconductors that are of interest for high-performance applications.

As seen from these papers, the task of selecting a process technology for a specific chip implementation can be challenging. There is no single process technology that can meet all of the systems' requirements. Hence as IC people, we do not have much choices except to invest time and effort to keep up with the fast pace of various semiconductor technologies. A long term view is that perhaps, many different IC technologies are needed, and this special issue is a good example of the technology race that has been going on since the early 1980's when the editor's career in microelectronics and integrated circuits was just started that he found fascinating and challenging and it is still true as today. This special issue and book volume will be useful for engineers, scientists, and graduate students working in high-speed electronics and systems in general, and in the area of compound semiconductor IC's in particular. It could be used as a supplementary text for graduate courses in this field as well.

Acknowledgments

The editor would like to thank Professor Michael S. Shur of Rensselaer Polytechnic Institute for inviting him to solicit and edit this special issue and book volume. As editor, he is indebted to all the contributors and reviewers. Their expertise, dedication, and promptness are greatly valued, especially during a global economy down turn in 2002 when the semiconductor and telecom markets reached a very low point. The expert reviewers include Christopher Bozada, Charles F. Campbell, Lewis M. Cohn, Jonathan D. Coker, William H. Davenport, Lester Eastman, Karl E. Fritz, Greg G. Freeman, Barry K. Gilbert, Hoang H. Hoang, Alvin Joseph, Sammy A. Kayali, Kevin Kobayashi, James Lyke, Steve Long, Keith Maile, Pierre Mandeville, Phil F. Marsh, Mounir Meghelli, Mehran Mokhtari, Phillip C. Nguyen, The'-Linh Nguyen, Linh Nuyen, Taiichi Otsuji,

Anh-Vu Pham, Barbara A. Randall, Leonard Reynolds, Marc Rocchi, Hisao Shigematsu, Michael S. Shur, Vladimir Sokolov, Sorin Voinigescu, Loc T. Vu, Jackie C. Vu, Tho T. Vu, Howard Walker, Keh-Chung Wang, Todd R. Weatherford, and Bradley D. Weaver. The editor appreciates the help of Mr. Yeow-Hwa Quek and Mr. Kwang-Wei Tjan extended by World Scientific Publishing Company in the production of this special issue and book volume. He also would like to thank his wife, Ngoc-Chi N. Vu, his three children, and the staff of Top-Vu Technology for their support during the work of this special issue and book volume.

THO TRUONG VU
Chip Design and Communication IC's
Top-Vu Technology, Inc. 2650 – 14th Street NW, Saint Paul, Minnesota 55112, USA
651-633-5925, tho@topvu.com, www.topvu.com

CONTENTS

International Journal of High Speed Electronics and Systems
Vol. 13, No. 1 (2003) 1–25

World Scientific
www.worldscientific.com

PRESENT AND FUTURE OF HIGH-SPEED COMPOUND SEMICONDUCTOR IC's

TAIICHI OTSUJI

Center for Microelectronic Systems, Faculty of Computer Science and Systems Engineering,
Kyushu Institute of Technology
680-4 Kawazu, Iizuka, Fukuoka, 820-8502, Japan

This paper describes state-of-the-art of high-speed electronic device and IC technologies for very high-speed lightwave communications systems. The technology of interest is for over 40-Gbit/s transmitter and receiver operations. Device technology including Si-Ge, GaAs-based, and InP-based heterostructure transistors as well as circuit design technology including analog/digital/mixed-signal and optoelectronic IC's are reviewed. The speed limiting factors are discussed to address the future trends toward 100 Gbit/s and beyond.

***Keywords*:** IC; compound semiconductor; HBT; HEMT; lightwave communication

1. Introduction

Explosion of the Internet multimedia communications has rapidly penetrated over the world, which urgently demands the expansion of transmission network capacity. The transmission throughput for the backbone network must enhance to a level at tens of terabits per second in the very near future. With the emergence of WDM (Wavelength-Division Multiplexing) and OTDM (Optical Time-Division Multiplexing) technologies, transmission throughput has, at the experimental level, exceeded 10 Tbit/s with electrical time-division multiplexing (ETDM) base rates of up to 42.7 Gbit/s.[1, 2] In terms of the system reliability, ease of administration, and cost, advancing the electronic integrated circuits (IC's) to achieve base rates of 40 Gbit/s and beyond is the most promising way to build practical terabit network systems.

Compound semiconductor devices appear to be the most promising for increasing the speed of electronic ICs. In particular, heterojunction bipolar transistors (HBT's) and high-electron mobility transistors (HEMT's) based on InP-based, GaAs-based, and SiGe-based material systems are the most attractive for breaking through the speed limit.[3] On going through the device scaling, the cutoff frequencies of those transistors are now exceeding 200 GHz. At the front line of development, 1-THz extrapolated maximum oscillating frequency has been achieved for an InP-based HBT, while over 400-GHz current-gain-cutoff frequency has been done for an InP-based HEMT.[4,5]

High-speed transistors can now offer near 100-Gbit/s or 100-GHz operation for primitive digital functional ICs.[6,7] As the signal wavelength approaches the physical dimension of circuit size, however, numerous distributed parasitic effects become serious in actual IC design, which in turn critically limits the circuit speed. In such an area, digital IC design can no longer exist in the sense of "digital". Consequently, circuit advances to relax the demand on device speed are very important.

This paper describes the present and future of high-speed electronic device and IC technologies for very high-speed lightwave communications systems. The technology of interest is for over 40-Gbit/s transmitter and receiver operations. The contents of this paper includes (1) Fundamentals of Lightwave Communication ICs, (2) Device Technology, (3) Fundamentals of High-Speed IC Design, (4) Mixed-Signal Design Approach for Ultra-High-Speed IC's, and (5) Future Trends and Technology Tradeoffs.

2. Fundamentals of Lightwave Communication ICs

Fig. 1 shows the basic transmitter and receiver block diagrams. The fundamental components of electronic devices are time-division multiplexer (MUX) ICs, modulator driver ICs, photodiodes, baseband amplifier ICs, decision (DEC) ICs consisting of a D-flip flop (D-FF) for retiming and regeneration, time-division demultiplexer (DEMUX) ICs, exclusive-OR ICs for clock recovery (CR), and frequency divider ICs for sub-harmonic clock generation. Simplified circuit block diagrams for MUX, DEC, CR, and DEMUX are shown in Fig. 2.

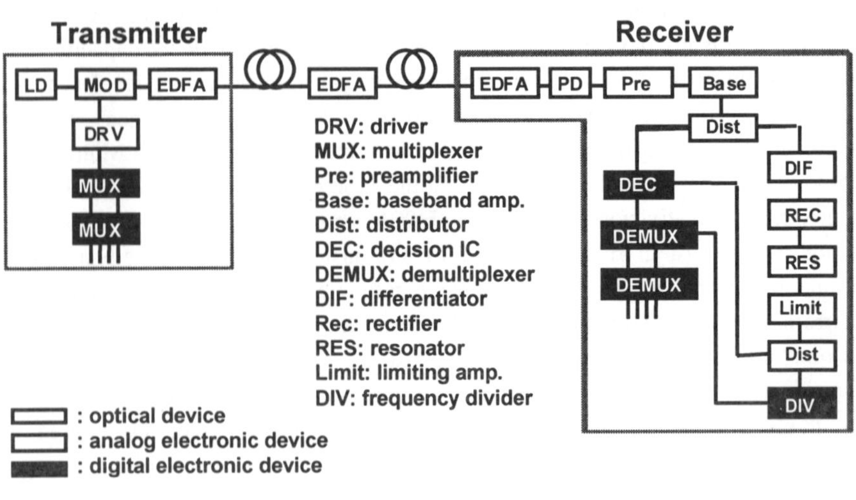

Fig. 1. Basic transmitter and receiver configurations for lightwave communications systems.

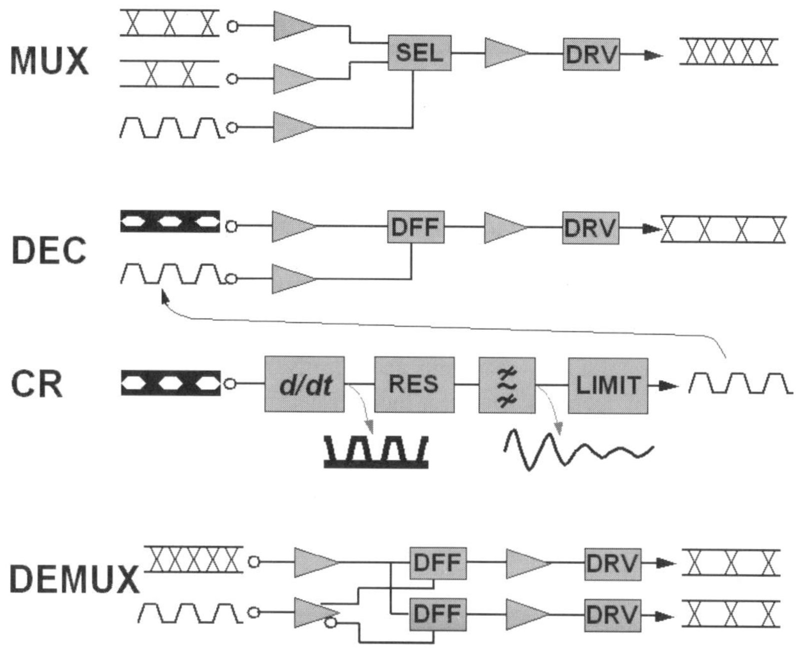

Fig. 2. Simplified circuit block diagrams of MUX, DEC, CR, and DEMUX.

Table 1. Record performance of lightwave communications ICs and their applications to ≥40-Gbit/s ETDM transmission experiments.

IC	FET / HEMT		HBT		
	GaAs	InP	SiGe	GaAs	InP
Multiplexer	45 Gbit/s	90 Gbit/s 70 Gbit/s#	50 Gbit/s	60 Gbit/#	50 Gbit/s
Demultiplexer	40 Gbit/s	50 Gbit/s#	40 Gbit/s	60 Gbit/s#	40 Gbit/s
Decision	40 Gbit/s	52 Gbit/s	40 Gbit/s	40 Gbit/s	46 Gbit/s
Freq. divider	0~29 GHz	5~60 GHz	0~40 GHz	0~67 GHz	0~75 GHz
EX-OR		40 GHz#			
Modulator driver	40 Gbit/s 6 Vpp		50 Gbit/s 4 Vpp	40 Gbit/s 2.5 Vpp	40 Gbit/s 2.3 Vpp
Baseband amp.	56 GHz	90 GHz 58 GHz#	55 GHz	45 GHz	85 GHz
ETDM transmission experiments		42.7Gx30ch. 3x125 km	40Gx273ch. 1x117 km	40Gx32ch. 3x100 km	40Gx82ch. 3x100 km

[#] results for packaged samples, others are on wafer.

For high-speed lightwave communications ICs at the level of 40 Gbit/s, there exist various transistors. In particular, SiGe HBT's and InP-based HBT's and HEMT's potentially offer affordable speed performance at 40 Gbit/s. The InP-based HBT's/HEMT's have another great advantage of optoelectronic circuit integration. The fabricated device/IC performances and results for 40-Gbit/s ETDM transmission experiments are summarized on Table 1.

The D-FF is the critical digital circuit that limits the TDM speed because it must offer clocking operation with feedback action at the highest system data rate. High speed digital ICs including the D-FFs basically consist of emitter-coupled logic (ECL) for bipolar junction transistors (BJT's) or source-coupled FET logic (SCFL) for field-effect transistors (FET's). E. Sano analytically revealed the simplified device figure of merit f_{fom} of operating speed for those circuits;

$$f_{fom} \approx 2\left[\frac{1}{f_T}+\left(\sqrt{\frac{2V_{SW}}{0.15}}+\left(0.5+\frac{V_{SW}}{0.6}\right)\frac{f_T}{f_{\max}}\right)\frac{1}{f_{\max}}\right]^{-1} \quad \text{for BJT's,} \tag{1}$$

$$f_{fom} \approx f_T \quad \text{for FET's,} \tag{2}$$

where f_T, f_{max}, V_{SW} are the current-gain cutoff frequency, the maximum oscillation frequency, and the logic swing.[8] Fig. 3 plots the D-FF speeds vs. f_{fom}. For FET's, the maximum speed of D-FFs stays one fourth to one fifth of f_{fom}, while it stays one third to one fourth of f_{fom} for BJT's. This implies that FET's having f_T of > 200 GHz or BJT's having balanced f_T and f_{max} of >120 GHz are required for 40-Gbit/s D-FF operation. Advanced high-speed circuit design technology should relax the demand on transistor speed.

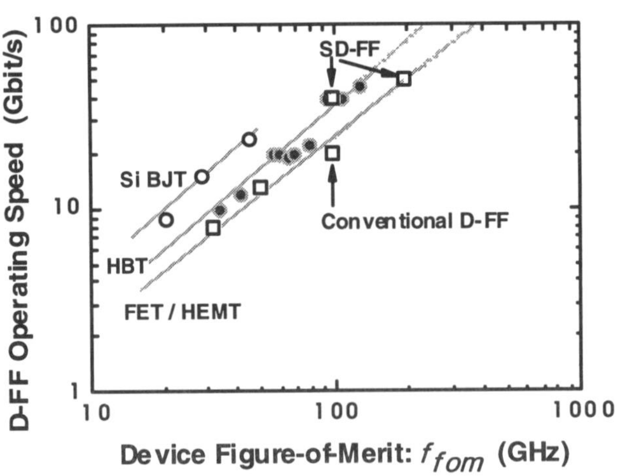

Fig. 3. D-FF speed versus device figure of merit f_{fom}.

3. Device Technology

For high-speed lightwave communications ICs at the level of 40 Gbit/s, there exist numerous transistors.[9] Historically, Si bipolar junction transistors (BJT's) have been continuing the longest effort to improve the speed performance. However, heterostructure and/or compound semiconductor devices including Si-Ge HBT's, GaAs metal-semiconductor FET's (MESFET's), GaAs-based HBT's and HEMT's, and InP-based HBT's and HEMTs can potentially offer much higher speeds of operation than can Si BJT's. The state-of-the-art of these device technologies are reviewed in this section. Typical device structures and band diagrams for HBT's and HEMT's are schematically shown in Fig. 4, which will be referred to in the subsections. Critical issues that limit the performance are discussed.

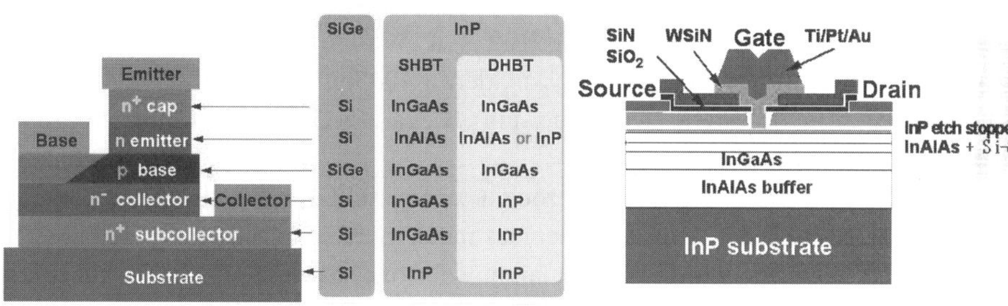

(a)

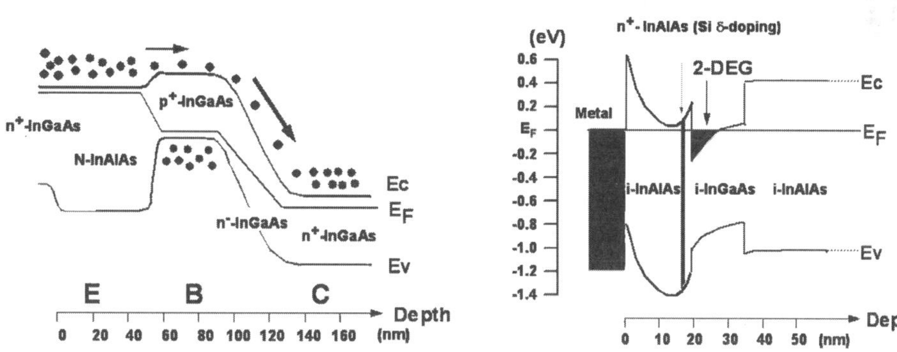

(b)

Fig. 4. Typical device structures (a) and energy band diagrams (b). The left side is for HBTs and the right side for HEMTs.

3. 1. *SiGe HBT's*

Si-Ge HBT's are the Si-based BJT's that replace a p-type Si base layer to a narrow bandgap p-type $Si_{1-x}Ge_x$ base (see Fig. 4). The selective epitaxial growth process stacking the constant Ge content layer followed by the graded Ge content layer toward the emitter can realize an extremely shallow base thickness, resulting in shortening the base transit time. Sophisticated self-aligned techniques are utilized to reduce the base resistance and collector-substrate capacitance. The record f_T and f_{max} of 207 GHz and 285 GHz have been achieved with a 0.12x2.5-μm^2 emitter self-aligned SiGe HBT by the group of IBM, while the superior f_T and f_{max} of 122 GHz and 163 GHz have been done with a 0.2x1.0-μm emitter self-aligned SiGe HBT by the group of Hitachi.[10-12] The speed improvement of beyond 100% from Si BJT's has been obtained at the sacrifice of lowering the breakdown voltage (~2 V), which might be a serious limit for use in modulator driver ICs.

In the early stage, H.M. Rein and co-workers of Ruhr Univ. and Siemens developed a 60-Gbit/s MUX and DEMUX ICs by using 0.3-μm emitter SiGe HBTs.[13] In the last three years, by using 0.2-μm emitter SiGe HBT's, K. Washio and co-workers of Hitachi developed a 67-GHz static 1/4 frequency divider, a 40-Gbit/s decision IC, a 40-Gbit/s 2:1 MUX IC, 1:4 DEMUX with decision IC, a 45-GHz bandwidth, 50-dBW transimpedance amplifier.[11, 12]

Scaled Si-Ge HBT's have now brought over 200-GHz f_T. However, as S. Sabbanna et al. suggested,[14] parasitics becomes much more serious limiting factors than those for III-V devices with semi-insulating substrate because of the Si-substrate conductivity.

3. 2. *GaAs MESFET's*

GaAs MESFET's are one of the most mature, mass-productive compound semiconductor transistors. By shortening the transistor feature size (the gate length) and suppressing the short-channel effects, GaAs MESFET's have been improving their speed. 0.15-μm class MESFET's having an f_T and f_{max} of ~100 GHz and a transconductance (G_m) of ~500 mS/mm were developed by the group of NTT.[15] So far, by using the 0.15-mm GaAs MESFET's, 40-Gbit/s class lightwave communication IC's have been developed by the group of NTT. These include a 20-40-Gbit/s super-dynamic decision IC and a 44-Gbit/s 2:1 MUX IC.[16]

The record f_T of 163 GHz was obtained with a 0.06-μm gate length.[17] Further shrinkage of the FET's gate length to below 0.1 μm, however, causes considerable short channel effect. This means the digital IC speed can no longer improve by further device scaling. Therefore, circuital advances to provide faster operation (as implemented in Ref. 15) and/or introduction of heterostructure to make thinner electron channel beneath the gate are essential to develop ≥40-Gbit/s ICs in the real world.

3. 3. *GaAs-based HBT's and HEMT's*

AlGaAs/GaAs material systems are widely accepted for GaAs-based HBT's, and were the first runner that realized 40-Gbit/s D-flip-flop IC operations.[18] Over the past 15 years, the AlGaAs/GaAs HBT's have improved their speed performance by incorporating smart device structures such as novel quasi-ballistic transport mechanism.[19] T. Ishibashi and co-workers of NTT developed the ballistic collection transistors (BCT's) having f_T and f_{max} of 170 GHz and 190 GHz with 2-μm emitter size at a collector current density (Jc) of 50 kA/cm^2.[20] However, due to serious reliability issues under high Jc, the high-speed performance has not been reflected onto the practical IC productions. K. Runge and co-workers of Rockwell developed 20-40-Gbit/s class analog/digital ICs including a 40-Gbit/s 4:1 MUX IC and a 30-Gbit/s 1:4 DEMUX IC by using a standard 0.13-mm emitter AlGaAs/GaAs HBT's.[21]

AlGaAs/InGaAs HBT's with p+ regrown GaAs extrinsic base layer are the other GaAs-based HBT's. The group of NEC developed a 40-Gbit/s receiver chip set by using 1.6-μm emitter HBT's having a f_T and f_{max} of 110 and 270 GHz.[22]

InGaP/GaAs HBT's have attracted much attention due to their superior speed and reliability performance. The large valence band discontinuity at the InGaP emitter/GaAs base interface improves the carrier injection efficiency compared to the AlGaAs/GaAs interfaces.[23] Significant improvements in the long-term reliability have been reported by replacing the AlGaAs emitter material with InGaP which assured the MTTF of $>10^5$ hours under relatively high Jc of $\gtrsim$50 kA/cm^2.[24, 25] The group of Hitachi developed a 39.5-GHz static frequency divider IC by using 0.5-μm emitter InGaP/GaAs HBT's having an f_T and f_{max} of 156 and 255 GHz.[26] The group of Yokogawa developed 50-Gbit/s modulator driver and 2:1 multiplexer ICs based on InGaP/InGaAs HBT's having an f_T and f_{max} of around 170 GHz.[27]

HEMT's are the heterostructure FET's having a two-dimensional electron gas (2-DEG) channel in an intrinsic narrow-bandgap layer in between wide-bandgap carrier-supplying buffer layers (see Fig. 4).[28] Because of their excellent electron transport and carrier confinement property of the 2-DEG, higher speed performance with less short channel effect can be achieved than comparable normal FET's. Typical GaAs-based HEMT is the AlGaAs/InGaAs/GaAs pseudomorphic HEMT (p-HEMT).

The group of Fraunhofer Institute developed a 0.15-μm gate-length p-HEMT having an f_T, f_{max} and G_m of 100 GHz, 180 GHz and 800 mS/mm, respectively.[29] For use in digital IC applications, 0.2-μm gate AlGaAs/GaAs/AlGaAs p-HEMT's have been utilized. They exhibited typical f_T and G_m of 60 GHz and 600 mS/mm providing a 45-Gbit/s 2:1 multiplexer IC, 40-Gbit/s, 2.5-$V_{p\text{-}p}$ modulator driver IC, 40-Gbit/s 1:4 demultiplexer IC and various 20-30-Gbit/s class analog/digital receiver ICs.[30, 31] On the other hand, the group of Fujitsu recently developed a record 54-GHz bandwidth, 6-$V_{p\text{-}p}$ distributed amplifier IC for a 40-Gbit/s modulator driver by using 0.15-μm gate-length p-HEMT's having an f_T and G_m of 90 GHz and 525 mS/mm.[32]

3. 4. *InP-based HBT's*

InP-based HBT's and HEMT's can potentially offer the highest speeds over currently available transistors, with an excellent f_{max} of 1000 GHz and f_T of 450 GHz, those of which are more than three times and twice as high as those of GaAs devices, respectively.[4, 5] In the early stage, the group of NTT improved the speeds of InP-based HBT's by incorporating the self-aligned InP/InGaAs composite collector structure with a hexagonal emitter geometry providing an f_T and f_{max} of 228 GHz and 227 GHz, respectively.[33] By using an improved reliable carbon-doped HBT's having relative low f_T and f_{max} of 116 and 169 GHz, the NTT group has developed various high-speed digital ICs including 40-Gbit/s decision IC and 1:4 demultiplexer IC.[34-36]

By using the novel substrate transfer process that allows very narrow collector stripes, M. Rodwell and co-workers of UCSB raised the f_{max} value to 1.08 THz.[4] The transferred-substrate HBT's were used to realize various over 40-Gbit/s analog/digital ICs including a 85-GHz distributed amplifier IC and a record 75-GHz static frequency divider IC.[37, 38] The key technology for those HBT's is the device structure that can effectively reduce the parasitic base resistance and the base-collector capacitance, which directly reflects on improvement of the f_{max}.

The production-level AlInAs/InGaAs single-HBT's have also gained their speed performance at the level of 130~250-GHz f_T and f_{max}.[39] They produced a 72.8-GHz 1/2 static frequency divider IC by the group of HRL and a 50-Gbit/s 4:1 MUX IC by the group of Bell Labs. and HRL.[7, 40] The group of OPTO+ developed a 40-Gbit/s 2:1 MUX IC and master-slave D-flip-flop IC using InGaAs/InGaAsP/InP double HBT's having f_T and f_{max} of up to 160 and 220 GHz.[41, 42] The group of UCSB and IQE has recently yielded the fastest production-level DHBT's exhibiting f_T and f_{max} of 282 GHz and ≥450 GHz.[43]

3. 5. *InP-based HEMT's*

Over the past ten years, HEMT's consisting of the InAlAs/InGaAs modulation-doped structure on InP substrates have become not only the fastest three-terminal devices, but also uniform devices that can be applied to wide variety of ICs for millimeter-wave and optical fiber communication systems. Compared to AlGaAs/InGaAs/GaAs p-HEMT's, InAlAs/InGaAs material systems have larger conduction-band discontinuity resulting in higher carrier confinement in the InGaAs channel (see Fig. 4).[44] This helps suppress the short-channel effect even for sub-100-nm gate-length devices.

T. Enoki and co-workers of NTT devised a highly-uniform InAlAs/InGaAs/InP HEMT structure,[45] which brought the big chance to integrate complicated digital functions onto an InP-based HEMT chip. The electron channel is formed with a 15-nm-thick InGaAs layer beneath a T-shaped gate with a 0.1-μm footprint. A novel InP gate-recess-etch stopper inserted into the InAlAs barrier layer dramatically improves the uniformity of transistor performance (the average threshold voltage (Vth): -0.65 V has a standard deviation of less than 40 mV over a 3-inch wafer), which is essential for the large-sale integration of digital logic circuits. For 0-1-μm gate HEMTs, the average transconductance (gm), f_T, and f_{max} are 1050 mS/mm, 195 GHz, and 230 GHz, respectively. Recently shortening the gate length has proceeded by the groups of NTT and Fujitsu.[5, 46, 47] The record f_T of 472 GHz has been realized with a 30-nm-gate length by K. Shinohara and co-workers of Fujitsu.[5]

By using the 0.1-μm gate HEMT's, the group of NTT developed 40-80-Gbit/s class various transmitter and receiver ICs, which included an 80-Gbit/s 2:1 MUX IC, a 10-dB, 90-GHz baseband amplifier IC, a 50-Gbit/s super-dynamic decision IC, a 50-Gbit/s 1:2 DEMUX IC, and a 43-Gbit/s clock and data recovery IC.[48-52] Very recently, the group of Fujitsu developed a record 90-Gbit/s 2:1 multiplexer IC, a 43-Gbit/s 4:1 multiplexer IC containing a 52-Gbit/s static D-flip-flop by using 0.13-μm gate HEMT's.[53, 54]

3. 6. *Optoelectronic ICs*

Photo receivers that monolithically integrate a pin-photodiode and a transimpedance amplifier are the fundamental optoelectronic IC's (OEIC's) for lightwave communications IC's. So far, over 40-GHz bandwidth photoreceiver OEIC's have been developed by several research groups using InP/InGaAs pin-PD-DHBTs, GaAs pin-PD-PHEMTs, InP/InGaAs pin-PD-HBTs, and InAlAs/InGaAs/InP waveguide-pin-PD HEMTs.[55-58] Big advantage of those OEIC's is the excellent interconnection-dependent parasitic-free characteristics which becomes very difficult to achieve in the high-speed region of 40-Gbit/s and beyond.

Recently, a new type of wide-band, high-output-saturation characteristic InP/InGaAs photodiodes, called a uni-traveling-carrier photodiode (UTC-PD), was realized by T. Ishibashi and co-workers of NTT.[59] Its extremely high output saturation voltage can eliminate the transimpedance and/or baseband amplifiers and can directly drive the decision ICs under the 50-Ω load impedance condition. Such an optical receiver was monolithically integrated using a UTC-PD and a HEMT super-dynamic decision IC, and error-free 40-Gbit/s return-to-zero-mode receiver operation was successfully demonstrated.[60]

4. Fundamentals of High-Speed Digital IC Design

A typical flowchart for high-speed digital IC design is shown in Fig. 5. It consists of model parameters extraction for active/passive circuit elements, circuit simulation, layout design, modeling parasitic elements, and feedback to the circuit design. For digital IC design, HSPICE™ simulator is frequently utilized. The accuracy of the simulation is dominantly relying upon the accuracy of the circuit model parameters.

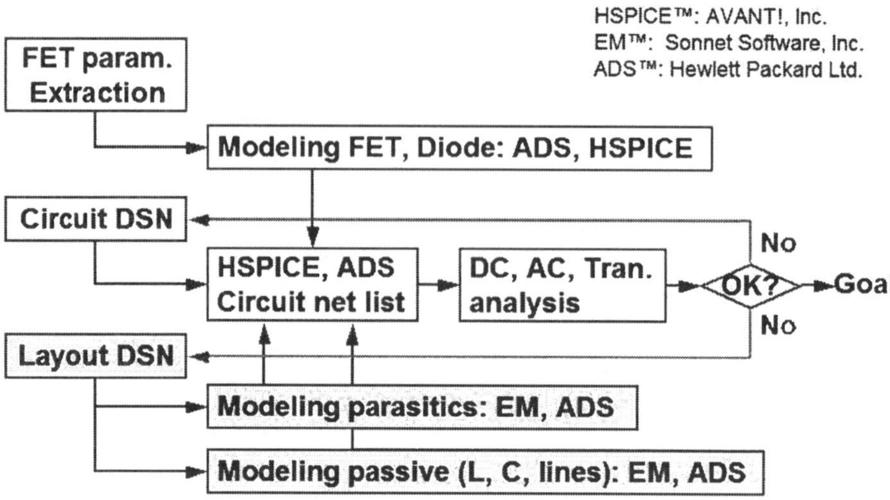

Fig. 5. Design flowchart of high-speed digital ICs.

The large signal equivalent circuit for a transistor includes more than 30 DC/AC parameters. The HSPICE™ circuit simulator and the ADS™ microwave design simulator are utilized for the parameter extraction. All of them are satisfactorily extracted according to the following well-considered but complicated procedure. (i) Optimizing the bias-independent DC parameters, (ii) optimizing the large-signal DC parameters (the gate diode, the drain current as functions of biases), (iii) optimizing the small-signal AC parameters at each bias points, and (iv) optimizing the large-signal AC parameters (bias-dependence of junction capacitances). In (i), (iii), and (iv), the parameters are optimized by fitting the simulated S parameters of the equivalent circuit to the measured ones.

The most important passive elements to be considered are interconnection lines and inductances. The simple way for those expressions is to obtain their frequency responses as sets of S parameters. However, they do not work in HSPICE™, so that they are to be modeled as the L-C-R networks by fitting their frequency responses to those obtained from the electromagnetic simulation for modeled structures. All the critical paths should be modeled.

According to the procedure mentioned above, a super-dynamic decision IC was designed and fabricated at NTT.[61, 62] Typical measured and simulated waveforms for PN-7 stage pseudorandom data input are shown in Fig. 6. At 40 Gbit/s in (a), the eye opening is severely degraded due to inter-bit distortion. When the data rate goes down to 36 Gbit/s in (b), the jitter is sufficiently suppressed. The simulated results perfectly express the reality.

There are important issues for modeling the HBT's. For accurate modeling, we must consider the following physical phenomena. Standard BJT models in HSPICE™ have some limitations: (i) high-injection effect which includes complicated mixture of

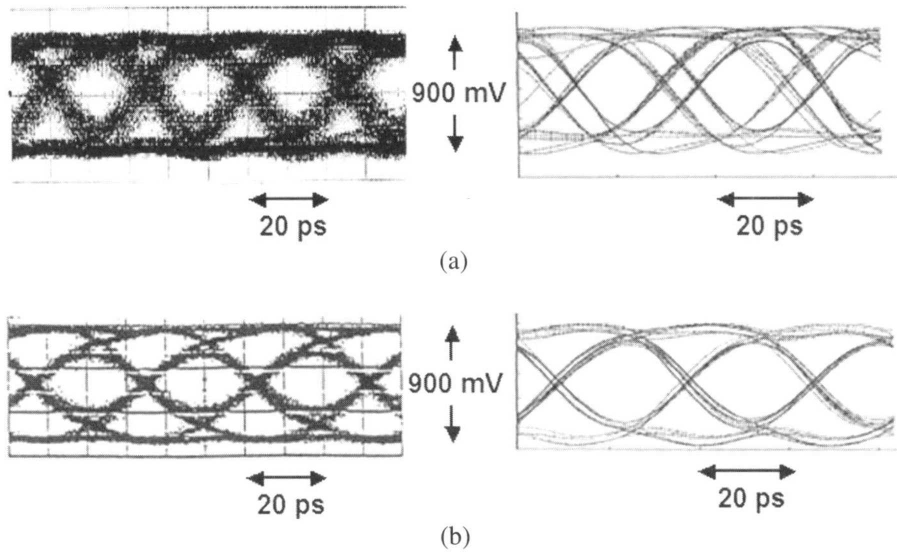

Fig. 6. Output waveforms of a decision IC for PN-7 pseudorandom data pattern input. Left: measured on wafer, right: simulated. (a) at 40 Gbit/s, (b) at 36 Gbit/s.

Kirk effect, emitter-crowding effect and thermal effect, (ii) the velocity overshoot, significantly contributing to the slow collector transit time, (iii) thinner collector structure making the bias dependence of the base-collector capacitance C_{bc} to be non-uniform. Designers, therefore, should take the base-collector bias (V_{bc}) dependence of C_{bc} into account. Large variation in C_{bc} at the low reverse biased V_{bc} region would cause dispersive switching, resulting in sever timing jitter.

5. Mixed Signal Issues in High Speed IC Design

As data rates increase to/beyond 40 Gbit/s, the wavelength of the signal approaches the physical size of IC chips, which gives rise to substantial difficulties in designing digital ICs: (i) numerous parasitics; (ii) cavity resonance; (iii) bandwidth limitation due to lumped-circuit treatment, and (iv) ultra-broadband operation from near-DC to the maximum bit rate. In such an area, digital IC design can no longer exist in the sense of "digital" but in the "mixed signals." Key technologies in high-speed IC design are (1) optimizing the process design, (2) minimizing the parasitic capacitance by optimizing operating bias conditions and/or by introducing distributed treatment, and (3) minimizing the effective logic swing to accelerate transitions by introducing dynamic operation. (1) is a process design issue, while (2) and (3) are the circuit design issues.

5. 1. *Distributed Design Approach*

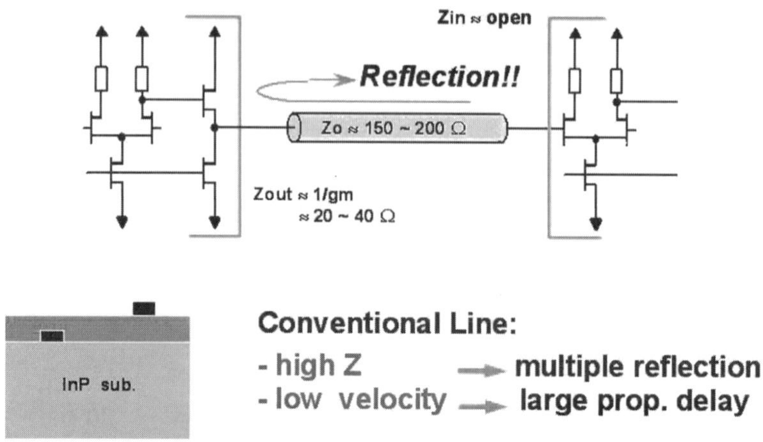

Fig. 7. Conventional interconnection with striplines: ciruit diagram and cross sectional structure.

Very high-speed digital ICs are generally based on the SCFL or ECL series-gated circuitry. The input node is the high impedance (in the order of tens of kΩ) gate/base, while the output is the low impedance ($\approx 1/(g_m+g_{ds})$, g_m: the transconductance, g_{ds}: the drain conductance) source/emitter follower. When the line length approaches one tenth of the signal wavelength (≈280 μm for ε_r of 7.0 at 40 GHz), the mismatch between the line impedance (Z_L) and the output impedance causes severe multiple reflections.

As shown in Fig. 7, single- or double-railed strip lines (SL's) are generally used for the digital IC layout. However, Z_L of the SL is relatively high (≈ 200 Ω). The double-railed SL's help reduce the Z_L, but still larger than the SCFL output impedance, and largely varies with surrounding environments because of the lack of steady ground. Therefore, severe distortion occurs and degrades speed performance.

A 2nd-layer microstrip line (MSL) with the 1st-layer ground, shown in Fig. 8, is best suited for high-speed, low-impedance signal transmission on ICs. The Z_L of the MSL can be lowered so as to match the SCFL output impedance (20~40 Ω). Its structure isolating from the substrate makes the effective dielectric constant be very low (≈2.6), and increase the propagation velocity by 50%. This makes for better matching to the output impedances of the SCFL's. Good example is the record 80-Gbit/s multiplexer IC[48] as shown in Fig. 9. A high-speed, impedance-matched transmission design was introduced for inter-cell connection of critical paths of 400-μm long. This drastically reduced the waveform distortion due to multiple reflections resulting in a speed improvement of 30% over the conventional lumped circuit design.

In case of ECL's, the output impedance is extremely low (< 10 Ω), so that you may choose another option where the inter-cell connection is made in between the collector node of the emitter-coupled differential pair and the emitter-follower input. Z_L, thus, becomes practically high so as to match the load impedance Z_{OUT} (see Fig. 10).

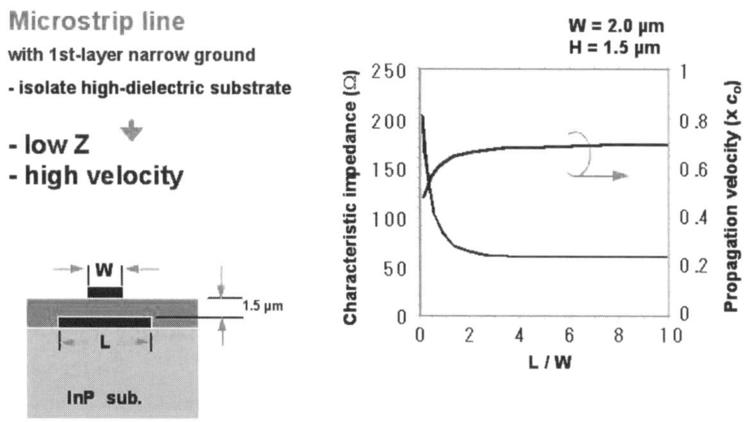

Fig. 8. Impedance-controlled microstrip lines: cross sectional structure and electrical property.

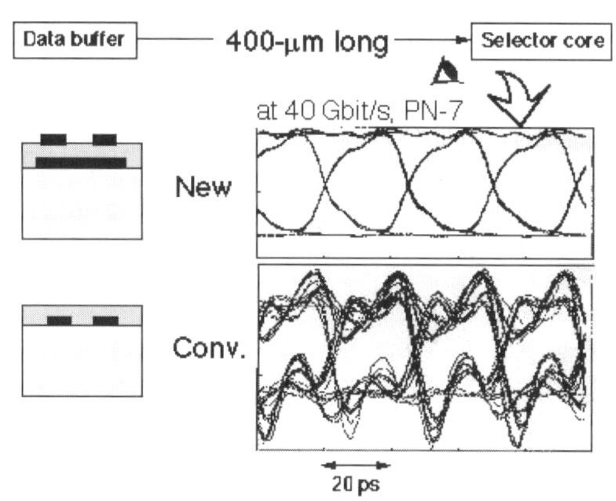

Fig. 9. Simulated 40-Gbit/s eye diagrams at the selector core input, upper: the MSL interconnection, lower: the conventional SL interconnection. (after Ref. 48.)

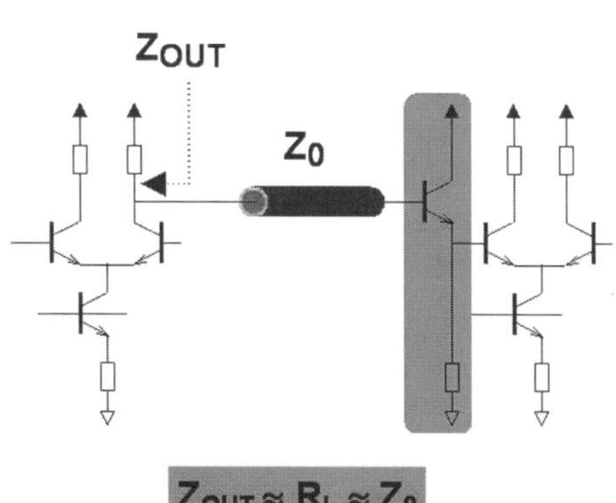

Fig. 10. Another option of impedance-matched interconnection for ECL circuits.

5. 2. *Dynamic Response Design Approach*

This approach is unique for digital IC designs based on large-signal, time-domain response. A typical example is the super-dynamic D-FF (see Fig. 11).[61, 62] The circuit features (i) a smaller latching current than the reading current, and (ii) a source-coupled negative feedback pair, which drastically reduces the effective logic swing with keeping fast signal transition slew rate. The circuit can operate about twice as fast as the conventional master-slave D-FF's can. Up to 50-Gbit/s DEC operation has been achieved using 0.1-μm InP-based HEMTs.[63] The basic idea of dynamical logic swing shrinkage to accelerate the switching speed was initially implemented on a D-FF by K. Murata et al. of NTT, and K. Ishii et al. of NTT.[64, 65]. So far various types of D-FF circuits have been proposed in the similar design manner.[11, 66, 67]

The key ideas in designing DEMUX's are (i) wider clock phase margin and (ii) higher input sensitivity, which is owing to the performance of the first-stage D-FF and to the clock-skew distribution. All the design techniques mentioned above reflect them.

5. 3. *Technology Tradeoffs*

In this subsection, the technology tradeoffs among the above mentioned transistor types

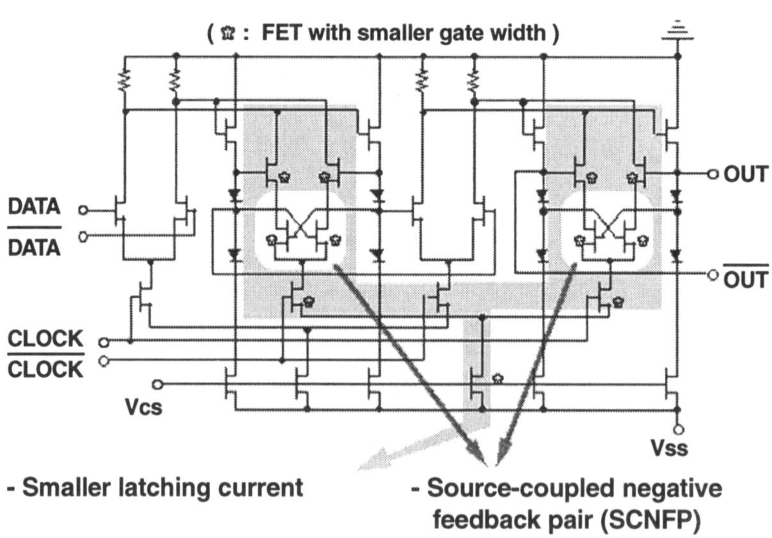

Fig. 11. Super-dynamic D-FF. (after Ref. 61.)

(HBT's vs. HEMT's) are to be discussed from the viewpoint of high-speed, low-power, digital circuit integration. Sokolich of HRL presented the power efficiency of the flip-flop speed for various process technologies,[68] which is shown in Fig. 12 with additional results for NTT's InP-based HEMT IC's. Obviously, HBT's can make lower power operation by almost one order of magnitude than the HEMT's.

According to the Sano's analytical expression for the device figure of merit f_{fom} (1) and (2), it is noted that the ECL's can accelerate the switching speed by reducing the logic swing, but the SCFL's cannot. This advantage for ECL's is mainly due to the rapid increase in f_T (or G_m) with increase in the base bias voltage, resulting in low-power and high-speed operation. As a result, in terms of the speed versus logic swing, HBT's can run faster with lower swing down to 300 or 250 mV, but HEMT's are very hard. As mentioned in Section 4, however, designers should take the V_{bc} dependence of the collector-base capacitance C_{bc} into account when shrinking the logic swing.

In terms of large-scale integration, HBT's are limited by thermal density, while HEMT's are limited by actual physical density. Thus, HBT IC's must be carefully integrated. In terms of clear eye opening, HBT's need extra care for the complexity of carrier motions. HEMT's have a big advantage of easy design due to their simple unipolar actions. One straightforward consequence may suggests that HEMT IC's are better suited for transmitter IC design, while HBT's are suited for receiver IC design.

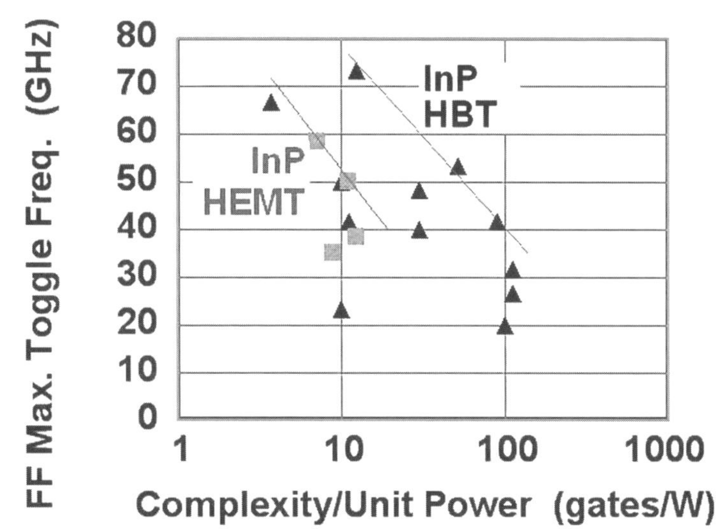

Fig. 12. Power efficiency of the flip-flop speed for various transistors. ■: for HEMTs, ▲: for HBTs. (add data to Ref. 68.)

6. Future Trends and Subjects

6. 1. *Device Scaling*

"How much faster can transistors and ICs operate?" For scaled Si-Ge HBT's, superior speed performance of over 200-GHz f_T and f_{max} has already been realized with an ultra-thin (≈10 nm) base layer.[10] As mentioned in 4.1, the device scaling sacrifices the breakdown property. Fig. 13 plots the breakdown voltages versus current-gain cutoff frequencies f_T for various transistors. Compared to the SiGe HBT's, III-V compound devices, in particular for InP-based DHBT's, have superior breakdown property. This weak breakdown property becomes a critical limiting factor for the speed improvement.

Another problem for SiGe HBT's is the parasitic effects due to the Si-substrate conductivity, which becomes much more serious than those for III-V devices with semi-insulating substrates. Therefore, reduction of the substrate effects becomes a critical issue. The transferred substrate or SOI (silicon on insulator) structure will be a solution.[14]

As to the transferred substrate InP-based HBT's, the f_{max} values can directly be improved, but the structure does not affect to the base transit time and the base-emitter capacitance. This means the f_T values cannot be improved. As mentioned in Section 2, in order to efficiently improve the IC speed, increase in f_T value is also necessary.

T. Enoki predicted that the speed performance limit of scaled InP-based HEMT's exists around 700 GHz due to tunneling leakage current and parasitic capacitance.[69] M. Matloubian et al. of HRL have proposed antimonide-based wider bandgap, larger conduction-band discontinuity material systems like InGaAs/AlGaPSb lattice-matched

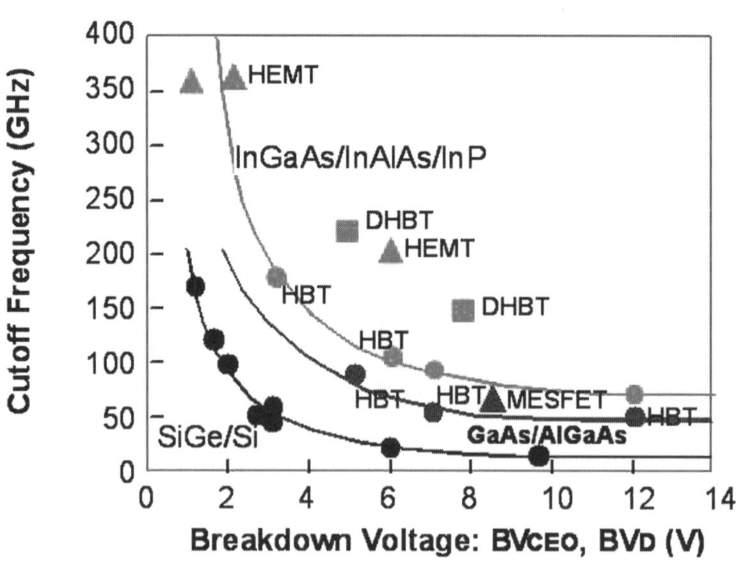

Fig. 13. Transistor cutoff frequency versus breakdown voltage for various transistors.

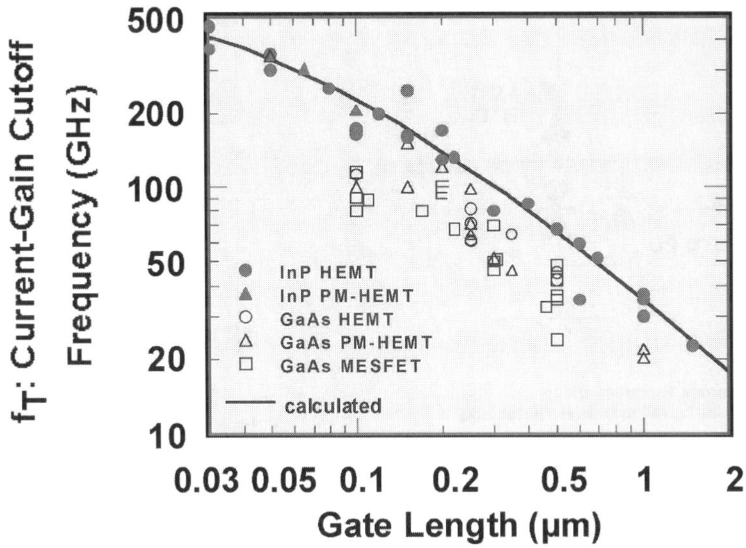

Fig. 14. Transistor current-gain cutoff frequency versus gate length for FET's and HEMT's. (add data to Ref. 69.)

to an InP substrate, which may realize terahertz operation for HEMT's with a high breakdown voltage.[70] C.R. Bolognesi et al. of Simon Fraser Univ. introduced a GaAsSb material for the base layer in place of InAlAs for InP-based DHBT's, exhibiting over 200-GHz f_T and f_{mac} with a 1-μm emitter width.[71] In both cases for HEMT's and HBT's, larger conduction or valence band discontinuity can make higher carrier confinement, resulting in higher speed operation for scaled devices. In a couple of years, antimonide-based compound semiconductors (ABCS) have extensively been investigated as the key material systems of the next generation of higher speed transistors.[72]

6. 2. *New Aspects of Optoelectronic IC Design Approach*

The above discussions lead to limiting the cutoff frequencies of InP-based HBT's and HEMT's at around terahertz. The trends also show that a scaled InP-based HBT's or HEMT's can potentially offer 100-Gbit/s static D-FF operation and 200-Gbit/s MUX operation. Is it feasible? In such an ultra-high-speed region, however, numerous parasitic effects like interconnection delay critically limit the circuit speed. In addition, conventional electrical I/O interfaces do not work anymore because allowable interconnection lengths for propagating 160-Gbit/s data via metallic coaxial cables would be as short as 10 cm at best.

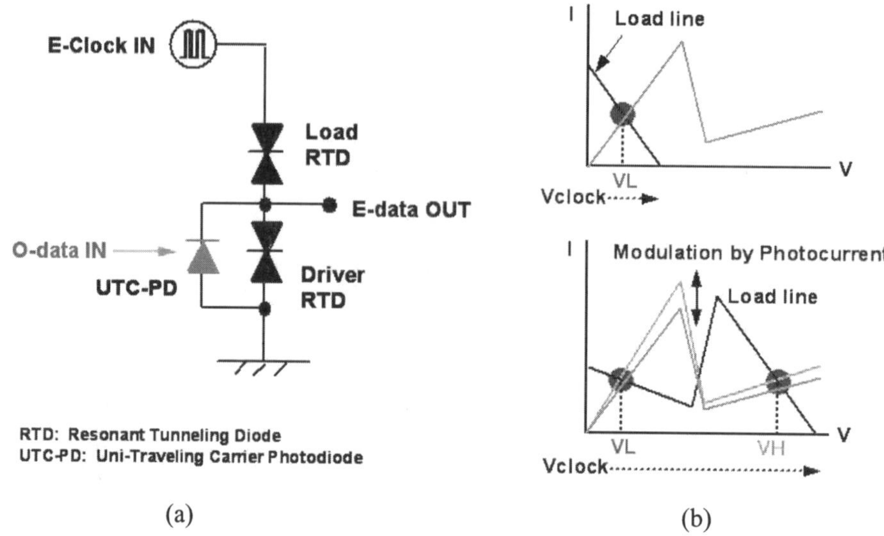

Fig. 15. Optoelectronic D-FF consisting of a MOBILE (two resonant tunneling diodes) and a uni-traveling-carrier photodiode. (a) circuit diagram, (b) operation principle, upper: when the clock voltage is low, lower: when the clock voltage is high. (after Ref. 74)

One direction is minimizing parasitics and power consumption to make sense of lumped circuit designs. Optoelectronic integration with highly functional devices will be a solution. A resonant tunneling diode (RTD) is a highly functional quantum effect device that exhibits a negative differential resistance yielding nonlinear, bistable operation. K. Maezawa devised monostable-bistable transition logic elements (MOBILEs) consisting of a serial connection of two RTD's, one for a load and the other for a driver (see Fig. 15).[73] The MOBILE can perform data latching operation equivalent to the D-FF function. The logic state can be controlled by modulating the drive current of the driver RTD. K. Sano introduced an optoelectronic scheme for modulating the driver, which is the monolithic integration of only one UTC-PD and a MOBILE on an InP substrate (see Fig. 15). This OEIC has demonstrated the direct demultiplexing of optical 80-Gbit/s data to electrical 40-Gbit/s data and decision operation for an 80-Gbit/s RZ data.[74, 75] An optoelectronic injection-locked oscillator integrating one UTC-PD and one RTD successfully extracted a sub-harmonic 11-GHz electrical clock from an optical 44-Gbit/s data input.[76]

As an alternative approach, pursuing traveling-wave or distributed circuit designs should also be respected. One unique concept transforms a sequential (feedback) FF circuit into an equivalent feed-forward one by making full use of distributed design topology and the management of the propagation modes of electromagnetic waves.[77] The speed limit of this type of the circuit has been estimated to far exceed 100 Gbit/s.

7. Conclusions

Recent advances in high-speed electronic device and IC technologies for very high-speed lightwave communications systems were reviewed. Now 40-Gbit/s ETDM field tests have been performed in Europe, USA, and Japan using SiGe HBT's, InP HBT's, and InP HEMT's. Si-Ge HBT's are attractive for mass-productive, low-power, highly-integrated applications at around 40 Gbit/s. Scaled InP-based HBT's and HEMT's can potentially offer the highest speed performance of all the transistors so that 100-Gbit/s logic ICs will be realized in the very near future. New material systems including antimonide-based compound semiconductors would enable further extension of the transistor speed limit. There exists numerous parasitics both inside and outside the chips, which critically limits the IC speed. Smart optoelectronic integration with highly functional devices will be a possible solution for over-100-Gbit/s world.

Acknowledgments

The author's contributions to the real IC fabrications were devoted when he was in NTT Laboratories. The author would like to acknowledge all the researchers who were involved in this work for their extensive contributions and valuable discussion.

References

1. K. Fukuchi, K. Kasamatsu, M. Morie, R. Ohhira, T. Ito, K. Sekiya, D. Ogawahara and T. Ono, "10.92-Tb/s (273 x 40-Gb/s) triple-band/ultra-dense WDM optical-repeated transmission experiment," in ***Tech. Dig. Opt. Fiber Comm. Conf.***, 2001, PD24-1-4.
2. S. Bigo, Y. Fregnac, G. Chorlet, S. Borne, P. Tran, C. Simonneou, D. Bayart, A. Joudan, J.-P. Hamaide, W. Idler, R. Dischler, G. Veith, H. Gross and W. Poehlmannk, "10.2 Tbit/s (256 x 42.7 Gbit/s PDM/WDM) transmission over 100 km TelaLight fiber with 1.28 bit/s/Hz spectral efficiency," in ***Tech. Dig. Opt. Fiber Comm. Conf.***, 2001, PD25-1-4.
3. M. Rodwell ed., ***Selected Topics in Electronics and Systems Vol. 21, High Speed Integrated Circuit Technology, Towards 100 GHz Logic***, World Scientific Publishing, Singapore, 2001.
4. Q. Lee, S.C. Martin, D. Mensa, R.P. Smith, J. Guthrie, S. Jaganathan, Y. Betser, T. Mathew, S. Krishnan, L. Samoska and M.J.W. Rodwell, " Submicron transferred-substrate heterojunction bipolar transistors with greater than 1 THz f_{max}," in *Postdeadline Paper* ***IEEE Dev. Res. Conf.***, 1999.
5. K. Shinohara, Y. Yamashita, A. Endoh, K. Hikosaka, T. Matsui, T. Mimura and S. Hiyamizu, "Extremely high-speed lattice-matched InGaAs/InAlAs high electron

mobility transistors with 472 GHz cutoff frequency," ***Jap. J. Appl. Phys.*** *Part 2-Lett.* **41** (2002) L437-L439.

6. T. Suzuki, Y. Nakasha, T. Takahashi, K. Makiyama, K. Imanishi, T. Hirose and Y. Watanabe, "A 90-Gb/s 2:1 multiplexer IC in InP-based HEMT technology," in ***Dig. Tech. Papers IEEE Int. Solid-State Circuits Conf.***, 2002, 11.7.
7. M. Sokolich, C. Fields, B. Shi, Y.K. Brown, M. Montes, R. Martinez, A.R. Kramer S. Thomas III and M. Madhav, "A low power 72.8 GHz static frequency divider implemented in AlInAs/InGaAs HBT IC technology," in ***Tech. Dig. IEEE GaAs IC Sym.***, 2000, pp. E.11.
8. E. Sano, Y. Matsuoka and T. Ishibashi, "Device figure-of-merit for high-speed digital ICs and baseband amplifiers," ***IEICE Trans. Electron.*** **E78-C** (1995) 1182-1188.
9. Keh-Chung Wang ed., ***Selected Topics in Electronics and Systems Vol. 13, High-Speed Circuits for Lightwave Communications***, World Scientific Publishing, Singapore, 1999.
10. B. Jagannathan, M. Khater, F. Pagette, J.-S. Rieh, D. Angell, H. Chen, J. Florkey, F. Golan, D.R. Greenberg, R. Groves, S.J. Jeng, J. Johnson, E. Mengistu, K.T. Schonenberg, C.M. Schnabel, P. Smith, A. Stricker, D. Ahlgren, G. Freeman, K. Stein and S. Subbanna, "Self-aligned SiGe NPN transistors with 285 GHz f_{max} and 207 GHz f_T in a manufacturable technology," ***IEEE Electron Dev. Lett.*** **23** (2002) 258-260.
11. K. Washio, R. Hayami, E. Ohue, K. Oda, M. Tanabe, H. Shimamoto and M. Kondo, "67-GHz static frequency divider using 0.2-μm self-aligned SiGe HBT," ***IEEE Trans. Microwave Theory and Tech.*** **49** (2001) 3-8.
12. K. Washio, "Self-aligned Si BJT/SiGe HBT technology and its application to high-speed circuits," ***Int. J. High Speed Electron. and Sys.*** **11** (2001) 77-114.
13. H.-M. Rein, "Si and SiGe bipolar ICs for 10 to 40 Gb/s optical-fiber TDM links," ***Int. J. High Speed Electron. and Sys.*** **9** (1998) 347-383.
14. S. Subbanna, J. Johnson, G. Freeman, R. Volant, R. Groves, D. Herman and B. Meyerson, "Prospects of silicon-germanium-based technology for very high-speed circuits," in ***Dig. IEEE MTT-S Int. Microwave Symp. Vol. 1***, 2000, pp. 361-364.
15. K. Nishimura, K. Onodera, S. Aoyama, M. Tokumitsu and K. Yamasaki, "High performance 0.1 μm self-aligned-gate GaAs MESFET technology," in ***Proc. European Solid-State Device Res. Conf. (ESSDERC) '96***, 1996, pp. 865-868.
16. T. Otsuji, K. Murata, K. Narahara, K. Sano and E. Sano, "20-40-Gbit/s-class GaAs MESFET digital ICs for future optical fiber communications systems," ***Int. J. High Speed Electron. and Sys.*** **9** (1998) 399-435.
17. M. Tokumitsu, M. Hirano, T. Otsuji, S. Yamaguchi and K. Yamasaki, "A 0.1-μm self-aligned-gate GaAs MESFET with multilayer interconnection structure for ultra-high-speed ICs," in ***Tech. Dig. Int. Electron Device Meeting (IEDM)***, 1996, pp. 211-214.

18. Y. Kuriyama, T. Sugiyama, S. Hongo, J. Akagi, K. Tsuda, N. Iizuka and M. Obara, "A 40GHz D-type flip-flop using AlGaAs/GaAs HBT's," in ***Tech. Dig. IEEE GaAs IC Symp.***, 1994, pp. 189-192.
19. T. Ishibashi and Y. Yamauchi, "A novel AlGaAs/GaAs HBT structure for near-ballistic collection," in ***Dig. 45th Annual Device Res. Conf.***, 1987, p. IVA-6.
20. T. Ishibashi, Y. Yamauchi, E. Sano, H Nakajima and Y. Matsuoka, "Balistic collection transistors and their applications," ***Int. J. High Speed Electron. and Sys.*** **5** (1994) 349-379.
21. K. Runge, P.J. Zampardi, R.L. Pierson, R. Yu, P.B. Thomas, S.M. Beccue and K.C. Wang, "AlGaAs/GaAs HBT circuits for optical TDM communications," ***Int. J. High Speed Electron. and Sys.*** **9** (1998) 473-503.
22. R. Ohhira, Y. Amamiya, T. Niwa, N. Nagano, T. Takeuchi, C. Kurioka, T. Chuzenji and K. Fukuchi, "AlGaAs/InGaAs HBT IC modules for 40-Gb/s optical receiver," ***IEICE Trans. Electron.*** **E82-C** (1999) 448-455.
23. H. Kroemer, "Heterostructure bipolar-transistors – what should we build," ***J. Vacuum Science and Tech. B*** **1** (1983) 126-130.
24. T.S. Low, C.P. Hutchinson, P.C. Canfield, T.S. Shirley, R.E. Yeats, J.S.C. Chang, G.K. Essilfie, M.K. Culver, W.C. Whiteley, D.C. D'Avanzo, N. Pan, J. Elliot and C. Lutz, "Migration from an AlGaAs to an InGaP emitter HBT IC process for improved reliability," in ***Tech. Dig. IEEE GaAs IC Symp.***, 1998, pp. 153-156.
25. K. Mochizuki, T. Oka, K. Ouchi and T. Tanoue, "Reliability investigation of heavily C-doped InGaP/GaAs HBTs operated under a very high current-density condition," ***Solid-State Electron.*** **43** (1999) 1425-1428.
26. T. Oka, K. Hirata, H. Suzuki, K. Ouchi, H. Uchiyama, T. Taniguchi, K. Mochizuki and T. Nakamura, "High-speed small-scale InGaP/GaAs HBT technology and its application to integrated circuits," ***IEEE Trans. Electron Devices*** **48** (2001) 2625-2630.
27. K. Uchida, H. Matsuura, T. Yakihara, S. Kobayashi, S. Oka, T. Fujita and A. Miura, "A series of InGaP/InGaAs HBT oscillators up to D-band," ***IEEE Trans. Microwave Theory and Tech.*** **49** (2001) 858-865.
28. T. Mimura, S. Hiyamizu, T. Fujii and K. Nanbu, "A new field-effect transistor with selectively doped GaAs–n-$Al_xGa_{1-x}As$ heterojunctions," ***Jpn. J. Appl. Phys.*** **19** (1980) L225-L227.
29. M. Schlechtweg, W.H. Haydl, A. Bangert, J. Braustein, P.J. Tasker, L. Verweyer, H. Massler, W. Bronner, A. Hulsmann and K. Kohler, "Coplanar millimeter-wave IC's for W-band applications using 0.15 μm pseudomorphic MODFET's," ***IEEE J. Solid State Circuits*** **31** (1996) 1426-1434.
30. Z. Lao, M. Lang, V. Hurm, Z. Wang, A. Thiede, M. Schlechtweg, W. Bronner, G. Kaufel, K. Kohler, A. Hulsmann, B. Raynor and T. Jakobus, "20-40-Gbit/s GaAs-HEMT chip set for optical data receiver," ***Int. J. High Speed Electron. and Sys.*** **9** (1998) 437-472.
31. Z. Lao, A. Thiede, U. Nowotny, H. Lienhart, V. Hurm, M. Schlechtweg, J. Hornung, W. Bronner, K. Kohler, A. Hulsmann, B. Raynor and T. Jakobus,

"40-Gb/s high-power modulator driver IC for lightwave communication systems," ***IEEE J. Solid State Circuits*** **33** (1998) 1520-1526.

32. H. Shigematsu, N. Yoshida, M. Sato, N. Hara, T. Hirose and Y. Watanabe, "45-GHz distributed amplifier with a linear 6-Vp-p output for a 40-Gb/s $LiNbO_3$ modulator driver circuit," in ***Tech. Dig. IEEE GaAs IC Symp.***, 2001, pp. 137-140.
33. S. Yamahata, K. Kurishima, H. Ito and Y. Matsuoka, "Over-220-GHz-f_T-and-f_{max} InP/InGaAs double-heterojunction bipolar transistors with a new hexagonal-shaped emitter," in ***Tech. Dig. IEEE GaAs IC Symp.***, 1995, pp. 163-166.
34. S. Yamahata, H. Nakajima, M. Ida, H. Niiyama, N. Watanabe, E. Sano and Y. Ishii, "Reliable carbon-doped InP/InGaAs HBT's technology for low-power 40-GHz static frequency divider," in ***Proc. Int. Conf. Solid State Devices and Materials***, 1999, 570-571.
35. E. Sano, H. Nakajima, N. Watanabe and S. Yamahata, "40 Gbit/s decision IC using InP/InGaAs composite-collector heterojunction bipolar transistors," ***IEE Electron. Lett.*** **35** (1999) 1194-1195.
36. E. Sano, H. Nakajima, N. Watanabe, S. Yamahata and Y. Ishii, "40 Gbit/s 1:4 demultiplexer IC using InP-based heterojunction bipolar transistors, " ***IEE Electron. Lett.*** **35** (1999) 2116-2117.
37. M.J.W. Rodwell, Miguel Urteaga, D. Mensa, Q. Lee, J. Guthrie, Y. Betser, S.C. Martin, R.P. Smith, S. Jaganathan, T. Mathew, P. Krishnan, S. Long, R. Pullela, B. Agarwal, U. Bhattacharya, L. Samoska, Denis Scott and Mattias Dahlstrom, "Submicron Scaling of HBTs." ***IEEE Trans. Electron Devices*** **48** (2001) 2606-2624.
38. T. Mathew, H.J. Kim, S. Jaganathan, D. Scott, S. Krishnan, Y. Wei, M. Urteaga, M.J.W. Rodwell and S. Long, "75 GHz ECL Static Frequency Divider Using InAIAs/InGaAs HBTs." ***IEE Electron. Lett.*** **37** (2001) 667-668.
39. C.H. Fields, M. Sokolich, S. Thomas, K. Elliot and J. Jensen, "Progress toward 100 GHz logic in InP HBT IC Technology," ***Int. J. High Speed Electron. and Sys.*** **11** (2001) 217-243.
40. J.P. Mattia, R. Pullela, G. Georgieu, Y. Baeyens, H.S. Tsai, Y.K. Chen, C. Dorschky, T. Winkler Von Mohrenfels, M. Reinhold, C. Groepper, M. Sokolich, L. Ngyuen and W. Stanchina, "High-speed multiplexers: a 50Gb/s 4:1 MUX in InP HBT technology," in ***Tech. Dig. IEEE GaAs IC Symp.***, 1999, pp. 189-192.
41. J. Godin, P. Andre, J.L. Benchimol, P. Berdaguer, S. Blayac, J.R. Burie, P. Desrousseaux, A.M. Duchenois, N. Kauffmann, A. Konczykowska and M. Riet, "40 Gbits optical communications: InP DHBT technology, ciucuits and system experiments," in ***Tech. Dig. IEEE GaAs IC Symp.***, 1999, pp. 185-188.
42. A. Kasbari, Ph. Andre, J. Godin and A. Konczykowska, "40 Gbit/s master-slave D-type flip-flop in InP DHBT technology," ***IEE Electron. Lett.*** **38** (2002) 330-331.
43. M. Dahlströml, M. Urteaga, S. Krishnan, N. Parthasarathy and M.J.W. Rodwell, "Ultra-Wideband DHBTs using a Graded Carbon-Doped InGaAs Base," in *Post-deadline Papers* ***Indium Phosphide and Related Materials Conf. (IPRM)***, 2002.

44. L.D. Nguyen, L.E. Larson and U.K. Mishra, "Ultra-high-speed modulation-doped field-effect transistors: a tutorial review," ***Proc. IEEE*** **80** (1992) 494-518.
45. T. Enoki, H. Ito, K. Ikuta, Y. Umeda and Y. Ishii, "0.1-μm InAlAs/InGaAs HEMTs with an InP-recess-etch stopper grown by MOCVD," ***Microwave and Optical Tech. Lett.*** **11** (1996) 135-139.
46. T. Suemitsu, T. Ishii and Y. Ishii, "Gate and recess engineering for ultrahigh-speed InP-based HEMTs," in ***Dig. Topical Workshop on Heterostructure Microelectronics***, 2000, 2-3.
47. Y. Yamashita, A. Endoh, K. Shinohara, M. Higashiwaki, K. Hikosaka, T. Mimura, S. Hiyamizu and T. Matsui, "Ultra-short 25-nm-gate lattice-matched InAlAs/InGaAs HEMTs within the range of 400 GHz cutoff frequency," ***IEEE Electron Dev. Lett.*** **22** (2001) 367-369.
48. T. Otsuji, K. Murata, T. Enoki and Y. Umeda, "An 80-Gbit/s multiplexer IC using InAlAs/InGaAs/InP HEMTs" ***IEEE J. Solid-State Circuits*** **33** (1998) 1321-1327.
49. S. Kimura, Y. Imai, Y. Umeda and T. Enoki, "0-90 GHz InAlAs/InGaAs/InP HEMT distributed amplifier IC, ***IEE Electron. Lett.*** **31** (1995) 1430-1431.
50. K. Murata, T. Otsuji, E. Sano, S. Kimura and Y. Yamane, "70-Gbit/s multiplexer and 50-Gbit/s decision IC modules using InAlAs/InGaAs/InP HEMTs," ***IEICE Trans. Electron.*** **E83-C** (2000) 1166-1169.
51. K. Sano, K. Murata and Y. Yamane, "50-Gbit/s demultiplexer IC module using InAlAs/InGaAs/InP HEMTs," ***IEICE Trans. Electron.*** **E83-C** (2000) 1786-1788.
52. K. Murata, K. Sano, E. Sano, S. Sugitani and T. Enoki, "Fully monolithic integrated 43 Gbit/s clock and data recovery circuit in InPHEMT technology," ***IEE Electron. Lett.*** **37** (2001) 1235-1237.
53. T. Suzuki, Y. Nakasha, T. Takahashi, K. Makiyama, K. Imanishi, T. Hirose and Y. Watanabe, "A 90Gb/s 2:1 multiplexer IC in InP-based HEMT technology," in ***Dig. Tech. Papers IEEE Int. Solid-State Circuits Conf. (ISSCC),*** 2002, 11.7.
54. Y. Nakasha, T. Suzuki, H. Kano, A. Ohya, K. Sawada, K. Makiyama, T. Takahashi, M. Nishi, T. Hirose and Y. Watanabe, "A 43Gb/s full-rate-clock 4:1 multiplexer in InP-based HEMT technology," in ***Dig. Tech. Papers IEEE Int. Solid-State Circuits Conf. (ISSCC),*** 2002, 11.6.
55. E. Sano, K. Sano, T. Otsuji, K. Kurishima and S. Yamahata, "Ultra-high speed low-power monolithic photoreceiver using InP/InGaAs double-heterojunction bipolar transistors," ***IEE Electron. Lett.*** **33** (1997) 1047-1048.
56. V. Hurm, W. Benz, W. Bronner, A. Hulsmann, T. Jakobus, K. Kohler, A. Leven, M. Ludwig, B. Raynor, J. Rosenzweig, M. Schlechtweg and A. Thiede, "40 Gbit/s 1.55 μm pin-HEMT photoreceiver monolithically integrated on 3in GaAs substrate," ***IEE Electron. Lett.*** **34** (1998) 2060-2062.
57. D. Huber, M. Bitter, T. Morf, C. Bergamashi, H. Melchior and H. Jackel, "46 GHz bandwidth monolithic InP/InGaAs pin/SHBT photoreceiver," ***IEE Electron. Lett.*** **35** (1999) 40-41.
58. K. Takahata, Y. Muramoto, H. Fukano, K. Kato, A. Kozen, S. Kimura, Y. Imai, Y. Miyamoto, O. Nakajima and Y. Matsuoka, "Ultrafast monolithic receiver OEIC

composed of multimode waveguide p-i-n photodiode and HEMT distributed amplifier," ***IEEE J. Selected Topics in Quantum Electron.*** **6** (2000) 31-37.

59. T. Ishibashi, N. Shimizu, S. Kodama, H. Ito, T. Nagatsuma and T. Furuta, "Uni-traveling-carrier photodiodes," in ***Tech. Dig. OSA Int. Topical Meeting on Ultrafast Electron. and Optoelectron.***, 1997, pp. 166-168.
60. K. Murata, H. Kitabayashi, S. Shimizu, S. Kimura, T. Furuta, N. Watanabe and E. Sano, "A 40-Gbit/s monolithic digital OEIC module composed of uni-traveling-carrier photodiode and InP HEMT decision circuit," in ***Dig. IEEE MTT-S Int. Microwave Sym.***, 2000, pp. 345-348.
61. T. Otsuji, M. Yoneyama, K. Murata and E. Sano, "A super-dynamic flip-flop circuit for broadband applications up to 24 Gb/s utilizing production-level GaAs MESFETs," ***IEEE J. Solid State Circuits*** **32** (1997) 1357-1362.
62. K. Murata, T. Otsuji, M. Yoneyama and M. Tokumitsu, "A 40-Gbit/s super-dynamic decision IC fabricated with 0.15-μm GaAs MESFETs," ***IEEE J. Solid-State Circuits*** **33** (1998) 1527-1535.
63. K. Murata, T. Otsuji, E. Sano, S. Kimura and Y. Yamane, "70-Gbit/s multiplexer and 50-Gbit/s decision IC modules using InAlAs/InGaAs/InP HEMTs," ***IEICE Trans. Electron.*** **E83-C** (2000) 1166-1169.
64. K. Murata, T. Otsuji, M. Ohhata, M. Togashi, E. Sano and M. Suzuki, "A novel high-speed latching operation flip-flop (HLO-FF) circuit and its application to a 19 Gb/s decision circuit using 0.2 μm GaAs MESFET," ***IEEE J. Solid State Circuit*** **30** (1995) 1101-1108.
65. K. Ishii, H. Ichino, M. Togashi, Y. Kobayashi and C. Yamaguchi, "Very-high-speed Si bipolar static frequency-divier with new T-type flip-flops," ***IEEE J. Solid-State Circuits*** **30** (1995) 19-24.
66. B. Tang, J. Northoff, A. Gutierres-Aitken, E. Kaneshiro and P. Chin, "InP DHBT 68 GHz frequency divider," in ***Tech. Dig. IEEE GaAs IC Symp.***, 1999, pp. 193-196.
67. T. Suzuki, H. Kano, Y. Nakasha, T. Takahashi, K. Imanishi, H. Ohnishi and Y. Watanabe, "40-Gbit/s D-type flip-flop and multiplexer circuits using InP HEMT," in ***Dig. IEEE MTT-S Int. Microwave Symp.***, 2001, pp. TUIF-50-1-4.
68. M. Sokolich, G. Raghavan, D.A. Hitko and Y. Brown, "InP HBT Production Technology for 100 Gbps Lightwave Communications," in ***GaAs ManTech Digest of Papers***, Vol. '99, 1999, pp. 169-172.
69. T. Enoki, H. Yokoyama, Y. Umeda and T. Orsuji, "Ultrahigh-speed integrated sircuits using InP-based HEMTs," ***Jpn. J. Appl. phys.*** **37** (1998) 1359-1346.
70. M. Matloubian, D. Docter, C. Nguyen, T. Liu, S. Bui and C. Ngo, "$Ga_{0.47}In_{0.53}As$/InP HEMTs with novel $GaP_{0.35}Pb_{0.65}$ Schottky layers grown by MOVPE, in ***Dig. 56th Annual Dev. Res. Conf.***, 1998, pp. 32-33.
71. C.R. Bolognesi, M. Matine, M.W. Dvorak, P. Yeo, X.G. Xu and S.P. Watkins, "InP/GaAsSb/InP Double HBTs: a new alternative for InP-based DHBTs," ***IEEE Trans. Electron Devices*** **48** (2001) pp. 2631-2639.

72. C. Marrian, "DARPA Antimonide Based Compound Semiconductors Program," in ***Third Workshop on the Fabrication, Characterization, and Applications of 6.1 Å III-V Semiconductors,*** 2001.
73. K. Maezawa and T. Mizutani, "A new resonant tunneling logic gate employing monostable-bistable transition," ***Jap. J. Appl. Phys.*** **37** (1993) 1359-1364.
74. K. Sano, K. Murata, T. Otsuji, T. Akeyoshi, N. Shimizu, M. Yamamoto, T. Ishibashi and E. Sano, "Ultra-fast optoelectronic decision circuit using resonant tunneling diodes and a uni-traveling-carrier photodiode," ***IEICE Trans. Electron.*** **E82-C** (1999) 1638-1646.
75. K. Sano, K. Murata, T. Otsuji, T. Akeyoshi, M. Shimizu and E. Sano, "An 80-Gb/s optoelectronic delayed flip-flop IC using resonant tunneking diodes and uni-traveling-carrier photodiode," ***IEEE J. Solid State Circuits*** **36** (2001) 281-289.
76. K. Murata, K. Sano, T. Akeyoshi, N. Shimizu, E. Sano, M. Yamamoto and T. Ishibashi, "An optoelectronic clock recovery circuit using a resonant tunneling diode and a uni-traveling-carrier photodiode," ***IEICE Trans. Electron.*** **E82-C** (1999) 1494-1501.
77. K. Narahara, T. Otsuji, N. Shimizu and E. Sano, "An optically-controlled phase shifter for traveling-wave retiming circuit," **OSA *Trends in Optics and Photonics Vol. 28, Ultrafast Electronics and Optoelectronics***, J.E. Bowers and W.H. Knox, eds., Optical Society of America, Washington, DC 1999, pp. 211-218.

International Journal of High Speed Electronics and Systems
Vol. 13, No. 1 (2003) 27–57

A COMPARISON OF SILICON AND III–V TECHNOLOGY PERFORMANCE AND BUILDING BLOCK IMPLEMENTATIONS FOR 10 AND 40 Gb/s OPTICAL NETWORKING ICs

S. P. VOINIGESCU, D. S. McPHERSON, F. PERA, S. SZILAGYI,
M. TAZLAUANU, and H. TRAN

Quake Technologies, 80 Hines Road,
Kanata, K2K 2T8, Ontario, Canada
sorinv@quaketech.com

Scalable models for both active and passive components are essential for the design of highly integrated fiber–optic physical layer ICs. This paper focuses on the various technology options available for 10 Gb/s and 40 Gb/s applications, on how their constituent components are modeled and what the characteristics and requirements are for the basic building blocks. As part of the technology comparison, an overview of the performance of leading edge Si CMOS, SiGe BiCMOS and III–V technologies is presented. Scalable models for SiGe HBTs and GaAs p–HEMTs are then compared with measured data for various device sizes. Inductors, varactors, transmission lines and isolation techniques on Si and III–V substrates are discussed next followed by technology–specific implementations of VCO and digital building blocks. Finally, Transimpedance Limiting Amplifier (TIALA) as well as laser and modulator driver designs in SiGe BiCMOS, InP HBT and GaAs p–HEMT processes using scalable device models are illustrated for 10 and 40 Gb/s fiber–optics applications.

Keywords: Heterostructure technolologies, transimpedance amplifier, output driver

1. Introduction

Bandwidth demand in local area networks (LANs) and over the Internet is growing rapidly due to applications such as video, multimedia, E–commerce, and advanced digital services. This ever–increasing need for bandwidth is creating huge demand on the short to medium reach applications. At the forefront of addressing this, the IEEE 802.3ae standards committee has approved Ethernet specifications for 10 Gb/s premises networking applications. Current generation Ethernet LANs are being deployed with 100 Mb/s connections at the desk, and 1 Gb/s on the LAN backbones. At present, 10 Gb/s physical layer (PHY) integrated circuits (ICs) can be found primarily in long haul applications such as transport or wide area networks (WAN). Unfortunately their power consumption, high cost, lower level of integration [1] and footprint make them unsuitable for Very–Short–Reach (VSR) links, Storage Area Networks (SAN), LAN, and MAN (Metropolitan Area Networks) networks.

Regardless of the reach and data rate of a typical application, fiber–optics modules have very similar architectures, containing both high speed digital, as well as analog building blocks such as TransImpedance Amplifier (TIA), post amplifier, driver, clock and data recovery (CDR), and transmit phase–lock–loop (PLL), as illustrated in Fig.1. In addition, depending on the application, a significant amount of logic gates might be required for traffic and protocol processing.

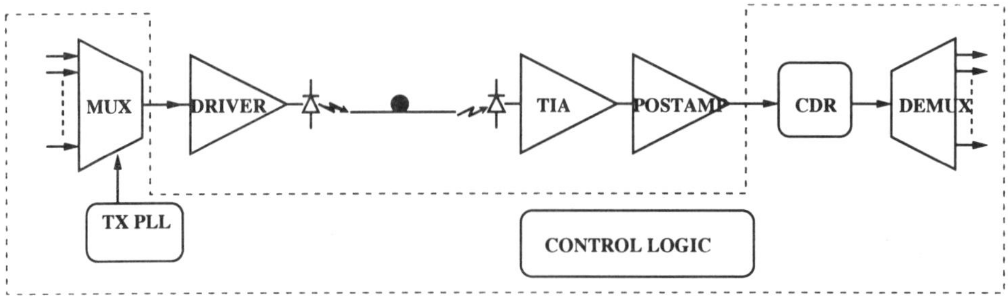

Fig.1. Typical PHY layer block diagram for 10 Gb/s and 40 Gb/s fiber systems.

High levels of IC integration in the 10 Gb/s and eventually 40 Gb/s markets are logical steps in optimizing such systems in terms of speed, power and cost performance. Indeed, several highly integrated SiGe BiCMOS (Bipolar Complementary Metal Oxide Semiconductor) and CMOS (Complementary Metal Oxide Semiconductor) transceivers in the 10 Gb/s range have been reported during the last three years [2–7].

This paper will start with an overview of the competing semiconductor technologies vying for 10 and 40 Gb/s applications. Next, basic building blocks such as Voltage–Controlled–Oscillators (VCOs), multiplexers (MUX), demultiplexers (DEMUX), output drivers and their technology requirements will be addressed. Finally, examples of 10 and 40 Gb/s transimpedance limiting amplifiers (TIALA) and laser and modulator drivers implemented in SiGe BiCMOS, InP Heterostructure Bipolar Transistor (HBT) and GaAs p–HEMT (pseudomorphic–High Electron Mobility Transistor) processes will be presented.

2. Semiconductor Technologies

2.1 *Technology requirements*

The first generation OC192 (10 Gb/s) systems of the mid 1990's [1] have relied exclusively on GaAs HBT, GaAs MESFET (Metal Semiconductor Field Effect Transistor) and GaAs p–HEMT technologies to deliver the performance required in fiber–optics modules. To date, they have low levels of integration and are dominated by a "mix–and–match" of discrete components and ICs. It is noteworthy that the raw transistor speed of these technologies, with typical f_T and f_{MAX} values of 60–70 GHz, far exceeded the capabilities of Si–based state–of–the–art devices at the time. This situation is now tending to favor SiGe npn and CMOS transistors as their cutoff and oscillation frequencies now exceeds 80 GHz.

The list of critical active and passive device requirements for successful implementation of highly integrated PHY ICs starts with the transistor speed. Historical data indicate that, at a minimum, the cutoff frequency, f_T, and the maximum oscillation frequency, f_{MAX}, should both exceed the system bit rate by a factor of at least 4. i.e. 43 GHz for 10.7 Gb/s systems, and 172 GHz for 43 Gb/s. These conditions must be met *over all process corners and the entire range of operating temperatures*. In some instances, as long as f_{MAX} is larger than four times the bit rate, the requirement for f_T can be relaxed, as in one circuit example discussed in Section 5. In order to successfully integrate VCOs and clock and data recovery circuits, the technology must also be able to support high quality passive components such as inductors, transmission lines, varactor diodes and MIM capacitors. All these passive components should have good quality factors and high resonant frequencies (f_{res}) beyond 20 GHz for 10 Gb/s systems, and beyond 80 GHz for 40 Gb/s applications. While in the past most of the research on model development focused on transistors, it has since been acknowledged that accurate models for passive components and interconnect are equally, if not more, important in the successful design of fiber–optic ICs. Finally, in order to maximize the integration level, minimize parasitic capacitance and to ensure good isolation between circuit blocks, a large number of metal layers with low–k inter–metal dielectric is desirable.

2.2. *Transistor performance*

The most relevant transistor performance for fiber–optic ICs is effectively captured by the following figures of merit, most of which are also summarized in Table 1:

- *peak f_T and f_{MAX}* values – determine the ultimate circuit speed;
- *peak f_T current density* J_{cpfT} – determines the power dissipation of a high speed circuit at a given data rate;
- $I_{cpfT}/C_{BC}(C_{GD})$, also known as the *intrinsic slew–rate* of the transistor, is defined as the ratio of the peak f_T current and the output capacitance (for completeness, C_{CS}/C_{DB} should be added to the output capacitance);
- BV_{CEO}/BV_{DG} – the *breakdown voltage* of the transistor which limits the maximum output swing achievable in the technology;
- *turn–on/threshold voltage*: V_{BE}/V_T – limits the minimum value of the power supply voltage and, usually, power dissipation of a circuit and favors CMOS over bipolar devices, and InP HBTs over SiGe HBTs;
- *minimum feature size* –affects power dissipation and integration levels and favors Si CMOS and SiGe BiCMOS over III–V technologies due to the more mature processing techniques;
- *thermal resistance* is closely related to the semiconductor material and to the minimum feature size and affects indirectly the integration levels and the transistor speed that can be achieved under reliable operating conditions.

Table 1. Performance of State–of–the–Art Production Semiconductor Technologies (best lab results shown in brackets)

Device/ Parameter	InP HBT @ Vce=1.5 V	SiGe HBT @ Vce=1.5V	CMOS @ Vds=1.2 V	GaAs p–HEMT @ Vds=1.5 V	InP HEMT @ Vds=1 V
feature size (μm)	1	0.18	0.13 (0.09)	0.15	0.1
f_T (GHz)	180 (300)	160 (200)	80 (120)	110	175
f_{MAX} (GHz)	200 (300)	150 (200)	80 (120?)	180	300
J_{cpfT} (mA/μm^2)	2	6	1.66 (2.5)	1.87	2.9
I_{CpfT}/C_{BC} (V/ps)	0.34	0.5	0.54	2.86	1.16
BV_{CEO}/BV_{DS} (V)	>2	>2	>1.5 (>1V)	>6	>2
V_{BE}/V_T (V)	0.75	0.95	0.35 (0.3)	–0.9	–0.6

Fig. 2 illustrates how the f_T of Si CMOS and Si/SiGe bipolar devices scales with process feature size. For the same generation of lithography, n–channel MOSFETs exhibit at best half the speed of SiGe HBTs with p–channel device performance half that again. SiGe HBTs have been aggressively scaled in the last two generations in order to retain their x2 and x4 speed advantage over n–channel and p–channel MOSFETs, respectively. In general, for the same f_T, the n–MOS feature size is two generations ahead of the bipolar one, thus offsetting the lower cost associated with the smaller number of masks in a CMOS process compared to a SiGe BiCMOS process.

SiGe HBT f_T and f_{MAX} values in the 160 to 200 GHz range, or even higher, have recently been announced by several foundries [8–11]. For 40 Gb/s applications, devices with over 160 GHz f_T and f_{MAX} values are required and such performance had, until the last year, been reached only by InP HBTs [12] and HEMTs [13]. Fig. 3(a) compares the f_T and f_{MAX} dependence on collector current density for state–of–the–art SiGe and InP HBTs. The InP devices enjoy some advantage in f_T (180 GHz versus 160 GHz), f_{MAX} (200 GHz versus 150 GHz), *peak f_T current density* (2 mA/μm^2 vs. 6 mA/μm^2) and *turn on voltage* (0.75 V as opposed to 0.95 V). The latter feature allows for InP ECL and even E^{2}CL circuits to operate with large margin from a 3.3V supply which is not achievable with SiGe HBT E^{2}CL circuits.

The lower *peak f_T current density* of InP HBTs is not typically exploited fully due to the coarser lithography employed in state of the art III–V HBTs compared to SiGe HBTs. For example, the minimum size 0.18 μm x 0.5 μm SiGe HBT has a peak f_T current of 0.6 mA, almost half that of a very "aggressive" minimum size 0.5 μm x 1 μm InP HBT featuring a peak f_T current of 1 mA. Reducing the emitter size in InP HBTs beyond 0.5 μm has proven very difficult because of increasing surface recombination, which reduces current gain below acceptable levels.

To facilitate the comparison between bipolar and FET performance, Fig. 3(b) shows the measured f_T and f_{MAX} dependence on drain current linear density (current per unit gate width) for 0.18 μm Si n–MOSFETs, 0.15 μm GaAs p–HEMTs and 0.1 μm InP HEMTs. While the f_T values of the GaAs p–HEMT and InP HEMT devices (110 GHz and 170 GHz, respectively) are lower or comparable to those of the SiGe and InP HBTs, the f_{MAX} values are higher (180 GHz and over 300 GHz, respectively.)

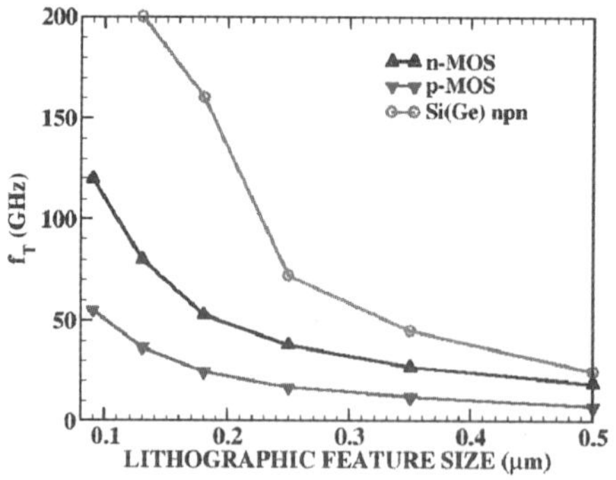

Fig.2. Scaling of Si CMOS and Si(Ge) npn devices as a function of feature size.

The f_{MAX} is higher in III–V HEMTs due to the very low resistance of the metal gate, when compared to HBT base resistance, and due to much lower drain–gate capacitance compared to HBT collector–base capacitance. The peak f_T current for all FETs shown in Fig. 3(b) occurs between 0.28 and 0.32 mA/μm. According to field effect transistor scaling rules [14], the current per unit gate width should remain constant, as the device gate length is reduced to 0.09 μm and beyond. As illustrated in Table 1, one can define a current density for FETs by dividing the drain current linear density by the gate length. In that case, the peak f_T current density for field effect transistors is comparable to that of the InP HBT and it scales as l_G^{-1}, where l_G is the gate length.

The *intrinsic slew rate* is an important figure of merit for digital circuits and output drivers. It reflects the capability of a device technology to operate with large voltage swing at very high speed. Typical values are 0.34 V/ps for InP HBT processes and 0.5 V/ps for the fastest SiGe HBTs and 0.13 μm n–MOSFETs. The best values are obtained for InP HEMTs and GaAs p–HEMTs. Even though the intrinsic slew rate is very high, the loaded slew rate (including interconnect capacitance) is severely degraded by the interconnect capacitance. This is particularly the case for Si MOSFETs and SiGe HBTs because the interconnect capacitance is higher in silicon technologies than on semi–insulating III–V substrates. As an example, a 3x0.13μmx2μm n–MOSFET or a 0.18μmx1μm SiGe HBT, both with a peak f_T current of about 1.1 mA driving a 2 fF interconnect capacitance will have a loaded slew rate of 0.27 V/ps, and 0.26 V/ps, respectively, half the value of the intrinsic slew rate. In most practical situations, the parasitic capacitance is larger than 2 fF and the slew rate degradation is more severe, unless larger devices with higher bias currents are deployed.

Unlike bipolar devices, which, for a given process, exhibit a fairly constant f_{MAX} for different emitter lengths, the f_{MAX} of field effect transistors is strongly dependent on unit gate finger width, as well as on gate access geometry (single side versus double side, T versus Π shape, etc.) The measured variation of f_{MAX} and f_T with unit gate finger width is illustrated in Fig.4 for various FET technologies.

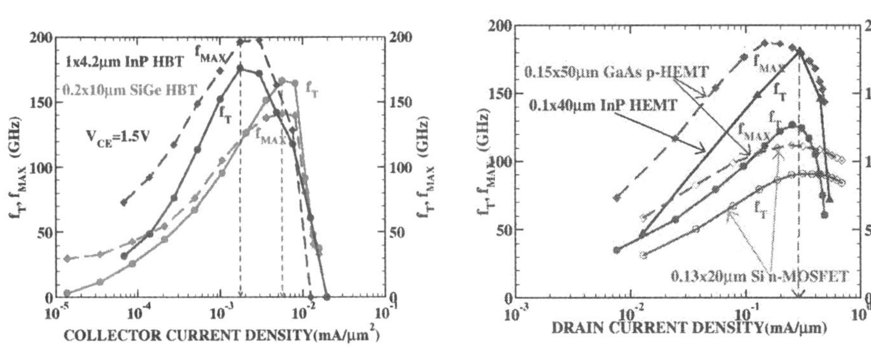

Fig.3. a) Measured 0.18x10μm² SiGe HBT vs. 1x4.2μm² InP HBT f_T and f_{MAX} characteristics as a function of the collector current density; b) measured 0.1μm InP HEMT vs. 0.15 μm GaAs p–HEMT vs. 0.13 μm Si n–MOSFET f_T and f_{MAX} characteristics as a function of drain current per unit gate width.

As the unit finger width is scaled down, fringing capacitance becomes dominant for all technologies degrading both f_{MAX} (in the case of III–V FETs) as well as f_T in the extreme case of very narrow gates. For Si MOSFETs, the polysilicon gate resistance, several orders of magnitude larger than the metal gate resistance of III–V HEMTs, is the dominant factor in f_{MAX} degradation [14–15]. f_{MAX} only scales as $l_G^{-1/2}$ [14] while f_T is proportional to $1/l_G$. Unless methods to significantly reduce the gate sheet resistance in MOSFETs are introduced, maintaining a higher f_{MAX} than f_T is going to be very problematic as gate lengths are shrunk to 0.09 μm and beyond.

With shrinking device dimensions, accurately predicting in simulation the $f_T(I_C)$ characteristics for a wide range of V_{CE}/V_{DS}, values, as well as for different device geometries (gate width and emitter length, respectively) becomes even more critical. Fig. 5 compares the modeled and measured characteristics for a family of 0.25 μm SiGe HBTs for emitter lengths varying between 3.2 and 25.6 μm. By carefully selecting a range of emitter sizes and using suitable de–embedding structures in transistor measurements and model extraction, excellent agreement is reached between measurements and simulations using one single scalable model for all SiGe HBTs. A foundry–proprietary, modified Spice Gummel Poon model with improved saturation region and breakdown behavior was employed. In Fig. 6, measured characteristics from an entire wafer are compared with best, worst and typical Spice–Gummel Poon model simulations. It is important to note that the measured peak f_T values, centered at 160 GHz, vary by +/–10% across the entire wafer, a range commensurate with the spread predicted by the best and worst case corner models. It becomes imperative that all process corners be used in circuit design to ensure good circuit yield in production.

Fig. 7 illustrates measured versus simulated $f_T(I_{DS})$ characteristics for single– and multi–finger 0.15 μm GaAs p–HEMTs with different unit gate finger widths. Again, given good process control and careful test and de–embedding structure design, it is possible to fit all device sizes with a single large signal scalable model. In this case a modified Agilent EE–HEMT model was used.

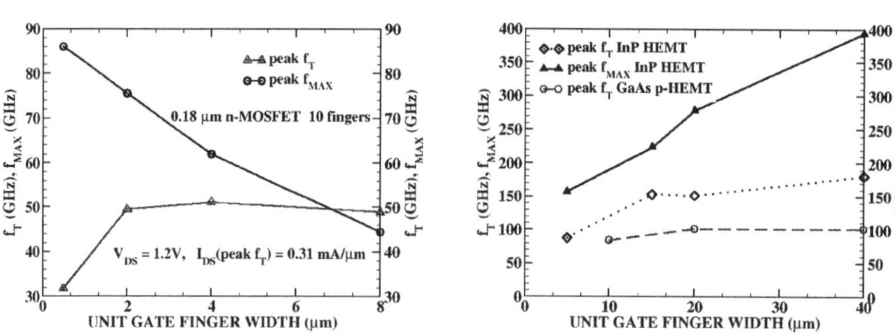

Fig.4. a) Measured f_T and f_{MAX} dependance on unit gate finger width for 0.18 μm n–channel MOSFETs. b) Measured f_T and f_{MAX} dependance on unit gate finger width for single–gate finger 0.1 μm InP HEMTs and 0.15 μm GaAs p–HEMTs.

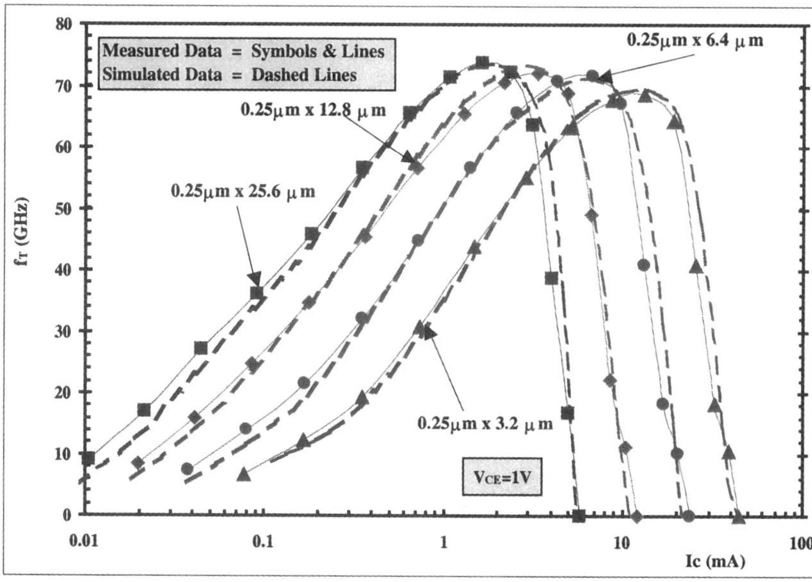

Fig.5. Measured vs. modeled f_T (I_C) characteristics as a function of emitter length for SiGe HBTs in a 0.25 μm SiGe BiCMOS process. A modified Spice Gummel Poon model was used.

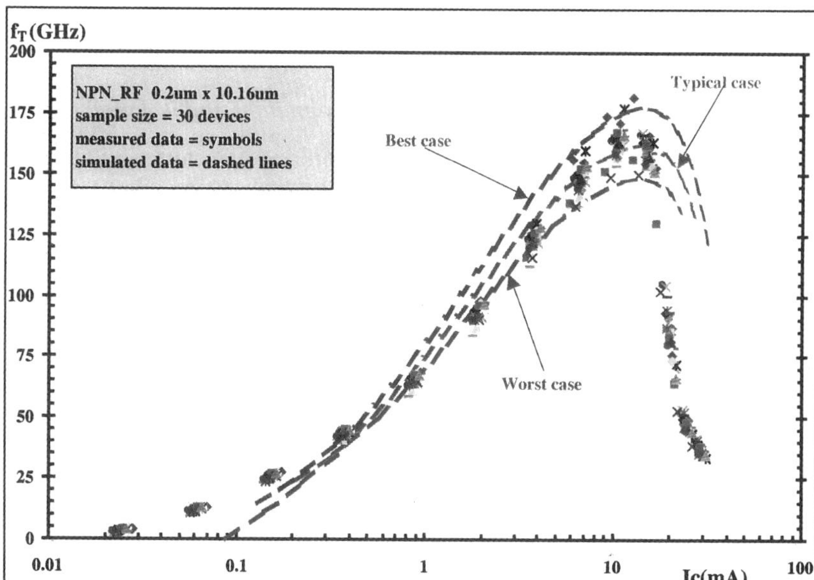

Fig.6. Measured across a wafer vs. best, typical and worst case model f_T (I_C) characteristics for SiGe HBTs in a 0.18 μm SiGe BiCMOS process. The Spice Gummel Poon model provided by the foundry was used in simulation.

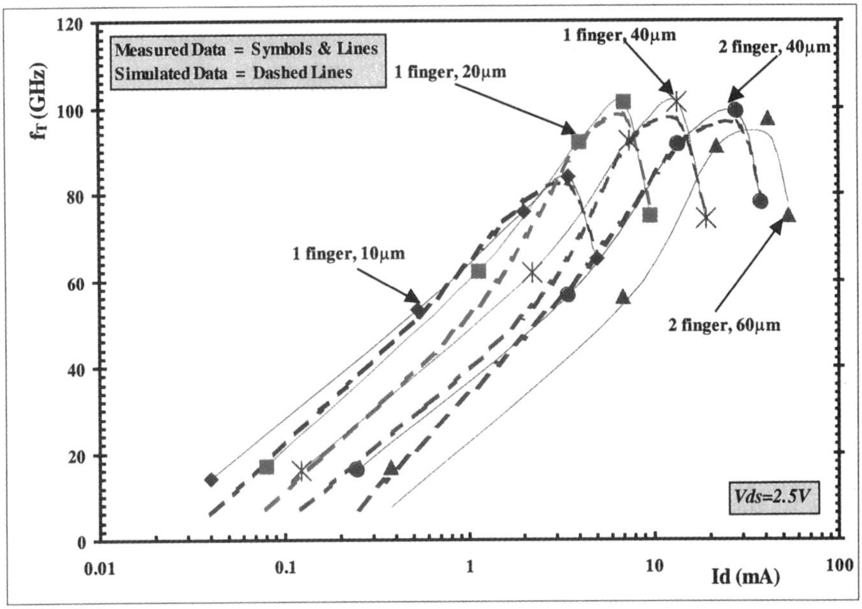

Fig.7. Measured vs. simulated f_T (I_{DS}) characteristics as a function of gate width (beyond 50 μm two or more gate fingers are used). The simulated characteristics are obtained using the large signal, scalable HEMT model.

2.3. *Passive devices: inductors, varactors, transmission lines and isolation techniques*

Recent results, shown in Fig. 8, indicate that inductors and varactors with Qs larger than 10 and resonant frequencies beyond 50 GHz can be realized on silicon substrates in conventional processes where only the top metal layer was thickened to 3 μm [18]. Measured characteristics of a 425 pH octagonal inductor are shown in Fig. 8(a) as a function of frequency. The peak Q is above 12 in the 20 to 50 GHz range. The frequency of operation was maximized by reducing the metal width to 5 μm and by minimizing the total inductor area, thus minimizing parasitic capacitance and substrate losses, the dominant loss mechanism beyond 5 GHz. Si varactor diode Q(f) characteristics, obtained from S parameter measurements, are shown in Fig. 8(b) for voltages between 0 and 5 V. Q is larger than 4 up to 50 GHz, even when the device is biased at 0 V. With a capacitance ratio of over 2.5, this device is superior to any varactor that can be realized in a conventional III–V technology that also integrates HBTs or HEMTs on the same die.

Computer programs are available today to accurately design and model two terminal and three-terminal (or differential) inductors. Fig. 9 compares measured vs. modeled apparent inductance and quality factor for a rectangular 150 pH inductor on an InP substrate and for a 650 pH three terminal (differential) inductor fabricated in a 0.35μm SiGe BiCMOS process. In both cases, good agreement between measurements and simulations is obtained.

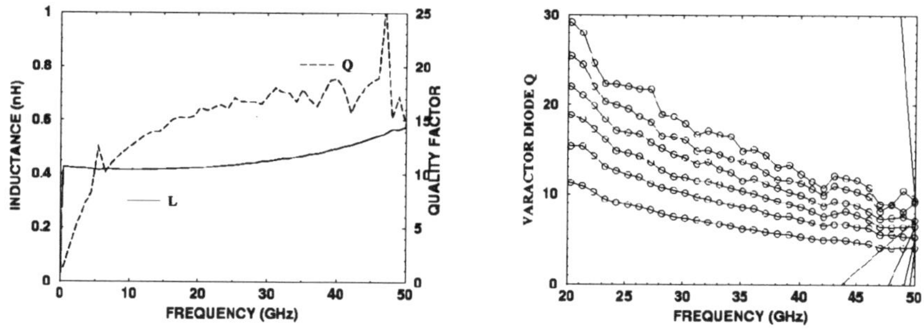

Fig.8. a) Measured inductance and quality factor of a 0.425 nH octagonal inductor in a 0.5μm SiGe HBT process; b) Measured quality factor of a 2x1.6μmx20μm multi–stripe varactor diode as a function of frequency and varactor voltage between 0 V and 5 V [18].

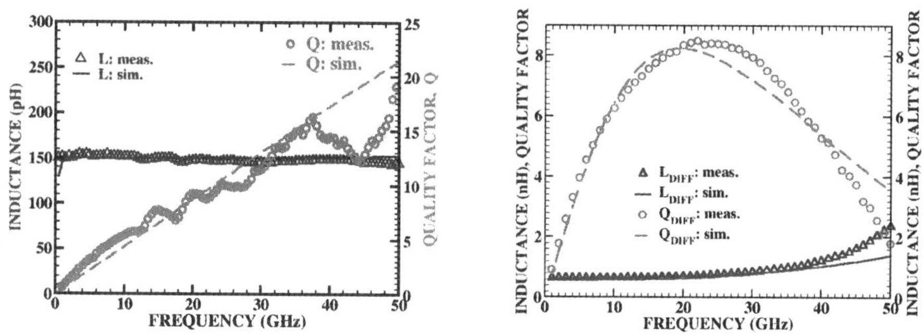

Fig.9. Measured vs. modeled inductance and quality factor for a) a rectangular 150 nH inductor in an InP process, and b) a 0.650 nH three terminal inductor in a SiGe HBT process

Traditionally, it has been assumed that silicon–based technologies lack the low–loss and good isolation properties of III–V technologies. Even with circuit techniques such as top metal over first metal ground planes and top metal over salicided polysilicon ground planes developed to overcome the disadvantages of the lossy silicon substrate [16–17], the loss and parasitic capacitance of interconnect over Si substrates continue to be slightly higher than those over semi–insulating III–V substrates.

The use of transmission lines is unavoidable in any highly integrated PHY IC at 10 or 40 Gb/s. At a minimum, clock signal distribution is typically implemented using controlled impedance, on–chip terminated transmission lines. In addition, it is common for the fiber side inputs and outputs of the PHY die to have on–chip matched 50 Ω transmission lines that conduct the signal between the pads and the circuit core. Transmission lines with a characteristic impedance larger than 50 Ω (typically up to 100 Ohm) are preferred for intra–chip signal distribution in order to minimize power dissipation. However, as data rates and frequencies increase, the bandwidth of the transmission line itself can become a high speed bottle–neck for the entire circuit and the value of the characteristic impedance must be lowered. By taking advantage of the large number of metal layers and of the high quality of low permitivity dielectrics typical of Si CMOS and SiGe BiCMOS technologies, transmission lines can be realized in Si processes with good control of the characteristic impedance and low loss even at 50 GHz. Fig. 10(a) compares the measured characteristic impedance and loss per mm of GaAs and SiGe microstrip lines as well as of InP coplanar waveguides (CPW). The 55Ω GaAs microstrip lines were fabricated on a 28 μm thick semi–insulating substrate with 5 μm thick Au interconnect. The 50 Ω SiGe transmission lines were realized using 3 μm thick Al top metal with 0.45 μm thick Al as the ground plane and 9.5 μm SiO_2 dielectric. The 50 Ω InP CPW line was implemented with 2 μm thick Au interconnect on a 600 μm thick semi–insulating substrate. The frequency dependence of the characteristic impedance, as well as of the attenuation of these microstrip and coplanar transmission lines, are well captured by the built–in models available in most common microwave and SPICE–like circuit simulators, as illustrated in Fig. 10(b). The agreement between the measured and simulated characteristic impedance for both SiGe microstrip and InP coplanar lines is better than 5%. However, to accurately model microstrip or coplanar lines over a lossy Si substrate, lumped RLC models based on measurements are still necessary [16].

It is not widely known that coupling between interconnect lines is significantly higher in III–V than in advanced Si technologies. Fig. 11 illustrates, using 3D EM simulation results, the cross–coupling between two adjacent metal interconnect lines on Si and InP substrates. As described above, the Si transmission line is realized using the top metal layer and the first metal layer as ground plane. In the InP case, the line is placed on top of the semi–insulating substrate while the ground is on the back side of the 100 μm thick wafer. In each case, the interconnect lines were designed to have a characteristic impedance of 66 Ohm when widely spaced apart. For a given line spacing, the cross–coupling is significantly weaker on the Si substrate. The thicker InP substrate seriously impedes high interconnect densities and integration levels. Ironically, this problem can only be solved in III–V technologies by resorting to either silicon–like interconnect with many metal layers and low–k dielectrics where the transmission lines have ground planes above the III–V substrate [14–15], or by thinning the semi–insulating GaAs or InP substrate. The latter situation is illustrated in a circuit example in Section 5.

On Si substrates, pn–junctions can be used in conjunction with deep n–wells and large first metal ground planes to reduce cross coupling through the substrate. Each circuit block is surrounded by a sufficiently wide guard–ring made of the above combination. 50–60 dB isolation is possible even above 10 GHz [2–5, 16]. In order to reduce the noise and loss associated with the substrate resistance under the signal pads, and in order to isolate the signal pads from the substrate, a salicided n–well is placed under the signal pad. The n–well is grounded outside the pad, thus forming a reverse–biased junction with the substrate. The pad behaves like an ideal high Q (>20 at 26 GHz) capacitor. This solution also provides very low pad capacitance. A typical 40x80 μm^2 pad has 12 fF capacitance, which is comparable to, or smaller than, that of a similar size pad on 75 μm thick GaAs or InP substrates, and is usable beyond 50 GHz [3, 16–17].

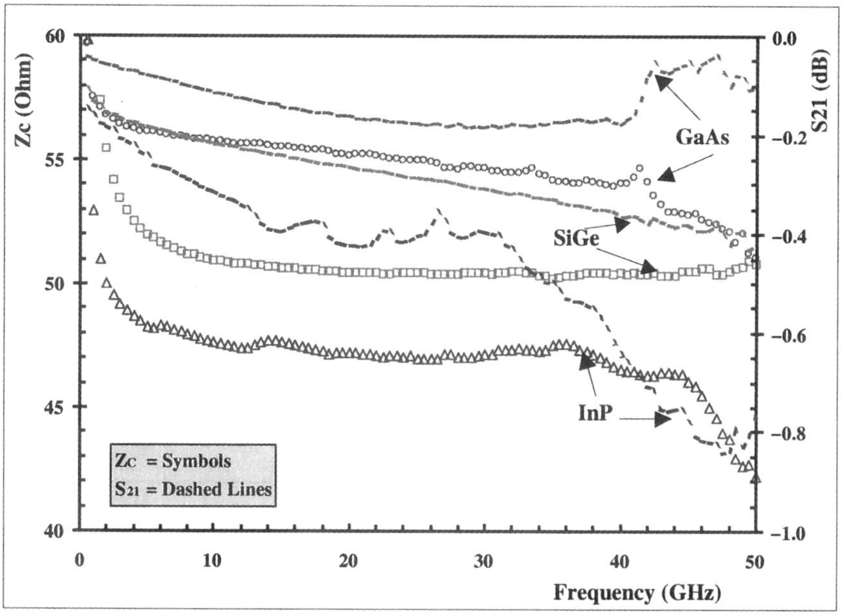

Fig.10. a) Characteristic impedance Z_C and attenuation S_{21} as a function of frequency for microstrip lines on Si and GaAs substrates and CPW lines on InP substrates.

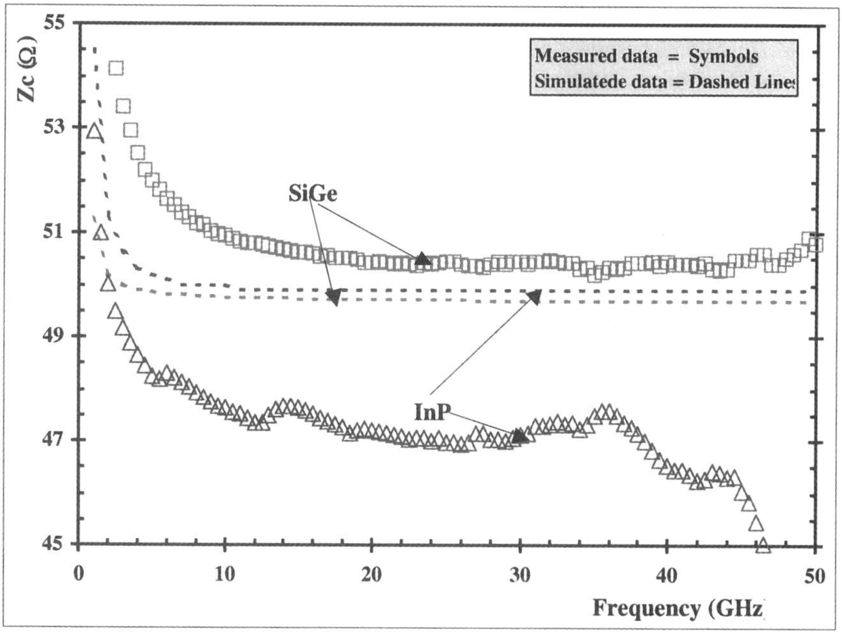

Fig.10. b) Measured vs. simulated characteristic impedance for SiGe and InP transmission lines.

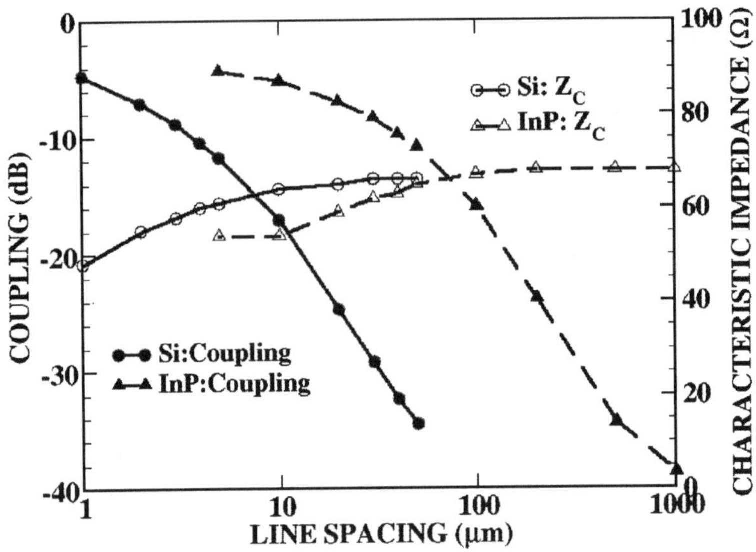

Fig.11. Coupling and characteristic impedance of two adjacent, infinitely long, microstrip lines on Si and InP substrates as a function of the spacing between lines.

3. Building Blocks

In terms of active and passive device performance, RF and fiber–optic ICs have similar requirements. However, unlike wireless applications which typically operate over a narrow frequency band and in which inductor–based tuned, narrow band circuit design is common place, PHY ICs for fiber–optics are broad band. Their frequency of operation typically extends from DC to a frequency equal to the bit rate. Also, while in most RF transceivers amplifiers and drivers tend to operate in linear mode and linearity is a critical system goal, in fiber–optics, with the exception of the transimpedance amplifier, drivers and post amplifiers are typically operated in limiting switching mode, similar to a digital circuit. For a given semiconductor device technology, the limiting switching mode of operation helps maximize the circuit bit rate and the voltage swing.

The requirements for various digital and analog blocks making up a fiber–optic system can be summarized as follows.

- Digital blocks need (i) high f_T/f_{MAX} for speed, (ii) low peak f_T current density to reduce power dissipation, (iii) low V_{BE} to reduce power supply and power consumption, (iv) small device size, and (v) fine metal pitch, the latter two being critical to reaching high levels of integration with low power dissipation.
- 50 Ω laser/modulator drivers require (i) large intrinsic slew–rate for bandwidth and S_{22} matching, (ii) large breakdown voltage for voltage swing, and (iii) high f_{MAX} to achieve the bandwidth.

- Transimpedance and post amplifiers are best realized with transistors that have (i) high f_{MAX} for bandwidth and (ii) low noise figure for good sensitivity.
- The VCO is the common block in fiber–optic and wireless ICs and it needs (i) high Q inductor for low noise, (ii) high Q varactor with large capacitance ratio to cover process spread, (iii) high Q, low parasitic capacitance MIM capacitor for low noise and large oscillation frequency and tuning range, (iv) high f_{MAX} transistor for large power and low–noise, and (v) low 1/f noise transistor.

Arguably, the analog functions place more demanding requirements on the speed of transistor technologies than do digital functions [19]. One exception is the master–slave D–type flip–flop in the decision circuit which is typically clocked at a frequency equal to the data rate [14]. Because of its simplicity and wide band operation, the differential inverter with resistive loads is the basic circuit topology – known as current–mode–logic (CML) in its bipolar implementation – employed in both digital blocks as well as in output drivers. Other functions such as multiplexing and de–multiplexing, pulse–width (duty–cycle) [26] and peaking control can be implemented using Gilbert–cell topologies based on such inverters, or by connecting inverters in parallel at the output node of the transistor, as in Fig. 12. Its bandwidth is limited by the value of the resistive load, the output capacitance of the differential transistor pair, as well as by the input capacitance of the next stage

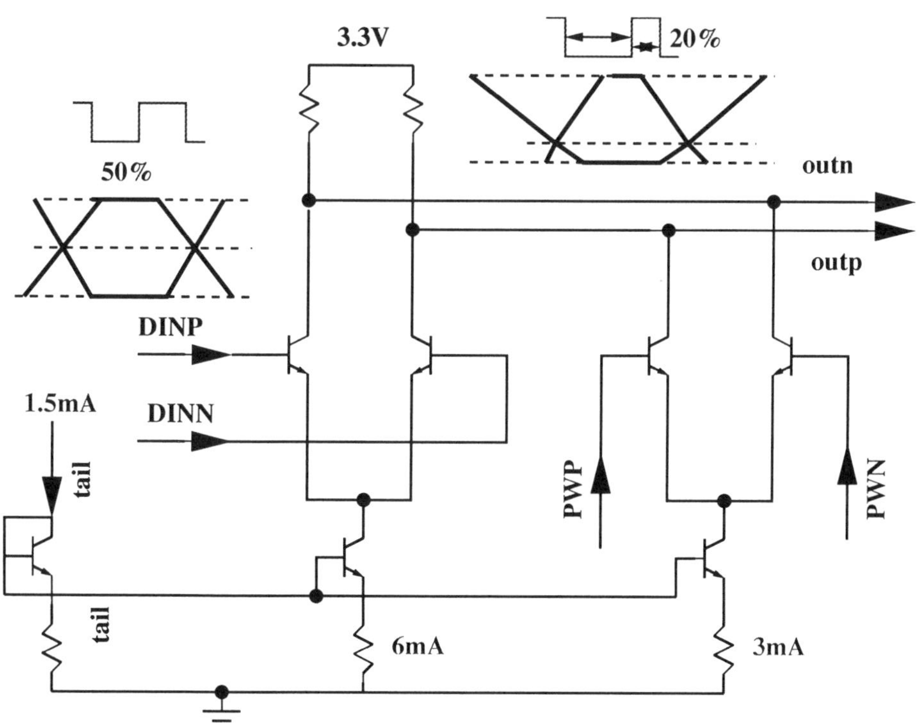

Fig.12. Circuit topology for a pulse–width control function implemented with bipolar circuits.

While active loads have the benefit of providing more gain with low DC voltage drop for a given current and power dissipation, they are seldom usable at the highest speed due to a smaller bandwidth when compared to a resistive load. If the cutoff frequency of the transistor is high enough, such that the current gain is adequate at the bit rate frequency, an inverter can drive directly another inverter, as shown in Fig. 13 for a 0.18 μm CMOS digital chain operating at 10 Gb/s [7]. The cutoff frequency of the n–channel MOSFET is 55 to 60 GHz. Also illustrated in Fig. 13 is how the bandwidth of a resistive load inverter can be extended by using inductive peaking [27]. This compensates to some extent for the low transconductance and gain of the MOSFET when operated with a resistive load and small supply voltage.

The distributed amplifier [1, 26, 27] and the distributed inverter [15] are extreme cases of inductor peaking, where, in each section of the distributed circuit, the resistive load is replaced by appropriately designed inductors to create an artificial transmission line together with the output capacitance of the transistor. This technique, which helps to push the circuit bandwidth closer to the maximum frequency of oscillation of the transistor, results in the highest bandwidth with the largest possible swing. It is primarily used in modulator drivers with voltage swings larger than 3 V at 10 Gb/s [1] and especially at 40 Gb/s [26]. Its main limitation, not an issue at 80 Gb/s [15], is the larger die area, as well as the requirement for a low–loss substrate as a transmission line medium. When the cutoff frequency of the transistor is not large enough to provide sufficient current gain at the desired bit rate, 43 Gb/s in the case of the GaAs p–HEMT implementation of Fig. 14(a) and of the SiGe HBT circuit of Fig. 14(b), the CML topology is modified by inserting one or two source/emitter follower stages between inverters [19,27]. Such f_T–doubling topologies, known as SCFL (source–coupled–FET–logic) for FET technologies and ECL (emitter–coupled–logic) or E^2CL for bipolar technologies, do not favor low supply voltage and low power dissipation.

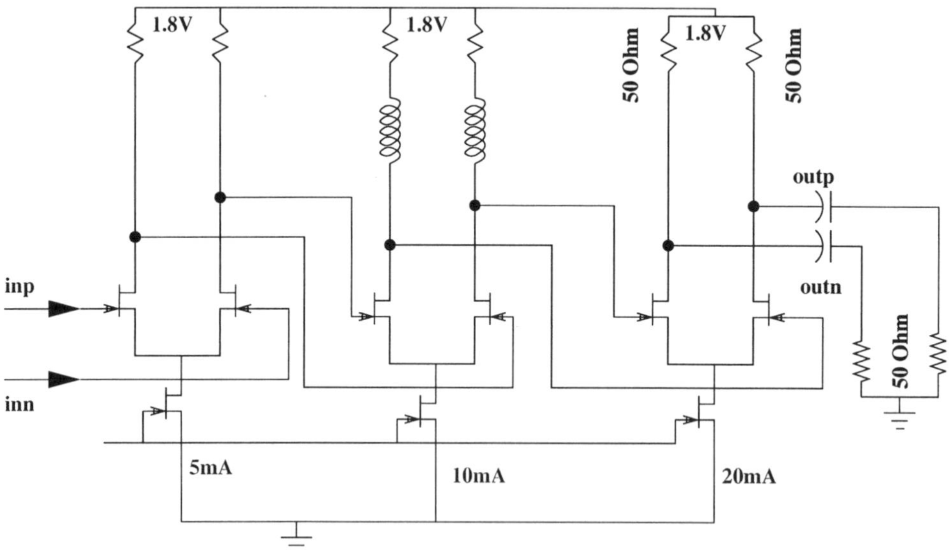

Fig.13. Schematics for CMOS–CML building block [7].

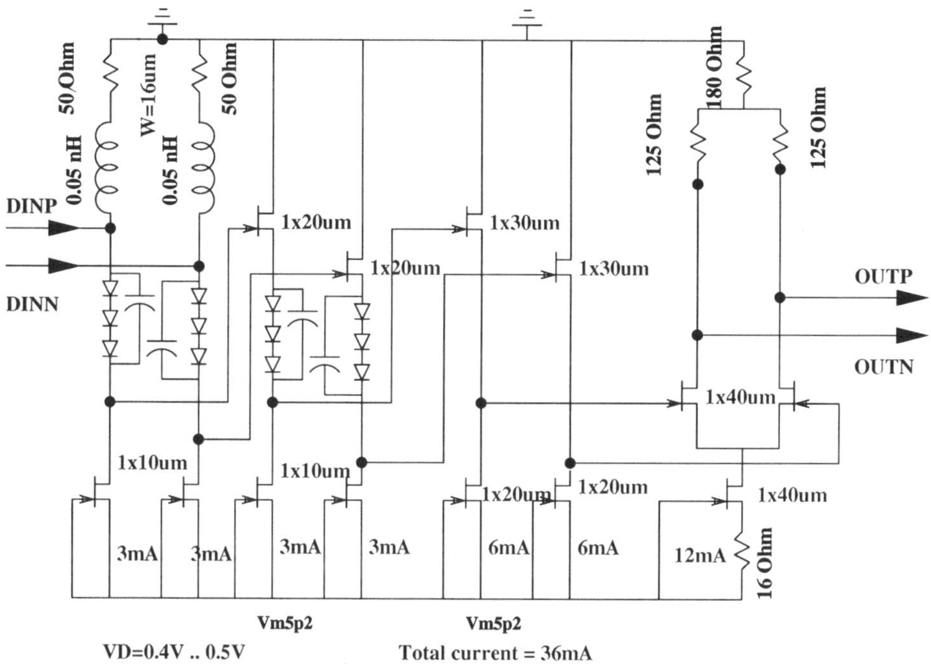

Fig. 14. a. Schematics for GaAs p–HEMT SCFL building block [26].

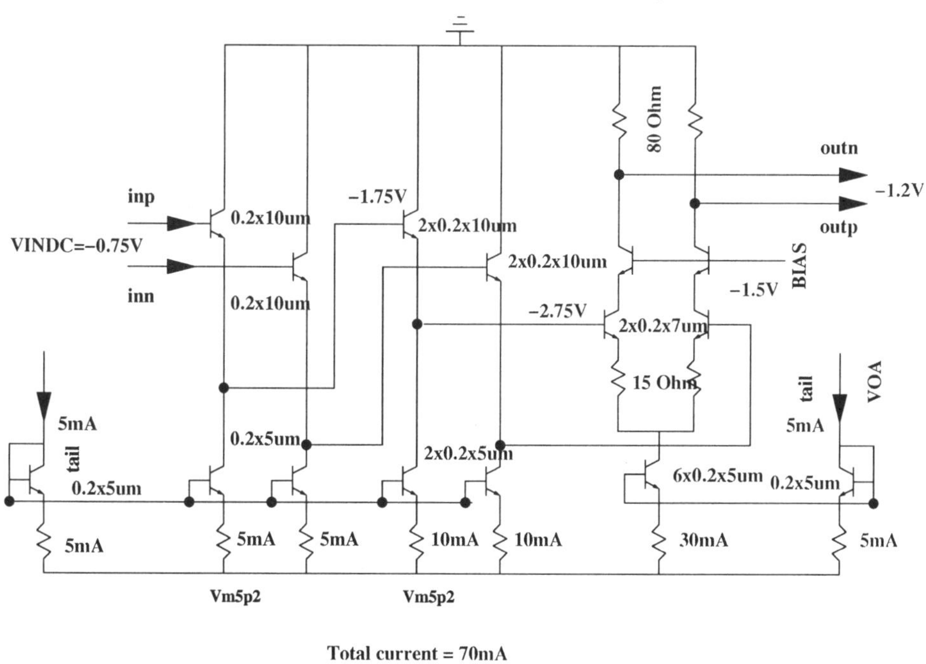

Fig.14. b. Schematics for SiGe HBT E^{2}CL building block.

Of all the building blocks in a fiber system, low phase noise VCOs with 15–20% tuning range have proven to be the most challenging to integrate at 10 Gb/s. Owing to the difficulty of integrating high quality inductors, varactors and fast transistors on the same substrate, it has not always been possible to achieve adequate performance. This is equally true for 20 GHz and 40 GHz VCOs. As an example, Fig. 15 shows the schematics of a 20 GHz VCO [18]. It has a differential varactor–tuned LC Colpitts topology in common–base configuration, with two inductors or a single, center–tapped 3–terminal inductor. This topology is scalable over a wide range of frequencies from 1.5 GHz to 120 GHz [3, 17–18, 24–25] and employs resistive emitter degeneration R_E to suppress harmonics and to reduce up(down) converted noise. The VCO can also operate on the second harmonic of the VCO tank, i.e. 40 GHz, when the signal is collected at node X. Fig. 16 illustrates the measured impedance as a function of control voltage and frequency for the half–circuit of the resonator, consisting of inductor L_B, MIM capacitor C_E and multi–stripe varactor diode D1. In order to characterize the resonator performance, the S parameters were measured between 10 GHz and 40 GHz with 50 MHz steps for each varactor bias. The resonant frequency was obtained from the peak, and the Q (larger than 4) was calculated from the 3 dB half–window, respectively, of the magnitude of the measured tank impedance. The resonant frequency of on–chip LC tanks is tunable over a 15% bandwidth and has low sensitivity to temperature variations [17–18]. The measured phase noise of the 20 GHz VCO is 100 dBc/Hz at 1 MHz from the carrier, as shown in Fig. 17. When operated on the second harmonic of the tank, the VCO frequency was tunable over the 40 to 45 GHz range but the phase noise was degraded due to the transistor operating beyond f_T (42 GHz) and close to f_{MAX}. (55 GHz).

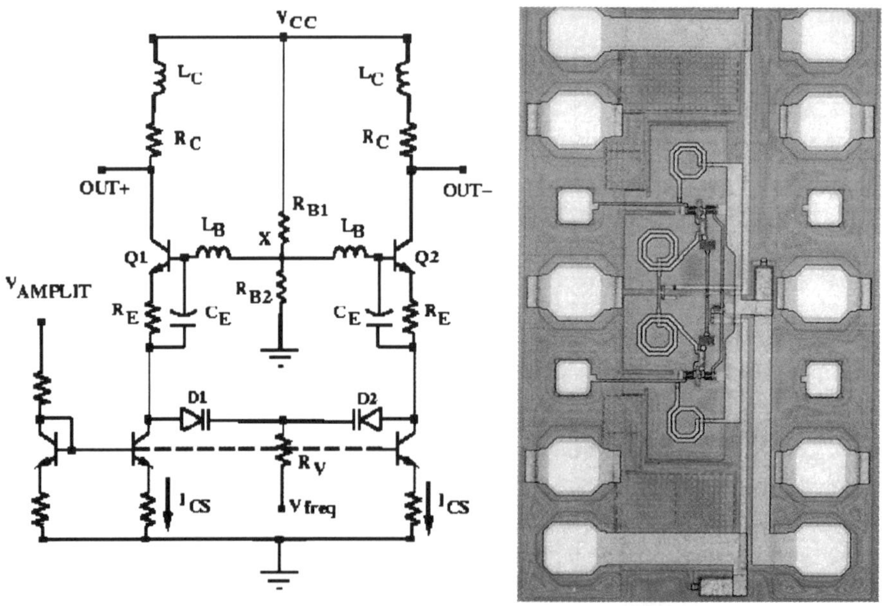

Fig.15. Schematic and layout of a family of L–C–varactor VCOs operating in the 20–40 GHz range [18].

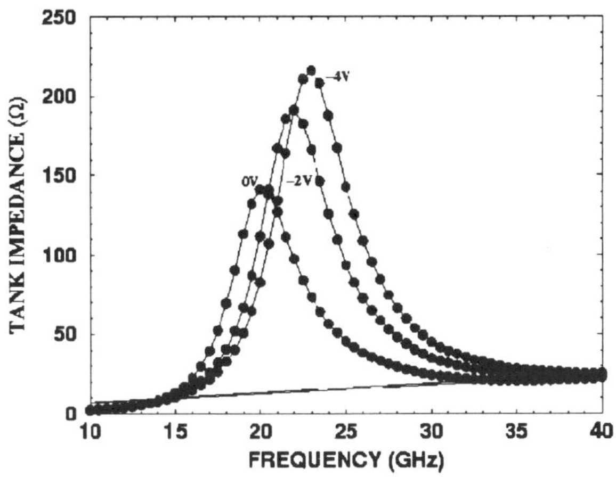

Fig.16. Measured impedance of a 20 GHz VCO L–C–varactor tank fabricated in a SiGe HBT process as a function of frequency and varactor voltage [18].

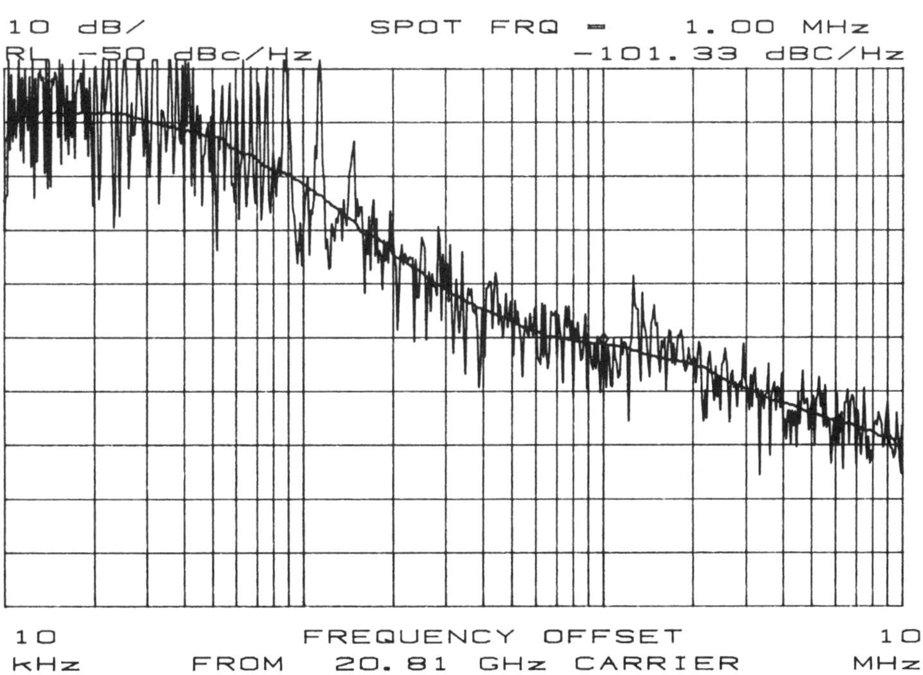

Fig.17. Measured phase noise of the 20 GHz VCO at 1 MHz from the carrier [18].

4. Technology Choices

At 10 Gb/s, it has been possible to implement most of these analog and digital functions, including 5V drivers, in a production GaAs HBT process [1]. In general, PLL and CDR circuits were initially implemented using discrete components. More recently they have been integrated in either Si CMOS or SiGe bipolar/BiCMOS processes, even for 40 Gb/s systems [2–4,20–21].

III–V HEMTs and MESFETs have been the technology of choice for modulator drivers at 10 Gb/s [1] and 40 Gb/s [26]. 10 Gb/s receivers and transceivers have been demonstrated in SiGe HBT processes and some are now commercially available [2,4–5]. Separate TIAs and 3 V laser/modulator drivers have been fabricated in Si bipolar or SiGe HBT technologies [19]. Receivers, transmitters and transceivers, clocked at 5 GHz [6] or at 10 GHz [7], have also been realized in 0.18 μm SOI–CMOS, or 0.18 μm CMOS, respectively, but it is highly probable that 10 Gb/s 3V or 5V drivers will not be realizable in present or future generation Si CMOS. GaAs HBTs and p–HEMTs will retain the 5V modulator driver markets in long–haul SONET/SDH applications.

For 10 Gb/s serializer–deserializer (SERDES) functions operating at or below 3.3V supply, 0.13/0.18 μm CMOS and second generation SiGe BiCMOS (f_T = 70 GHz, f_{MAX} = 80 GHz) have become the technologies of choice. 10 Gb/s short to medium reach applications require the most cost effective high performance technology available. As illustrated in Figs 18–19, for f_T values beyond 80 GHz, the impact of the transistor speed on the rise/fall time of the 10 Gb/s eye diagram at the output of a packaged SERDES diminishes. However, the improvement in the overall deterministic and random jitter is still noticeable with every new technology generation.

SiGe BiCMOS is a technology that has been in high volume production for the last 3 years, driven by consumer wireless applications and given the more relaxed lithography, is very cost effective. It is perfectly suited for analog functions at 10 Gb/s because of the high speed (SiGe) bipolar transistors and CMOS for some control function implementation. From the point of view of implementing large amounts of digital logic, CMOS is crucial. For more complex applications where the SERDES function is integrated with large digital processing functions such as Forward Error Correction (FEC), 0.13 μm CMOS is needed to keep die size and power consumption as low as possible. However, such a CMOS process is higher cost than conventional "digital" CMOS processes because it also has to integrate high quality varactor, MIM capacitor and thick top metal inductors.

OC–768 circuits require rise/fall times below 10 ps and random jitter values below 0.25 ps rms. The integration of low–phase noise VCOs with very wide tuning range and adequate output power at 40 GHz continues to be a problem, irrespective of the technology. The first commercial 40 Gb/s systems will have lower levels of integration and reduced functionality using 20 GHz rather than 40 GHz clocking schemes [20–21,23]. A combination of technologies, such as InP HBTs for the highest speed sections and CMOS or SiGe BiCMOS for the lower speed sections of the system, is the most likely implementation. Third generation SiGe BiCMOS (> 170 GHz f_T, f_{MAX}) will take over the SERDES function as the technology matures and volumes pick up. Modulator driver functions will continue to be implemented in GaAs p/m–HEMT technologies, the only ones that have proven capable of accommodating large voltage swings reliably at 40 Gb/s and beyond.

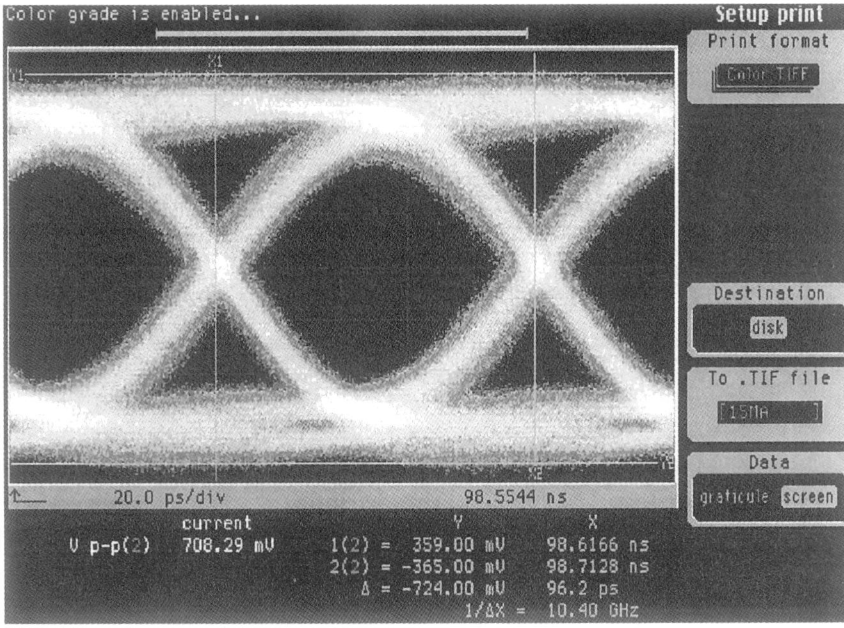

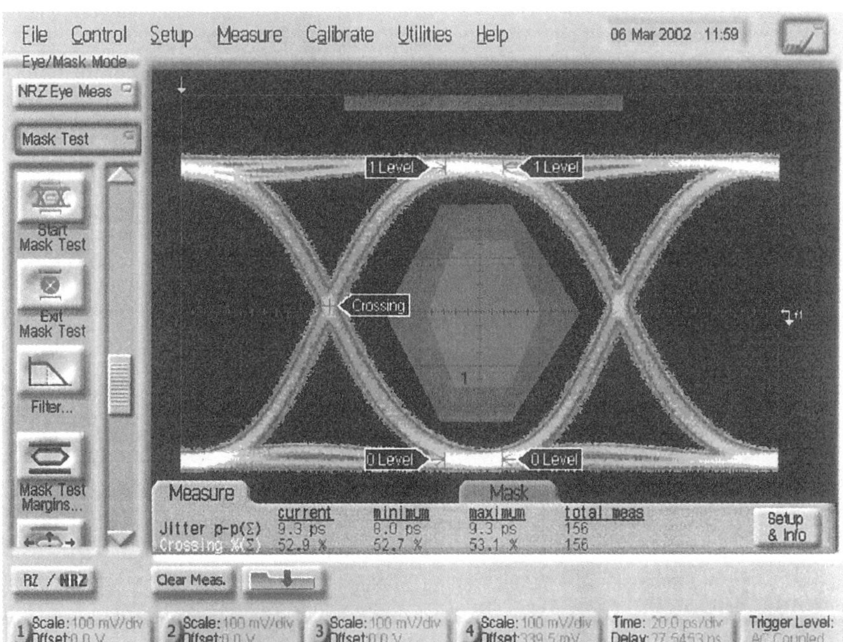

Fig.18. Measured 10.3 Gb/s eye–diagrams at the transmitter output for a packaged SERDES fabricated in a SiGe BiCMOS process with: a) npn f_T = 45 GHz and f_{MAX} = 45 GHz; b) npn f_T = 70 GHz and f_{MAX} = 80 GHz.

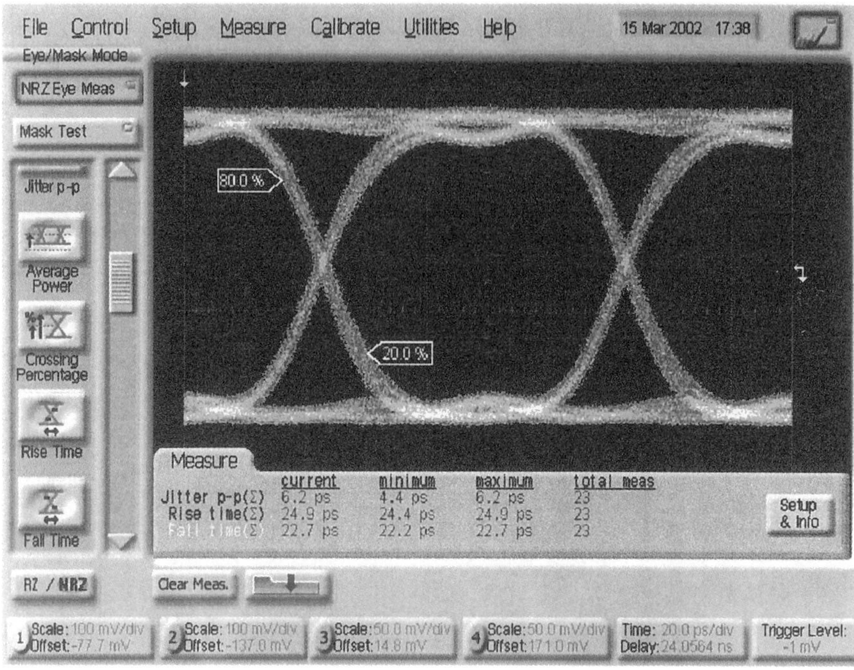

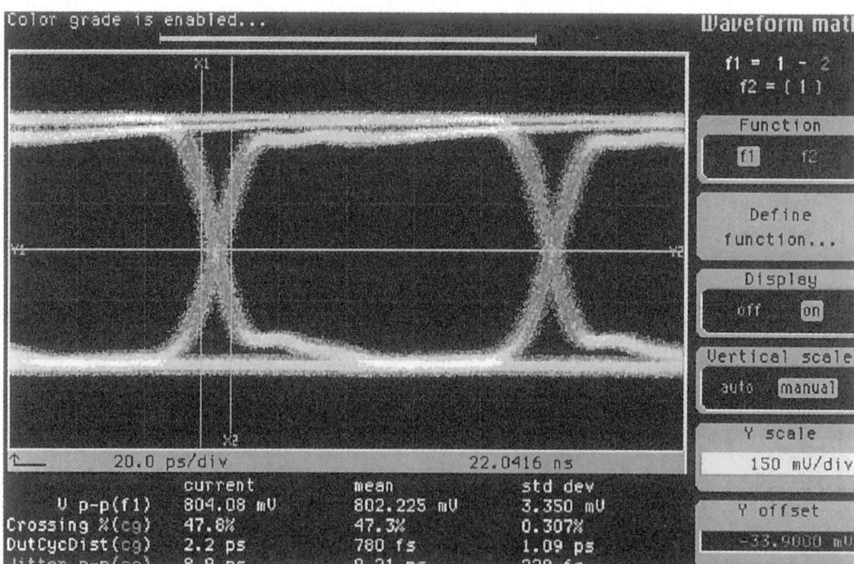

Fig.19. Measured 10.7 Gb/s eye–diagrams for an output driver fabricated in:
a) a SiGe BiCMOS process with npn f_T = 160 GHz and f_{MAX} = 150 GHz and mounted in a PBGA package
b) an InP HBT process with f_T = 170 GHz and f_{MAX} = 150 GHz and measured on wafer (no package).

5. Examples of 10 and 40 Gb/s Circuits

5.1. *10 Gb/s Transimpedance–limiting amplifier*

Fig. 20 shows the block diagram of a single–chip transimpedance limiting amplifier (TIALA) for OC–192 SONET/SDH STM–64 and 10 Gb/s Ethernet (IEEE802.3ae, 10GE) applications with data rates up to 12.5 Gb/s. The block diagram combines a single–ended TIA stage, a multi–stage limiting amplifier, an output buffer with a gain of 1.5, a feedback amplifier for dc offset cancellation, a peak detector for monitoring the input signal, a loss–of–signal (LOS) circuit with adjustable threshold, and a bandgap reference. The TIA stage has 500 Ω transimpedance gain and was designed to minimize the equivalent input noise, while maintaining good linearity for the entire range of input currents: 25 μAp–p to 2 mAp–p. It operates in linear mode and has a bandwidth greater than 8 GHz over all process, supply and temperature corners. The limiting amplifier stages consist of differential inverters with optional three–terminal inductor peaking and emitter followers. The CAZ1 pad can also be used to externally adjust the slicing level [27] at the input of the limiting amplifier between 20% and 80% of the eye height. The LOS block features a hysteresis comparator with low–voltage TTL outputs. The threshold of the LOS circuit is adjustable via an external potentiometer placed between the VREF, VSET and GND pads. The entire circuit draws 60 mA from a single 3.3 V supply.

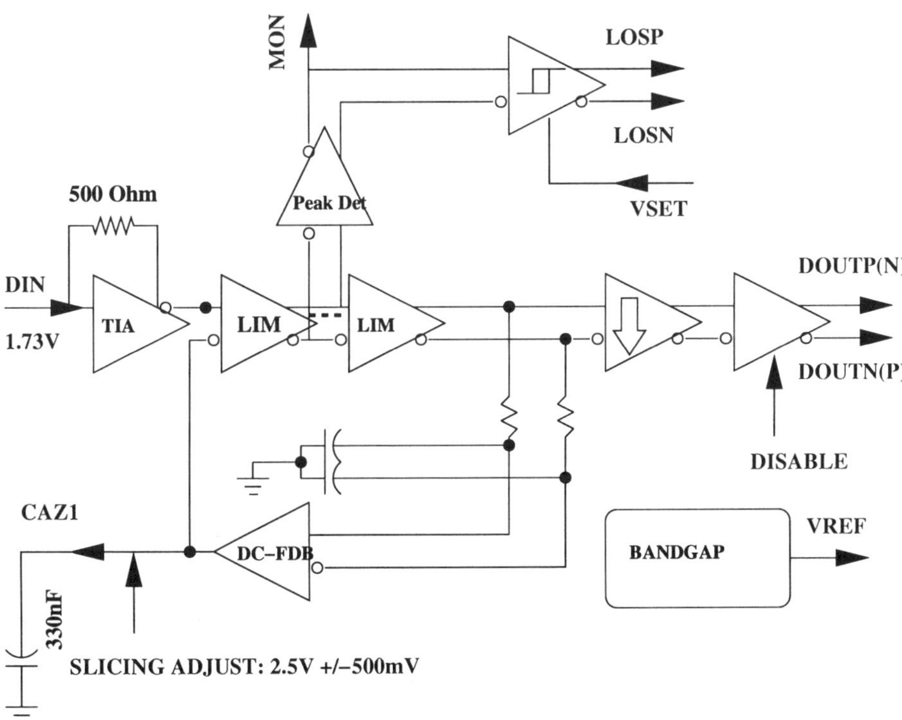

Fig.20. Block Diagram of the TIALA. The gain block with an arrow pointing down indicates an emitter–follower stage.

The circuit was fabricated in two successive generations of SiGe BiCMOS technologies with 0.35 μm and 0.25 μm feature sizes, respectively. The main features of the two processes are described in Table 2, together with the TIALA performance. The 1.1x1.3mm^2 TIALA die is shown in Fig. 21 and has over 400 npn/MOS transistors, inductors for peaking, and MIM capacitors for bias de–coupling filters. On–chip resistively matched high impedance transmission lines and pn junctions [5,17] are employed to ensure high input–output isolation (S_{12} < −60 dB) as required by the small die size and high gain (S_{21} > 40 dB) of the circuit. Because of the very large gain, the TIALA operates in limiting mode even at input currents as low as 30 μAp–p.

Table 2. 10Gb/s TIALA Technology and Circuit Performance

Parameter	***0.35 μm BiCMOS***	***0.25 μm BiCMOS***
SiGe npn peak f_T @ V_{CE} = 1 V	45 GHz	75 GHz
SiGe npn peak f_{MAX} @ V_{CE} = 1 V	50 GHz	80 GHz
peak f_T current density	1 mA/μm^2	3 mA/μm^2
BV_{CEO}	3.5 V	2.8 V
TIALA rise/fall time	30 ps	25 ps
Imin @BER 1E–12, 10.7 Gb/s R = 0.9	37 uAp–p −16.5dBm	25 uAp–p −18.5 dBm
Imax @BER 1E–12, 10.7 Gb/s	3 mAp–p	2 mAp–p
Duty Cycle Distortion	<8 ps	<8 ps
TIA bandwidth	>8 GHz	>8 GHz
TIALA small signal bandwidth	5.5 GHz	9 GHz

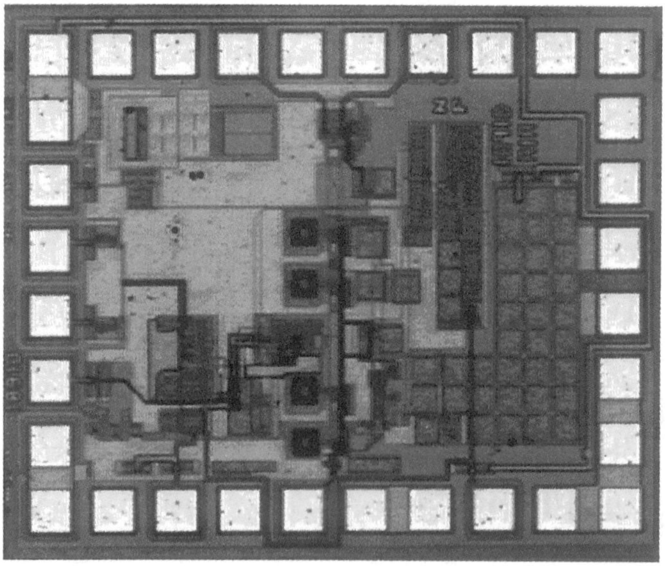

Fig.21. Layout the of the 0.35 μm SiGe BiCMOS TIALA.

Fig. 22 shows the measured single–ended small signal power gain. Also plotted is the output return loss for different loads on the input pad. It is better than −10 dB up to 20 GHz. Because the input impedance of the TIALA is almost 50 Ω, S_{11} is better than −10 dB. This allows for the circuit to be fully testable, including sensitivity, through electrical measurements only, by directly applying the signal from the Bit–Error–Rate–Tester (BERT).

Fig. 23 reproduces the on–wafer measured eye diagrams at 10.7 Gb/s for an input current level of 30 μAp–p and a pseudo–random–bit–signal PRBS pattern of $2^{31}-1$. The output swing is limited to 300 mVp–p per side for input currents between 25 μAp–p and 3 mAp–p. The performance parameters summarized in Table 2 indicate that, as long as the TIA bandwidth remains unchanged, the process speed and small signal bandwidth of the entire TIALA only impact the sensitivity and, due to the limiting mode of operation, to a less extent the rise/fall times and data rate of the circuits. The peaking inductors were eliminated from the limiting amplifier stages in the case of the faster process implementation.

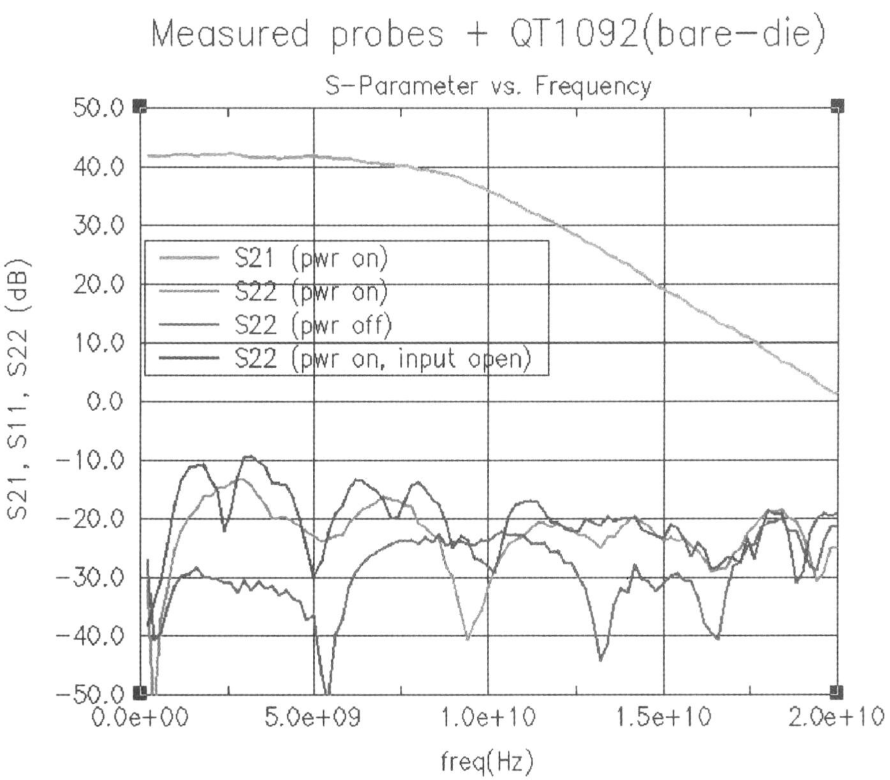

Fig.22. Measured small signal (Pin=–65 dBm) bandwidth and S parameters for the 0.25 μm SiGe BiCMOS implementation.

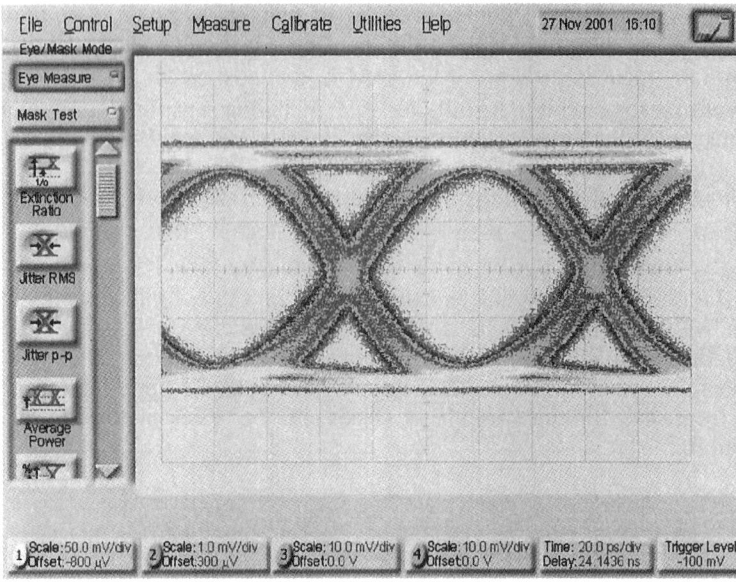

Fig.23. Measured 10.7 Gb/s eye–diagram with 30 μAp–p input for the 0.25 μm SiGe implementation. The scales are 20 ps/div. and 50 mV/div, respectively.

5.2. *40 Gb/s Transimpedance–limiting amplifier*

Figs. 24 and 25 show the block diagram and layout, respectively, of a 40 Gb/s TIALA implemented in an InP HBT process. The circuit integrates more than 300 HBTs on a single 1x1.8mm^2 die, while consuming only 550 mW from a 3.3 V supply. The measured on–wafer differential eye diagram at 42 Gb/s is illustrated in Fig. 26 for an input current level of 500 μAp–p and a $2^{31}-1$ PRBS pattern. The output swing is limited to 400 mVp–p per side for input currents between 100 μAp–p and 3 mAp–p.

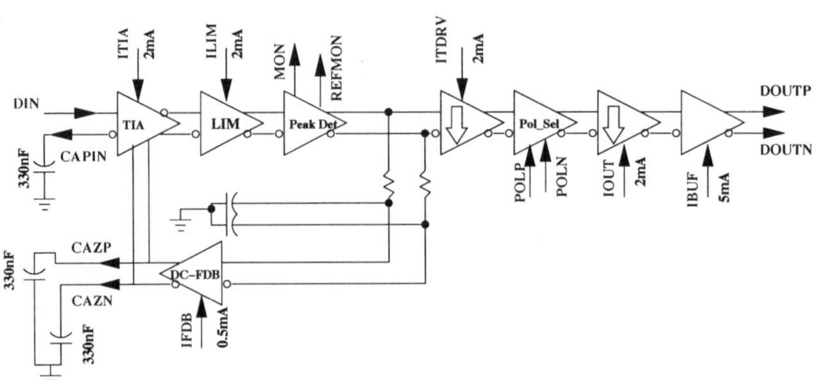

Fig.24. Block diagram of the 40 Gb/s InP HBT TIALA

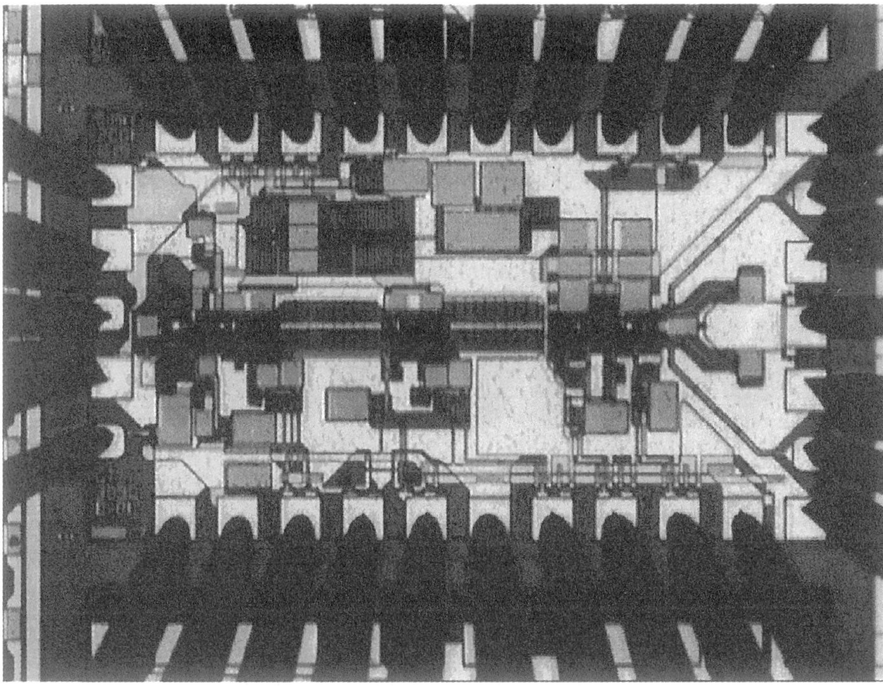

Fig.25. Layout of the 40 Gb/s InP HBT TIALA

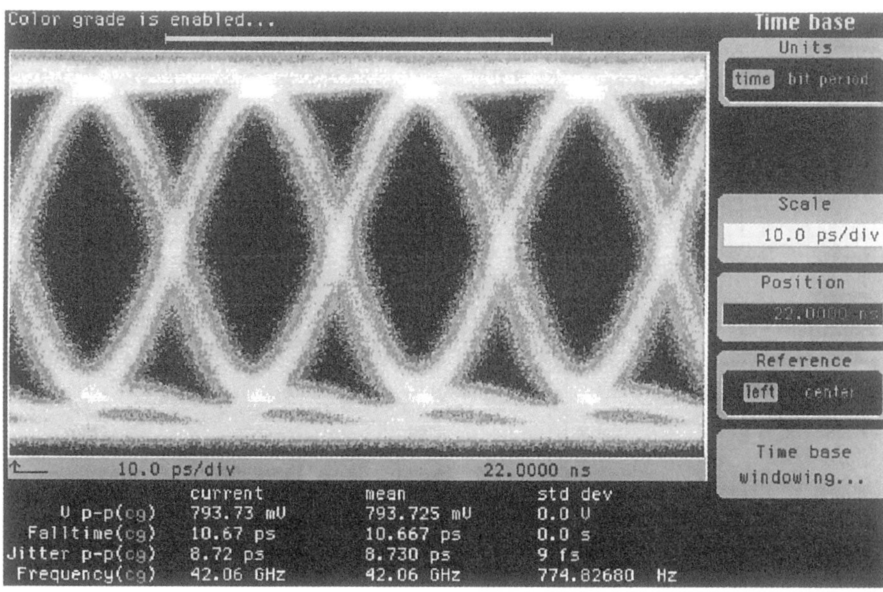

Fig.26. Measured TIALA 42 Gb/s differential eye–diagram with 0.5 mAp–p input. The scales are 10 ps/div. and 100 mV/div, respectively.

5.3. *10 Gb/s Driver.*

The block diagram of a differential output driver with programmable wave shape is illustrated in Fig. 27. It features output swing control between 250 mVp–p per side and 800 mVp–p per side into 50 Ω loads, up to 30% separate positive and negative peaking control, as well as pulse–width (or duty–cycle) control. As in the case of the 10 Gb/s SiGe BiCMOS TIALA, simple building blocks using differential inverters and emitter followers are employed.

The pulse–width (DCD) control block is implemented as shown in Fig. 12. In order to separately control the output signal amplitude, the amount of positive, and the amount of negative pre–emphasis (or peaking), the signal is split in three parallel paths. The two bottom digitally differentiate the rising and falling edge, respectively, of the data signal, as illustrated with dotted triangles in the block diagram. Each path terminates with an open–collector differential inverter featuring an adjustable bias current source which controls the amplitude of the signal for that path. The collectors of all three inverters (one for each path) are tied together in a summing 50 Ohm resistive load. To ensure the proper alignment of the signals, the delays along each path must be matched across process corners and temperature.

The chip operates from a single positive 3.3 V supply with a power consumption of 0.5 W for a midrange output swing of 500 mVp–p per side. It has over 3000 HBTs and MOSFETs and is fabricated in a 0.35 μm SiGe BiCMOS process. Fig. 28 presents the on–wafer measured 10.3 Gb/s eye–diagrams with a 2^{31}–1 PRBS pattern at a single–ended output demonstrating 800 mVp–p per side and peaking control.

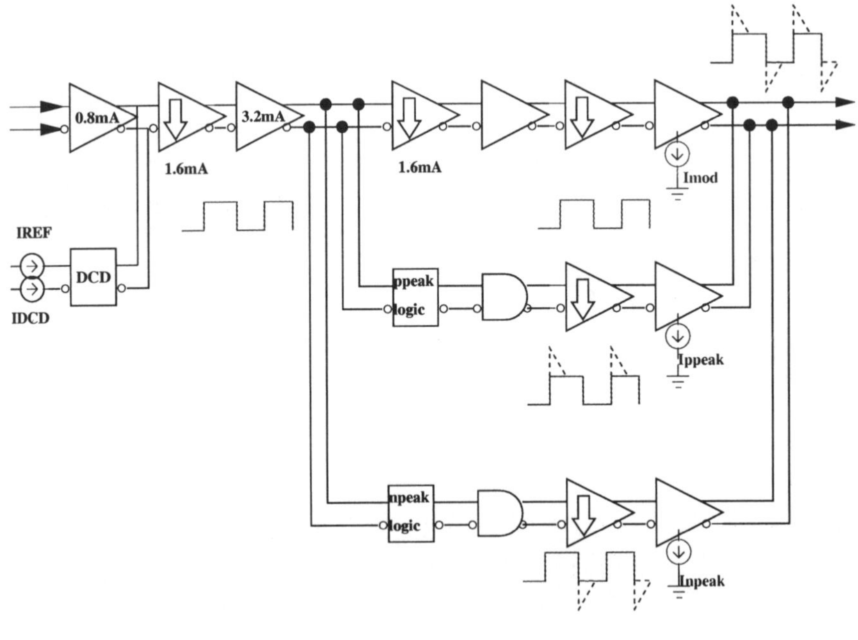

Fig.27. 10 Gb/s output driver schematic. The gain block with an arrow pointing down indicates an emitter–follower stage. The signal shape at various stages is also illustrated.

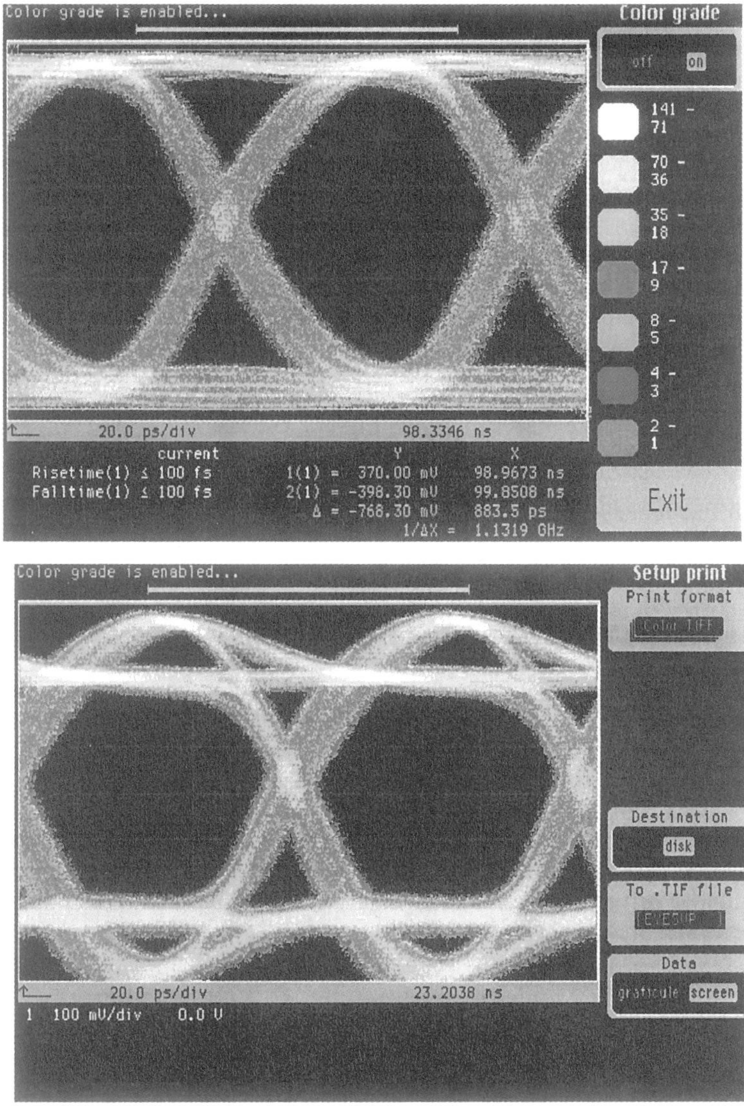

Fig.28. On wafer measured 10.3 Gb/s eye–diagrams for a driver fabricated in a 0.35 μm SiGe BiCMOS process with npn f_T = 45 GHz and f_{MAX} = 45 GHz. a) without pre–emphasis, b) with both positive and negative pre–emphasis. The scales are 20 ps/div. and 100 mV/div, respectively.

5.4. *40 Gb/s Driver.*

The block diagram of a 40 Gb/s Electro–Absorbtion (EA) modulator driver, shown in Fig. 29, features three limiting amplifier blocks: a lumped input block, a middle lumped limiting amplifier with adjustable duty cycle (pulse width) and a 5–stage distributed limiting amplifier block. The lumped limiting amplifier blocks, as well as the sections of the distributed amplifier have a similar topology, consisting of double source follower stages and a differential inverter, as in Fig. 14(a) More details can be found in [26]. The entire driver operates from a single –5.2 V supply with a gain of at least 16dB and limits the output swing to 3 Vp–p per side for input signals larger than 0.7 V p–p per side. The differential outputs are DC–coupled to the EA modulator and to an external 50 Ω load, respectively. The DC level at the two differential outputs, also called the DC output offset, is adjustable between –0.75 V and –2.25 V. This provides flexibility in biasing the EA modulator. A differential drive Mach–Zender modulator can also be DC–coupled to the differential outputs of the driver.

The die, shown in Fig. 30, is 60 μm x 1.95 mm x 3.99 mm and is fabricated in a 0.15 μm GaAs p–HEMT process with f_T of 110 GHz and f_{MAX} of 180 GHz.. The process features one metal layer, airbridge and back–side via, interdigitated p–HEMTs with more than 6 V breakdown, interdigitated Schottky diodes, epitaxial resistor (110 Ω/sq) and MIM capacitor. The maximum permissible junction temperature is 120 °C. The measured 42 Gb/s differential eye–diagram with 3 Vp–p per side is shown in Fig. 31.

6. Conclusions

The main semiconductor technologies which vie for the implementation of highly integrated 10 Gb/s and 40 Gb/s PHY ICs have been compared based on system and circuit block requirements. Important aspects related to active and passive device scaling and modeling were addressed demonstrating excellent accuracy when compared to experiments. Specific trade–offs in the implementation of digital and analog building blocks in silicon and III–V technologies were discussed. In the case of basic digital building blocks, circuit topologies which best take advantage of the features of a particular technology were illustrated using FETs as well as bipolar transistors. Finally, fully featured 10 Gb/s and 40 Gb/s transimpedance limiting amplifier and laser/modulator driver circuits fabricated in SiGe BiCMOS, InP HBT and GaAs p–HEMT processes were described.

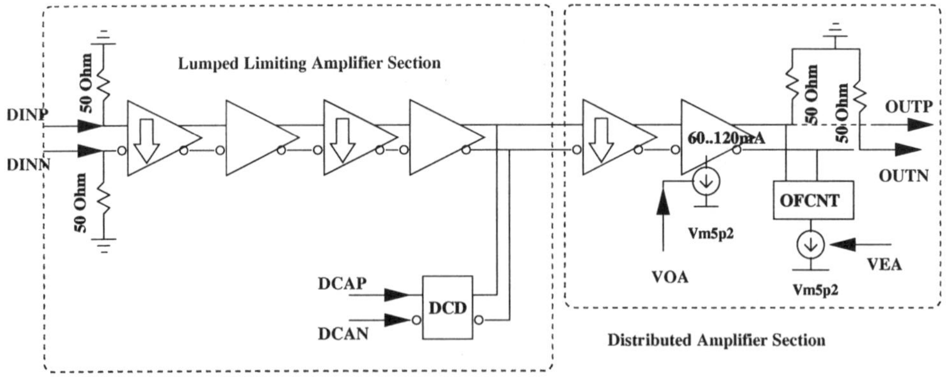

Fig.29. Block diagram of a 40 Gb/s SERDES EA modulator driver with pulse–width, output amplitude and dc offset control. The gain block with an arrow pointing down indicates a source–follower stage.

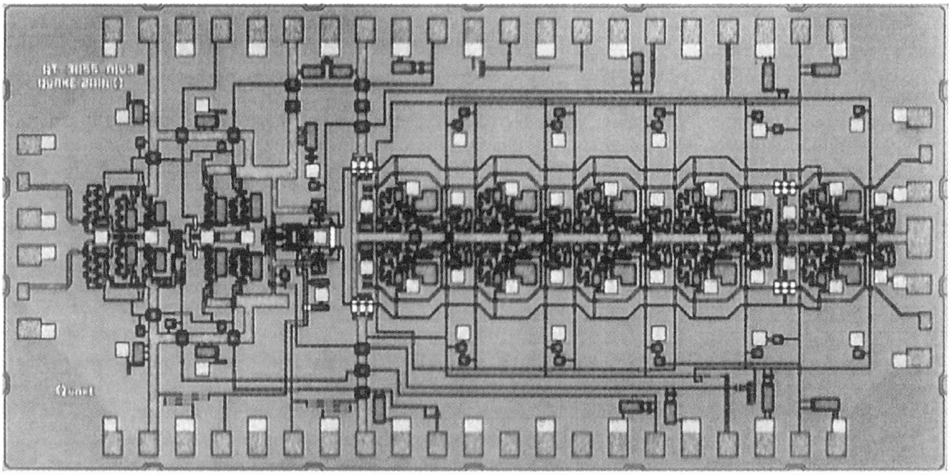

Fig.30. Layout of the 40 Gb/s EA modulator driver.

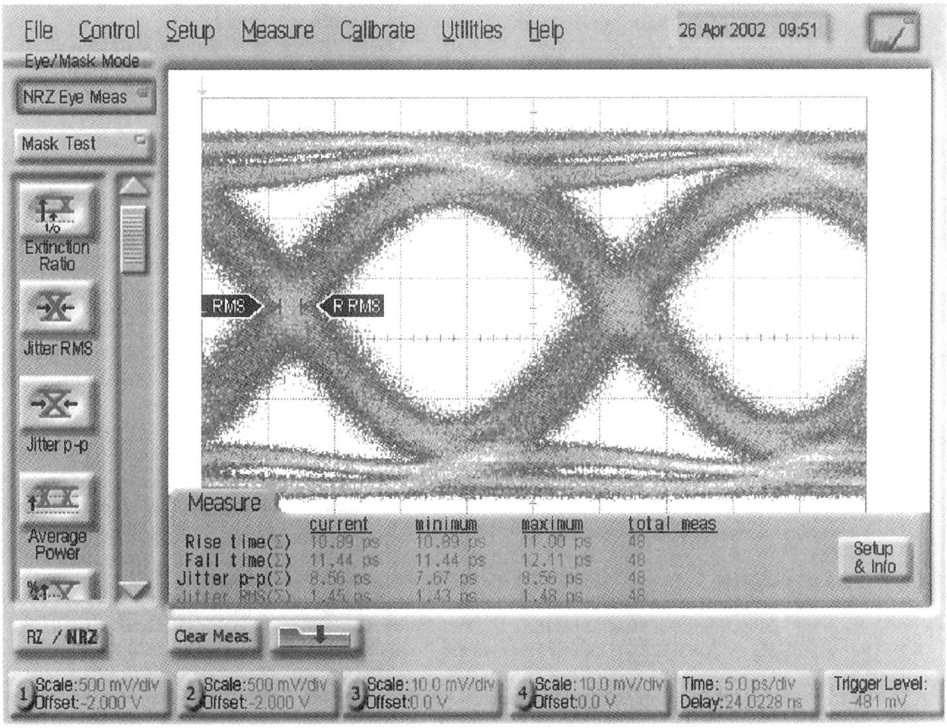

Fig.31. Measured 42 Gb/s differential eye–diagram with 3 Vp–p output swing per side. The scales are 5 ps/div and 500 mV/div, respectively.

References

1. B. Beggs, "GaAs HBT 10 Gb/s Product," IEEE MTT–S International Microwave Symposium Workshop, June 13–19, 1999, Anaheim, Ca.
2. R.Hadaway, P.Schvan, S.Voinigescu, Y.Greshishchev, D.Marchesan, J.Ojha, P.Dawe, "Application of SiGe HBTs to Datacom and Wireless", MRS, San Francisco, April, 1998.
3. Y. Greshishchev, P.Schvan, J. Showell, M–L Xu, J.Ojha, and J.Rogers, "A Fully Integrated SiGe Receiver IC for 10 Gb/s Data Rate", ISSCC Digest, pp.52–53, 2000.
4. M.Menghelli, B. Parker, H.Ainspan, and M.Soyuer, "SiGe BiCMOS 3.3V Clock and Data Recovery Circuits for 10 Gb/s Serial Transmission Systems", ISSCC Digest, pp.56–57, 2000.
5. S.Voinigescu, P.Popescu, P.Banens, M.Copeland, G.Fortier, K.Howlett, M.Herod, D. Marchesan, J.Showell, S.Szilagyi, H.Tran, and J.Weng, "Circuits and Technologies for Highly Integrated Optical Networking IC's at 10 Gb/s to 40 Gb/s," Proc. of the IEEE CICC, pp. 331–338, 2001.
6. T. Yoshimura, K. Ueda, J. Takasoh, Y. Wada, T. Oka, H. Kondoh, O Chiba, Y. Azekawa and M. Ishiwaki, "A 10Gbase Ethernet Transceiver (LAN PHY) in a 1.8V, 0.18 μm SOI/CMOS Technology.", Proc. of the IEEE CICC, pp.355–358 , 2002.
7. M.M. Green, A. Momtaz, K. Vakilian, X. Wang, K–C. Jen, D. Chung, J. Cao, M. Caresosa, A. Hairapetian, I. Fujimori, and Y. Cai, " OC–192 Transmitter in Standard 0.18 μm CMOS," and J. Cao, A. Momtaz, K. Vakilian, M. Green, D. Chung, K–C. Jen, M. Caresosa, B. Tan, I. Fujimori, and A. Hairapetian, OC–192 Receiver in Standard 0.18 μm CMOS," ISSCC Digest, pp.248–251, 2002.
8. T.Masuda, K–I Ohhata, F. Arakawa, N. Shiramizu, E. Ohue, K.Oda, R. Hayami, M. Tanabe, H. Shimamoto, M. Kondo, T. Harada, and K. Washio, "45 GHz Transimpedance 32 dB Limiting Amplifier and 40 Gb/s 1:4 High–Sensitivity Demultiplexer with Decision Circuit using SiGe HBTs for 40 Gb/s Optical Receiver", ISSCC Digest, pp.60–61, 2000.
9. R. Tang et al., "A low–Cost Modulator SiGe BiCMOS Technology and Analog Passives for High–Performance RF and Wide–Band Applications", IEEE BCTM, pp.102–105, 2000.
10. S. Subana, L. Larson, G. Freeman, D. Ahlgren, K. Stein, C. Dickey, J. Mecke, A. Rincon, P. Bacon, R. Groves, M. Soyuer, D. Harame, J. Dunn, D. Rowe, W. Chon, D. Herman, and B.Meyerson, "Silicon–Germanium BiCMOS Technology and a CAD Environment for 2–40 Ghz VLSI Mixed–Signal ICs", Proc. of the IEEE CICC, pp.559–566, 2001.
11. M. Racanelli, K. Schuengraf, A. Kalburge, A. Kar–Roy, B. Shen, C. Hu, D. Chapek, D. Howard, D. Quon, F.Wang, G. U'ren, L. Lao, H. Tu, J. Zheng, J. Zhang, K. Bell, K. Yin, P. Joshi, S. Akhtar, S. Vo, T. Lee, W. Shi, and P. Kempf, "Ultra High Speed SiGe NPN for Advanced BiCMOS Technology," IEDM Techn. Digest, pp.336–339, 2001
12. J.Godin, M.Riet, S.Blayac, Ph.Berdaguer, J.–L.Benchimol, A.Konczykowska, A.Kasbari, Ph.Andre, N.Kauffman, "Improved InGaAs/InP DHBT Technology for 40 Gbit/s Optical Communication Circuits", IEEE GaAs IC Symposium Technical Digest, pp.77–80, 2000.
13. T.Otsuji, K.Murata, T.Enoki, and Y.Umeda, "An 80–Gbit/s Multiplexer IC Using InAlAs/InGaAs/InP HEMTs", IEEE JSSC, vol.33, pp.1321–1326, 1999.
14. S.P.Voinigescu, P.B. Rabkin, C.A.T. Salama, and P.A. Blakey, "2D Numerical Investigation of the Impact of Compositional Grading on the Performance of Submicrometer Si–SiGe MOSFETs", IEEE Trans. ED, vol.42, pp.1039–1046, 1995.
15. S.P.Voinigescu, S.W.Tarasewicz, T.MacElwee, and J.Ilowski, "An Assessment of the State–of–the–Art 0.5μm CMOS Technology for RF Applications,", IEDM Techn. Digest, pp.721–724, 1995.
16. S.P.Voinigescu, D.Marchesan, J.L.Showell, M.C.Maliepaard, M.Cudnoch, M.G.M.Schumacher, M.Herod, G.E.Babcock, P.Schvan,and R.A.Hadaway, "Process– and Geometry Scalable Bipolar Transistor and Transmission Line Models for Si and SiGe MMICs in the 5– 22 GHz Range," Proc.of IEDM, pp.307–310, 1998.
17. S.P. Voinigescu, M.A. Copeland, D. Marchesan, P. Popescu, and M.C. Maliepaard, "5 GHz SiGe HBT Monolithic Radio Transceiver With Tunable Filtering", Digest of Papers, RFIC Symp., pp.131–134 1999.
18. S.Voinigescu, D.Marchesan, M.A.Copeland, "A Family of Monolithic Inductor–Varactor SiGe–HBT VCOs for 20 GHz to 30 GHz LMDS and Fiber–Optic Receiver Applications,", Proc. 2000 RF IC Symp, Boston.
19. H.–M. Rein, "Si and SiGe Bipolar IC's for 10 to 40 Gb/s Optical–Fiber TDM Links", Int.J.of High–Speed Electron. Syst., vol.9, pp.1–37. 1998.
20. M. Wurzer, J. Bock, H. Knapp, W. Zirwas, F. Schumann, and A. Felder, "A 40 Gb/s Integrated Clock and Data Recovery Circuit in a 50–GHz f_T Silicon Bipolar Technology", IEEE JSSC, vol.34, pp.1320–1324,

1999.
21. M. Wurzer, T.F. Meister, H. Schafer, H. Knapp, J. Bock, R. Stengl, R. Aufinger, M. Franosch, M. Rest, H.–M. Rein, and A. Felder, "42 GHz Static Frequency Divider in a Si/SiGe Bipolar Technology", ISSCC Digest, pp.122–123, 1997.
22. Y.Miyamoto, M.Yoneyama, T. Otsuji, K.Yonenaga, and N.Shimizu, "40–Gbit/s TDM Transmission Technologies Based on Ultra–High–Speed IC's", IEEE JSSC, vol.34, pp.1246–1253, 1999.
23. J.P. Mattia, R. Pullela, Y. Baeyens, Y–K Chen, H–S Tsai, G. Georgiou, T.W. Von Mohrenfels, M. Reinhold, C. Groepper, C. Dorschky, and C. Schulien, "A 1:4 Demultiplexer for 40 Gb/s Fiber–Optic Applications", ISSCC Digest, pp.64–65, 2000.
24. L.Dauphinee, M.A.Copeland, and P.Schvan, "A Balanced 1.5 GHz Voltage Controlled Oscillator with an Integrated LC Resonator," IEEE ISSCC Digest, pp.390–391, 1997.
25. Y. Baeyens, R.Pullella, C. Dorschky, J.–P. Mattia, R.Knopf, H.–S. Tsai, G. Georgiou, R. Hamm, Y.C. Wang, Q. Lee, Y.–K. Chen, "Compact Differential InP–Based HBT VCO's with Wide Tuning Range at W–Band", IEEE MTT–Symposium Technical Digest, 2000.
26. D. S.McPherson, F. Pera, M.Tazlauanu, and S.P.Voinigescu, "A 3V Fully Differential Distributed Limiting Driver for 40 Gb/s Optical Transmission Systems," accepted at the IEEE GaAs IC Symp., Oct. 2002
27. E.Sackinger, "Broadband Circuits for Optical Fiber Communication," MEAD Microelectronics lecture notes, March 2002.

International Journal of High Speed Electronics and Systems
Vol. 13, No. 1 (2003) 59–64

World Scientific
www.worldscientific.com

THE TRANSFORMING MMIC

EDGAR J. MARTINEZ
Defense Advanced Research Projects Agency
3701 N. Fairfax Drive
Arlington, VA 22203
emartinez@darpa.mil

In this paper, we describe two new DARPA initiatives addressing new concepts in compound semiconductor materials and architectures that will radically transform monolithic microwave integrated circuits (MMICs) technology to address future requirements for military and commercial sensors and mobile communication networks.

Keywords: DARPA, wide-bandgap semiconductor materials, intelligent RF front-ends, intelligent Microsystems, MMICs.

1. Introduction

Since the invention of the transistor 55 years ago, microelectronics technologies have enabled the way that we communicate, acquire, and process information, playing one of the most important roles in today's information era. Over the last three decades, the Defense Advanced Research Projects Agency (DARPA) has funded and conducted many programs that have served as catalysts for innovation in the compound semiconductor industry. These include the Gallium Arsenide Insertion Initiatives, the Microwave Monolithic Integrated Circuit (MIMIC), and Microwave Analog Front End Technology (MAFET). Many of the technology advances from these programs have been incorporated into today's highly capable communication, radar, electronic warfare, and information systems.

While DARPA has always focused on military unique problems, the community has also benefited from the economies of scale that the commercial markets provide. Today, the increasing demands for commercial electronics have resulted in a national semiconductor industry that remains strong and robust. Even with last year's recession, sales of semiconductor products exceeded $40 billion in the United States alone - with an estimated global market of $141 billion.

While commercial technology developments are meeting the demands of the evolving consumer markets, they still fall short of satisfying many emerging requirements critical to future military systems. For example, the twenty-first century battlefield scenarios are anticipated to be asymmetric, unpredictable, and will experience continued and rapid changes imposing a new set of requirements that will be hard to meet with current weapons and sensor system concepts. Likewise, the increasing demand for mobile communication-network solutions is anticipated to continue stressing the performance of commercial RFIC and mixed-signal components while targeting high manufacturing volumes and low cost solutions. To meet the demands of future military and commercial markets, a new paradigm shift in MMIC and RFIC technologies is anticipated in which component performance will be enhanced by new materials, device, and architectural concepts. Therefore, DARPA has started two new initiatives focusing on radical transformation in microwave and millimeter-wave component technologies.

2. Transforming MMICs through Innovations in Compound Semiconductor Materials

The first of these initiatives is aimed at the exploitation of the properties of wide bandgap compound semiconductor (WBGS) materials and devices to improve the analog performance of future microwave and millimeter-wave integrated circuits by orders of magnitude of current GaAs MMIC technology. WBGS materials are defined as semiconductor materials with conduction to valence band discontinuities larger than 3.0 eV – mainly silicon carbide, gallium nitride, and related compound semiconductors. In addition to WBGS, many of these materials exhibit higher electron velocity and thermal conductivity than any other compound semiconductor in their most pure form. Table 1 summarizes some of the material properties of wide bandgap semiconductors and compares them against other semiconductor materials. All these material properties make WBG semiconductors ideal for the demonstration of high-frequency analog integrated circuits in which high voltage and high temperature operation could lead to higher power density per unit cell and higher linearity MMICs.

Table 1: Comparison of Compound Semiconductor material properties

	Si (----)	GaAs (AlGaAs/ InGaAs)	InP (InAlAs/ InGaAs)	4H SiC (----)	GaN (AlGaN/ GaN)
Bandgap (eV)	1.1	1.42	1.35	3.26	3.49
Electron mobility (cm ^{2}/Vs)	1500	8500	10000	700	900
Saturated (peak) electron velocity (x10 ^{7} cm/s)	1	2.1	2.3	2	2.7
2DEG sheet electron density (cm $^{-2}$)	NA	<4 x 10^{12}	<4 x 10^{12}	NA	20 x 10^{12}
Critical breakdown field (MV/cm)	0.3	0.4	0.5	2	3.3
Thermal conductivity (W/cm-K)	1.5	0.5	0.7	4.5	>1.7
Relative dielectric constant (εr)	11.8	12.8	12.5	10	9.0

$$P_{max} \propto E_g^4$$

The Wide Bandgap Semiconductor Technology Initiative is currently focusing in material and device technology activities to achieve affordable MMIC fabrication processes. The initiative is being executed in three phases: (a) materials development phase, (b) device technology phase, and (c) WBG MMIC demonstration phase.

The focus of the material development phase is in the demonstration of robust fabrication processes for the manufacturing of large area substrates and reproducible epitaxial processes that will lead to devices with optimum performance.

To date, silicon carbide is the most mature wide bandgap semiconductor material. It is also the only wide-bandgap material available in the form of high quality, single crystalline, moderate diameter substrate – two-inch substrates are commercially available. It is also the preferred substrate material for all wide bandgap RF devices due to its high thermal conductivity and high crystalline quality. A specific objective of the Wide Bandgap Semiconductor Technology Initiative is the realization of four-inch substrates for the realization of affordable MMICs. Because, there is a 4% lattice mismatch between GaN and SiC, other substrate materials currently under consideration include aluminum nitride, and gallium nitride. Like SiC, in order for these alternate substrates to be considered for MMIC fabrication, they must be made reproducibly with properties and uniformity suitable for the production of large area devices and components.

For microwave and millimeter-wave devices, semi-insulating substrates are essential and particularly challenging. Due to impurities in the starting materials and the complexity of the fabrication processes, today's wide bandgap substrates are conductive and highly populated with dislocations and killer defects limiting the production of affordable RF components. Therefore, the current focus is in developing new fabrication processes for the manufacturing of highly pure semi-insulating substrates suitable for the fabrication of large area microwave and millimeter-wave integrated circuits.

While the substrate is the semiconductor's foundation, upon which engineers build devices and circuits, the epitaxial layers define the electrical and/or optical characteristics of the device. The crystalline structure of silicon carbide does not match that of gallium nitride or of most other candidate epitaxial materials. Therefore, special attention must be paid to producing interface layers that provide a good transition between silicon carbide substrates and the epitaxial layers grown on them. In the area of epitaxial material technologies, our focus is currently investigating and assessing the potential of several material growth techniques for the demonstration of highly reproducible epitaxial layers that will lead to high yield MMIC fabrication processes. It is currently a race between two techniques, Molecular Beam Epitaxy and Chemical Vapor Depositions, to demonstrate the best uniformity, reproducibility, and superior electronic performance.

The second phase of this effort will focus on device technologies. The goal of the second phase will be to establish fabrication processes for microwave and millimeter-wave MMICs. The approach will be in device-performance optimization, understanding device physics and failure mechanisms, establishment of appropriate device models for MMIC design, and advanced thermal management and packaging concepts. The last phase of this initiative will focus on the demonstration of reliable and cost effective high power microwave and millimeter-wave MMICs.

This effort not only will transform the state-of-practice of MMIC technologies but also will have a profound transformation at the system level. It is anticipated that MMIC with ten times the power density per unit cell, linearity, and bandwidth than their current GaAs counterparts will be demonstrated in the near future. This will allow MMICs to be a viable option for high-power, high-linearity applications in which microwave power modules, based on vacuum electronics, are currently the only option. This will enable lighter, cheaper, and more efficient high-discrimination sensors and mobile communication networks.

3. Transforming Architectures for Real-time Adaptability

The second DARPA initiative focuses on new component architectures in which digital and micro-electro-mechanical technologies are introduced into MMIC processes to achieve real-time optimization of the analog functions. The Intelligent RF Front End (IRFFE) Initiative is aiming at the demonstration of a new class of highly integrated RF and mixed-signal components that through self-assessment and optimization can adapt to rapid changes in the environment or operational demands. Our vision is to enable a new generation of MMIC technologies with the ability to exploit embedded information and convert it into knowledge to achieve superior levels of adaptability and performances. This new class of Intelligent MMIC will provide system designers with new degrees of freedom necessary to meet future network-centric architecture requirements.

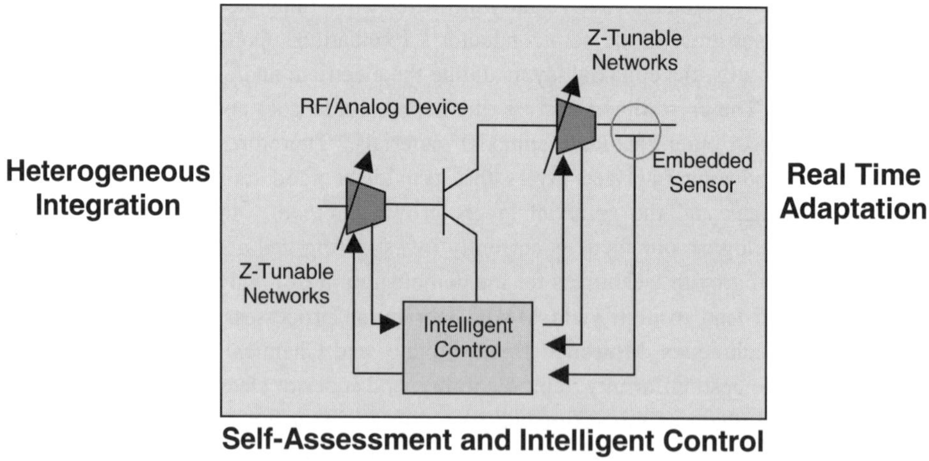

Figure 1: Intelligent RF Front End architecture.

Figure 1 shows a schematic of an IRFFE. Like in MMICs, the core of an IRFFE is a conventional RF microwave or millimeter-wave device. These devices will be able to reconfigure themselves through the implementation of tunable impedance-matched networks placed at their input and output terminals. We anticipate that these tunable impedance-matched networks will be enabled principally by the integration of MEMS technologies with conventional MMIC technologies.

To achieve self-assessment and the intelligent control of the reconfiguration process, embedded sensors, mixed-signal, and digital technologies will be introduced into the MMIC architecture. Intelligent adaptability will enable components capable of meeting diverse system operational requirements effectively and efficiently. The digital control or intelligence in the integrated circuit will be capable of optimizing the component architecture and performance at any given frequency within the operating bandwidth of their core active device.

For example, one can envision a single component capable of meeting the performance requirements of communication, radar, and/or base station enabling multifunctional operation through a single aperture. This will be possible as components are capable of self-optimizing, in real-time, their efficiency, analog performance, and bandwidth to meet the operational demands of all of these applications. Our goal is for RF designers to devise means for achieving circuits with self-optimization capabilities creating a revolutionary transformation in MMIC technology – chips that can think.

These "Intelligent Microsystems" represent a paradigm shift in analog component design. They will transform future system architectures to provide the ability needed to rapidly and effectively respond and react to unforeseeable, continual changes in operational demands.

One particular characteristic of these Intelligent Microsystems is their ability to self-optimize in real time their efficient-performance over broad bandwidth, overcoming the design trade-offs that current analog components must endure. This unique ability will lead to very efficient systems with reduced size, weight, power, and cost.

We anticipate that the IRFFE concept will also have a major impact in future mobile network configurations. At the component level, other benefits could include device auto-calibration to compensate for aging effects or transient faults and auto-correction to compensate for design and manufacturing errors. Collectively these capabilities will result in higher fabrication yields, lower component costs, and enhanced system reliability.

Current technical challenges include:

- Development of embedded analog sensors and algorithms for self-assessment and intelligent control of the analog performance
- Demonstration of real-time adaptation through the implementation of RF MEMs technology to innovative MMIC designs
- Demonstration of heterogeneous integration of mixed-signal technologies

4. Conclusions

While the concept of monolithic microwave ICs is still the workhorse of wireless sensors and communication systems, new paradigms in compound semiconductor materials and architectures are expected to transform MMIC technology to improve performance, efficiency, and their ability to adapt to operational changes. Wide bandgap semiconductors will have a major impact in future MMIC technology by improving the power performance per unit cell of MMIC by at least a factor of ten. In order to achieve the maturity level required for the manufacturing of WBG MMICs, the community must shift its focus from scientific experiments to activities that will lead to technology innovation and the eventual establishment in manufacturing environments. During the next decade, MMIC architectures and fabrication processes will experience a radical change as digital, and actuation functions are integrated in the circuit architecture to achieve higher levels of adaptability and intelligent self-optimization. All these ideas will influence future RF system architectures as MMIC technology experience radical transformation through innovation in compound semiconductor materials and architectures.

Reference

1. Semiconductor Industry Association, June 2002, Semiconductor Forecast Summary.

International Journal of High Speed Electronics and Systems
Vol. 13, No. 1 (2003) 65–89

METAMORPHIC LOW NOISE AMPLIFIERS AND OPTICAL COMPONENTS

C. S. WHELAN, P. F. MARSH, R. E. LEONI III, W. E. HOKE, S. M. LARDIZABAL, S. J. LICHWALA, Y. ZHANG, P. BALAS AND T. E. KAZIOR

Raytheon RF Components, 362 Lowell Street, Andover MA 01810
Phone: 978-684-5456, e-mail: cwhelan@rrfc.raytheon.com

GaAs based metamorphic HEMT (MHEMT) technology has emerged as an attractive, low cost alternative to InP HEMTs. The strain-induced imperfections caused by high indium content layers on GaAs is eliminated in metamorphic devices by providing a properly grown lattice-matching buffer between the substrate and active device layers. With this limitation overcome, it is now possible to provide the superior performance of InP-based devices with the cost advantages of highly manufacturable 4- and 6-inch GaAs wafers that can easily be integrated on existing GaAs fabrication lines. This paper will review device performance as well as state-of-the-art low noise amplifiers fabricated with this technology operating from 1 to 100 GHz. Fiber optic receiver components such as 40 Gb/s optical-to-electrical photodiodes and traveling wave amplifiers fabricated metamorphically will also be discussed. Finally, device and circuit reliability data will be presented demonstrating median-time-to-failure of more than 30 years at 125 C.

Keywords: Metamorphic; LNA; HEMT; Optical; Noise.

1. Introduction

The need for bandwidth, driven primarily by the demand for information transfer and communication services, has necessitated the push to microwave and millimeter wave frequencies. While silicon based semiconductor devices dominate lower frequency electronics, high speed compound semiconductor technologies such as gallium arsenide (GaAs) are the enablers for such capabilities as military radar and communication, wireless commercial communications and high speed data transfer (e.g. fiber optics). The next generation of these systems, as well as new systems such as automotive radar, will require even higher transistor frequency responses.

One of the work horses in GaAs technology, the pseudomorphic high electron mobility transistor (pHEMT), uses a layer consisting of indium, gallium, and arsenic (InGaAs) to provide electrons with high speed conduction. Unfortunately, the indium that gives the PHEMT its enhanced transport properties also introduces imperfections due to strain. Therefore, channels with indium mole fractions greater than approximately 25% are not typically fabricated on GaAs substrates (Figure 1). Historically, indium phosphide (InP) met the small primarily military and space based performance needs required of higher indium content channels. InP overcomes the strain induced performance limitation since it is lattice matched with 53% mole fraction InGaAs channels. However, the technology is expensive due to manufacturability issues associated with fabrication of the InP HEMTs, along with the lack of high quality, large diameter (6 inch or greater) InP substrates.

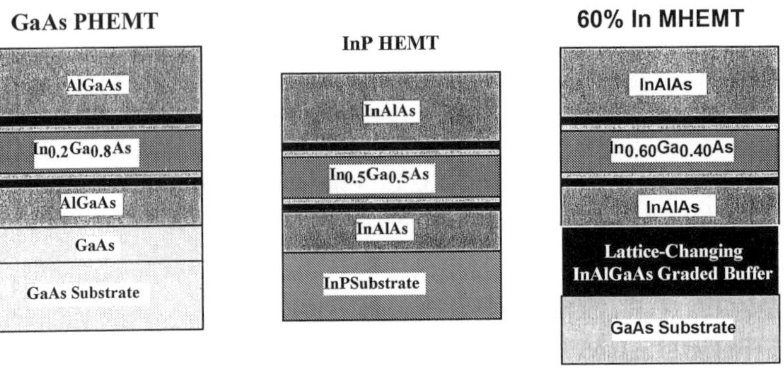

Fig. 1. Three schematic cross-sections of a 20% In GaAs pHEMT, a 53% In InP HEMT and a 60% In MHEMT detailing the compositions of the top Schottky barrier layer, the InGaAs channel and the substrate.

Recently, GaAs based metamorphic HEMT (MHEMT) technology has emerged as an attractive, low cost alternative to InP HEMTs. The strain-induced imperfections caused by high indium content channels on GaAs are eliminated in MHEMTs by providing a properly grown buffer between the substrate and active device layers. With this limitation overcome, it is now possible to provide the superior performance of InP based HEMTs (high speed, low noise, and low dc power dissipation) with the cost advantages of highly manufacturable 4 and 6 inch GaAs wafers. For a 4" HEMT structure, the less expensive metamorphic HEMT on a GaAs substrate translates to a cost per wafer savings of nearly $1,000 (Figure 2) for just the starting material. The ability to fabricated InP-like, metamorphic devices on existing GaAs tool sets and seamlessly integrate the process into an existing GaAs pHEMT fabrication facility is the largest factor reducing cost.

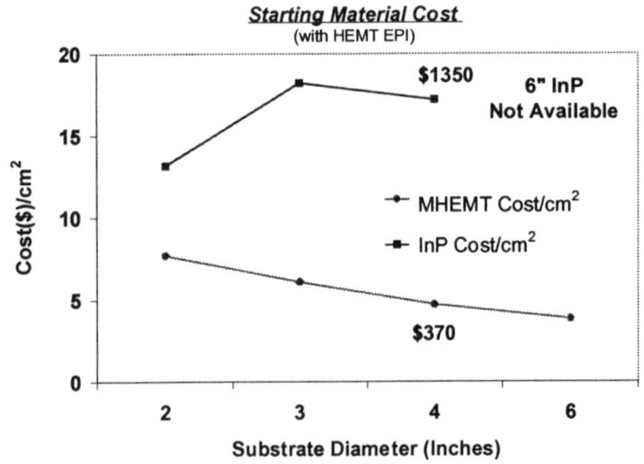

Fig. 2. The starting cost per square centimeter of a HEMT structure on GaAs is significantly lower than InP, providing significant financial incentive for pursuing metamorphic growth.

In the MHEMT, the device active layers are grown on a strain relaxed, compositionally graded, metamorphic buffer layers which transforms the lattice constant from GaAs up through InAs, allowing high In content channel layer to be grown on low cost GaAs substrates. Additionally, the buffer layer provides the ability to tailor the lattice constant to any indium content channel desired (20% to >80%), and therefore allows the device designer an additional degree of freedom to optimize transistors for high frequency gain, power, linearity and low noise. For example, using a metamorphic buffer layer, InP-based high electron mobility transistors (53% In) can be grown on GaAs substrates for a substantial cost reduction and manufacturability improvement over a InP-substrate based devices [1]-[5].

High In content $In_xGa_{1-x}As$ channel MHEMTs (x = 53%-60%) have shown impressive results, achieving noise performance comparable to InP HEMTs, along with excellent linearity. These devices have been used to fabricate low noise amplifiers through 91 GHz, including state-of-the-art X-, Ka- and W-band LNAs. Using the unique ability of metamorphic technology to fabricate devices with intermediate (33-43%) In contents, high on- and off-state breakdown voltage, large power densities and 1.4W of output power PAs at 44 GHz have been realized.

In this paper the current status of metamorphic HEMT technology is reviewed, including low noise, power and optical devices, as well as amplifiers fabricated from these components. This paper will review the material properties, device and circuit performance, and finally the reliability of metamorphic HEMTs, with a focus on work done at Raytheon RF Components.

2. Material Properties

The metamorphic buffer layer [6]-[8] serves two purposes: to transform the lattice constant from that of the GaAs substrate to that of the high In content device active layers, and to trap lattice dislocations and prevent them from propagating into the device's channel layer. The Transmission Electron Microscope (TEM) cross-sectional image in Figure 3 illustrates the growth of an intermediate indium content HEMT (33% In) directly on GaAs using a metamorphic grading layer. The dislocations (shown as dark lines in the TEM) are indeed trapped in the grading layer and propagate parallel to the substrate's surface, not vertically into the device's active layers. The MHEMT's layer interfaces remain atomically smooth despite having twice the indium content and five times the total thickness as a typical GaAs pHEMT, demonstrating that the metamorphic buffer layer technology can successfully transform the lattice constant from GaAs.

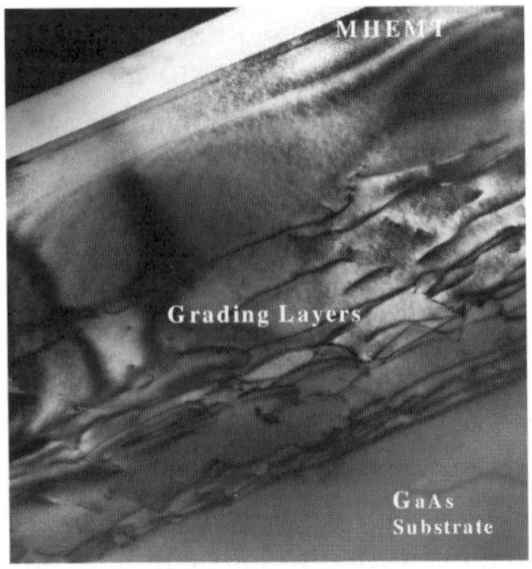

Fig. 3. Cross section TEM of a MHEMT on a GaAs substrate.

Figures 4 and 5 illustrate the distinct tradeoff that occurs as one increases the In content of the channel of a (metamorphic) HEMT. In Figure 5, the mobility of an electron traveling in the In_xGaAs channel is plotted using measured room temperature Hall data, and the quantum well's depth (conduction band discontinuity) is calculated from photoluminescence measurements. The 300K mobility of $In_{0.56}GaAs$ HEMT grown on a GaAs substrate measured within 5% of the mobility of an identical structure on an InP substrate [7], demonstrating the high quality of the MHEMT channel. Channel electron mobility increases with indium in the InGaAs channel, due mainly to a fall in effective mass. This reduction in mass results in higher channel electron velocity, which increases F_t (frequency at which current gain goes to 0 dB) for a fixed 0.18 micron gate length (Figure 6).

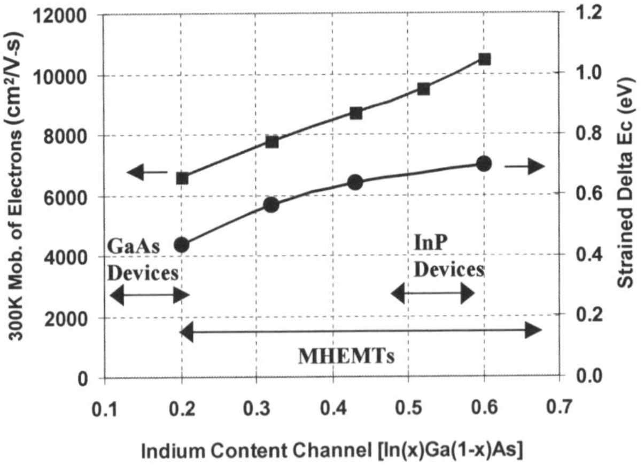

Fig. 4. As indium is added to the channel, both the mobility and well depth increase (when a strained InAlAs layer is used).

Figure 5 shows the drop in on-state breakdown with increasing indium, due to a reduction in the channel's band gap. A larger conduction band discontinuity can be engineered through the use of a strained In(Ga)AlAs Schottky layer with a band gap of greater than 1.7 eV on the intermediate content metamorphic devices. This provides even better quantum well confinement of the electrons and less parallel conduction, as well as improved off-state breakdown. (The calculations shown in Figure 4 are using this strained In(Ga)AlAs layer to calculate delta E_c.) The combined effects of improved channel mobility and larger conduction band discontinuity result in lower overall noise figure, especially at higher frequencies.

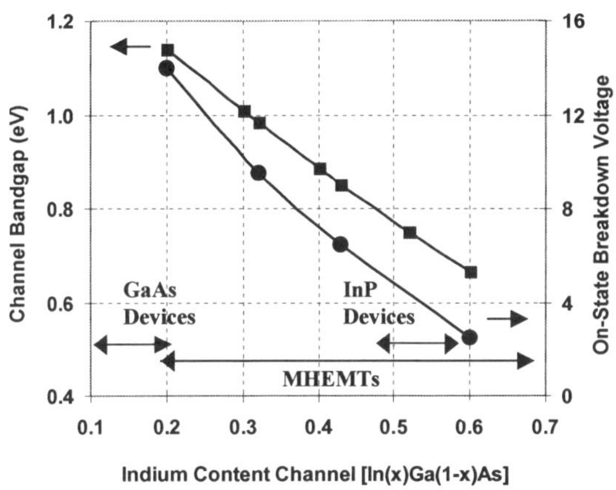

Fig. 5. As indium is added to the channel, the channel band gap falls, reducing on-state breakdown voltage.

While GaAs pHEMTs and InP HEMTs are limited to In compositions near their lattice spacing, MHEMTs have a wide range of lattice constants that are available, and therefore enable the customization of the device's properties specific to each application. Although thin strained channels can be grown on InP, the metamorphic devices grown here contain nearly 1000 angstroms of strained material, making their growth impossible without a graded buffer layer. The data points in Figures 4 and 5, some of which lie within the pHEMT's and InP HEMT's forbidden channel indium content regions, are devices grown at Raytheon which exploit this additional degree of freedom.

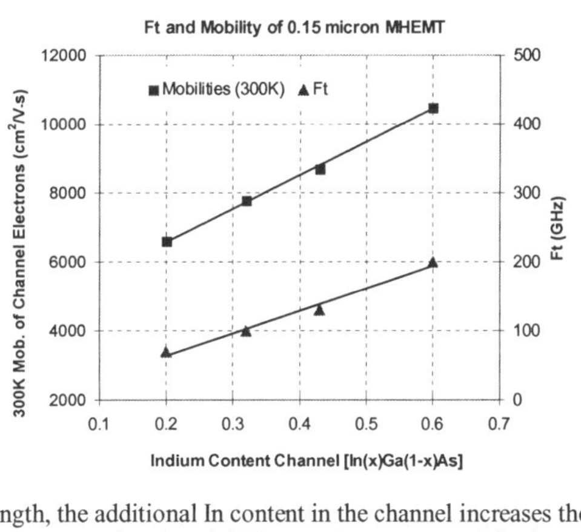

Fig. 6. For a fixed gate length, the additional In content in the channel increases the mobility and therefore the F_t (extrapolated at -6 dB/octave from measured data).

3. Device Processing and Low Noise DC Performance

MHEMT devices are typically mesa-etched for isolation using a sulfuric or phosphoric based etchant. A series of metals containing Au/Ge are evaporated and annealed on the semiconductor to form an ohmic contact, with contact resistance numbers in the range of 0.06-0.08 Ohm-mm. Following ohmic formation, gate etching is performed by selectively removing the InGaAs cap layer and stopping on the InAl(Ga)As barrier layer. Ti/Pt/Au gates are then evaporated. Finally, silicon nitride is used to passivate the device. The device processing is nearly identical to our GaAs pHEMT process and has allowed for easy integration into our 4" GaAs production line.

The DC performance data for a typical $In_{0.60}Ga_{0.40}As$ MHEMT device shows an extrinsic G_m of 850 mS/mm, an I_{max} of 680 mA/mm at V_{ds}= 1V, a V_{po} of –0.80V and a two terminal drain-gate breakdown of 8V (Figure 7). An excellent uniformity of less than 3.0% standard deviation for all parameters across a 4" wafer is due to both the high selectivity of the gate etch process, and the precision of the MBE growth process.

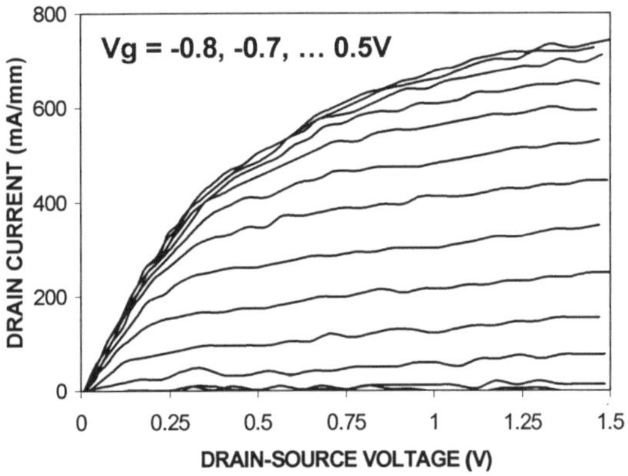

Fig. 7. Normalized I-V curve for a 0.18 micron gate, 60% indium MHEMT.

4. Low Noise Devices And Circuits

Raytheon's MHEMT low noise results [9]-[10] rival the best published MHEMTs [11], as well as the best InP HEMTs [12]. A 0.18 micron $In_{0.60}GaAs$ Raytheon MHEMT biased at 1V and 90 mA/mm showed 0.24 dB F_{min} with 16.2 dB associated gain at 12 GHz, and 0.61 dB F_{min} with 13.8 dB G_{assoc} at 26 GHz [9]-[10]. Rohdin et al [11] showed 0.25 dB with 15 dB of associated gain at 12 GHz, using a 0.1 micron $In_{0.53}GaAs$ MHEMT discrete device.

Figure 8 shows the results of a comprehensive study on noise figure versus the channel In content of MHEMTs for four gate-finger, 75 micron gate-width devices biased at V_{ds} = 1V. As the channel In content is increased from a pHEMT's 19% In to a substrate-forbidden 33% In MHEMT, the 25 GHz minimum noise figure is reduced from 1.4 dB to 1.2 dB. Further increasing In content to 43% In results in a substantial drop in minimum noise figure to 0.75 dB. Beyond 43% In, only a moderate improvement in noise figure is achieved at 25 GHz, as demonstrated by the 53% In MHEMT with a F_{min}= 0.85 dB and the

60% In MHEMT with a F_{min}= 0.65 dB. Devices fabricated with shorter unit gate-widths resulted in lower gate resistance and F_{min}s as low as 0.5 dB at 25 GHz with the 43% In MHEMT.

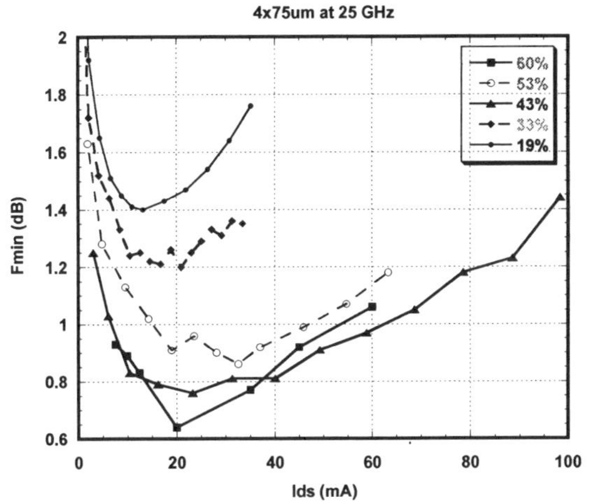

Fig. 8. F_{min} versus drain current at 25 GHz for 0.15 micron gate length, 4x75 micron periphery, devices.

The associated gain at the minimum noise match for the same 300 micron device is plotted versus drain bias at 25 GHz (Figure 9). The pHEMT shows approximately 8 dB of associated gain near its minimum F_{min}, in contrast to the 33% In MHEMT's 10 dB. The 43% In MHEMT has ~10-11.5 dB of G_{assoc} over the large current range where noise figure remains quite low. The 53% and 60% devices reach their peak associated gain of 12-12.5 dB quickly, demonstrating the clear advantage of high indium channels at very low currents. This particular property is often exploited when building very low DC power consumption low noise amplifiers, especially for space-based applications.

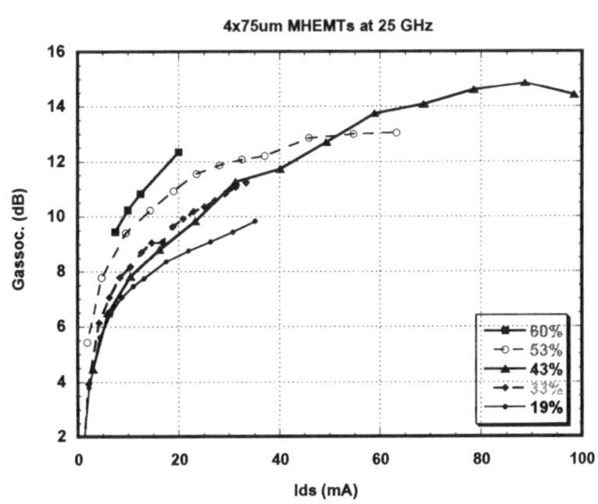

Fig. 9. Associated gain versus drain current at 25 GHz for 0.15 micron gate length, 4x75 micron periphery, devices.

The measured result of a 2-stage X-band MHEMT Low Noise Amplifier (LNA) that utilize these low noise device characteristics is shown in Figure 10. The LNA has 24 dB of gain from 7-11 GHz and a very low noise figure of less than 0.6 dB from 8-11 GHz.

The self-biased LNA consumes 30 mW of DC power consumption and can survive up to 1W of input power before degrading.

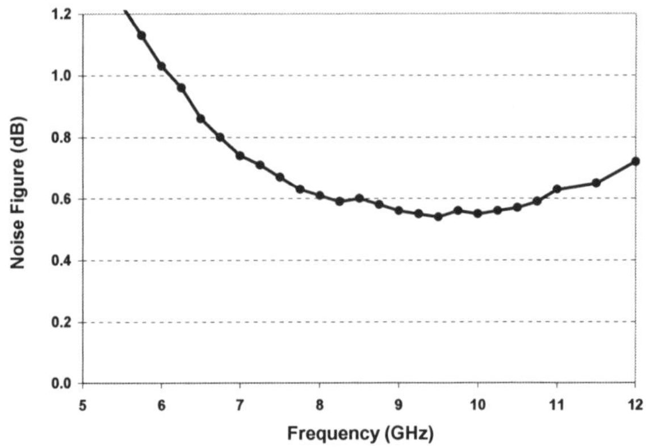

Fig. 10. Simulated (black) and measured (red, blue) noise figure for multiple 2-stage X-band LNAs.

A second LNA designed for a high linearity and low noise operation demonstrates 24 dB of gain from 15-25 GHz and less than 1.6 dB noise figure from 20-25 GHz (Figure 11). The output stage of the LNA has a periphery of 600 microns, allowing it to handle high input powers with low distortion.

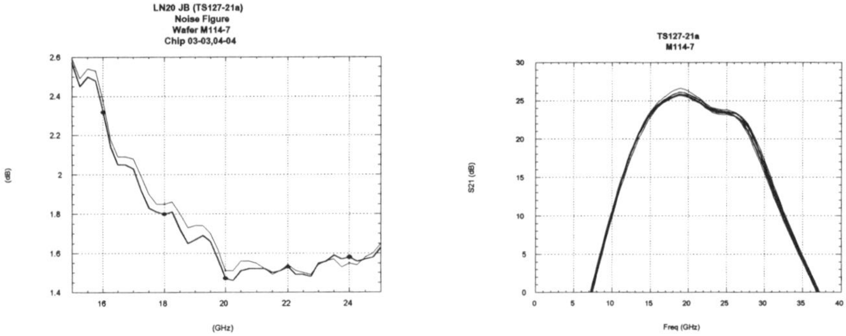

Fig. 11. A 2-stage high linearity MHEMT LNAs shows 24 dB of gain and less than 1.6 dB noise figure from 20-25 GHz.

Figure 12 plots the noise figure and gain of a narrow band 3-stage MHEMT LNA with less than 1.5 dB NF and greater than 23 dB of associated gain from 31-32 GHz [13]. Particularly impressive is the 15 mW of total DC power consumed by this 3-stage LNA. Due to the high mobility (10,800 cm^2/V-s) of the transistors, the drain bias can be reduced from 1V to 0.35V with no increase in noise figure, resulting in only 5mW of power consumption.

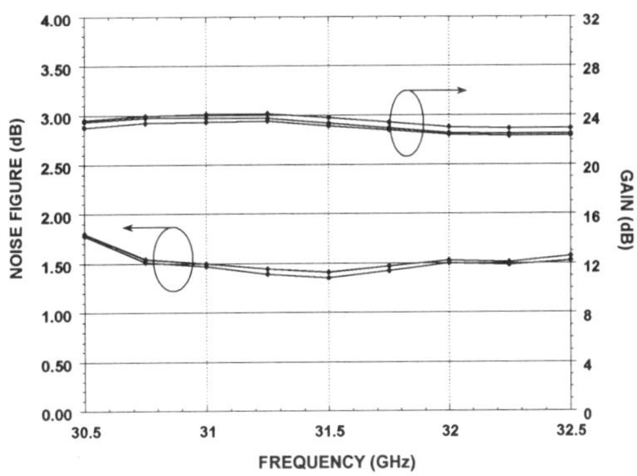

Fig. 12. The 3-stage MHEMT LNAs show 1.5 dB noise figure and 23 dB of associated gain from 31-32 GHz

Another wider bandwidth, Ka-band LNA is shown in Figure 13. This small 1.2 mm by 2.2 mm chip uses a single supply self-biased design that consumes 30 mW of DC power. The LNA shows 20 dB of gain and approximately 1.6 dB NF over a very large 20% bandwidth from 31-37 GHz.

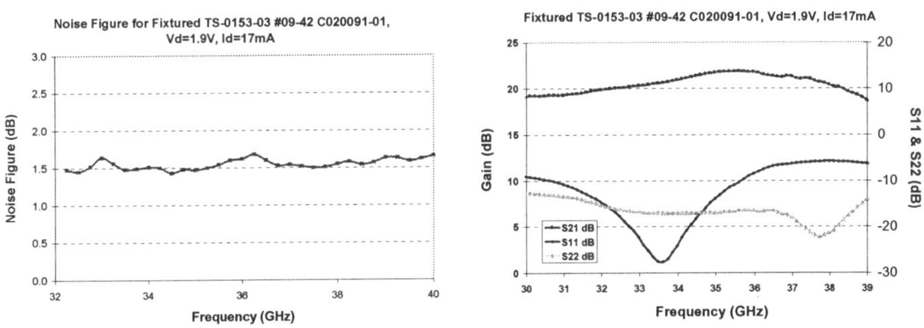

Fig. 13. This 20% bandwidth Ka-band MHEMT LNA shows 20 dB of gain and 1.6-1.7 dB NF.

A number of very wide bandwidth amplifiers [14-15] have also been demonstrated using this technology. The wide bandwidth traveling wave amplifier (TWA) [15] is of particular interest for high-speed data links up to 40 Gb/s. The amplifier consists of 6 cascode pair cells (Figure 14) and uses purely resistive loads in order to extend the on-chip low frequency range of operation. Figure 15 summarizes the s-parameters of 63 chips from a typical wafer. The amplifier has a high frequency 3 dB cutoff of approximately 45 GHz and a KHz-range low frequency cutoff. Gain in a 50 Ohm system is typically 16 dB with a 1.5 dB peak-to-valley variation to 40 GHz. RMS gain variation over this range is typically 0.25 dB. The chip is capable of providing over 3 dBm over the majority of the band, which is approximately 0.9V peak-to-peak in a 50 Ohm system.

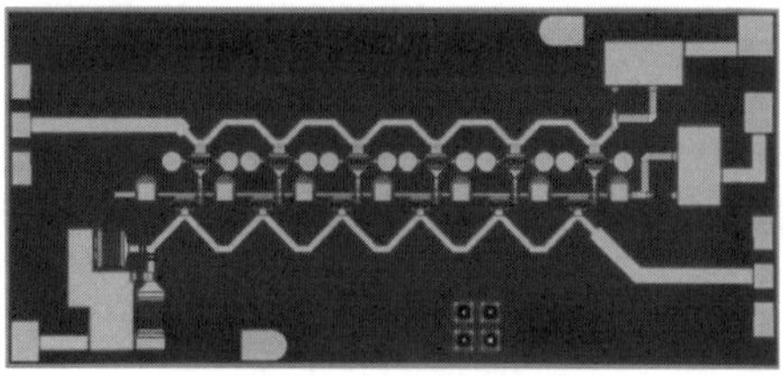

Fig. 14. Photograph of a dc – 45 GHz TWA. The amplifier consists of 6 cascode pair cells and occupies an area of 6.3 mm^2.

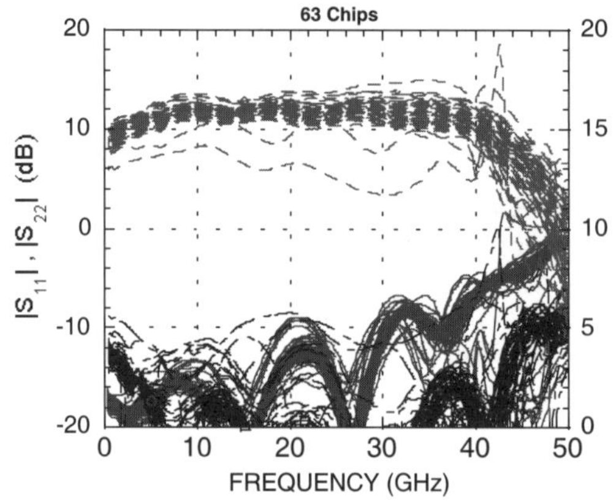

Fig. 15. S-parameters results of 63 MHEMT TWAs from a single wafer. The superimposed S_{11}, S_{21}, and S_{22} of the amplifiers are displayed.

Continuing up the frequency spectrum, an MHEMT LNA designed for W-band frequency operation is shown in Figure 16. It provides a small signal gain of 19 dB and noise figure less than 6 dB while maintaining input and output return losses less than –10 dB. The MMIC uses a compact microstripline design measuring 105 by 45 mils and avoids the need for series MIM capacitors in the matching networks. Bias is applied to a single contact for both the gate and drain.

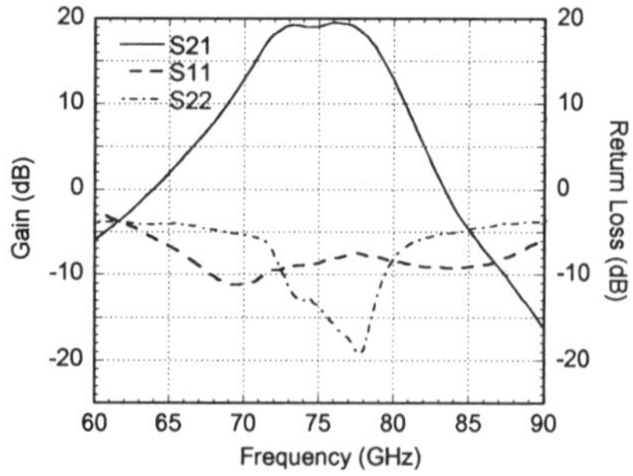

Fig. 16. The three-stage MHEMT LNA provides 20 dB of gain, centered at 75 GHz.

Similar in design to the 77 GHz MMIC, results for a 91 GHz amplifier are shown in Figure 17. This three-stage MHEMT LNA achieves a small signal gain of 14.5 dB with less than 6.5 dB noise figure and a saturated output power of +5 dBm. It is designed using a 2 mil thinned substrate and individual source vias to reduce parasitic inductance at this high frequency.

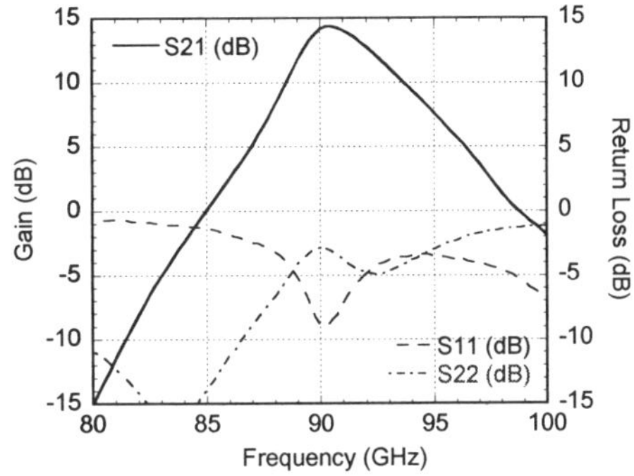

Fig. 17. This three-stage W-band LNA features 14.5 dB of gain, 6 dB noise figure and +5 dBm saturated output power.

These numerous state-of-the-art X- through W-band circuit results demonstrate the performance and maturity of metamorphic technology. The performance of InP LNAs can now be achieved with metamorphic technology, using existing GaAs production lines.

5. Power Devices

Drain bias limits of approximately 3.0V have hampered the output power density of some MHEMT devices and most InP HEMTs, predominately due to low on-state breakdown governed by the small band gap of their high indium content channels. Even so, output power(s) (densities) of 1W at 950 MHz [16], 509 mW/mm at 20 GHz [17] and 240 mW/mm at 60 GHz [18] have been achieved with 53% In MHEMT devices. Figure 18 compares various published MHEMT results versus a number of InP HEMT power devices (> 150 microns in total gate periphery), most of which consist of composite channels ($In_{0.5}GaAs/InP$) to improve on-state breakdown. The 30-45% In MHEMT devices lie considerably above the trend line, and show promise as a high power mm-wave device alternative to InP HEMTs. The 43% In device has also demonstrated 1.5 dB improved G_{assoc} at the same power output density as a GaAs pHEMT. Both devices were biased at 5V, class AB and power tuned at 35 GHz.

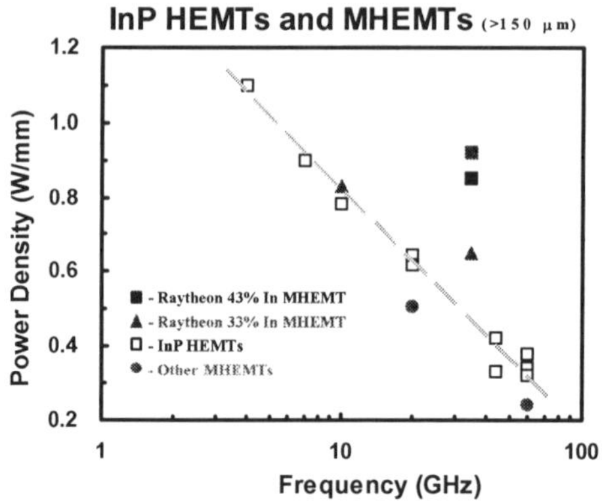

Fig. 18. Power density of InP HEMTs and MHEMTs versus frequency for devices of greater than 150 microns total periphery.

Exploration of lower indium contents has resulted in higher on-state breakdown, enabling 5V and 6V drain biasing to be used for higher power density. Using a 33% In channel, > 820 mW/mm at 10 GHz [19] and > 640 mW/mm at 35 GHz [20] have been achieved with 6V drain biasing. With a 43% In, 12x50 micron (600 micron) MHEMT, over 900 mW/mm was achieved at 35 GHz using a double recess structure for improved breakdown [21]. Interestingly, this same 43% In device (with lower electron sheet density) displayed ~0.7 dB F_{min} at 25 GHz due to its high channel mobility and excellent confinement. A 8.16 mm periphery 3-stage 44 GHz power amplifier fabricated using a 33% In MHEMT demonstrated 1.4W output power, 18% PAE and 14 dB of gain.

6. Metamorphic Optical Devices and Circuits

In addition to low noise and power transistors, top illuminated metamorphic PIN (p-doped/ intrinsic/ n-doped) photodiodes that absorb 1.55 micron wavelength light have been reported [22, 23, 24]. Raytheon's PIN photodiode consists of $In_{0.53}Ga_{0.47}As$ P+ anode and thick undoped drift layers grown on top of an N+ $In_{0.52}Al_{0.48}As$ cathode layer. A metal stack consisting of Au/Ge/Ni is used for N+ contacts and a Ti/Pt/Au is deposited for P+ contacts. Air bridges are formed to connect the anode probe-pad to the anode ring. Devices with optical diameters of 10, 20 … 50 micron diameter are typically fabricated using this material. A 10 micron diode has a total capacitance of approximately 80 fF, an inductance of 85 pH and a on-resistance of < 12 Ohms.

The active layers in these structures are much thicker than those found in HEMTs and therefore are affected by residual strain in the buffer much more than the HEMTs. Any stress-induced dislocations that propagate into the device active layers will electrically show themselves as increased leakage current. Figure 19 displays typical leakage currents of both Raytheon's InP and metamorphic PIN photodiodes as a function of diode diameter. A quadratic dependence on diode diameter infers bulk-related leakage while a linear dependence indicates periphery-related leakage. Both technologies exhibit a roughly linear dependence on diameter and have about the same amount of dark current, indicating the high quality of the metamorphic material.

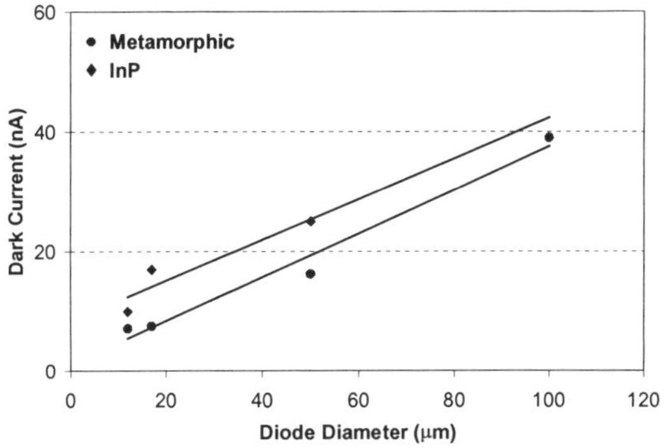

Fig. 19. The leakage current of metamorphic GaAs and InP 1.55 micron wavelength photodiodes exhibit approximately the same levels of dark current. Leakage current for both technologies shows an approximately linear dependence on device diameter, indicating little bulk leakage.

The metamorphic photodiodes demonstrate frequency responses and responsivities adequate for high data rate applications. Figure 20 shows the DC photocurrent for a 10 micron device. The data acquisition system used for this measurement has a 20 micron spot size and therefore is not capable of responsivity measurements. However, from the figure we see a linear relationship between current and optical power. From a comparison of relative DC responsivity and 20 GHz modulated light as a function of reverse bias, we see that the devices require at minimum 3V reverse bias (Figure 21). A packaged 10 micron diameter metamorphic photodiode, pigtailed with a 9 micron fiber and biased at -5 V, has greater than 50 GHz of bandwidth with 0.5 A/W responsivity (Figure 22).

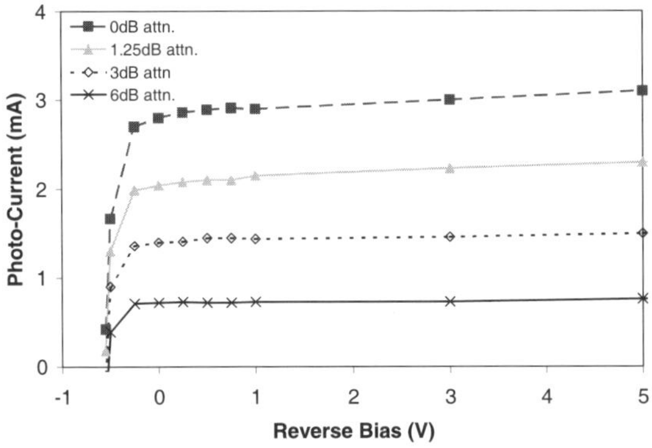

Fig. 20. Photocurrent as a function of optical power and reverse bias. The optical power is increased from 2.5 to 10 mW in 2.5 mW steps.

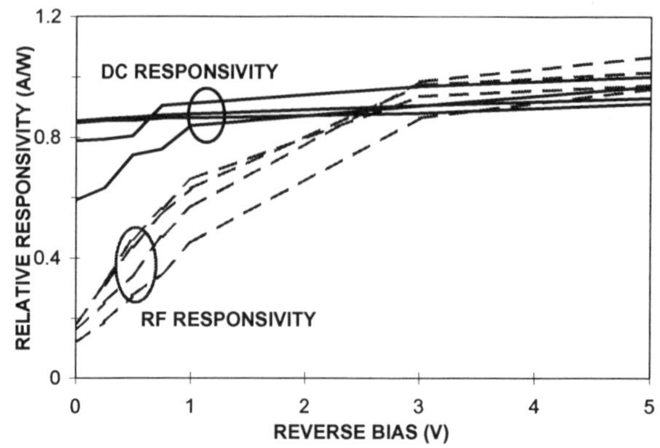

Fig. 21. Relative responsivity of a 10 micron photodiode tested at dc (solid lines) and 20 GHz responsivity calculated from the power delivered to a 50 W load. RF results show that a minimum of 3 V reverse bias is necessary.

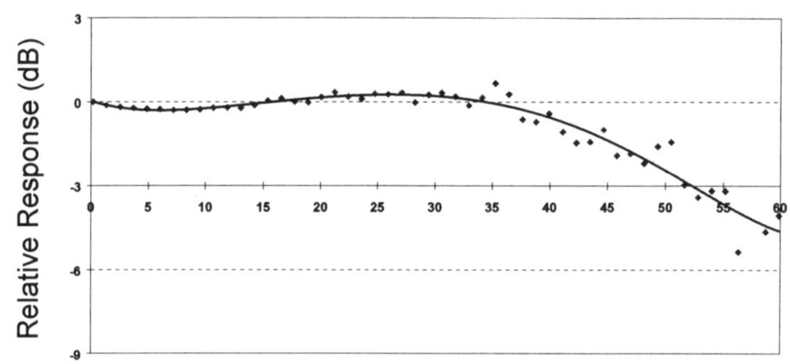

Fig. 22. Frequency response of a fixtured 10 micron metamorphic photodiode. The device provides greater than 50 GHz of bandwidth with a responsivity of 0.5 A/W.

A photodiode and a wide bandwidth amplifier make up the front end of a photoreceiver. Using our TWA and InP photodiodes we have been able to build optical receivers for both analog and digital applications. The frequency response of a 20 GHz bandwidth receiver is shown in Figure 23. The module has an average responsivity of 90 V/W with a variation of +/- 15 V/W over its operational frequency and temperature ranges. Also, a hybrid photoreceiver with a 3 dB cutoff of 32 GHz has been demonstrated (Figure 24). This combination of components showed an open eye diagram with utilizing a 40 Gb/s PBRS non-return-to-zero digital signal.

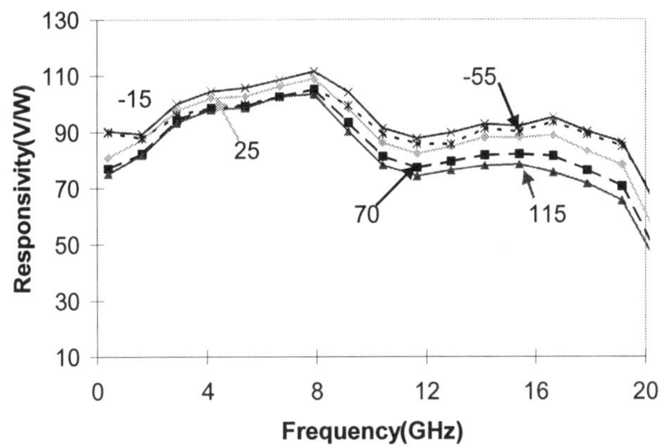

Fig. 23. Frequency response of a hybrid photoreceiver covering the majority of the frequency range up to 20 GHz. The receiver operates with little gain variation from –55 to 115°C.

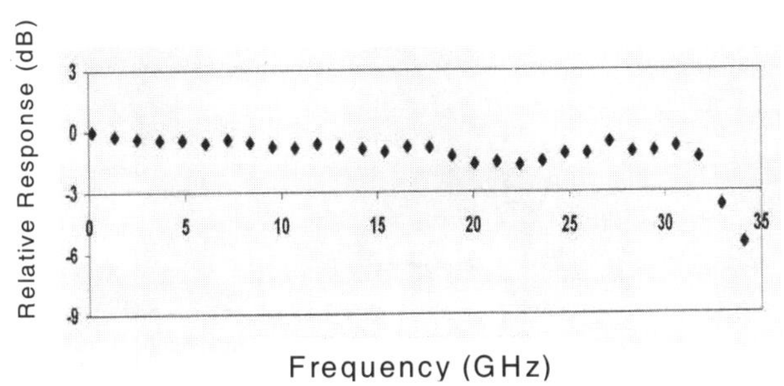

Fig. 24. Frequency response of a hybrid photoreceiver composed of a metamorphic TWA and InP photodiode.

The interface between the photodiode and wide bandwidth gain stage greatly affects the overall frequency response of a photoreceiver due to the high degree of mismatch between the photodiode, which electrically is a capacitively loaded current source, and the amplifier. Therefore, for high data rate optical links there are great benefits to monolithic integration of photodiode and amplifier. With the ability to grow high indium content active layers on GaAs, Raytheon and others are working towards the development of monolithically integrated metamorphic opto-electronic integrated circuits [25].

7. Reliability

An extensive study of metamorphic reliability has been carried out to investigate the long-term performance of first the material, then the devices and the fabrication process, and finally the circuits. The first experiments focused on the stability of the metamorphic material, since it is at the base of device reliability.

Under a 250 C bake in nitrogen, unprocessed MHEMT layers showed excellent stability over a 250 hour time frame with the sheet charge and mobility varying between $3x10^{12}$ to $3.1x10^{12}$ cm^{-3} and 10,800 to 11,100 cm^2/V-s, respectively, with no clear trend.

This and other data indicate that the thermal stability of MHEMT material itself is not a significant factor in device reliability.

The MHEMT gate process leaves a region of exposed InAlAs Schottky material near the gate metal that is later covered by a subsequent silicon nitride (SiN_x) passivation. Before passivation, this InAlAs surface can absorb fluorine and oxygen from the air. Fluorine diffuses rapidly in InAlAs [26] where it is gettered by the Si pulse doping of this layer [27] but is blocked by the InGaAs contact layer. Therefore, this is a concern for only InP and MHEMT devices and warranted investigation. In the first experiment, we confirmed that fluorine diffusion could occur in an InP wafer composed of: 400 Å Si pulse-doped InAlAs layer on top followed by 40 Å undoped In InAlAs then 500 Å undoped InGaAs then 3000 Å InAlAs, and finally the InP substrate. The wafer had been exposed for several weeks to lab air and was then subjected to a tube furnace bake under nitrogen for 5 minutes at approximately 370 C, simulating an ohmic alloy. Figure 25 shows that the thermal stress caused O and F to diffuse into the InAlAs Schottky and InGaAs layers, with the F collecting near the Si pulse. Diffused fluorine is thought [27] to bind electrons (donor passivation) which would be expected to increase the sheet resistance (R_{sheet}) of the channel.

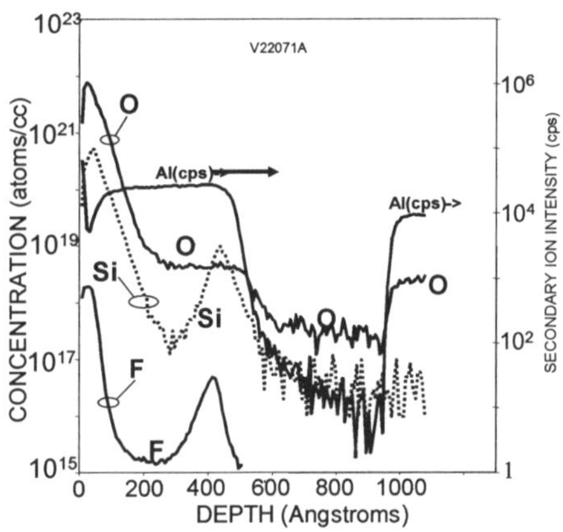

Fig. 25. Fluorine and oxygen diffusion into InAlAs and InGaAs device layers grown on an InP wafer when annealed in a tube furnace under N_2. During anneal, the sample's temperature is ramped from 349 C to 391 C in 5 minutes and is then removed.

With F diffusion in InP-based layers confirmed, a second experiment was performed to investigate the sensitivity of MHEMT layers' R_{sheet} to this effect. First, two MHEMT wafers, one grown in our research MBE (molecular beam epitaxy) system and another grown in our production MBE, had their InGaAs cap layers selectively removed, while leaving the InAlAs Schottky layers intact. Both wafers had a 2000Å silicon nitride passivation layer deposited at 300 C to protect the InAlAs Schottky layers from oxidation. Figure 26 shows that baking these wafers at 240 C under N_2 resulted in an actual decrease of R_{sheet} over time. One would have expected R_{sheet} to increase due to fluorine donor passivation. However, a change in surface potential of the SiNx-InAlAs interface during the bake could have increased the channel electron density in spite of some donor passivation. A similar effect, noted in some MHEMT devices after a ~250 C stabilization at the completion of device fabrication, appears as a slight decrease in source resistance and might also be due to a similar change in the SiNx-InAlAs interface near the gate. More experiments will be needed to isolate the effects of fluorine from those of the SiNx-InAlAs interface changes, but nevertheless the fluorine passivation effects seem to be minimal in the MHEMT structure.

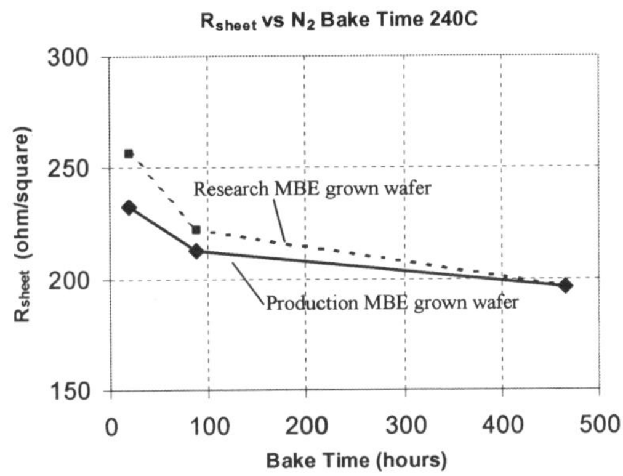

Fig. 26. Sheet resistance stability of MHEMT wafers vs. time at 240 C under N_2. In these tests, the InGaAs cap layer was removed and silicon nitride was deposited directly over the InAlAs Schottky layer.

MHEMT reliability instead faces challenges mainly arising from the choice of gate and ohmic metallization, as well as device passivation, all of which occur during the transistor fabrication process. Ohmic contacts must be stable over time to ensure device reliability, and during the alloy process, interaction of the metal-semiconductor contact must be sufficient to dope the semiconductor. When ohmic metal diffusion, during the alloy, was excessive enough to destroy the HEMT channel, contact resistance, R_o, dramatically increased. For instance, [28] and [29] discuss the use of ohmic contacts containing Ni, Ge, and Au or Pd, Ge, and Au. However, such conventional contacts were found here to yield mean time to failures (MTTFs) of only 1000 hours at 125 C due to excessive metal diffusion into the device layers as shown in Figure 27. Non-alloyed ohmic contacts using Mo/Au and other refractory metals next to the semiconductor were tried, but resulted in severe Au diffusion at the contact edges because the refractory metal became thin there. Improved metal layers, developed here, greatly reduced metal diffusion and gave very stable ohmic contacts as shown in Figure 28. The authors' final MHEMT ohmic contact design gives a low source resistance (R_s) of 0.28-0.33 Ohm-mm combined with an excellent MTTF (10% increase in R_o) of 9×10^6 hours at 240 C increasing to 3×10^{15} hours at 125 C. Therefore, MHEMT ohmic contacts show no significant degradation at 125 C and below.

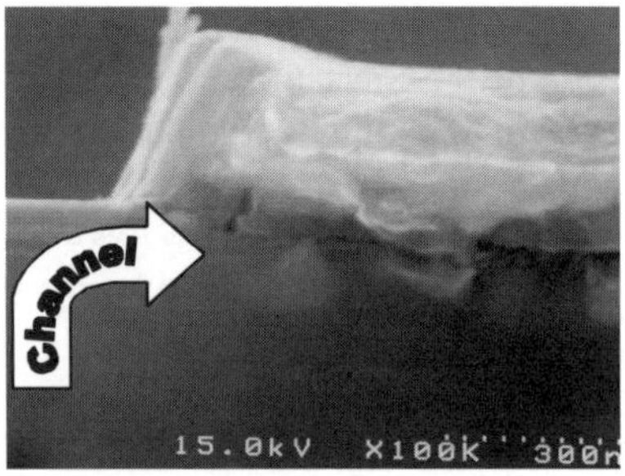

Fig. 27. Cross-section of a Pd, Au ohmic on MHEMT device layers after 72 hours at 250 C. Note the excess metal diffusion penetrating deep beneath the InGaAs channel layer.

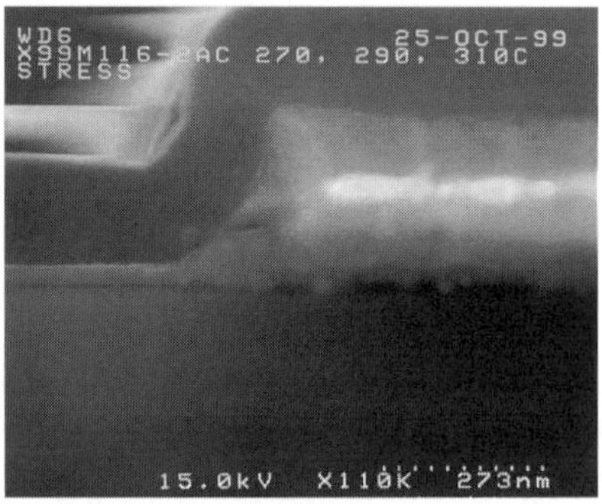

Fig. 28. Cross-section of an MHEMT sample having an improved ohmic metal stack after a series of stress bakes at 270, 290, 310 C for 24 hours each. This ohmic stack increases the optimum alloy temperature over that of the earlier, obsolete MHEMT ohmic design and eliminates excess diffusion.

Metal-semiconductor interactions and the application of currents and electric fields [30]-[31] introduce reliability mechanisms in finished devices that add to those seen in unprocessed material.

Overall, MHEMT degradation manifests itself in reductions of I_{max}, I_{dss}, and G_m, and increases in R_s and R_d, i.e. source and drain resistance, respectively. Various new failure mechanisms can occur under simultaneous bias and thermal stress such as metal electromigration, impact ionization under high electric fields, and gate sinking.

The following equation (Eq. (1)) is used to extract device reliability at the operating temperatures from shifts in device parameters resulting from high temperature stress. Most of the device degradation is assumed to related to the inter-diffusion of elements and hence can modeled at a temperature of T and activation energy of E_a with the equation [32]:

$$\ln(MTTF) = \frac{E_a}{K}\left(\frac{1}{T}\right) + A$$

MTTF is the median of the device failure times of a group of devices and A is a constant related to the maximum allowed degradation of the device characteristic in question, e.g. the peak transconductance G_m. Performing a least-squares linear fit of MTTF data vs. 1/T allows estimation of E_a and MTTF for all temperatures. While E_a and

MTTF can, in principle, be determined from two points, it is recommended to use a minimum of three temperature MTTF points for accuracy.

To investigate reliability, Raytheon's MHEMTs were stressed with a drain bias of 1V and 75 mA/mm at 210, 230, and 250 C ambient temperatures in lab air. Each of these devices had two gate fingers of 0.18 micron length and 50 micron width. Approximately twelve devices were bonded into each header and stressed, with each header at a constant temperature. Channel temperature was calculated using Cooke's law [33] which gave a temperature rise of 7.5 C above ambient. For this work, channel temperature was conservatively estimated at 4.5 C above ambient.

One MTTF data point was obtained for each header of devices according to the following procedure. As the devices in a header degraded, their times to failure were noted for each device. In several cases, the MTTF, at a temperature point, was calculated from an estimate of the cumlative probability density function (CDF) of failures vs. stress time. To obtain the MTTF this way, the cumulative number of failed devices, divided by total number of devices, was plotted as a function of stress time on a log scale. Since one expects the cumulative failure data points to fall on a straight line, the CDF is formed as a least-squares linear fit to the data set. The MTTF is then defined to be the stress time where the CDF crosses over the 50% point, i.e. the expected failure time of half the devices. After several headers are stressed and their MTTF points are calculated, E_a and MTTF estimates, for all temperatures, are then extrapolated from a least-squares linear fit of *ln(MTTF)* vs 1/T as shown in Figure 29.

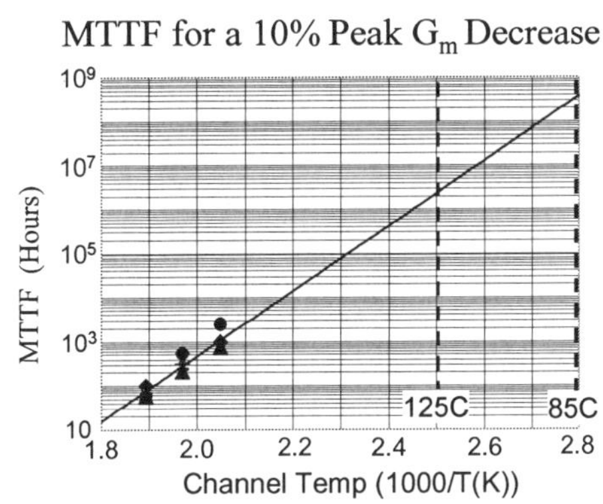

Fig. 29. MTTF of Raytheon research MHEMTs with respect to 10% peak G_m degradation. Each data point represents the MTTF of 8-15 devices having gate length of 0.15 micronand gate width of 100 micron (2 x 50 micron). Separate wafers are designated by the different types of data point symbols. The line is a least-squares fit of the plotted MTTF data points.

A subset of many hundreds of stressed devices is shown in Figure 28 and Table 1. With a failure criterion of a 10% reduction in peak G_m, our development fabrication lineand subsequently transferred production line MHEMTs show MTTFs of $3.8x10^8$ and $9x10^{10}$ hours at 85 C, and MTTFs of $2.3x10^6$ and $8x10^7$ hours at 125 C, respectively. These results compare very favorably to others' MHEMT and InP HEMTs [34]-[38].

Type	Peak G_m			I_{dss}		
	MTTF (hrs)		E_a (eV)	MTTF (hrs)		E_a (eV)
	85 C	125 C		85 C	125 C	
Raytheon Research MHEMT	$3.8\text{x}10^8$	$2.3\text{x}10^6$	1.5	$1.7\text{x}10^9$	$9\text{x}10^6$	1.6
Raytheon Production MHEMT	$9\text{x}10^{10}$	$8\text{x}10^7$	1.9	$9\text{x}10^{10}$	$8\text{x}10^7$	1.9
InP HEMT [35]	$1.8\text{x}10^9$	$5\text{x}10^6$	1.8			
InP HEMT [36]	$1.1\text{x}10^{10}$	$3\text{x}10^7$	1.8			
MHEMT [37]-[38]				$1.9\text{x}10^9$	$7.5\text{x}10^6$	1.7

Table 1. Raytheon development line and production line MHEMT reliability as compared to other work. Failure is considered a 15% decrease in I_{dss} and a 10% degradation of peak G_m. MTTF data were derived from MHEMTs stressed at $V_d = 1$ V and $I_{ds} = 75$ mA/mm at 210, 230, and 250 C.

Ambient hydrogen can cause gradual parameter shifts in InP and GaAs-based HEMTs [39]-[41] affecting reliability. This hydrogen exposure is the result of outgassing inside hermetic circuit packages. The Pt contained within the HEMT gate metal layers can split H_2 into atomic H. Atomic hydrogen is believed to affect InP-based HEMTs via its combining with titanium in the gate metal to form Ti-H [39]. Since gate Ti expands as it changes to Ti-H, hydrogen exposure results in stress applied to the semiconductor just under the gate metal [39]. As more hydrogen is absorbed, the piezoelectric property of the III-V semiconductor then causes in a drift of the total gate metal-semiconductor interface potential that can alter the HEMT pinchoff voltage over time. Investigations, Figure 30, show that a 6 hour 270 C bake in forming gas caused our MHEMTs to undergo a 100 mV positive shift in the peak of G_m with no change in I_{max} and a very slight gain in peak G_m. For comparison, pHEMTs having Pt-containing gates showed a 300 mV shift of the G_m curve in less than 1 hour of 270 C forming gas bake [40]-[41]. In Figure 30, the V_{gs} where G_m peaks suffered most of its shift, 90 mV, in the first 3 hours of the forming gas bake and leveled off thereafter. By contrast, a control device, from the same MHEMT wafer, was baked at 270 C under nitrogen, and showed no significant shift in G_m characteristics. This indicates that hydrogen was necessary to shift the device characteristics. Overall, our MHEMTs showed considerably less sensitivity to hydrogen than pHEMTs.

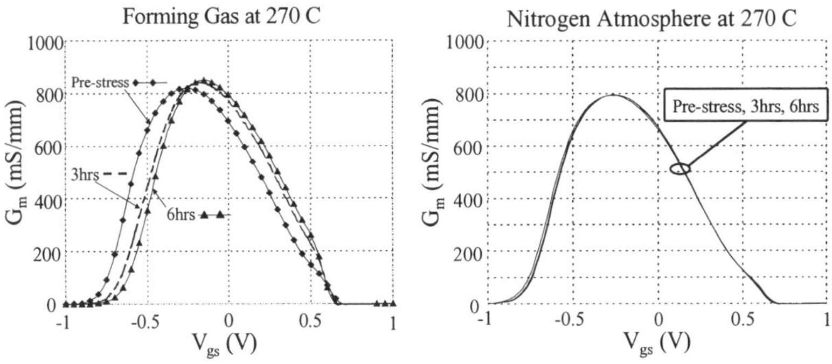

Fig. 30. Effect of hydrogen on the G_m vs. V_{gs} curves. A bake at 270 C in forming gas causes a 90 and 100 mV total shift in G_m characteristics for a 3 and 6 hour respective bake time. A control device, from the same wafer, showed little change under a 270 C nitrogen bake with the pre-stress, 3 hour, and 6 hour G_m characteristics nearly on top of one another.

Hydrogen exposure had very little impact on the maximum drain current ($I_{ds} = I_{max}$ at $V_{ds} = 1V$ and $I_g = 1$ mA/mm) or G_m of the MHEMT devices except for a shifting of the curves; as one would expect from a purely piezoelectric effect. If hydrogen donor passivation were involved, one would also expect a reduction of I_{max}. The stability of I_{max} under the forming gas bake indicated an insignificant amount of donor passivation.

In certain applications, amplifiers might be exposed to moderate levels of RF drive. Hence, reliability testing must look for evidence of degradation under moderate levels of RF input. Here, a sample of MHEMT-based travelling wave amplifiers (TWAs) designed for 40 Gb/sec fiber systems were driven at approximately 1 dB compression at 40 C and 85 C. Under this drive, the TWAs showed essentially no change in gain and RF output power, P_{out}, in the 1300 hour duration of the test.

These experiments have measured the reliability and stability of MHEMT material, devices and circuits with excellent results. They have demonstrated that MHEMT reliability is comparabe to mature InP and GaAs pHEMT technology, and is sufficient for system implementation.

8. Conclusion

Metamorphic HEMTs have distinct advantages over the existing GaAs and InP HEMT technology: the freedom to choose virtually any indium content InGaAs channel provides for application specific device optimization and high frequency performance, while the GaAs manufacturing economies of scale (high volume, large wafer size) reduce the cost. With low noise and high frequency gain comparable to InP HEMTs, metamorphic HEMTs show promise as the next RF technology of choice to fill the emerging need for high-speed communication. Metamorphic technology provides the capability for fabricating other electronic devices such as optical devices onto GaAs substrates, allowing for higher order integration of components and device technologies onto a single chip.

Acknowledgments

The authors would like to acknowledge the support of Chris McCarroll, Dave Laighton and Steve Kaspar, who are critical to the success of Raytheon's MHEMT program. We also would like to thank J. Bettencourt, M. White, C. Laighton, T. Kennedy, J. Sactuary, P. Lyman, K. Herrick, J. Mosca, N. Pisiello, J. Hunt and M. Grigas, for their significant contributions to this work. Also, special thanks to Abhay Joshi at Discovery Semiconductors for his invaluable optical packaging and measurement expertise.

References

1. A. Cappy, Y. Cordier, S. Boollaert and M. Zaknoune, "Status of Metamorphic $In_xAl_{1-x}As/In_xGa_{1-x}As$ HEMTs," *1999 GaAs Integrated Circuit Symposium*, pp. 217-220.

2. M. Kawano, T. Kuzuhara, H. Kawasaki, F. Sasaki, and H. Tokuda, "InAlAs/InGaAs Metamorphic Low-Noise HEMT", *IEEE Microwave and Guided Wave Letters*, Vol. 7, No. 1, pp. 6-8, 1997.

3. M. Chertouk, H. Heiss, D. Xu, S. Kraus, W. Klein, G. Bohm, G. Trankle and G. Weimann, "Metamorphic InAlAs/InGaAs HEMT's on GaAs Substrates with a Novel Composite Channels Design," *IEEE Electron Device Lett.*, vol. 17, no. 6, pp. 273-275, 1996.

4. D. Gill, B. Kane, S. Svensson, D. Tu, P. Uppal and N. Byer, "High Performance, 0.1micronInAlAs/InGaAs High Electron Mobility Transistors on GaAs," *IEEE Electron Device Lett.*, vol. 17, no. 7, pp. 228-230, 1996.

5. S. Bollaert, Y. Cordier, V. Hoel, M. Zaknoune, H. Happy, S. Lepilliet and A. Cappy, "Metamorphic $In_{0.40}Al_{0.60}/In_{0.40}Ga_{0.60}As$ HEMT's on GaAs Substrate," *IEEE Electron Device Lett.*, vol. 20, no. 3, pp. 123-125, 1999.

6. W. E. Hoke, P. S. Lyman, J. Mosca, C. S. Whelan, and A. Torabi, "Growth and Characterization of Metamorphic $In_x(AlGa)_{1-x}As/In_xGa_{1-x}As$ High Electron Mobility Transistor Structures with X=0.3-0.4," *J. Vac. Sci. & Tech. B, May/June 2000.*

7. W. E. Hoke, P. J. Lemonias, J. J. Mosca, P. S. Lyman, A. Torabi, P. F. Marsh, R. A. McTaggart, S. M. Lardizabal and K. Hetzler, "MBE Growth and Device Performance of Metamorphic High Electron Mobility Structures Fabricated on GaAs Substrates," *J. Vac. Sci. & Tech. B,* 17 (3) 1999, pp. 1131-1135.

8. Y. Cordier, S. Bollaert, M. Zaknoune, J. diPersio, and D. Ferre, "AlInAs/GaInAs Metamorphic HEMT's on GaAs Substrate: From Material to Device," *10th International Conference on Indium Phosphide and Related Materials*, pp. 211-214, May 1998.

9. C. S. Whelan, S.M. Lardizabal, W.E. Hoke, P.F. Marsh, R.A. McTaggart and T.E. Kazior, "A Comparison of Channel Indium Content in Low Noise Metamorphic HEMTs with $In_xGa_{1-x}As$ ($0.3<X\leq 0.6$)," *InP and Related Materials 2000.*

10. P. F. Marsh, S. L. G. Chu, S. M. Lardizabal, R. E. Leoni III, S . Kang, R. Wohlert, A.M. Bowlby, W. E. Hoke, R. A. McTaggart, C. S. Whelan, P. J. Lemonias, P. M. McIntosh, and T. E. Kazior, "Low Noise Metamorphic HEMT Devices and Amplifiers on GaAs Substrates," *1999 Microwave Theory and Techniques Symposium,* Vol. 1, pp. 105-108.

11. H. Rohdin, A. Nagy, V. Robbins, C. Su, C. Madden, A. Wakita, J. Raggio, and J. Seeger, "Low-Noise High-Speed $Ga_{.47}In_{.53}As$/ $Al_{.48}In_{.52}As$ 0.1-micronMODFETS and High-Gain/Bandwidth Three-Stage Amplifier Fabricated on GaAs Substrate", *Seventh International Conference on Indium Phosphide and Related Materials*, pp. 73-76, 1995.

12. P. C. Chao, A. J. Tessmer, K. G. Duh, P. Ho, M. Kao, P. M. Smith, J. M. Ballingall, S. M. J. Liu and A. A. Jabra, " W-Band Low-Noise InAlAs/InGaAs Lattice-Matched HEMTs," *IEEE Elect. Dev. Lett.,* Vol. 11 (1), 1990.

13. P. F. Marsh, S . Kang, R. Wohlert, P. M. McIntosh, W. E. Hoke, R. A. McTaggart, S. M. Lardizabal, R. E. Leoni III, C. S. Whelan, P. J. Lemonias, and T. E. Kazior, "Millimeter-Wave Low Noise Metamorphic HEMT Amplifiers and Devices on GaAs Substrates," *1999 GaAs IC Symposium*, pp. 225-228.

14. C. S. Whelan, P. F. Marsh, W. E. Hoke, R. A. McTaggart, P. S. Lyman, P. J. Lemonias, S. M. Lardizabal, R. E. Leoni III, S. J. Lichwala, T. E. Kazior, "Millimeter-wave low-noise and high-power metamorphic HEMT amplifiers and devices on GaAs substrates," *IEEE Journal of Solid State Circuits,* Vol. 35, No. 9, Sept. 2000, pp. 1307-1311.

15. R. E. Leoni III, S. J. Lichwala, J. G. Hunt, C. S. Whelan, P. F. Marsh, W. E. Hoke, T. E. Kazior, " A Dc- 45 GHz metamorphic HEMT traveling wave amplifier," *2001 GaAs IC Symposium*, pp. 133-136.

16. W. Contrata and N. Iwata, "Double-Doped $In_{0.35}Al_{0.65}As$/ $In_{0.35}Ga_{0.65}As$ Power Heterojunction FET on GaAs Substrate with 1 W Output Power," IEEE Electron Device Lett., vol. 20, no. 7, pp. 369-371, 1999.

17. D. Tu, S. Wang, J. S. M. Liu, K. C. Hwang, W. Kong, P. C. Chao and K. Nichols, "High-Performance Double-Recessed InAlAs/InGaAs Power Metamorphic HEMT on GaAs Substrate," *IEEE Guided Wave Lett.,* vol. 9, no. 11, pp. 458-460.

18. M. Zaknoune, B. Bonte, C. Gaquiere, Y. Cordier, Y. Druelle, D. Theron and Y. Crosnier, "InAlAs/InGaAs Metamorphic HEMT with High Current Density and High Breakdown Voltage," *IEEE Elect. Dev. Lett.*, vol. 19, no. 9, pp. 345-347, 1998.

19. C. S. Whelan, W. E. Hoke, R. A. McTaggart, P. S. Lyman, P. F. Marsh, R. E. Leoni III, S. J. Lichwala, and T. E. Kazior, "High Breakdown Voltage InAlGaAs/$In_{0.32}Ga_{0.68}As$ Metamorphic HEMT for Microwave and mm-wave Power Applications," *1999 MTT,* vol. 3, pp. 1187-1190.

20. C. S. Whelan, W. E. Hoke, R. A. McTaggart, P. S. Lyman, P. F. Marsh, S. J. Lichwala, and T. E. Kazior, "High Power and Gain at 35 GHz Utilizing an InAlGaAs/$In_{0.32}Ga_{0.68}As$ Metamorphic HEMT," *1999 GaAs IC*, pp. 221-223.

21. C. S. Whelan, W. E. Hoke, R. A. McTaggart, S. M. Lardizabal, P. S. Lyman, P. F. Marsh and T. E. Kazior, "Low Noise $In_{0.32}(AlGa)_{0.68}As/In_{0.43}Ga_{0.57}As$ Metamorphic HEMT on GaAs Substrate with 850 mW/mm Output Power Density," *IEEE Elect. Dev. Lett.,* January 2000.

22. W. E. Hoke, P. J. Leomonias, T. D. Kennedy, A. Torabi, E. K. Tong, R. J. Bourque, J. H. Jang, G. Cueva, D. C. Dumka, I. Adesida, K. L. Chang, K. C. Hsieh, "Metamorphic heterojunction bipolar transistors and P-I-N photodiodes on GaAs substrates prepared by molecular beam epitaxy," *Journal of Vacuum Science and Technology B*, Vol. 19, July/Aug. 2001, pp. 1505-1509.

23. C. S. Whelan, P. F. Marsh, R. E. Leoni III, J. G. Hunt, M. Grigas, W. E. Hoke, K. C. Hwang, T. E. Kazior, A. M. Joshi, X. Wang, "Metamorphic PIN photodiodes for the 40 Gb/s fiber market," *2001 GaAs IC Symposium*, pp. 251-254.

24. J. H. Jang, G. Cueva, D. C. Dumka, W. E. Hoke, P. J. Leomonias, P. Fay, and I. Adesida, "Long wavelength metamorphic double heterojunction In0.53Ga0.47As/InAlGaAs/In0.52Al0.48As photodiodes on GaAs substrates," *Electronics Letters*, Vol. 37, No. 11, 24th May 2001, pp. 707-708.

25. Discovery Semiconductor Inc. "OEIC's: reality or laboratory curiosity?", presented during *Panel Session 4 of 2001 GaAs IC Symposium.*

26. T. Ishida, Y. Yamamoto, N. Hayafuji, S. Miyakuni, R. Hattori, T. Ishikawa, and Y. Mitsui, "Fluorine Limited Reliability of AlInAs/InGaAs High Electron Mobility Transistors with Molybdenum Gates," *1997 IEEE InP and Related Materials* pp. 201-204.

27. A. Wakejima, K. Onda, A. Fujihara, E. Mizuki, and M. Kanamori, "Fluorine Diffusion and Accumulation in Si Step-Doped InAlAs Layers," *Applied Physics Letters*, Vol. 73, No. 17, October 1998, pp. 2459-2461.

28. M. Van Hove, T. Skrabka, H. Bender, W. De Raedt, M. Van Rossum, and Y. Baeyens, "Optimization of Ohmic Contacts on Lattice-Matched and Pseudomorphic AlInAs/InGaAs/InP," *Materials Research Society Symposium Proceedings*, Vol. 326, 1994, pp. 463-468.

29. W. Hong, K. Seo, P. Bhattacharya, and H. Lee, "Low-Resistance Ohmic Contacts to AlGaAs/GaAs and $In_{0.52}Al_{0.48}As/In_{0.53}Ga_{0.47}As$ Modulation-Doped Structures Obtained by Halogen Lamp Annealing," *IEEE Electron Device Letters*, Vol. EDL-7, No. 5, May 1986, pp. 320-322.

30. N. Hayafuji, Y. Yamamoto, T. Ishida, and K. Sato, "Degradation Mechanism of the AlInAs/GaInAs High Electron Mobility Transistor due to Fluorine Incorporation," *Applied Physics Letters*, 69, (26), 23 December, 1996, pp 4075-4077.

31. T. Ishida, Y. Yamamoto, N. Hayafuji, S. Miyakuni, R. Hattori, T. Ishikawa, and Y. Mitsui, "Fluorine Limited Reliability of AlInAs/InGaAs High Electron Mobility Transistors with Molybdenum Gates," *1997 IEEE InP and Related Materials,* pp. 201-204.

32. D. LaCombe, W. Hu, F. Bardsley, "Reliability of 0.1 micron InP HEMTs, *International Reliability Physics Symposiu, 1993*, pp. 364-371.

33. H. Cooke, *Microwave & RF*, August 1986.

34. M. Chertouk, F. Benkhelifa, M. Dammann, M. Walther, K. Kohler, and G. Weimann, "Metamorphic InAlAs/InGaAs HEMT MMIC Technology on GaAs Substrate: From Promise to Reality," *GaAs Mantech 2000*, pp. 233-236.

35. M. Dammann, M. Chertouk, W. Jantz, K. Kohler, K. Schmidt, G. Weimann, "Reliability of Passivated 0.15mm InAlAs-InGaAs HEMTs with Pseudomorphic Channel," *Reliability Physics Symposium Proceedings*, pp. 99 –102.

36. M. Chertouk, M. Maśsler, M. Dammann, K. Kohler, G. Weimann, "Manufacturable 0.15mm InP-Based HEMT MMIC Process with High Yield and Reliability on 2" InP Substrate," *1999 GaAs Manufacturing Technology Symposium*, pp. 25-28.

37. A. Wakita, H. Rohdin, V. Robbins, N. Moll, C-Y Su, A. Nagy, and D. Basile, "Low-Noise Bias Reliability of AlInAs/GaInAs MODFETs with Linearly Graded Low-Temperature Buffer Layers Grown on GaAs Substrates," *10th International Conference on Indium Phosphide and Related Materials*, May 1998, pp. 223-226.

38. A. Wakita, H. Rohdin, V. Robbins, N. Moll and C. Su, "Low Noise Bias Reliability of AlInAs/InGaAs Modulation Doped Field Effect Transistors with Linearly Graded Low-Temperature Buffer Layers Grown on GaAs Substrates," *Japanese Journal of Applied Physics*, Vol. 38 (1) 2B, 1999. pp. 1186-1189.

39. R. Blanchard, J. del Alamo, P. Chao, and S. Adams, "Hydrogen Degradation in InP HEMTs," *1998 International Devices Meeting*, pp 231-234.

40. R. Gaddi, R. Menozzi, D. Dieci, C. Lanzieri, G. Meneghesso, C. Canali, E. Zanoni, "Bulk and Surface Effects of Hydrogen Treatment on Al/Ti-Gate AlGaAs/GaAs Power HFETs," *Reliability Physics Symposium Proceedings,* 1999. 37th Annual.1999 IEEE International , 1999 , pp. 110 –115.

41. P. Chao, W. Hu, H. DeOrio, A. Swanson, W. Hoffmann, and W. Taft, "Ti-Gate Metal Induced pHEMT Degradation in Hydrogen," *IEEE Electron Device Letters*, Vol. 18, No. 9, September 1997, pp. 441-443.

International Journal of High Speed Electronics and Systems
Vol. 13, No. 1 (2003) 91–109

GaAs PHEMT CHIP SETS AND IC PROCESSES FOR HIGH-END FIBER OPTIC APPLICATIONS

REMY LEBLANC, AHMED GASMI, MESSAOUD ZAHZOUH, DEREK SMITH, FRANCIS AUVRAY, JOEL MORON, JEAN HOURANY, DOMINIQUE DEMANGE, ANDREAS THIEDE*, MARC ROCCHI, Senior Member IEEE

OMMIC
22, avenue Descartes, 94453 Limeil-Brévannes, France
m.rocchi@ommic.com

** Paderborn University – FB14/160 – 33095 Paderborn, Germany*
Thiede@uni-paderborn.de

Continuous improvement of high bit rate fiber optic networking up to 40 Gb/s, results in increased demand for III/V ICs offering ever lower current input noise, higher output voltage swing, lower jitter and lower DC power consumption. Using production E/D and power sub 0,2 μm GaAs PHEMT processes, transimpedance amplifiers have been developed enabling optical input sensitivity as good as -25 dBm at 2.5 Gb/s and -20 dBm at 40 Gb/s. Similarly laser and modulator drivers have been fabricated with jitter as low as 1,8 ps rms at 10 Gb/s and over 7 V output swing up to 40 GHz. Future generations of products will be implemented using sub 0,1 μm processes based on E/D and power GaAs MHEMTs as well as InP HEMTs maintaining a long term performance edge over the Silicon counterparts.

Key words : fiber optic ICs, PHEMT, MHEMT, transimpedance amplifiers, drivers

1. Introduction

Since the mid 90's, the system requirements of advanced telecommunication infrastructure have been progressing faster than Silicon IC's performance offering new design-in opportunities for III/V processes especially for high-end applications in mobile terminals, cellular base stations and long haul fiber optic repeaters. Each new generation of fiber optic modules for high-bit rate transmission standards such as OC48, OC192 and OC768 requires smaller amplifier ICs with ever lower noise for post photo-diode receivers and higher voltage swing with minimum jitter and group delay for laser and modulator drivers. From 2.5 Gb/s to 40 Gb/s GaAs, it has been proven that using GaAs PHEMT processes particularly optimized for these applications, the toughest IC specifications can be met, well superseding the performance of the Si CMOS and SiGe BiCMOS counterparts [1-2].

The differences in performance between the III/V and Si technologies, have consistently resulted from the intrinsic transport properties in each respective material. GaAs PHEMT technologies and quite recently GaAs MHEMT, InP HEMT and HBT technologies exploit the very high saturation velocity of electrons in high indium content InGaAs channel or base [3]. This enables to achieve state of the art ft's up to 400 GHz which is about 2 times higher than the best Si/Ge results[4]. In addition, III/V technologies exhibit breakdown voltages that are 2 to 3 times higher than all Si technologies [3] with the same ft max.

The results presented here, are the starting point that will enable in the coming years to continously improve the Tx/Rx functions for fiber optic networking. This will be achieved by scaling the critical dimensions of active devices while optimizing the channel conductivity by increasing the In content on GaAs or InP substrates, before eventually shifting to Sb based solutions [5].

In this paper, we first review in sections 2 and 3, power and low noise IC processes optimized to meet the fiber optic networking requirements. In particular the high potential of E/D PHEMT processes is described in terms of IC design optimization for the input stage of a 2.,5 Gb/s transimpedance amplifier.

A detailed analysis of the system requirements for long haul and metropolitan applications, clearly show that transimpedance amplifiers from 2.5 Gb/s up and laser or modulator drivers from 10 Gb/s up are the main functions where advanced III/V IC processes will give real advantages.

Input sensitivity is a key parameter of receiving modules for long haul and metropolitan fiber optic links. With the best photodiodes available sensitivity as good as - 25 dBm at 2.5 Gb/s and better than - 20 dBm at 40 Gb/s have been achieved with the transimpedance amplifiers reported in sections 4 and 5.

Output voltage swing into a 10 or 50 ohm load resistance, DC power dissipation and low jitter are key parameters of laser and modulator drivers. 10 Gb/s 3 V pp laser drivers with jitter less than 1.8 ps rms and 45 GHz single and dual input modulator drivers with output swing higher than 7 V are described in sections 6 and 7 respectively. A very compact RC reshaping circuit is also reported to improve the saturation behaviour of the 45 GHz drivers based on a classical distributed architecture.

Finally, the design and measured performance of 50 Gb/s multiplexers and demultiplexers are given in section 8 with excellent eye diagrams up to the limit of the test equipment at 43 Gb/s This shows that the key factor of merit in large signal operation is not the ft max of the process but the mean ft across the logic swing. In this respect E/D PHEMT processes offer a unique compromise for the integration of high speed digital and analogue functions on the same chip.

In section 9, the potential improvement of the next generations of transimpedance amplifiers and drivers in the near future is shown to be fully supported by the microwave performance of scaled down GaAs MHEMTs and InP HEMTs processes. The long term edge of III/V solutions over silicon counterparts will be essential in meeting ever more stringent system specifications.

2. PHEMT and MHEMT Processes

The PHEMT and MHEMT processes optimized for fiber optic applications (see table 1) are developed along two main tracks :

- High frequency low noise processes, mixing enhancement and depletion mode transistors ("ED") : ED02AH, ED01AH and ED01IH.

- High Frequency, high breakdown process : D01PH, D01MH.

The main features of all these processes are (see Figure 1) : air bridges, via holes, GaAs and NiCr resistors, level shifting and varactor diodes, 1 μm and 2 μm thick interconnect layers, 400 pF/mm^2 and 50 pF/mm^2 MIM capacitors, spiral inductors, full SiN passivation, full SiO2 protection.

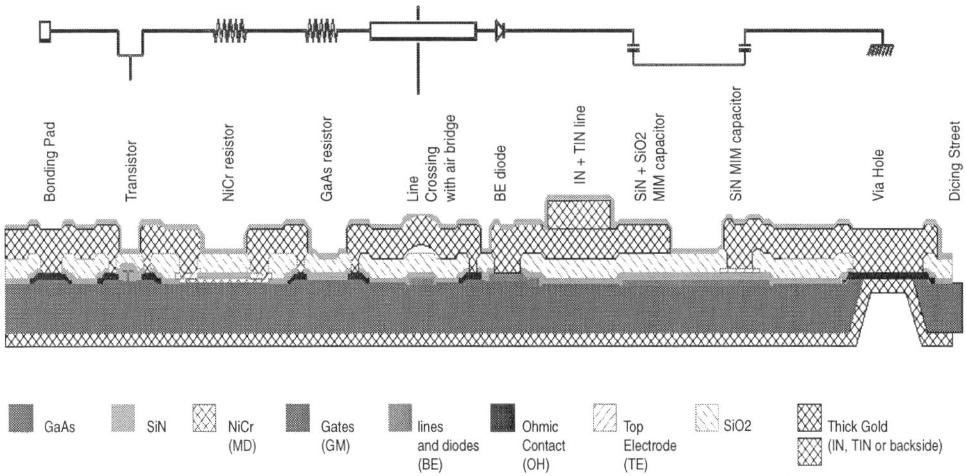

Figure 1 : Cross section of the D01PH process

The high frequencies achieved by these processes can make off-chip interconnections difficult : a 0.25 nH bonding wire creates at 50 GHz a series impedance close to 80 Ω, which is hard to match reproducibly in production. For this reason, we have developed micro bumps allowing very short connections and hence very low inductances. These bumps have a 60 μm diameter and a 35 μm height (Figure 2). No undesirable proximity effects are observed when mounting the circuit on the main board.

Process	PHEMT		MHEMT	
	ED02 / 01AH	D01PH	D01MH	ED01IH
Gate length	0.18 - 0.15 μm	0.13 μm	0.13 μm	0.13 μm
ft (GHz) @ 2 V	60 / 75 GHz	100 GHz	150 GHz	150 GHz
NF (dB)	< 0.5 dB at 12 GHz			< 3 dB at 120 GHz
Vbgd (V)	8	12	10	7
Vt off & on (V)	+ 0.2 / - 0.9 V	- 0.9 V	- 0.9 V	+ 0.1 / - 0.9 V
Status	Production	Production	Prototyping	Development

Table 1 : Pseudomorphic and Metamorphic HEMT process main characteristics
(ft is the extrinsic current gain cut-off frequency including the capacitance of the input pad used for probing, V_{bgd} is the gate to drain voltage corresponding to a gate current of 1 mA/mm, V_t is the gate to source voltage at I_{DS} = 1 mA/mm and V_{DS} = 2 V)

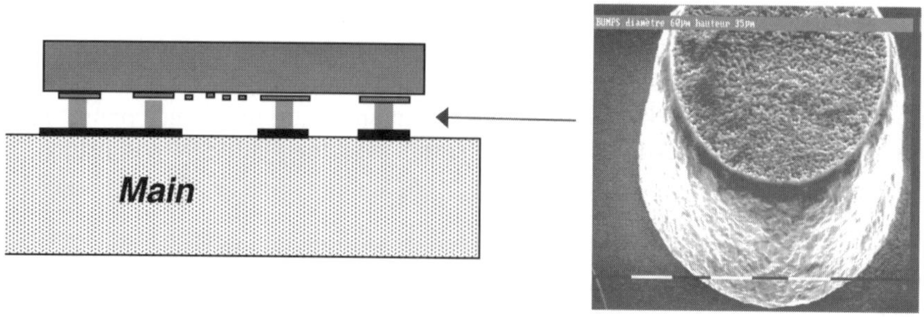

Figure 2 : Micro bumps

The ED02AH process is a 0.18 μm multipurpose pseudomorphic HEMT process. The two available threshold voltages (+ 0.2 and – 0.9 V), the high ft (Figure 3) and the low noise make it suitable for many different applications : low noise amplifiers for GSM base stations, 2.5 or 10 Gb/s transimpedance amplifiers and drivers, microwave amplifiers, mixers, attenuators and phase shifters including digital control, military or space components. With the 0.15 μm gate length option, 50 Gb/s multiplexers have been demonstrated.

Presenting at the same time a 100 GHz ft and a 12 V gate drain breakdown voltage, the D01PH pseudomorphic HEMT 0.13 μm process is an excellent tool for the fabrication of 40 Gb/s transimpedance amplifiers or modulator drivers, and also for some other applications such as microwave power amplifiers.

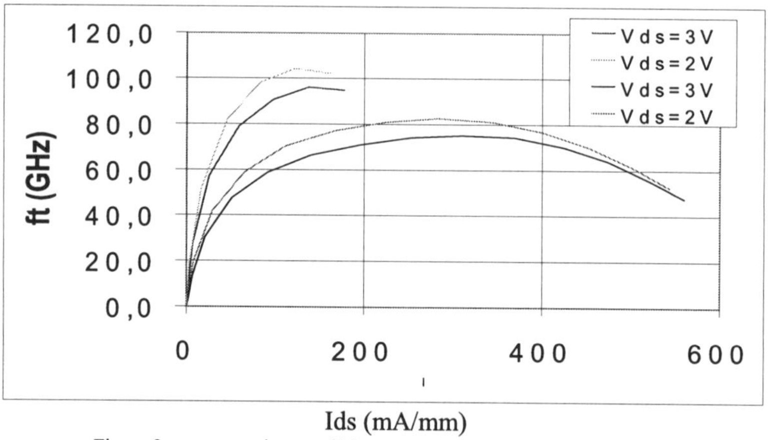

Figure 3 : current gain cut-off frequency ft versus drain current for N-off (upper two curves) and N-on transistors (lower two curves) in the ED02AH process

(ft in Figure 3 is the available extrinsic current gain cut-off frequency excluding the capacitance of the input pad used for probing)

The next process generation, already emerging, follows the same two tracks, but with improved performances obtained by a higher indium content, either from a metamorphic active layer on GaAs, or processing on InP. An example of this is the depletion mode metamorphic (D01MH) process which exhibits a 175 GHz ft at Vds =1V and 145 GHz at Vds = 2 V (see Figures 4a and 4b) .

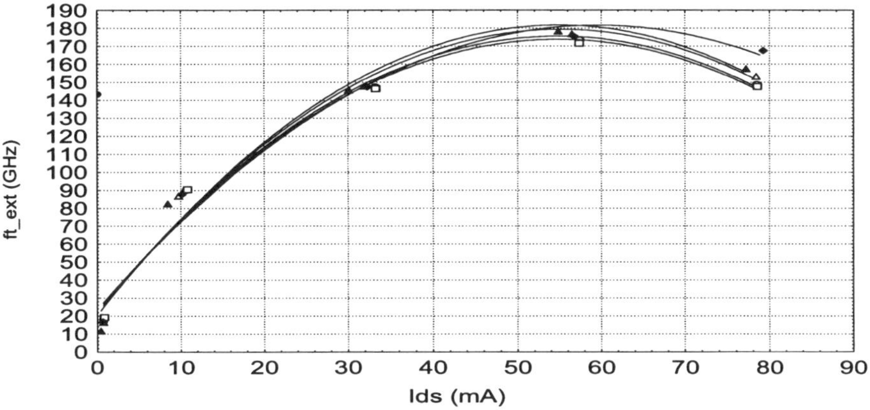

Figure 4 a : ft of the Metamorphic D01MH process at 1 V

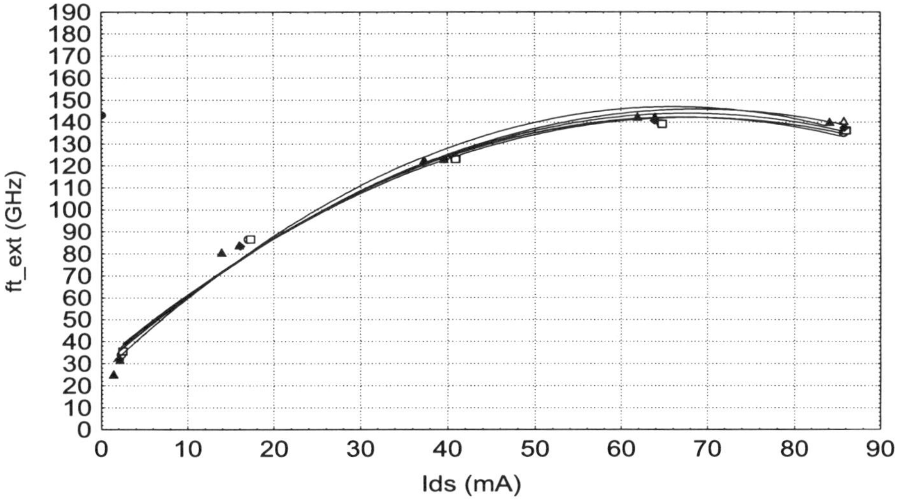

Figure 4 b : ft of the Metamorphic D01MH process at 2 V

3. Optical Fibre Interface Circuits

The results presented below are examples of circuits fabricated with the ED02AH, ED01AH or D01PH processes. These circuits are dedicated to optical fiber interface communication systems, at the 2.5, 10 or 40 Gb/s bit rates. They all make use of the specific process features to obtain either very high sensitivity or very low DC consumption.

4. 2.5 Gb/s Transimpedance Amplifiers

This circuit has been fabricated with the ED02AH process, using both enhancement mode and depletion mode transistors. It is intended to be used as a low noise preamplifier for light-wave receiver modules in optical fiber with an exceptionally good sensitivity and high gain. The circuit features a very small chip size, differential outputs, low power consumption, very low noise and built-in Automatic Gain Control (AGC) to allow operation over a wide dynamic range.

In the design of a transimpedance amplifier (TIA) with a fixed gain, there is a trade-off between input current noise and the maximum input current (overload current) via the value of the shunt feedback resistor used in the signal path. The use of a PHEMT process with very low noise and high frequency cut-off helps to find a better trade off, while the availability of enhancement mode devices allows a single 3.3 V biasing voltage and a very low DC consumption for what is essentially a DC to 2 GHz amplifier [6].

The Figure 5 shows the functional block diagram of the TIA. The device consists of a low noise transimpedance stage (T1) followed by a differential amplifier (A1). Figure 6 shows a photograph of the TIA chip. The circuit was designed to be as compact as possible. The size of the circuit is 0.76 mm^2, on which about 40 transistors are integrated.

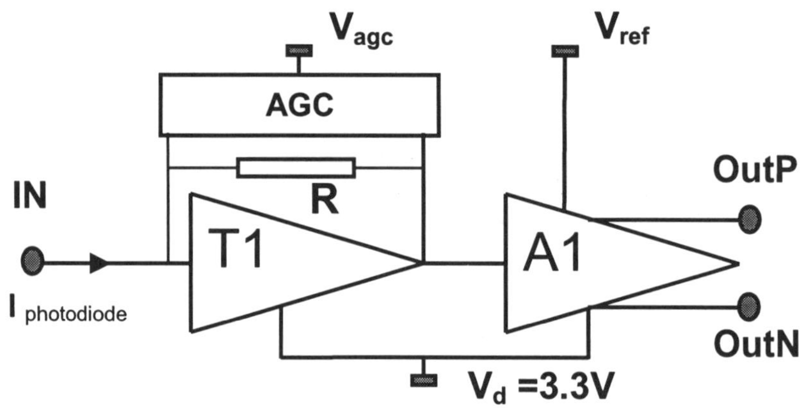

Figure 5 : 2.5Gb/s TIA schematic

The circuit presents a transimpedance gain (single-ended) of 75 dBΩ with a 3 dB cut-off frequency of 2.8 GHz (Figure 7). For the frequencies below 3 GHz, the output return loss (S_{22}) is better than 10 dB.

To be able to perform electrical-electrical eye diagram measurements, the transimpedance amplifier has been mounted on a test board with CMS components, and the photodiode is modelled by a 1 kΩ resistor (to transform the voltage generator into current generator) and a shunt 0.3 pF capacitor. The output eye diagram of the amplifier at 2.5 Gb/s is shown in figure 7. With an input data level of 6.5 μA, an output amplitude of 40 mV was obtained with a good eye opening (Figure 7).

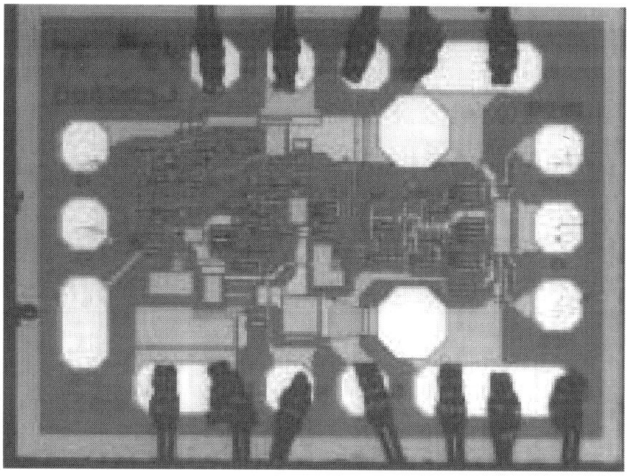

Figure 6 : Chip photograph of the 2.5Gb/s transimpedance amplifier.

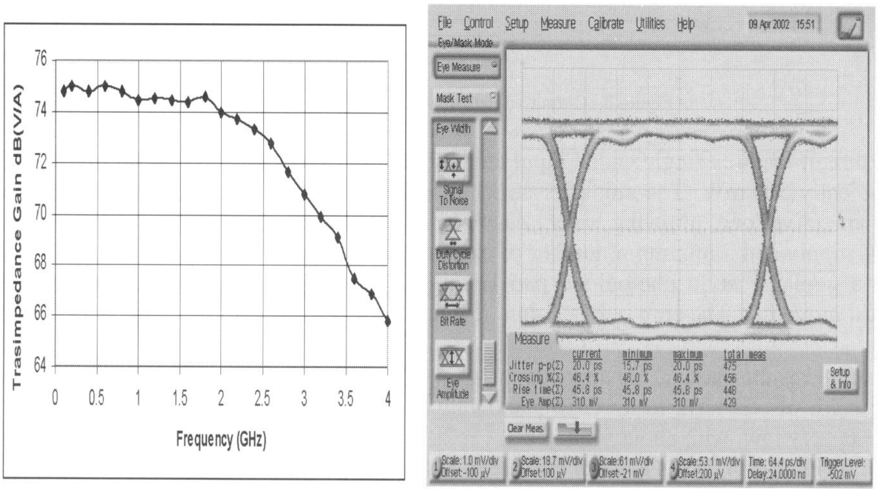

Figure 7 : Measured Transimpedance Gain versus Frequency
and output eye diagram at 2.5 Gb/s with an input current swing of 6.5 μA

The equivalent input noise current from 100 MHz to 1700 MHz is better than 3 pA/Hz^{½}. The power consumption of the TIA is 40 mA with a single V_d = 3.3 V power supply. The TIA uses an integrated automatic gain control circuit to operate with higher input photo-current up to 2.2 mA peak to peak. (Figure 8). -25 dBm sensitivity has readily been achieved at 10^{-10} BER using conventional PIN diodes.

5. 40Gb/s Transimpedance Amplifier

For 40 Gb/s transimpedance amplifiers, the D01PH process has been used because of its 100 GHz extrinsic current gain cut-off frequency required by the broad bandwidth. Typical use of such a circuit is as a Low Noise Preamplifier for Lightwave Receiver Modules in optical fibre networks.

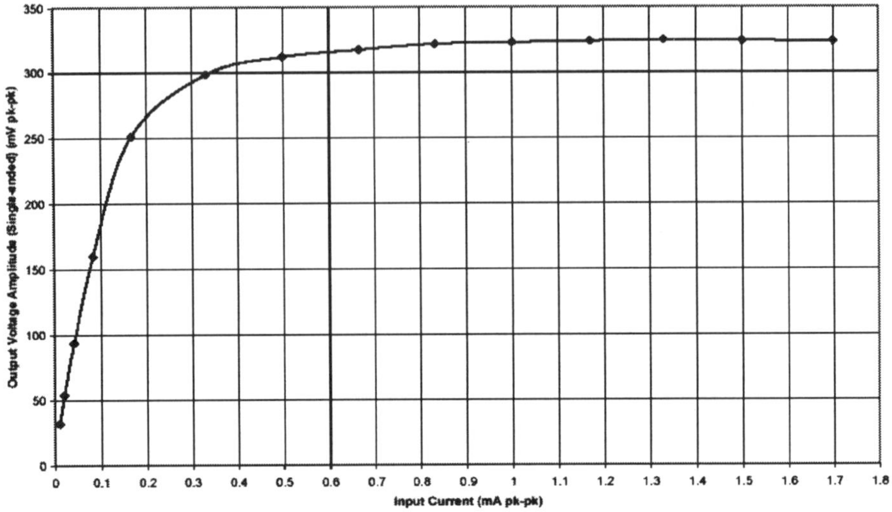

Figure 8 : Output voltage of the 2,5Gb/s TIA versus input current

This circuit features single ended input and output and operates with a power consumption of typically 500 mW. The amplifier structure is based on a distributed cascode amplifier, with an active load, allowing a very easy biasing of the circuit, requiring only a single + 5.0 V supply and a minimum number of external components. The input pad is compatible with a coplanar style photodiode pad layout. This offers extended bandwidth and less critical mounting. The chip size is 3 mm^2 (Figure 9).

Measured results show a transimpedance gain of 250 Ω (48 dBΩ), with a very wide 3 dB bandwidth of 25 KHz to 45 GHz , allowing data rates above 45 Gb/s (Figure 10). The noise is very low, with about 6 pA/(Hz)$^{1/2}$ at 20 GHz (carrier frequency) and 9 pA/(Hz)$^{1/2}$ at 40 GHz (Figure 11). This is a key parameter for very high sensitivity links.

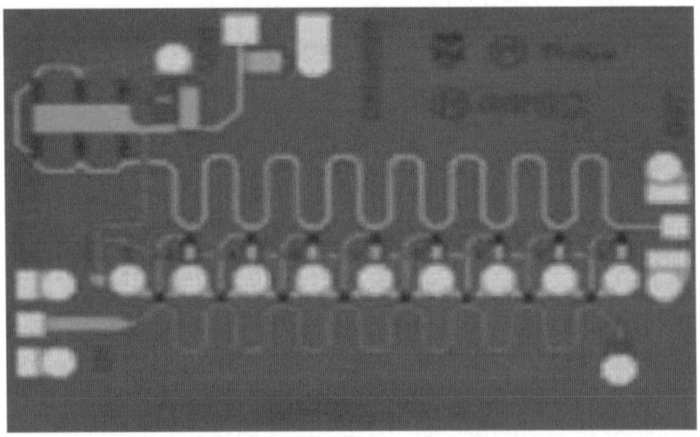

Figure 9 : 40 Gb/s TIA photograph

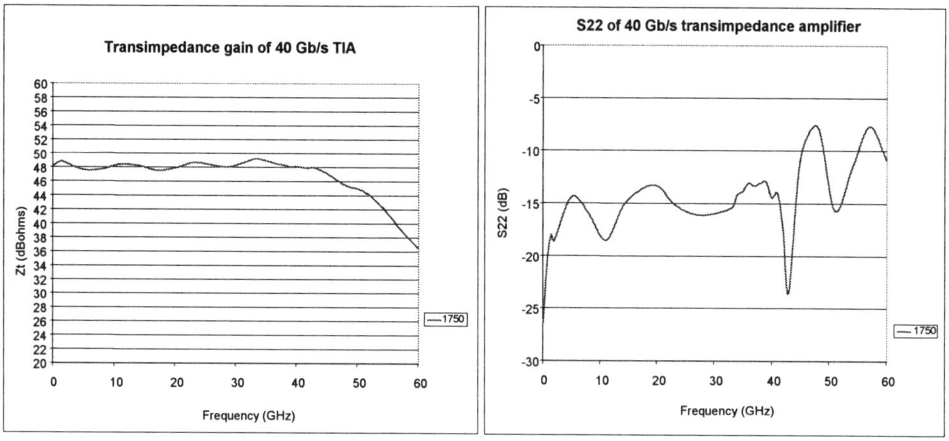

Figure 10 : Transimpedance gain in dBΩ (left) and output return loss in dB (right)
(diode parameters are : C =50 fF, R=10 Ω, L=0.3nH)

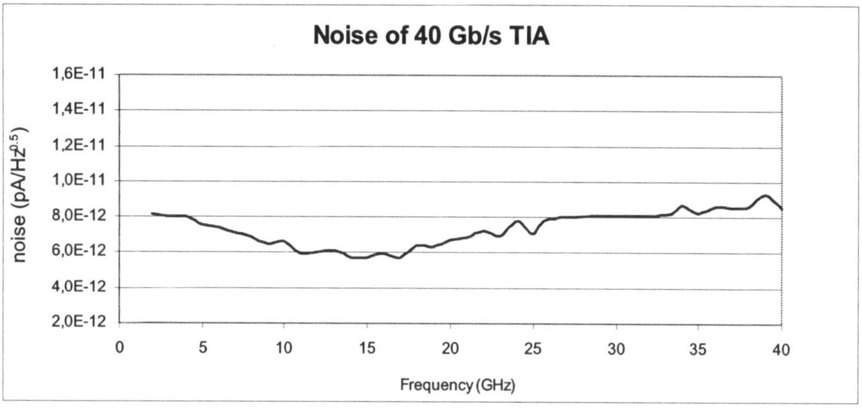

Figure 11 : Input current noise in $pA/(Hz)^{1/2}$

The combination of the distributed amplifier approach and the use of the power process D01PH leads to a very high input current overload capability (> 6 mA)

6. 10 Gb/s Laser Drivers

The same ED02AH process has been used for a 10 Gb/s modulator/driver. Typical use of such a circuit is as a current driver to directly drive a laser or as a voltage driver to drive an electro-absorption modulator for Lightwave Transmitter Modules in optical fibre networks. The architecture of this amplifier is based on a RC design approach, allowing at the same time dense layout and easy signal control and monitoring.

As with a transimpedance amplifier, the driver functions from almost DC to the maximum frequency required by the data rate. The use of enhancement mode transistors (associated with depletion mode devices) allows the easy implementation of DC coupling with optimum power consumption and RF performance.

The circuit, measured in its ceramic package, presents an output swing of 3 Vpp, with 30 ps rise and fall time, together with a very low jitter of 1.8 ps RMS. It operates using a single supply voltage of - 5.2 V, and the enhancement mode devices allow a low DC power consumption, less than 1400 mW. Input and output are single ended, and are DC coupled avoiding low frequency cut-off, and removing the need for an external bias T. The driver includes output DC offset adjust, a cross point and independent output swing controls (Figure 12).

The electrical performance of the circuit remains excellent at high temperature, without any increase of the DC consumption (Figure 13). Chip size is 1.47 mm^2.

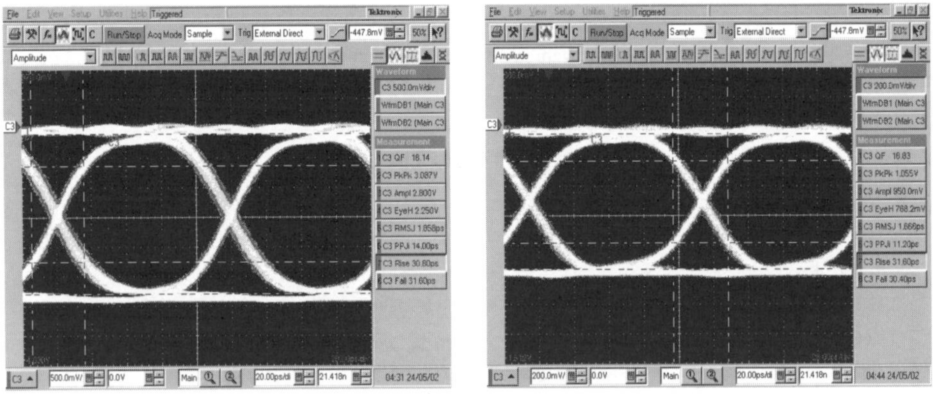

Figure 12 : 10 Gb/s eye diagrams at 3 Vpp, and 1 Vpp

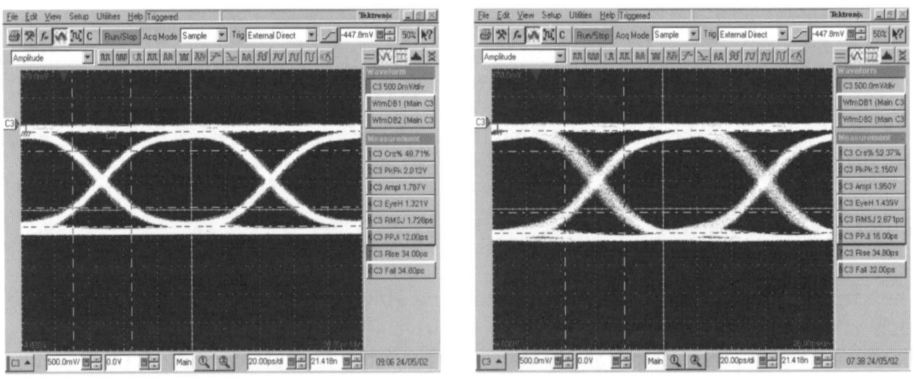

Figure 13 : 10 Gb/s eye diagrams at 25°C and 85°C
Including the effect of the long cables required for temperature measurements

7. 40Gb/s Modulator Drivers

The D01PH process combining high 100 GHz ft and 12 V gate to drain breakdown voltage is very well suited for the fabrication of high bite rate drivers. For large output voltage swings, distributed amplifiers are still the best solution to maintain the required bandwidth. On the other hand, they have a limited gain and give little opportunity for the control of the signal. This can be solved by using a reshaping circuit as described in section 7.3

7.1. Single input driver

The first amplifier designed and fabricated is a single input/single output driver. It is based on a cascode distributed amplifier, designed to produce output signals from 3 Vpp to 8 Vpp depending on the amplitude of the input swing. This chip has a very low DC consumption of less than 1 W under 5 V. The chip size is 2.2 mm^2.
Figure 14 shows the very large small signal bandwidth obtained with this amplifier, while Figure 15 shows the available output swing versus frequency for 0.5 Vpp, 1 Vpp, and 2 Vpp input signals. The output return loss (S_{22}) is as low as - 15 dB up to 42 GHz and better than - 10 dB up to 50 GHz.
Figure 16 shows the 7.5 Vpp eye diagram under 1.25 Vpp 12 Gb/s input signal.

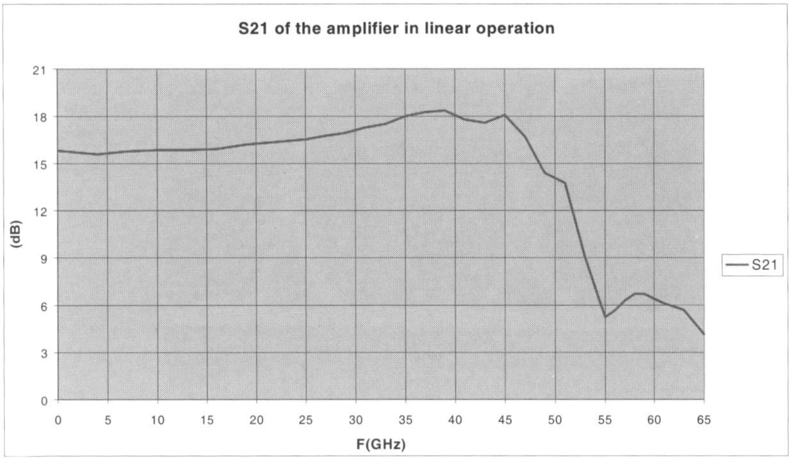

Figure 14 : S21 of the single input driver (dB)

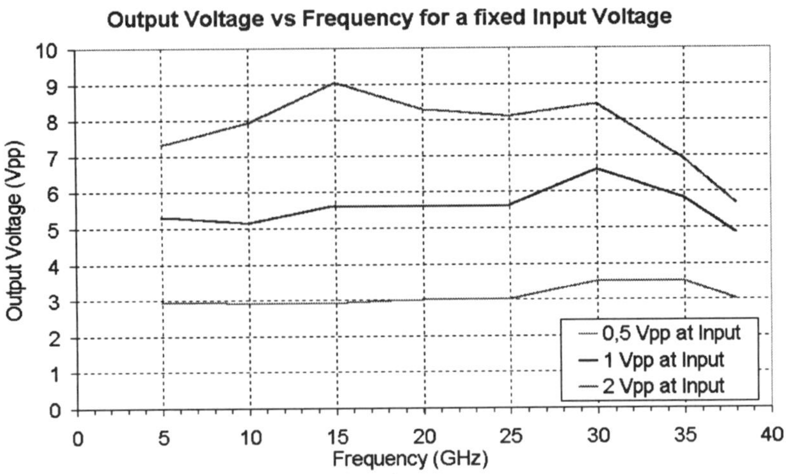

Figure 15 : Output swing of the single input driver (V)

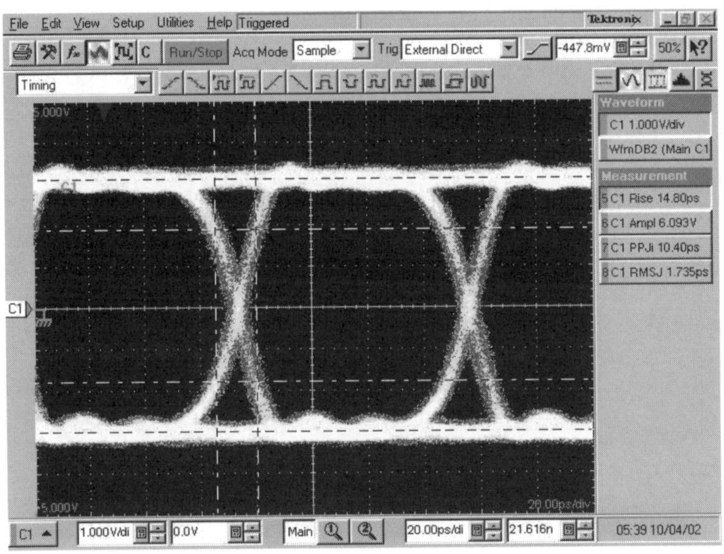

Figure 16 : 12 Gb/s eye diagram of the single input driver

7.2. Dual input driver

The design concept of this second 40 Gb/s driver[7] is based on two identical travelling wave amplifiers sharing a common drain line on which their output power is combined. The additional capacitive loading of the drain line enables comparable length in respect of the gate lines simplifying layout. Each half of the combiner consists of 5 stages in coplanar wave guide technology. A single stage contains a two finger HEMT cascode consisting of two FETs, each having a gate width of 30 μm. The cascode is designed as a single cell to save space and reduce parasitics. Figure 17 shows a chip micrograph of the combiner with probes.

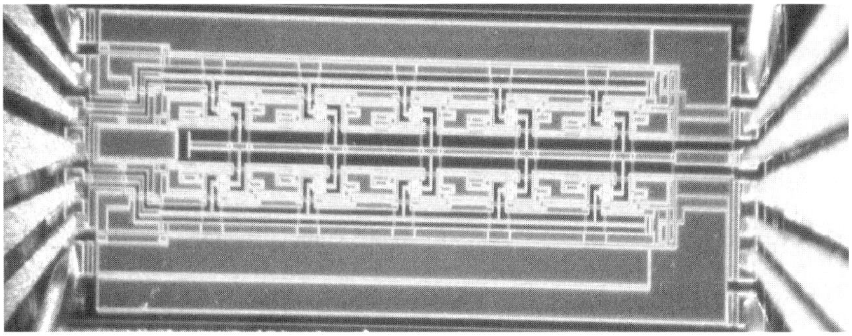

Figure 17 : Chip micrograph of the dual input driver

Since a 40 Gbit/s PRBS generator was not yet available, large signal 20 GHz sine wave measurements were carried out. All losses were first measured at 20 GHz and were compensated either by applying additional power at the input or by setting the attenuation parameter of the oscilloscope. Moreover, only the 10 MHz reference could be used as trigger signal for the scope, requiring averaging to get a clear signal. A passive power splitter was employed to deliver two equal input signals.

As shown in figure 18, maximum output voltages well above 7 V at input voltages of 1.5 V_{PP} were achieved.

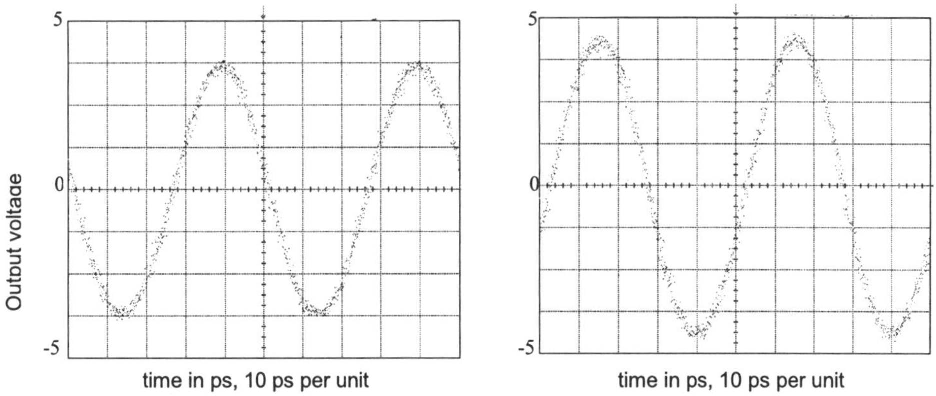

Figure 18 : Output Voltage of the dual input driver
for Input Voltages of 1 V_{PP} (left) and 1.5 V_{PP} (right)

7.3 Reshaping pre-amplifier

The distributed amplifier structure used to achieve the very wide bandwidth has some drawbacks :

- Little opportunity for the control of the signal, such as cross point controls
- Linear operation, leading to output signal shape always worse, or in the best case equivalent to the input signal.

To overcome these drawbacks, a reshaping pre-amplifier has been designed with the D01PH process. In spite of the extremely wide bandwidth required, this amplifier is based on a push-pull « RC » architecture, indeed similar to the one used for the 2.5 Gb/s or 10 Gb/s circuits reported above. This is made possible by a strong saturation at the output, creating an important reshaping of the signal. In addition, the differential structure allows the controls normally required for drivers, such as cross point tuning.

This first circuit has been designed to amplify input signals down to 0.3 Vpp, and give at the output a reshaped stabilised saturated signal of 0.5 Vpp. The chip size is 1.6 mm^2 for a DC consumption of 200 mA under 5 V. The circuit has a single ended input with DC level adjustment, and a 50 Ω differential output with cross point control.

Measuring such an amplifier is challenging, as most of the improvement of the 40 Gb/s signal lies on the 3rd harmonic, at 60 GHz, which cannot be captured by the oscilloscope available at that time, limited to 50 GHz. Figure 19 shows the comparison of the input and output signals at 20 GHz, which is the fundamental frequency of 40 Gb/s bit rates. input signal was a sine-wave with an amplitude of 0.3 V, a rise time of 15.2 ps and a fall time of 15.01 ps (at 10 % - 90 %). The output signal has an amplitude of 0.52 V, with a rise time of 13.57 ps and a fall time of 12.64 ps. We expect the actual improvement will be more marked when measuring up to the 3rd harmonic.

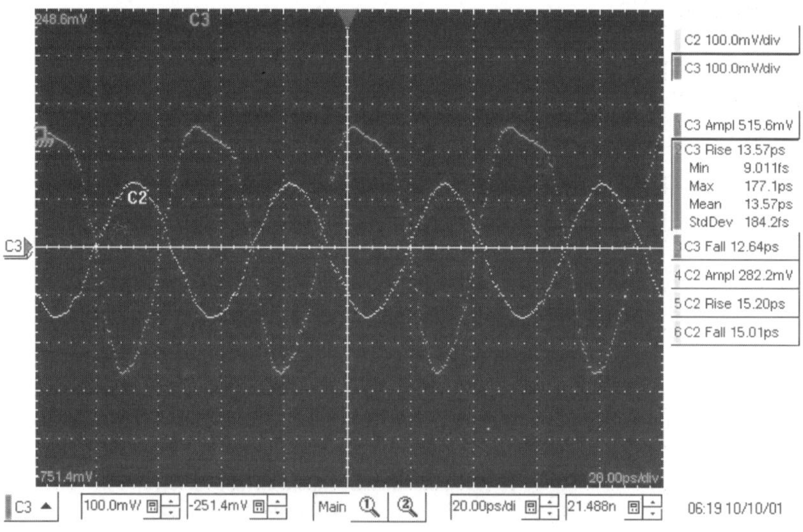

Figure 19 : input and output signals of the reshaping pre-amplifier

8. 43 Gb/s Multiplexers and Demultiplexers

The design of such digital circuits is eased by the availability of enhancement and depletion mode transistors in the ED02AH process. Full fledged SDFL logic gates are used with push-pull output buffers. Although we already demonstrated some years ago 40 Gb/s multiplexers and demultiplexers with this process, the 15 GHz better ft of the ED01AH process allows a better margin in the design.

The 2:1 fully retimed multiplexer combines two 0.5 Vpp signals into a 0.5 Vpp differential output signal at twice the bit rate. This circuit is able to operate up to 50 Gb/s, and shows a very good eye opening at 43 Gb/s, as presented in Figure 20. Rise and fall times are better than 7 ps. The 1:2 demultiplexer has been tested up to 43 Gb/s input signal, which was the maximum bit rate of our test equipment. Figure 21 shows also the two output signals obtained from a 40 Gb/s input signal.

The multiplexer chip size is 1.5 x 1.5 mm^2, while the demultiplexer chip size is 2 x 1.5mm^2 (Figure 22). Both circuits operate under - 5.2 V.

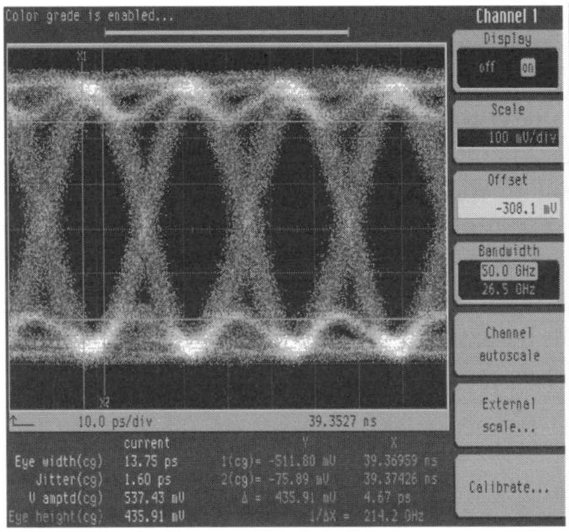

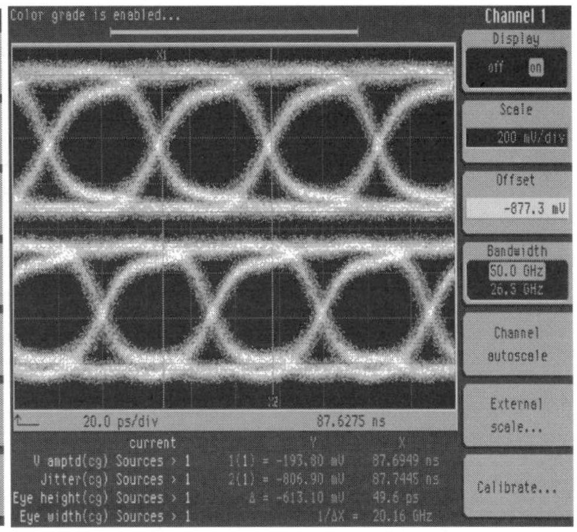

Figure 20 : 43 Gb/s Multiplexer output and Figure 21 : 20 Gb/s Demultiplexer outputs

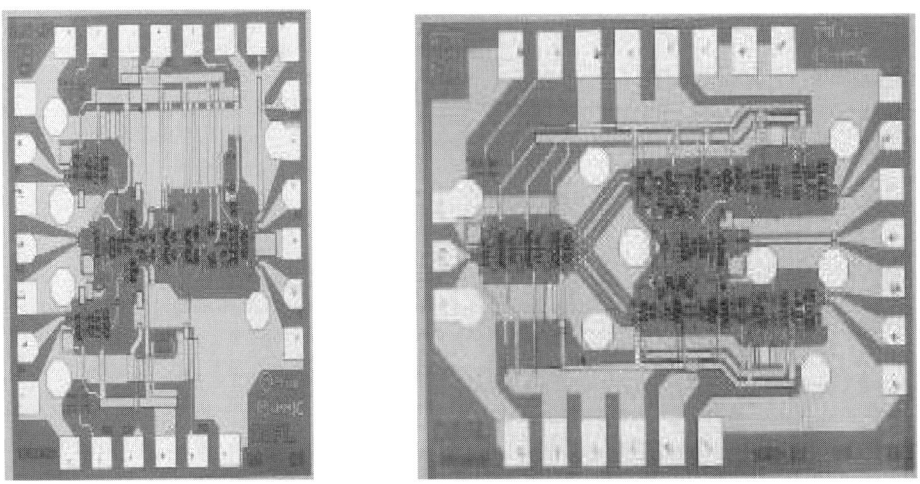

Figure 22 : Photograph of Multiplexer and Demultiplexer Die

9. Future Performance of Fiber Optic IC's and Related Processes

As for Silicon processes, there is ample room for further scaling down and optimization of the key III/V IC processes in order to achieve very high ft and fmax while preserving the highest breakdown voltages possible.

The two key parameters for process optimization are :

- The gate length of transistors which can be scaled down in medium volume production to 30 nm using the latest generation of e-beam machines.
- The transport properties in the InGaAs channel under the gate which can be further improved by increasing the In content. For flight times shorter than the energy relaxation time (1 ps), the velocity of electrons exhibits an overshoot with respect to the steady state velocity under a constant electric field [8]. The velocity overshoot is directly related to the low field mobility (momentum relaxation) which increases with the In content in the InGaAs channel. Electron velocity versus electric field under such non-stationary conditions can be derived from flight time measurements [9]. Figure 23 shows the electron velocity versus electric field for various semiconductors across a distance of 0.05 μm. For 100 % In, a velocity as high as $11*10^7$ cm/s can be expected, which is about 5 times higher than with SiGe or even strained Silicon.

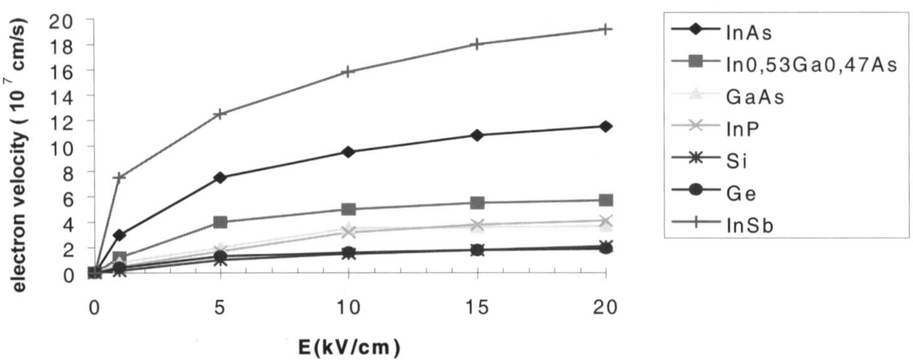

Figure 23 : Electron velocity versus electric field for high purity semiconductors under non-stationary transport conditions (across a 0.05 μm channel)

Scaling down of the gate length and increasing the In content in the channel, result in very high ft and fmax. With 45 nm gate length and 70 % In , ft as high as 400 GHz have been reported with an associated fmax of 470 GHz [4] as shown in figure 24. The PHEMT curve corresponds to 20 % In content in the channel. The slope change in the MHEMT / InP HEMT curve comes from the higher In content as the gate length is decreased.

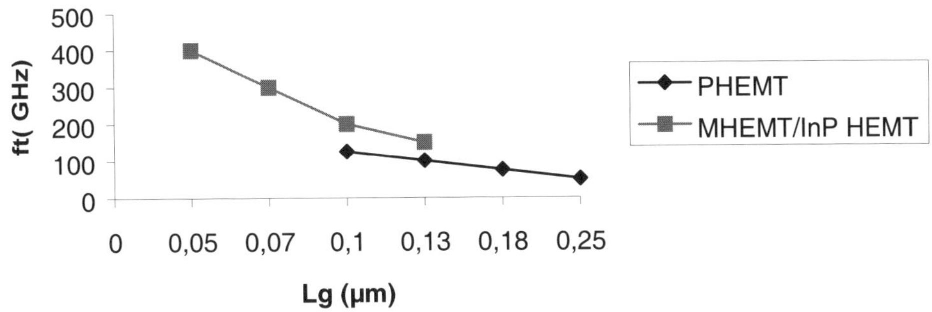

Figure 24 : current gain cut-off frequency versus gate length for GaAs P/MHEMTs and InP HEMTs

Increasing the In content means higher impact ionization at high electric field which results in lower gate to drain breakdown voltage, while decreasing the gate length without proper scaling of the hetero-structure thicknesses, results in short channel effects and lower drain to source breakdown voltage.

Although a significant amount of optimization is required below 0,1 µm gate length, excellent transistor breakdown (about two times better than with Si) has already been proven up to 150 GHz ft; as shown in Figure 25 [10]. This is achieved without degrading the manufacturability and reliability of the IC processes [11].

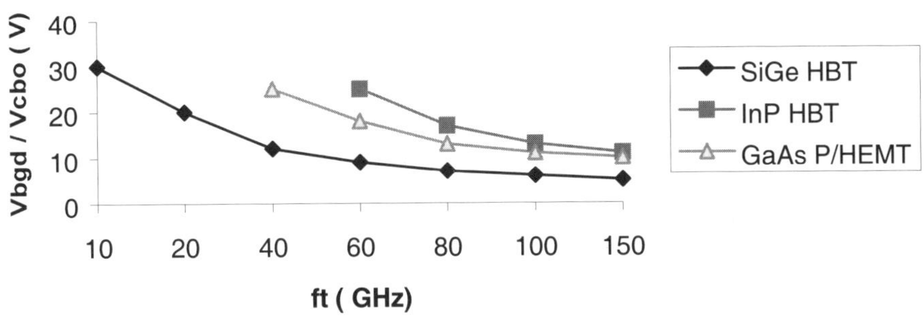

Figure 25 : drain to gate and collector to base breakdown voltages versus ft for the main III/V processes

In large signal operation, only the average ft across the voltage swing counts and not the maximum ft . For collector / drain current around 0.1 mA/mm, this is a clear advantage of P/MHEMT processes as long as E/D transistors are available on the same die (Figure 26).

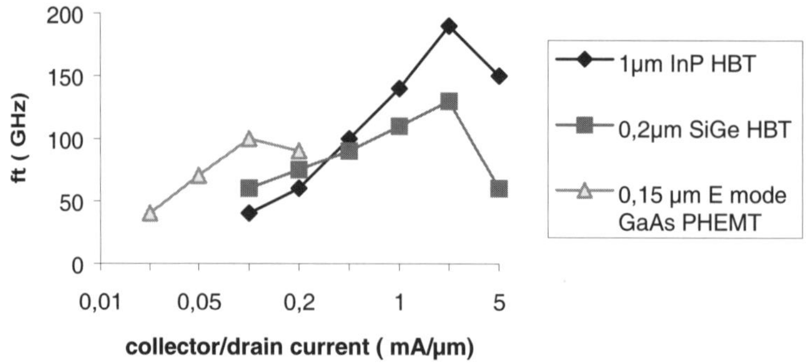

Figure 26 : ft versus collector / drain current per unit width for key III/V processes.

A direct consequence of the III/V P/MHEMT process roadmap described above, is that all the low frequency lumped element RC design of 2.5 Gb/s and 10Gb/s transimpedance amplifiers and drivers will be extended to 40 Gb/s and higher. Using 300 GHz GaAs HEMT or InP HEMT processes and based on proven scaling rules, 40 Gb/s transimpedance amplifiers are expected to exhibit 5 $pA/(Hz)^{1/2}$ input noise with 65 dB ohm gain. Similarly lumped element drivers, are expected to yield less than 1 ps rms jitter with 5 V output swing.

10. Conclusion

Advanced telecommunication infrastructures, continue to be very demanding in terms of IC performance, especially for high bit rate fiber optic interfaces. We have developed key interface functions such transimpedance amplifiers and laser/modulator drivers with electrical performance significantly superseding the best in class Silicon results (3 times lower noise input current at 40 GHz)[12]. The price difference between III/V solutions and Si counterparts is largely outweighed, by cost saving resulting from system optimization using high-end parts. Aggressive III/V process roadmaps, are in place to maintain this performance edge over the 10 years to come.

References

1. G. Freeman et al. "40 Gb/sec circuits built from a 120 GHz ft technology", GaAs IC Symposium, Digest 2001
2. S. Georgian, "SiGe circuits for wireless applications up to 40 Gb/s", Diget of Gorham conference on compound semiconductors for wireless & optical communications Europe, June 2002
3. M. Rocchi, III-V MMICs will continue to complement SI MMICs for advanced telecommunication applications", Digest of Gorham conference on compound semiconductors for wireless & optical communications Europe, June 2002.
4. K. Shinohara et al. "Ultrahigh speed pseudomorphic InGAs/InAlAs HEMTs with 400 GHz cut-off frequency", IEEE Electron Device Letters, vol 22, n° 11, november 2001

5. M.W. Dorak et al., IEEE Electron Device Letters, vol 22, n° 8, pp 361-363, 2001
6. A. Gasmi et al. "Ultra low noise 2.5 Gbit/s 3.3 V transimpedance ampifier with automatic gain control", 2001 IEEE GaAs IC Conference, Baltimore, Maryland, USA, October 21-24, 2001.
7. Results to be presented at the European GaAs application Symposium, 2002, Milano, Italy
8. Jacques G. Ruch, "Electron Dynamics in short channel field effect transistors", IEEE Electron Devices, ED-19, n° 5, May 1972
9. S. Laval, C. Bru et al., "Experimental evidence for velocity overshoot in GaAs from photoconduction measurements". Int Symposium of GaAs and related compounds, Sept. 1980.
10. D.C. Streit, "High volume Indium Phosphide production for wireless and opto-electronic applications". Digest of Gorham conference on Compound Semiconductors for wireless and optical communications Europe, June 2002.
11. T. Sonoda et al. "Manufacturability and reliability of InP HEMTs", Solid State Electronics, vol 41, n° 11, pp. 1621-1623, 1997.
12. Results to be presented at the GaAs IC Symposium 2002, Monterey, CA.

International Journal of High Speed Electronics and Systems
Vol. 13, No. 1 (2003) 111–139

DISTRIBUTED AMPLIFIER FOR FIBER-OPTIC COMMUNICATION SYSTEMS

H. SHIGEMATSU, M. SATO, M. TAKECHI*, T. TAKAHASHI, AND T. HIROSE

FUJITSU LABORATORIES LTD.
10-1 Morinosato-Wakamiya, Atsugi 243-0197, Japan.
Tel.: +81-46-250-8244 Fax: +81-46-250-4337
E-mail: shigematsu.hisa@jp.fujitsu.com

**FUJITSU QUANTUM DEVICES LTD.*
1000 Kamisukiahara, Showa-cho, Nakakomagun, Yamanashi 409-3883, Japan

We have developed distributed amplifiers for high bit-rate fiber-optic communication systems. We have designed a $LiNbO_3$ modulator driver using 0.15-μm double-doped AlGaAs/InGaAs/AlGaAs pseudomorphic HEMTs. We obtained a 15-dB gain and 54-GHz bandwidth with GCPW lines, and a fairly clear 6-V_{p-p} output at 40 Gb/s. We also developed preamplifiers for both NRZ and RZ applications. Our preamplifier for the NRZ-data format has a 52-dBΩ gain and a 49-GHz bandwidth. It consists of a lumped element circuit and a distributed amplifier with InP-based HEMTs. We fabricated a module that consisted of the preamplifier connected to a photodiode, and confirmed a clear NRZ-output waveform at 40-Gb/s. The preamplifier with 0.15-μm double-doped AlGaAs/InGaAs/AlGaAs pseudomorphic HEMTs for the RZ-data format has an 11-dB gain and a 57-GHz bandwidth. We also developed a PIN/preamp module that has a 44-dBΩ gain and a 45-GHz bandwidth. The module showed a superior phase characteristic and produced a clear RZ eye-pattern. Furthermore, we fabricated an ultra-broadband amplifier for over-40-Gb/s applications. We obtained a 9-dB gain and 70-GHz bandwidth. These results demonstrate that a distributed amplifier is the best candidate for over-40-Gb/s fiber optic communication systems.manuscript

Keywords: HEMT; distributed amplifier; preamplifier.

1. Introduction

The rapid development of multimedia communications in recent years has generated stronger demands for high-speed and high-capacity transmission systems. Wavelength-division multiplexing (WDM) and time-division multiplexing (TDM) have been developed to realize next-generation-transmission systems. Ultra-massive capacity transmission experiments using a WDM system with data-rate performances of 40 Gb/s per channel have been reported.[1,2] A 40-Gb/s-transmission system will make animated-data communication feasible. High-speed integrated circuit (IC) technologies capable of 40-Gb/s data rates are required for both WDM and TDM systems, and ICs with various devices have been developed for practical use.[3-11]

Figure 1 shows a block diagram of a 40-Gb/s fiber optic transmission system.

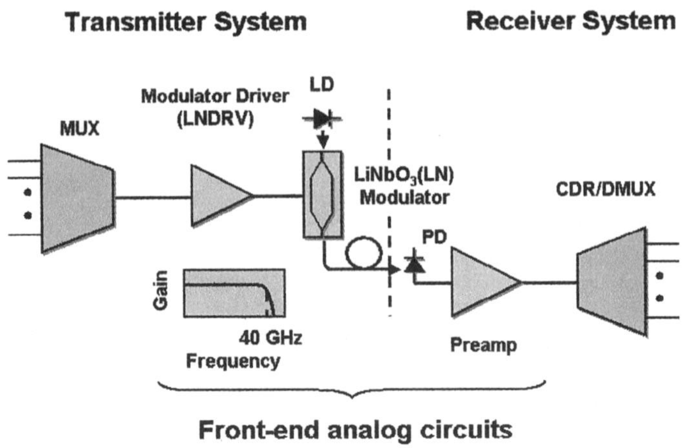

Fig. 1. Block diagram of a 40-Gb/s fiber optic transmission system

A $LiNbO_3$ modulator driver and a preamplifier are key components of the system. The $LiNbO_3$ modulator driver, which is located at the front-end of the transmitter system, plays a role in amplifying the output signal from the multiplexer to more than the 5-$V_{p\text{-}p}$ output needed to drive a $LiNbO_3$ modulator. The preamplifier is connected to a pin photodiode (PD) and is located at the front-end of the optic-electric conversion. It amplifies a very small photo-current to a voltage large enough to drive the subsequent ICs. Though the 40-Gb/s data input has various frequency components, a bandwidth over 40 GHz and a flat gain are required to amplify this waveform accurately. High-speed IC technology for over 40 Gb/s, for example 80 Gb/s, will also be needed in the future. In particular, we need to consider what kind of circuit structure meets this high-speed requirement. Figure 2 shows a conventional lumped amplifier and a distributed amplifier.

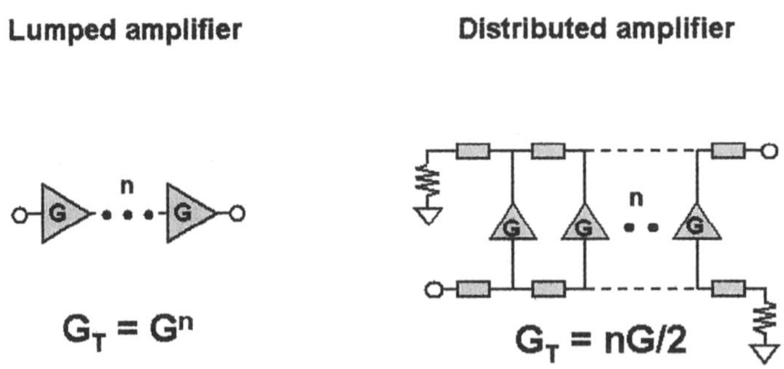

Fig. 2. Lumped amplifier and distributed amplifier.

The total gain of a conventional lumped amplifier, which has a cascaded configuration, is one if the gains of the individual transistors are close to one as frequencies rise. Therefore, a conventional lumped amplifier cannot amplify an input signal at high frequencies. The total gain of a distributed amplifier represents the sum-form of each stage, nG/2, not the product-form. The total gain of a distributed amplifier becomes n/2 if the gains of the transistors are close to one. This means a distributed amplifier has a wider bandwidth than a lumped amplifier. In truth, the distributed amplifiers with broad bandwidth have long been reported.[12, 13]

Figure 3 shows the dependence of the lumped and distributed amplifier gains on the frequency.

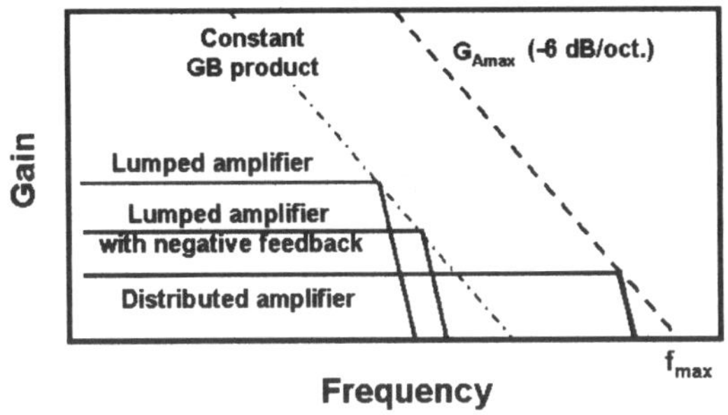

Fig. 3. Dependence of lumped and distributed amplifier gains on frequency.

The potential of the transistors is utilized in a distributed amplifier design because its matching circuits are made of loss-less circuits with LC components. The cut-off frequency of a lumped amplifier is much lower than that defined by the maximum available gain of a transistor, because resistors, which are loss elements, are used in it. Furthermore, the distributed amplifier consists of transistors connected in parallel, and so can drive large currents and produce a large output power. We therefore chose to use a distributed amplifier for the modulator driver circuit.

Though a distributed amplifier is promising for this application, the electrical parameters must be finely optimized to meet strict requirements including high output power, high gain, flat gain, and broad bandwidth. Simply using several stages to obtain high output power would increase the gain ripple and degrade stability. In this paper, we describe a distributed amplifier designed with stabilization, negative resistance control, and broadband impedance matching for a 40-Gb/s $LiNbO_3$ modulator driver circuit. The design enables a 54-GHz bandwidth and a 6-V_{p-p} output at 40 Gb/s. We also discuss preamplifiers for both the RZ- and NRZ- data formats to which this design was applied. Moreover, in the last section, we present an over 40-Gb/s amplifier with 70-GHz bandwidth.

We used InP-based and GaAs-based high electron mobility transistors (HEMTs) to fabricate the distributed amplifiers because of their superior material characteristics, such as a high saturated electron velocity that results in a high cut-off frequency f_T and a high maximum oscillation frequency f_{max}. Moreover, shortening the gate length raises the cut-off frequency. Recently, f_T of more than 560 GHz has been reported.[14] The input capacitance of an HEMT is lower than that of a heterojunction bipolar transistor (HBT) at the same driving current. This is an important advantage in designing a distributed amplifier, especially for a driver amplifier, as will be described in the next section.

2. Conventional Distributed Amplifier Design

First, we will outline the basic design of a conventional distributed amplifier (Fig. 4). The procedure for designing a distributed amplifier is to first add L_{twa} to an input transmission line to satisfy the following equation in a 50-Ω system:

$$\sqrt{\frac{L_{twa}}{C_{in}}} = \sqrt{\frac{L_{twa}}{C_{gs} + (1 + A)C_{gd}}} = 50\Omega \qquad (1)$$

Here, A is the amplification of a transistor. The output capacitance C_o is continuously added to the drain terminal of the transistor to satisfy the following equation, $C_{in} = C_{ds} + C_o$. Here, C_o is designed by the transmission lines. Moreover, L_{twa} is also added to the output transmission line. This allows the phase velocity and characteristic impedance conditions to be satisfied. The cut-off frequency is defined approximately by

$$f_c = \frac{1}{\pi\sqrt{L_{twa} C_{in}}} = \frac{1}{\pi\sqrt{L_{twa}\left[C_{gs} + (1 + A)C_{gd}\right]}} \qquad (2)$$

This equation indicates that a transistor with a low input capacitance is needed to extend the bandwidth. If a transistor with a high input capacitance is used, a large L_{twa} is required to satisfy the characteristic impedance condition. However, this would degrade the cut-off frequency because the product of L_{twa} and C_{in} would become large. Therefore, a transistor with a low C_{gs} should be used in an ultra-broadband amplifier. Thus, we used HEMTs.

To obtain a wider bandwidth, a cascode configuration is suitable. The C_{in} equals C_{gs} because we can ignore the Miller-capacitance, $(1+A)C_{gd}$, in a cascode configuration. We can therefore achieve an ultra-broad bandwidth. However, there are two problems with using a conventional distributed amplifier in a fiber-optic communication system. One is a forward gain slope and the other is a gain ripple at high frequencies. Gain slope, which is caused by losses in the input and output transmission lines, must be compensated for. Gain ripples are caused by degradation of the return loss at high frequencies. To suppress these ripples, good impedance matching within the bandwidth is necessary.

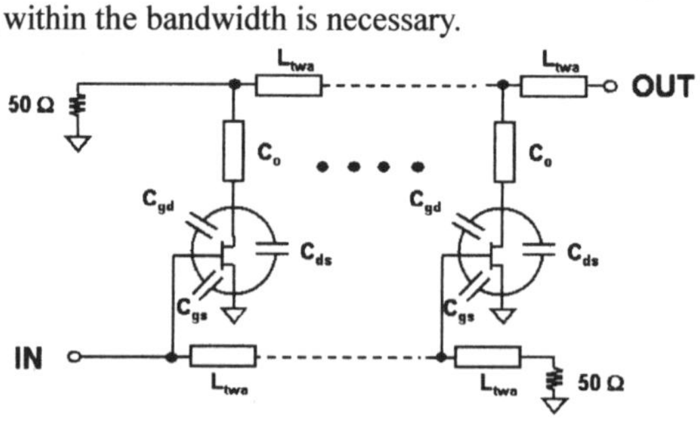

Fig. 4. Schematic of a conventional distributed amplifier.

From the next section, we will describe the individual circuits. We will explain in detail the techniques we applied to solve these problems in the following section. In particular, the design of the unit cell is quite important in solving the first problem; that is, in compensating for the gain slope.

3. $LiNbO_3$ Modulator Driver (LNDRV)

3.1. *Circuit Design*

Figure 5 shows the equivalent circuit of a cascode-pair of transistors. In a conventional cascode configuration, the output impedance of Ql seen from the source terminal of Q2 without a source inductance of l_s is always capacitive because of the drain-source capacitance (C_{ds1}) of Q1.

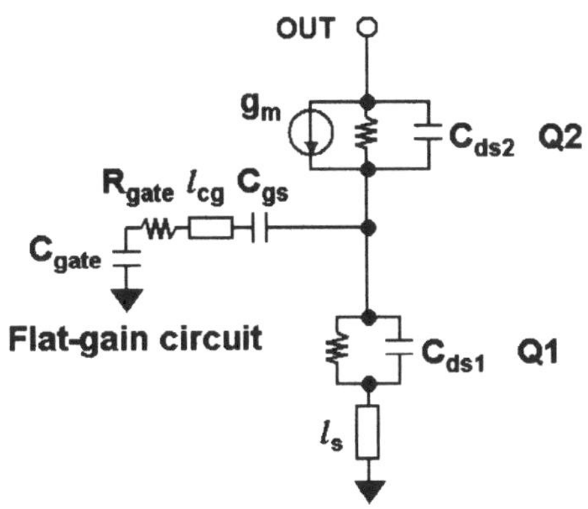

Fig. 5. Equivalent circuit of a cascode-pair of transistors.

Therefore, a positive feedback circuit is formed. This configuration generates negative resistance, thus compensating for the loss of the transmission lines. However, a drastic increase in negative resistance also degrades the stability of the amplifier when it is operating near the cut-off frequency. To suppress negative resistance, we inserted a source inductance of l_s between the source terminal of Q1 and the ground, and connected a circuit consisting of l_{cg}, R_{gate}, and C_{gate} to the gate terminal of Q2. There is an option that allows us to change the gain profile of the circuit so that it produces a flat gain profile.[15]

Figure 6 shows the dependence of the gain profile on the frequency. The inclusion of l_s, where l_s was 20 pH and the other parameters were fixed, preserved gain flatness and improved stability and bandwidth. The K-factor of the amplifier without l_s was below 1.0 and the circuit was unstable. The K-factor was increased to above 1.4 by inserting l_s.

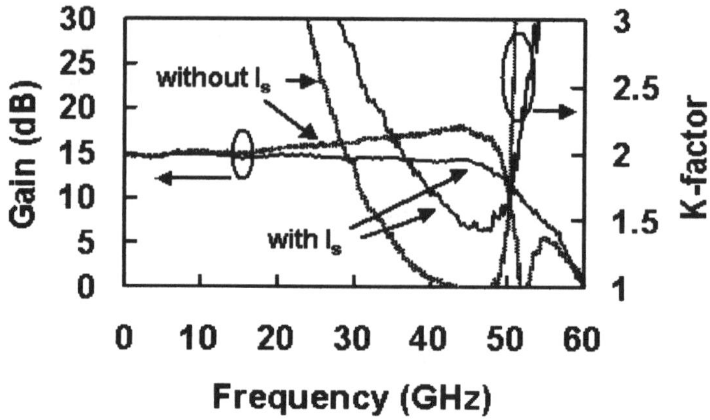

Fig. 6. Simulated gain profiles with and without I_s.

Gain ripples can be suppressed as follows. Figure 7 shows a schematic of a distributed amplifier with an input matching circuit.

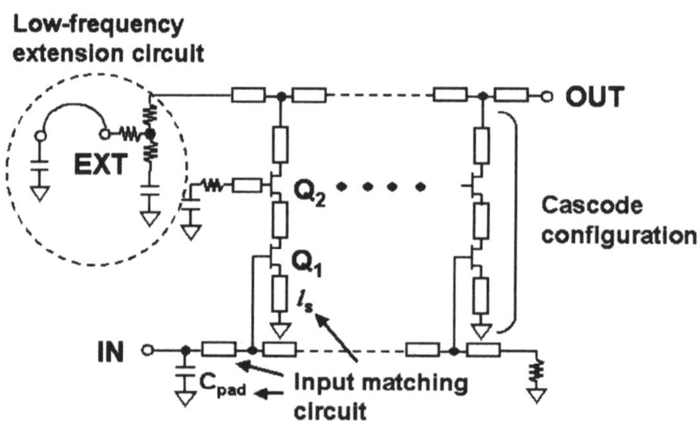

Fig. 7. Distributed amplifier with an input matching circuit.

The return loss of a conventional distributed amplifier deteriorates as the frequency increases. We therefore formed an impedance matching circuit to reduce the input return loss at around 40 GHz by using pad capacitance, a source inductance of Q1, and a line between the input pad and the first stage.[16] In a conventional broadband amplifier design, narrow band impedance matching is not used. However, we tried to realize broadband impedance matching between near DC and about 50 GHz through this circuit. Figure 8 shows the calculated gain profile with and without an impedance matching circuit. The input return loss with an input matching circuit is lower than that of an amplifier without the circuit between 35 and 45 GHz. The circuit also suppressed the gain ripple at high frequencies.

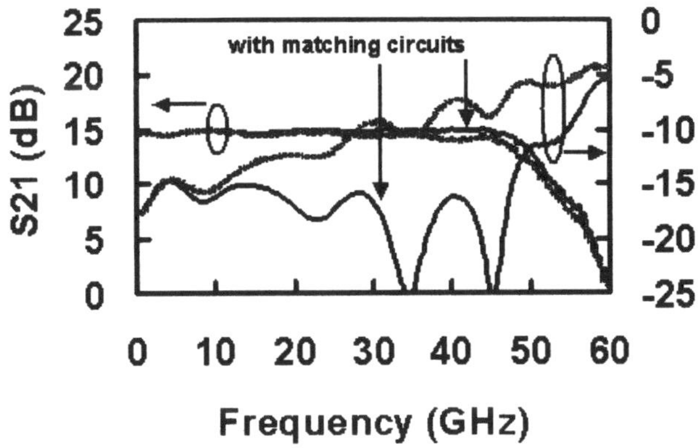

Fig. 8. Calculated S-parameters with and without the input matching circuit.

Here, we will also briefly explain a low-frequency extension circuit we used. Our output termination circuit consists of an on-chip capacitance and a large off-chip capacitance. The chip has damping resistance to remove the unwanted resonance caused by the inductance of the bonding wire. This structure has the advantage of being easy to assemble. In our circuit design, we used grounded coplanar waveguided (GCPW) lines with a 75-μm-thick substrate for the layout designs to make the chip easy to assemble and to facilitate on-wafer DC and RF measurements. Since a conventional micro-strip line (MSL) chip needs viaholes for normal operation, it cannot be used for on-wafer measurement. Also, a CPW structure enables a lower chip-cost than would be the case with a MSL structure. We calculated the S-parameters of coplanar waveguided (CPW) lines on a 600-μm-thick substrate and GCPW lines on a 75-μm-thick substrate; in both cases, the lines were 1 mm long, and the line width and the gap were both 20 μm. The CPW and GCPW structures had similar characteristics. The characteristic impedance of the 600-μm-thick CPW structure was 52 Ω and that of the 75-μm-thick GCPW structure was 50 Ω. The loss in both lines was –0.3 dB/mm. The simulated gain profile of distributed amplifiers with GCPW lines was the same as for one with CPW lines. To obtain an output swing of over-5-$V_{p\text{-}p}$, our distributed amplifier consists of eight stages with cascode-configured unit cells. The gate width of each transistor is 80 μm.

3.2. *Device Fabrication*

We fabricated the distributed amplifiers using double-doped AlGaAs/InGaAs/AlGaAs -pseudomorphic HEMTs (p-HEMTs), because a high breakdown voltage is required for a driver circuit. The gate was a T-shaped structure made of Al with a 0.15-μm gate length and was formed by electron beam lithography. The source and drain electrodes were made of AuGe/Ni/Au. The device consisted of a channel layer sandwiched between two electron supply layers. The transconductance g_m was 525 mS/mm and the maximum drain current If_{max} was 560 mA/mm. The cut-off frequency at V_{gs} of -0.1 V and V_{ds} of 3.0 V was 90 GHz. The resistors were made from epitaxial layers that had a sheet resistance of 120 Ω per square. The MIM capacitors were made from SiN.

Figure 9 shows a microphotograph of the $LiNbO_3$ modulator driver amplifier.

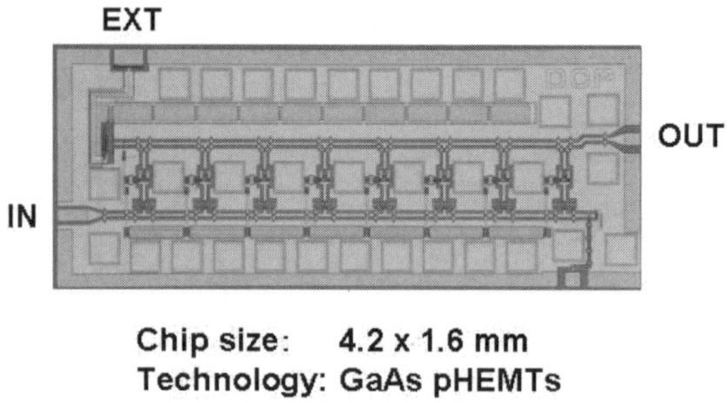

Fig. 9. Chip photograph of the distributed amplifier for the $LiNbO_3$ modulator driver.

We used GCPW lines with via holes. We placed ground via holes at the source terminals to achieve accurate l_s as required for our design. This structure is suitable for face-up assembly and enables RF-performance to be checked without having to use the via holes. Ground-bridge interconnects were also used to avoid any undesired mode. It is important to make the grounds of the circuit stable when designing high-frequency circuits, because the grounds can not fulfill their function if there are small inductive components at high frequencies. The GCPW structure also suppresses the influence of adjacent lines resulting from the side-walls of the grounds. The chip was 4.2 x 1.6 mm and power consumption was 1.1 W.

3.3. *Circuit Performance*

We measured the S-parameters using an HP8510XF vector network analyzer. RF-probes were calibrated using the line reflect reflect match (LRRM) method.

Figure 10 shows the frequency characteristics of the 75-μm GCPW chip.

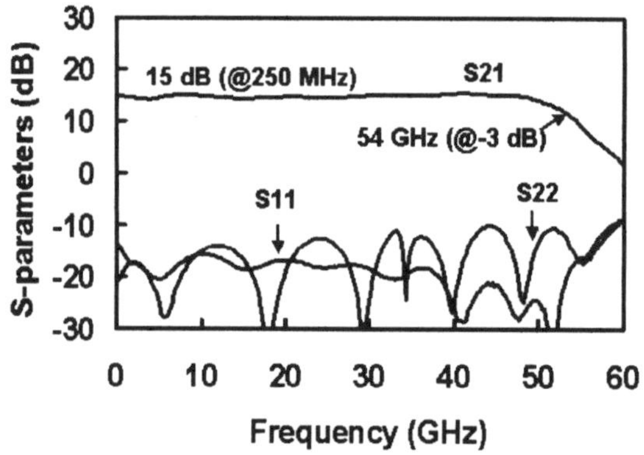

Fig. 10. Frequency characteristics of the distributed amplifier.

We achieved very good performance with a 15-dB gain and a 54-GHz bandwidth at a drain-line bias of 6 V and a gate-line bias of –0.1 V. This is the highest gain bandwidth product yet reported for baseband amplifiers fabricated using GaAs-based FETs and that can be applied to an optical transmission system. This result indicates that our circuit design is suitable for extending bandwidth and stabilizing an amplifier. The gain ripple was within ±0.5 dB. The deviations in group delay were about ±10 ps up to 40 GHz. The input-return loss was below –15 dB and the output was below –10 dB up to 50 GHz. The input impedance matching circuit provided a low input-return loss up to 50 GHz as we expected. The bandwidth of the 600-μm-thick substrate was 45 GHz. The bandwidth was wider on the GCPW chip because of the stable source via-grounds. The source inductance of the 600-μm-thick substrate was greater than that of the chip with via holes. The source ground of a CPW-distributed amplifier is important to achieve wide bandwidth when using flip-chip bonding or viaholes.

Figure 11 shows the low-frequency characteristics between 100 kHz and 200 MHz of a distributed amplifier connected by a 2.2-nF and a 1.0-uF capacitor to the EXT terminal.

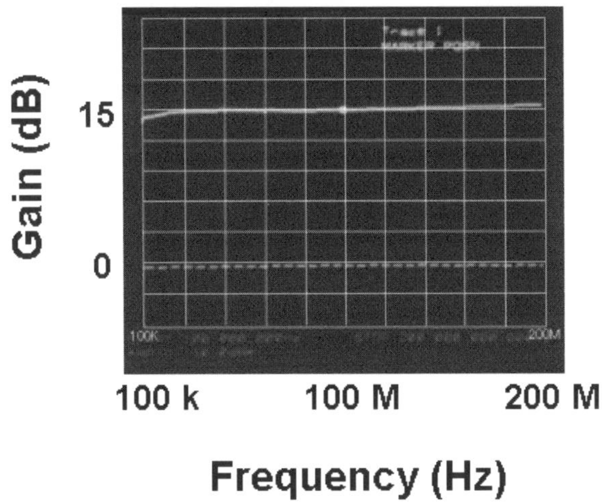

Fig. 11. Low-frequency characteristics of the distributed amplifier.

The gain profile was quite flat and no ripple or peaking was observed. We observed a –2-dB-low frequency cut-off at 50 kHz.

Figure 12 shows the $P_{in} - P_{out}$ characteristics at 20 GHz (and a drain bias of 6.0 V), which is a fundamental frequency of 40-Gb/s data. A P_{1dB} of 20 dBm was obtained, which corresponds to 6.3 $V_{p\text{-}p}$ at a 50-Ω load. The saturated power of P_{sat} was 23 dBm. We also obtained a P_{sat} of 20 dBm at 40 GHz. Figure 13 shows an eye-diagram of the output at 40 Gb/s. We observed a linear and clear 6-$V_{p\text{-}p}$ output-voltage swing. We also observed output above 7.6 $V_{p\text{-}p}$. These results indicate that our circuit is promising for use as a $LiNbO_3$ modulator driver. We can fabricate a module with a 30-dB gain using only two chips, and a two-chip configuration will reduce the module's power consumption.

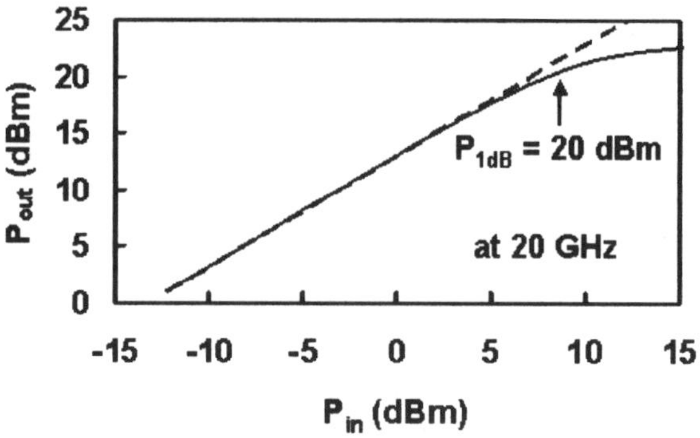

Fig. 12. $P_{in} - P_{out}$ characteristics at 20 GHz.

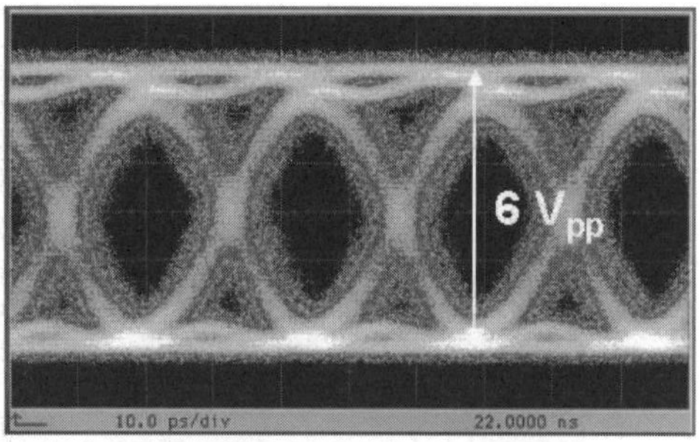

Fig. 13. Output waveform of the LNDRV.

We also fabricated a test fixture to investigate the driver-module configuration. Figure 14 shows a block diagram of the $LiNbO_3$ modulator driver test fixture.

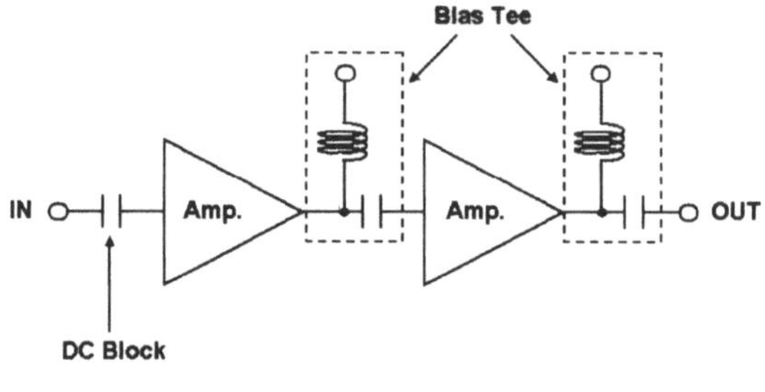

Fig. 14. Block diagram of the $LiNbO_3$ modulator driver test fixture.

It consisted of two chips, two bias-tees, and a DC-block capacitor. An important parasitic arose from the wire bond connection between the chip and the transmission line for the DC block capacitor or the bias-tees. This degraded the input and output return losses. Both the input and the output return losses from near DC to 20 GHz are important when amplifying a non-return-to zero (NRZ) 40-Gb/s input signal because 20 GHz is a fundamental frequency for a NRZ 40-Gb/s signal. Considering the outer bonding wire inductance, we designed circuits that would reduce this parasitic. Figure 15 shows the frequency characteristics of a chip with bonding wires for both a signal and grounds.

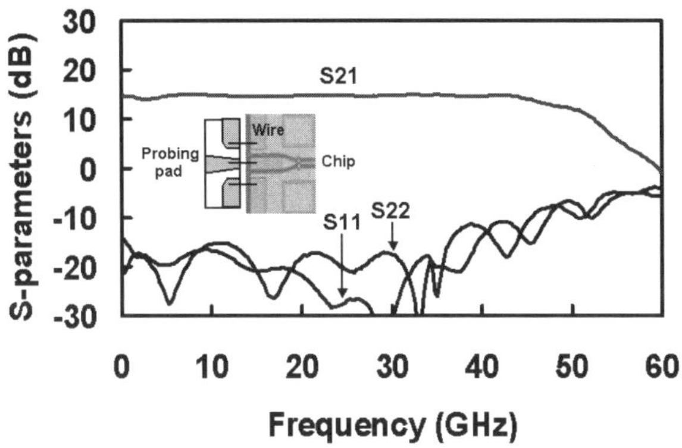

Fig. 15. Frequency characteristics of a chip with bonding wires for both a signal and grounds.

Both the input and the output return losses were less than –15 dB up to 35 GHz and –10 dB up to 45 GHz. The bandwidth was 51 GHz and no degradation in the form of gain ripples was observed. This confirmed the effectiveness of our design for minimizing degradation by parasitic inductances. The GCPW structure minimizes the effect of ground wire inductances because of the chip's stable grounds. These grounds are important for such an ultra-broadband amplifier.

The most important passive component used to make the driver module is the DC block capacitor. A broadband capacitor for 40-Gb/s fiber-optic communications is difficult to realize because one would require a wide bandwidth from 50 kHz to more than 50 GHz. We designed our DC-block to minimize resonance, ripples, insertion losses, and return losses. The loss for a DC block capacitor was 0.8 dB at 40 GHz. We used a conical coil inductor as a bias choke coil at a drain terminal of the individual chips. Also, V-connectors were used for the input and output ports of the test fixture. The small signal gain was 28 dB and the bandwidth was 40 GHz. The decreases in the gain and bandwidth were caused by insertion losses from the V-connectors, DC block capacitors, and transmission lines. Optimization of these parts will enable better performance. The gain ripple was within 1 dB. We reduced the drain current of the chip at the first stage to reduce its power dissipation. As a result, the power dissipation was less than 2 W. The configuration with only two chips also suppressed the power dissipation and will permit a small module size without a heat sink. We directly connected this test fixture and an attenuator to a sampling head of an oscilloscope and measured the output waveform without using a cable for the output port. Figure 16 shows a 2^{31}-1 PRBS output eye-diagram at 40 Gb/s. We observed a clearer 6-$V_{p\text{-}p}$ output-voltage swing than that of Fig. 13 because the measurement was done without a cable. The rise time was below 10 ps. The rms- and p-p- jitter were 1.1 ps and 5.6 ps, respectively. Cable less measurements enables us to achieve a clear eye-opening. Use of a cable degrades the waveform at such high bit rates. We also observed output of above 7.6 $V_{p\text{-}p}$ by increasing the input voltage. These results confirmed the promise of our circuit for use as a $LiNbO_3$ modulator driver.

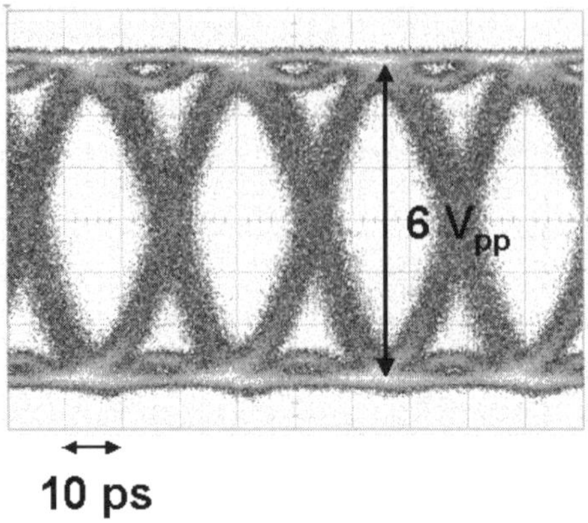

Fig. 16. Output waveform of the test fixture.

4. NRZ-Preamplifier

4.1. *Circuit Design*

Next, we will describe the preamplifier ICs that the above design techniques were applied to. There are several data formats for fiber-optic communication systems. First, we describe a preamplifier for the NRZ-data format, which is often used for a very short reach (VSR) system. A high gain is required for a preamplifier in a VSR system because optical amplifiers are not used in VSR systems and the minimum input power is low.

Figure 17 shows a block diagram of our preamplifier. It has two stages – a transimpedance stage and a gain stage. We use a lumped element amplifier for the transimpedance (first) stage and a distributed amplifier for the gain (second) stage. The distributed amplifier compensates for the narrow bandwidth of the lumped element amplifier.

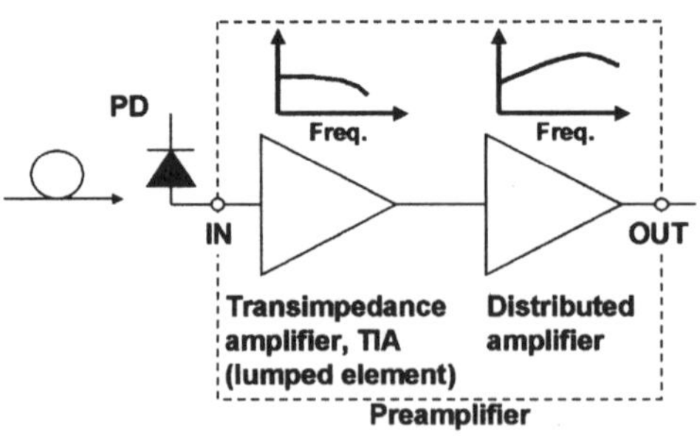

Fig. 17. Preamplifier block diagram.

The transimpedance amplifier consists of the input stage and a two-stage source follower used as an inter-stage buffer (Fig. 18). These were designed with lumped element components using a conventional circuit topology. To reduce the input impedance of the input stage, we used a common gate FET. This also provides broadband characteristics due to a reduced Miller capacitance.

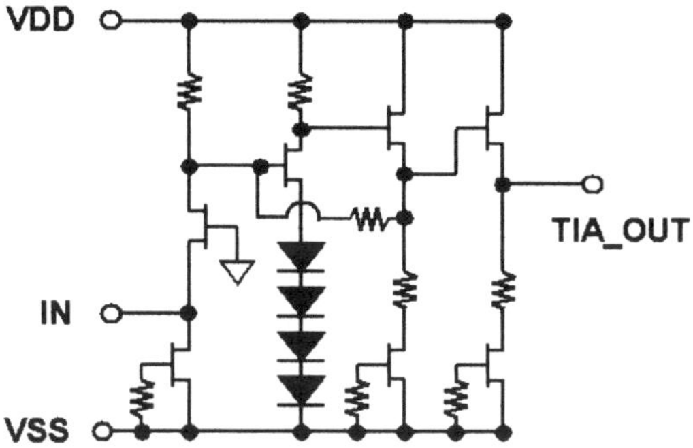

Fig. 18. Schematic of the transimpedance amplifier.

For the gain stage, we used a seven-stage distributed amplifier with cascode-configured unit cells (the same configuration shown in Figs. 5 and 7). Here, the most important parameter is R_{gate} to control the gain profile of the distributed amplifier.[15] We have to design a distributed amplifier with a backward gain slope to compensate for the lumped-element amplifier's lack of bandwidth. However, the backward gain slope degrades the stability of the amplifier near the cut-off frequency, and this was a problem we would have to solve.

Figure 19 shows the calculated gain profile of the distributed amplifier for the second stage with and without R_{gate}. Here, all parameters except R_{gate} were set so that the gain profile was slightly increased from DC to 60 GHz. As can be seen, the gain profile without R_{gate} drastically increased above 70 GHz. On the other hand, the proposed circuit was well stabilized and also provided an ultra-broad bandwidth.

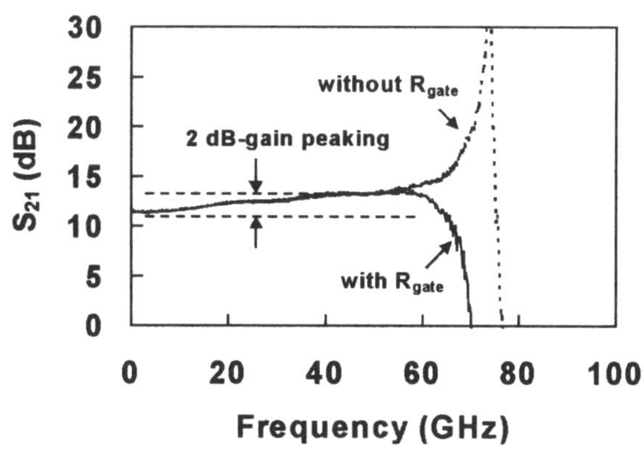

Fig. 19. Calculated gain profile of the distributed amplifier with and without R_{gate}.

4.2. *Device Fabrication*

We fabricated the preamplifiers using 0.15-μm-gate InP HEMTs. We explain our InP-based HEMTs before presenting circuit performance. Figure 20 shows a cross-section of our InP-based HEMT.

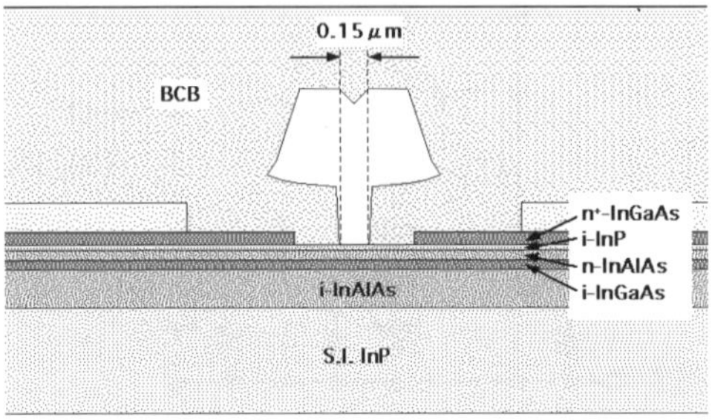

Fig. 20. Cross-section of an InP-based HEMT.

We used a benzocyclobutene (BCB) film, which has low permittivity (ε_r = 2.8), as an interlevel dielectric. The resistors were made from thin films of NiCr with a sheet resistance of 50 Ω per square. The MIM capacitors were made from SiN. The measured transconductance of the HEMT was 900 mS/mm and the cutoff frequency was 167 GHz. Our HEMTs have good uniformity and perform well.[17] Although HEMTs have been considered unsuitable for highly integrated circuits, we overcome their shortcomings regarding such applications in the following ways. We used an InP recess layer for selective wet chemical etching, and this structure resulted in a uniform V_{th} and a carrier supply layer with little damage. The InP surface in the recess region was covered with a thin SiN dielectric film formed by plasma-enhanced CVD so that it would be stable enough to prevent carrier depletion when the device was subjected to high temperatures during thermal processing. A gate electrode consisting of Ti/Pt/Au was evaporated onto the InP layer and lifted off. To realize integrated circuits with FETs, the uniformity of the threshold voltage, the level-shift voltage of diodes, and the offset voltage in the differential amplifiers are all important.

Figure 21 shows a threshold voltage (V_{th}) histogram for HEMTs fabricated on a 3-inch wafer. The average V_{th} is -0.633 V and the 13-mV standard deviation indicates that the uniformity of the gate and the recess region was better than that reported previously.[18] It also shows that the thickness and doping concentration of the epitaxial layers were uniform across the 3-inch wafer.

Figure 22 shows a histogram for the level-shift voltage of diodes, which we define as the voltage at a current of 50 mA/mm. The small standard deviation, 5 mV, indicates that the Schottky diodes consisting of the gate metal are suitable for use in ICs.

Differential amplifiers are often used in digital circuits based on source-coupled FET logic (SCFL). We therefore investigated the effects on V_{th} of the offset-voltages in the differential amplifiers.

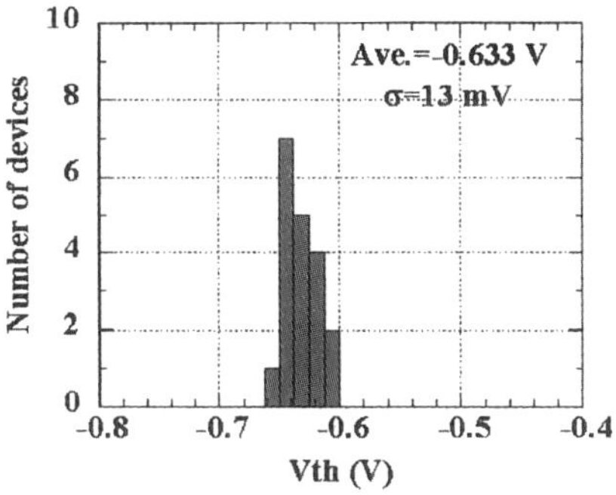

Fig. 21. Threshold-voltage histogram.

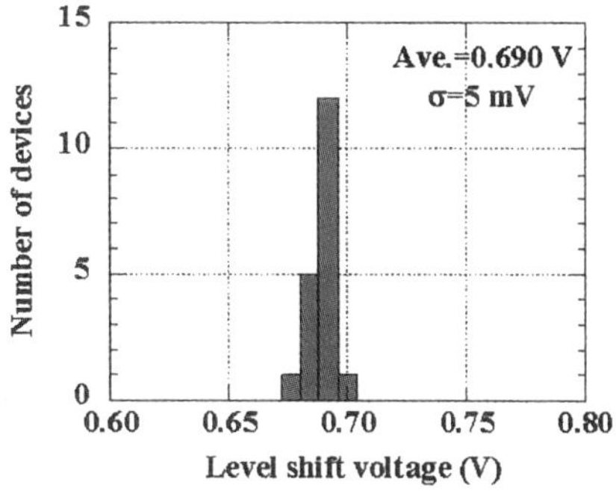

Fig. 22. Level-shift voltage histogram.

Figure 23 shows a histogram of the offset-voltage in differential amplifiers. Even though this data was gathered from several wafers of different lots, the small standard deviation (6.2 mV) indicates InAlAs/InGaAs HEMTs are suitable for use in ICs based on SCFL.

Thermal stability is also important and the dependence of V_{th} on the annealing time is shown in Fig. 24. After 60 minutes at 330°C the V_{th} had changed by only 11 mV. Figure 25 shows that Idss was decreased by about 5% after annealing at 330°C, but we consider this change negligible and believe the InP-based HEMTs will be thermally stable in practical use. Furthermore, we believe that it is essential to passivate the InP etch-stop layer in the recess region entirely by using a dielectric film.

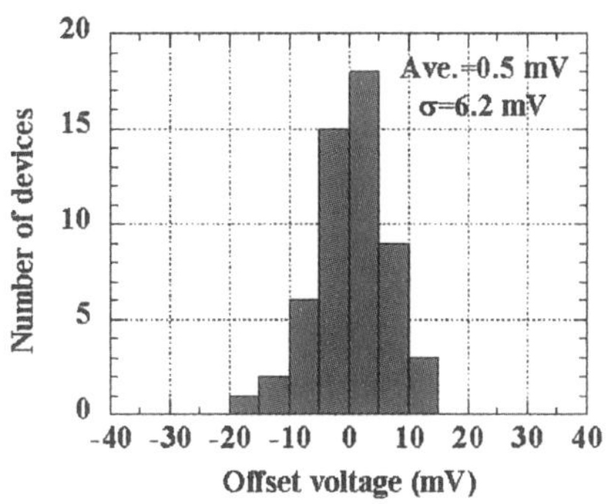

Fig. 23. Histogram for the offset voltage in differential amplifiers.

Therefore, we can realize highly integrated circuits with this device that are suitable for digital applications. We developed a full rate 40-Gb/s 4:1 multiplexer,[19] a 50-Gb/s 1:4 demultiplexer,[20] and a 90-Gb/s 2:1 multiplexer[21] using these devices.

We also fabricated a preamplifier using HEMTs with good uniformity and capable of high performance.

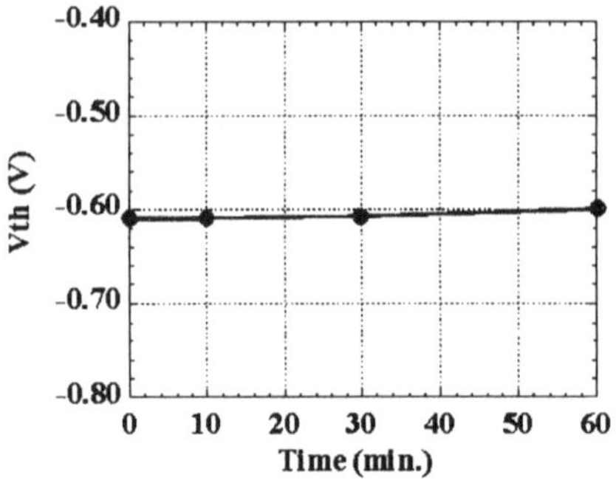

Fig. 24. V_{th} dependence on annealing time.

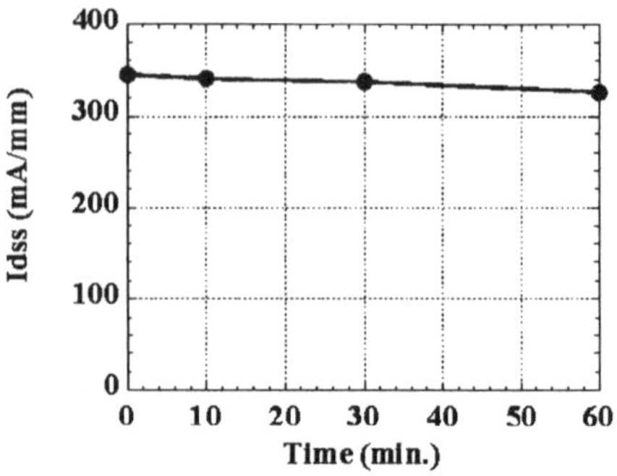

Fig. 25. Idss dependence on annealing time.

4.3. *Circuit Performance*

Figure 26 is a microphotograph of the preamplifier. The chip size was 3.09 x 1.75 mm. We used HEMTs with a gate width of 40 μm in the distributed amplifier. Coplanar waveguide lines with ground-bridge interconnects were used in the distributed amplifier to avoid any undesired mode. The coplanar waveguide lines were 20 um wide. The characteristic impedance was 68 Ω. The power dissipation of the designed IC was 0.52 W. Figure 27 shows the frequency characteristics of the distributed amplifier. We obtained a 12-dB gain and a 76-GHz bandwidth. The transconductance of an InP-based HEMT is greater than that of a GaAs HEMT. We can thus use smaller transistors for a distributed amplifier design and achieve a wider bandwidth.

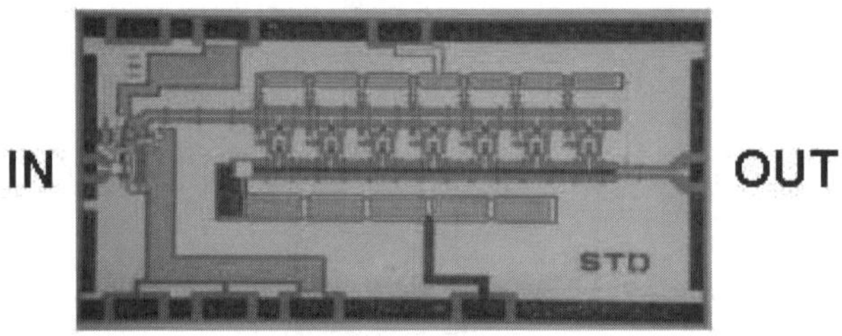

Fig. 26. Microphotograph of the preamplifier.

Figure 28 shows the frequency characteristics of the preamplifier. We achieved a transimpedance gain of 52 dBΩ and a bandwidth of 49 GHz. As far as we know, this is the widest bandwidth yet reported for a preamplifier with such a high gain. The gain of S_{21} was 20.4 dB. The input impedance was 31 Ω, and the output return loss (S_{22}) was below -10 dB up to 50 GHz. Low input impedance plays an important role in widening the total bandwidth of a preamplifier connected to a pin-photodiode. These results also indicate that our circuit design is suitable not only for 40 Gb/s-preamplifiers but also for other wide bandwidth amplifiers.

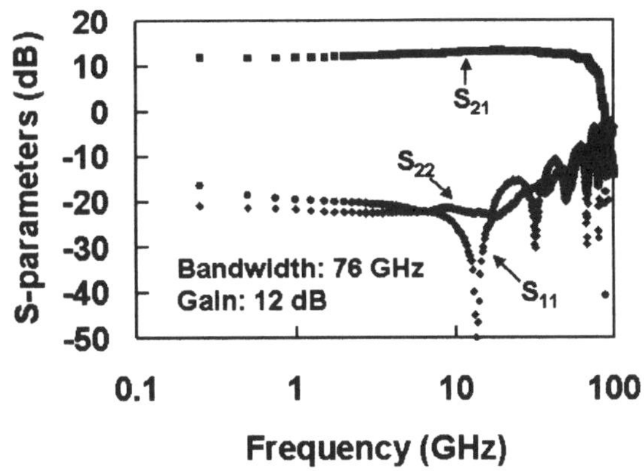

Fig. 27. Frequency characteristics of the distributed amplifier.

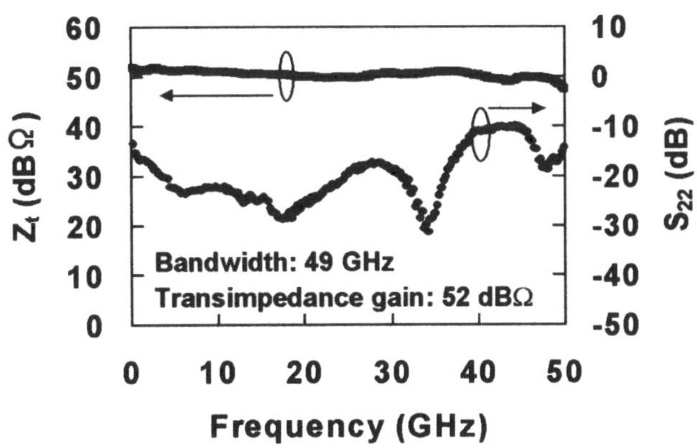

Fig. 28. Frequency characteristics of the preamplifier.

The group delay characteristic is important for broadband amplifiers in fiber-optic communication systems, because 40-Gb/s data input contains various frequency components. Figure 29 shows the group delay characteristic of the preamplifier. The group delay can be described as

$$T_g = -\frac{d\varphi}{d\omega} = -\frac{1}{360} \cdot \frac{d\varphi_{degrees}}{df} \qquad (3)$$

where

φ = phase (radians)

ω = angular frequency (radians/second)

$\varphi_{degrees}$ = phase (degrees)

f = frequency (Hz)

A flat group delay profile versus frequency over a given bandwidth implies a linear phase. The group delay is important for a broadband amplifier in a fiber-optic communication system because a large deviation of the group delay will degrade waveforms. The group delay deviation in this preamplifier was within ±15 ps up to 40 GHz. This shows that our proposed circuit has good phase linearity for 40-Gb/s NRZ-preamplifiers.

However, this is insufficient for the RZ data format, which requires stricter phase characteristics than those for the NRZ-data format. We will discuss this in a later section.

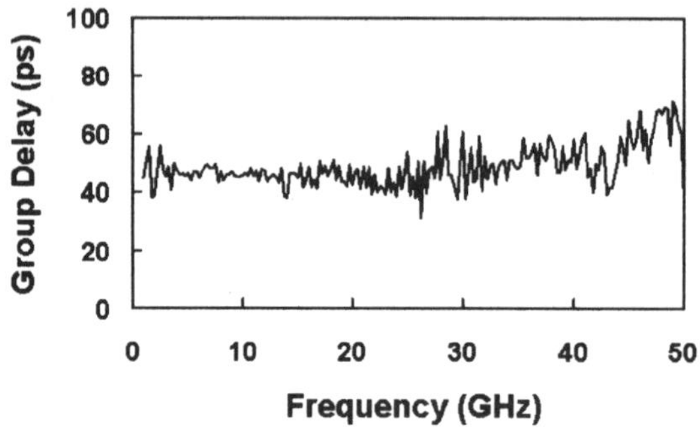

Fig. 29. Group delay characteristic of the preamplifier.

We fabricated an O/E-module, consisting of a preamplifier connected to a pin-photodiode, which had a bandwidth greater than 35 GHz under a 50-Ω load, responsivity greater than 0.85 A/W, and polarization dependent loss (PDL) of less than 0.2 dB. (The details of this structure and the photodiode characteristics have been reported elsewhere.[22]) Figure 30 shows the output waveform of the pin-preamp module at 40 Gb/s. We obtained a clear eye-pattern.

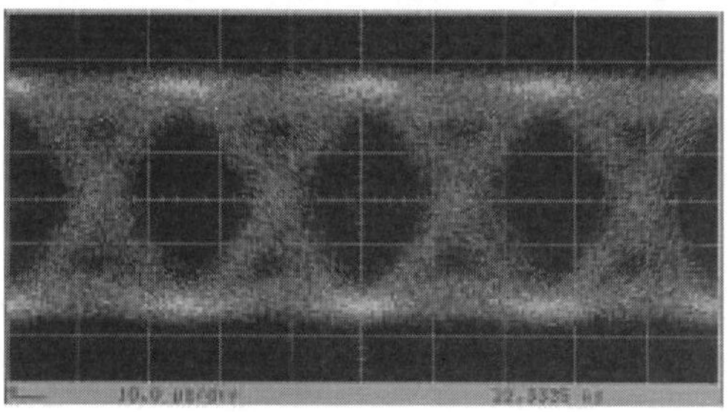

Fig. 30. Output waveform of the preamplifier.

5. RZ-Preamplifier

Next, we will explain the preamplifier for the RZ data format. The initial market for 40-Gb/s systems is likely to be long-haul DWDM systems, where over-1-Tb/s transmission capacity can be realized. However, fiber dispersion is a critical problem in such long-haul systems. To manage fiber dispersion, the RZ data format is widely used, because it has higher immunity to polarization mode dispersion (PMD). Therefore, longer transmission span can be achieved. On the other hand, the RZ data format requires a much wider bandwidth; thus, it requires stricter phase linearity for optical receivers than the NRZ data format, which is conventionally used in systems below 10 Gb/s. A typical receiver block diagram for a 40-Gb/s long-haul optical communication system is shown in Fig. 31.

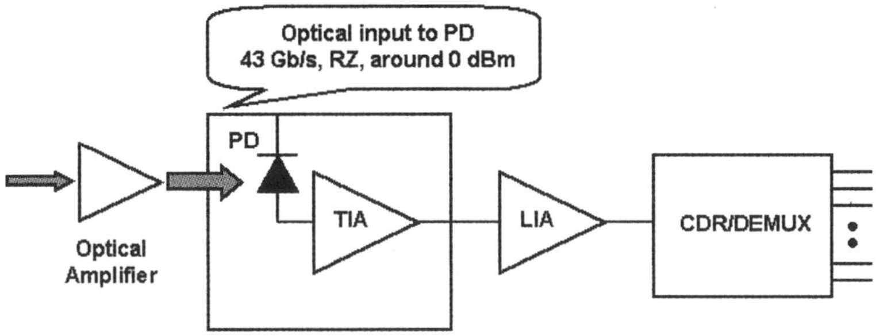

Fig. 31. Typical receiver block diagram for a 40-Gb/s long-haul optical communication system.

An optical preamplifier is used to feed enough optical power to a receiver. PIN photodiode modules have often been used as optical receivers; however, a fairly high optical input power, usually more than 10 dBm, is needed to obtain a voltage swing sufficient for the following circuits. This requirement restricts flexibility in the system design of 40-Gb/s long-haul optical communication systems. Therefore, high performance PIN/preamp modules that have a wide bandwidth and superior phase linearity – to avoid distorting the 43-Gb/s RZ signal – are strongly desired. We used a distributed amplifier with a low input impedance to make our module. A distributed amplifier can be designed with input impedance not necessarily equal to output impedance. A conventional distributed amplifier has both input impedance and output impedance close to 50 Ω. We designed an input impedance of 35 Ω to extend the cut-off frequency, which is defined by the pin diode capacitance, and the input impedance of the amplifier.

We fabricated the distributed amplifiers using the double-doped AlGaAs/InGaAs/AlGaAs-pseudomorphic HEMTs (p-HEMTs) described in section 3. The gate length was 0.15 μm.

Figure 32 shows the frequency characteristics of the distributed amplifier. We obtained a 10-dB gain and 57-GHz bandwidth. The phase deviation from the linear region was only 5°. This amplifier showed the Bessel roll-off characteristic, indicating that the slope within the 3-dB cut-off region was quite gentle.

We fabricated a PIN/preamp module with this circuit. (Figure 33 shows a photograph of the receiver module.) The V-connector was used for an output port. We used a conventional wire/ribbon bonding method for the interconnections and did not use a flip-chip technique or OEIC because of productivity and reliability considerations. We confirmed that the parasitic inductance of the interconnection between the TIA and PD did not result in significant gain ripples and did not restrict the bandwidth, because our TIA does not significantly degrade the input and output return loss up to nearly 50 GHz.

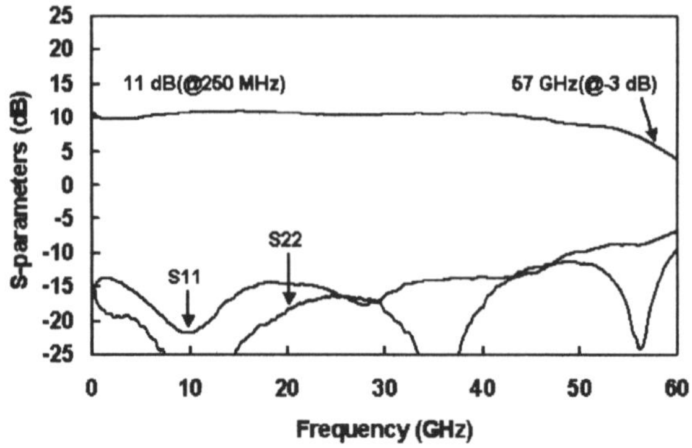

Fig. 32. Frequency characteristics of a distributed amplifier for a TIA.

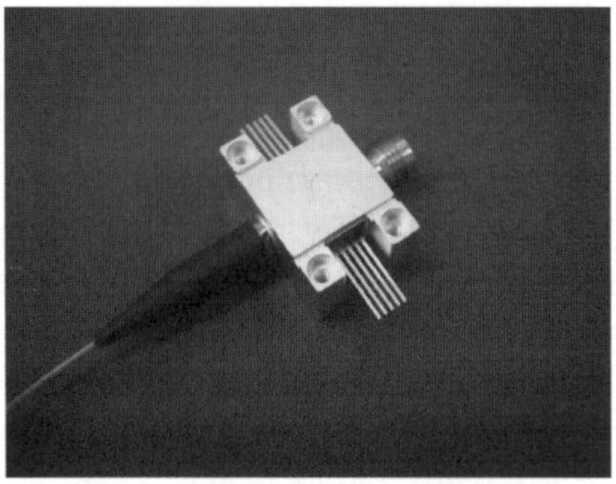

Fig. 33. Photograph of the PIN/preamp module.

Figure 34 shows the frequency characteristics of the PIN/preamp module. We applied +7 V and –2 V to the TIA and +3 V to the PIN photodiode.

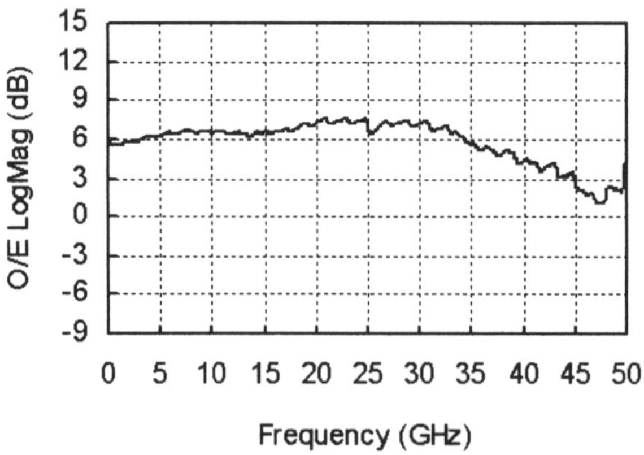

Fig. 34. Frequency characteristics of the PIN/preamp module.

The frequency response was measured with a lightwave component analyzer. We obtained a

44-dBΩ transimpedance gain and a 45-GHz bandwidth. An output return loss better than -11 dB was also obtained up to 50 GHz. The most remarkable point of this module, though, is its phase characteristic.

The deviation from the linear phase was within ±7 degrees up to 43 GHz (Fig. 35), which was the same as that of the distributed amplifier. The phase linearity plays a crucial role, especially when the RZ data format is used. A large phase deviation at high frequencies causes the eye-pattern to show a large ringing "0" level even if the bandwidth is greater than 40 GHz. On the other hand, the eye-pattern with the NRZ data format shows little degradation if the phase deviation below a fundamental frequency of 20 GHz is small. This difference is due to the many high frequency components in the RZ data format.

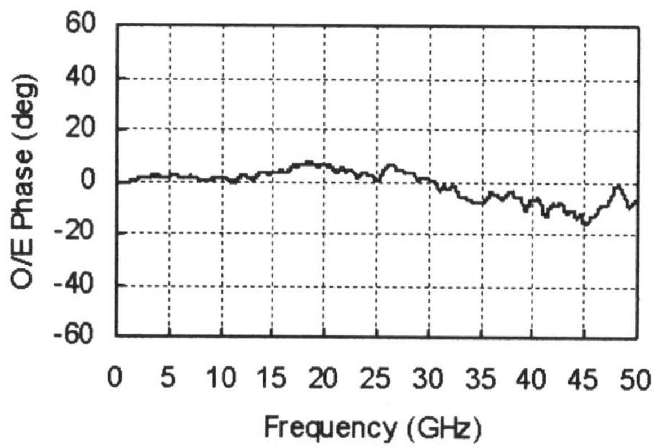

Fig. 35. Deviation from the linear phase of the PIN/preamp module.

Figure 36 shows the group delay characteristics of the PIN/preamp module. The group delay was quite flat up to a fundamental frequency of 20 GHz and was within ±6 ps up to 50 GHz. Thus, we realized good group delay performance as a PIN/preamp module.

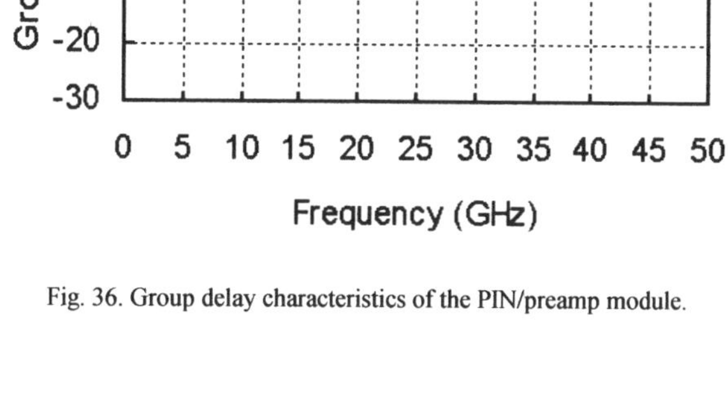

Fig. 36. Group delay characteristics of the PIN/preamp module.

The average input equivalent noise current density up to 19 GHz (Fig. 37) was within 15 pA/sqrtHz. This indicates that our module is suitable for both long-haul and VSR systems.

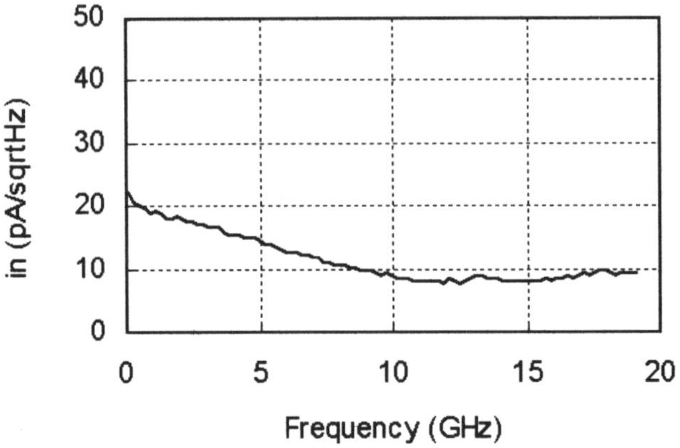

Fig. 37. Input equivalent noise current density.

Figure 38 shows the eye-pattern at 40-Gb/s and 2^{31}-1 (PRBS) in the RZ data format with an average optical input power of 0 dBm. A fairly large eye opening, low jitter, and low "0" level ringing were obtained.

The NRZ eye-pattern at 40-Gb/s and 2^{31}-1 (PRBS) with an average optical input power of 2 dBm was also quite large with low jitter of 1.0 ps (Fig. 39).

These results demonstrate that our distributed low-input-impedance amplifier is suitable for use as an RZ-preamplifier. Its superior phase characteristic ensures a clear waveform.

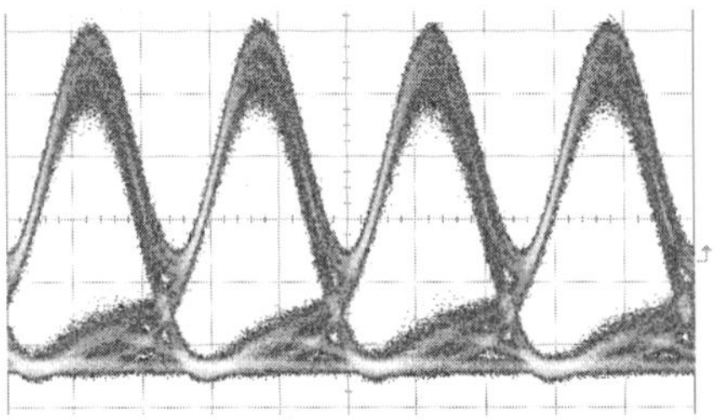

Fig. 38. 40-Gb/s RZ eye-pattern at 0 dBm.

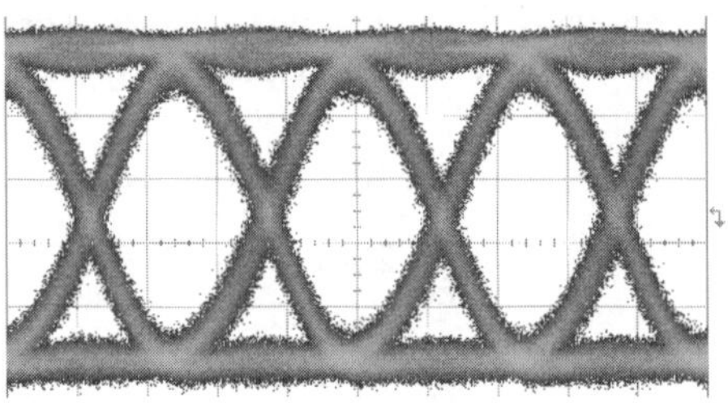

Fig. 39. 40-Gb/s NRZ eye-pattern at 2 dBm.

6. Over-40-Gb/s amplifier

As explained above, we have demonstrated the high performance of front-end analog ICs at 40 Gb/s. However, transmission system capacity will increase in the future and higher bit-rates, possibly 80 Gb/s, will be needed for next-generation fiber optic communication systems. Thus, we have fabricated an ultra-broadband amplifier through attenuation-factor optimization.[23] This technique is suitable for extending the 3-dB bandwidth, because an amplifier with an optimized attenuation factor does not have R_{gate} (as explained in section 2), and makes the most of the transistor potential. However, with this method, a slight gain slope results from the losses of the input and output transmission lines. Increasing the negative resistance degrades the stability of the amplifier at high frequencies to compensate these losses. Therefore, we have to refine the preamplifier design.

6.1. *Circuit Design*

Figure 40 is a schematic diagram of the distributed amplifier for over 40-Gb/s applications.

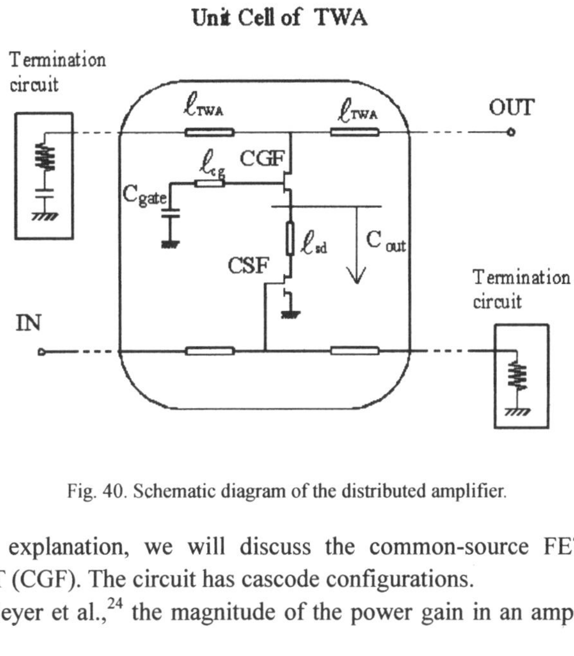

Fig. 40. Schematic diagram of the distributed amplifier.

In the following explanation, we will discuss the common-source FET (CSF) and the common-gate FET (CGF). The circuit has cascode configurations.

According to Beyer et al.,[24] the magnitude of the power gain in an amplifier stage can be described as

$$G = \frac{g_m^2 R_{01} R_{02} \cdot \sinh^2[n(A_d - A_g)/2] \cdot e^{-n(A_d + A_g)}}{4[1 + (\omega/\omega_g)^2][1 - (\omega/\omega_c)^2]\sinh^2[(A_d - A_g)/2]} \quad (4)$$

$$\omega_g = 1/R_i C_{gs}, \omega_c = 2/\sqrt{L_{TWA} C_{gs}}$$

$$L_{TWA} \approx \frac{Z_0 \cdot \ell_{TWA}}{\lambda_g \cdot f}.$$

Here, R_{01} and R_{02} are the image impedances of the input and output transmission lines, respectively, n is the number of unit cells in the amplifier, and A_g and A_d are the attenuation factors on the input and output transmission lines, respectively.

In Eq. (4), $e^{-n(A_d + A_g)}$ is the dominant factor determining the gain-flatness and circuit stability of the distributed amplifier. A_g and A_d are expressed as [25]

$$A_g = \frac{(\omega_c / \omega_g) X_k^2}{\sqrt{1 - [1 - (\omega_c / \omega_g)^2] X_k^2}} \quad (5)$$

$$A_d = \mathrm{Im}\left[1 + \frac{j\omega L_{TWA}}{2Z_{out}}\right] \Big/ \sqrt{1 - \mathrm{Re}\left[1 + \frac{j\omega L_{TWA}}{2Z_{out}}\right]^2} \quad , (6)$$

where Z_{out} is the output impedance of a cascode HEMT.[26]

$$Z_{out} = \frac{Z_{ds} + j\omega L_{sd}}{Z_{ds} + j\omega(L_{cg} + L_{sd}) + 1/j\omega C_{gs} + 1/j\omega C_{gate}} \cdot \left(\frac{g_m Z_{ds}}{j\omega C_{gs}} + j\omega L_{cg} + \frac{1}{j\omega C_{gs}} + \frac{1}{j\omega C_{gate}} \right) + Z_{ds} \quad (7)$$

$$L_{cg,sd} \approx \frac{Z_0 \cdot \ell_{cg,sd}}{\lambda_g \cdot f}$$

ω_g: gate-circuit radian cutoff frequency
ω_c: radian cutoff frequency of the line
C_{gs}: gate-to-source capacitance
R_i: effective gate-to-source resistance
ℓ_{TWA}: length of series transmission line for unit cell
Z_0: characteristic impedance of transmission line
f: frequency
λ_g: wavelength in the transmission line
$X_k = \omega/\omega_c$: normalized frequency
C_{gate}: capacitance connected at the CGF gate
Z_{ds}: drain-to-source impedance of transistor
$L_{TWA,cg,sd}$: inductance of $\ell_{TWA,cg,sd}$

The frequency response of A_g depends on parameters ℓ_{TWA} and C_{gs} of the CSF. For A_d, the dominant parameters are ℓ_{sd}, ℓ_{cg}, and C_{gate}. In addition, these parameters also determine the stable conditions for the distributed amplifier. When A_g+A_d equals a large negative value, the gain described in Eq. (4) would diverge since the term $e^{-n(A_d+A_g)}$ diverges. As shown in Eq. (5), A_g is positive at any frequency. Thus, determination of the A_d profile is the key to improving the gain-flatness and achieving stable operation over a wide frequency range. Therefore, we focused on how to determine the parameters ℓ_{sd}, ℓ_{cg}, and C_{gate} of the distributed amplifier.

C_{gate} and ℓ_{cg} make up a resonant circuit. Since the impedance looking into the CSF at the source terminal of the CGF (Fig. 34) is always capacitive, C_{out}, the CGF including the resonant circuit makes up a positive feedback circuit – which will be an unstable circuit. This resonant frequency is determined by C_{gate} and ℓ_{cg} and C_{out}. To obtain greater stability, we set

C_{gate} and ℓ_{cg} to small values to raise the resonant frequency. Thus, we can determine the gain flatness and circuit stability with ℓ_{sd}.

Figure 41 shows plots of A_g and A_d as a function of frequency with various ℓ_{sd} values.

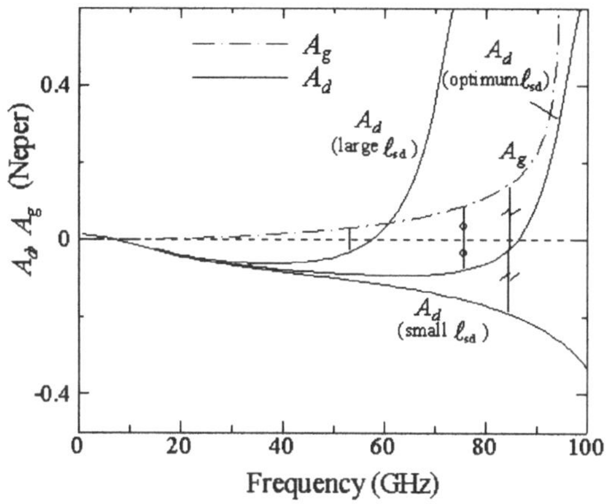

Fig. 41. Calculated attenuation factor of input and output transmission lines (A_g, A_d).

The A_g profile did not change with ℓ_{sd}, because the Miller capacitance of a cascode HEMT is very small. On the other hand, the A_d profile depended on the ℓ_{sd} value. We also looked at the frequency, f_x, where A_d+A_g=0. If f_x was high, the gain variation was small over a wide frequency range. For a large ℓ_{sd}, A_d drastically increased at a relatively low frequency, causing a low f_x. For a small ℓ_{sd}, however, A_d diverged into a large negative value, and A_g+A_d also became negative. This behavior will cause unstable operation. Consequently, large variations occur in the gain profile at high frequencies. The optimum conditions for ℓ_{sd} to obtain gain-flatness and stable operation in the distributed amplifier is for A_g+A_d to equal a small value over a wide frequency range and for f_x to be set within a higher region. We obtained 60 μm as the optimum value for ℓ_{sd} for our HEMTs.

6.2. *Circuit performance*

Figure 42 shows the measured frequency characteristics of a distributed amplifier. We obtained a gain of 9 dB and a bandwidth of 70 GHz. The measured noise figure averaged 5 dB ±1 dB across a frequency band of 2 to 40 GHz (Fig. 43). These results show that our technique significantly improved the 3-dB bandwidth of a distributed amplifier.

We have also been developing a 0.9-$V_{p\text{-}p}$-operation $LiNbO_3$ modulator.[27] Using this modulator with our amplifier, we should be able to realize 80-Gb/s NRZ-transmitter systems.

Moreover, we can realize over-100-GHz amplifiers using InP-based HEMTs.[28] Therefore, we should be able to realize 80-Gb/s RZ- and over-80-Gb/s NRZ-transmitter systems.

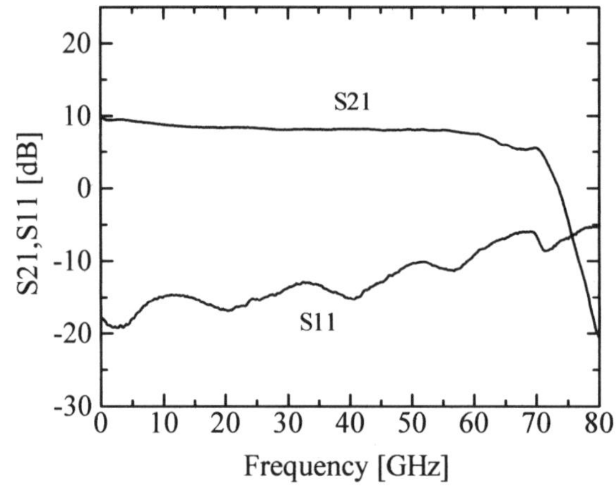

Fig. 42. Frequency characteristics of a 70-GHz distributed amplifier.

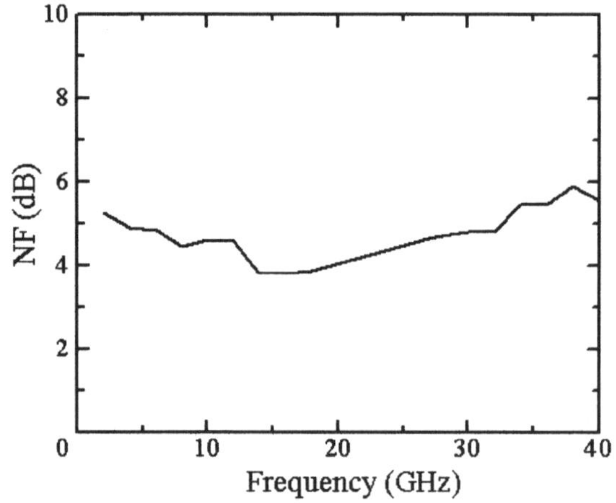

Fig. 43. Measured noise figure of a 70-GHz distributed amplifier.

7. Conclusion

We have developed a distributed amplifier for high bit-rate fiber-optic communication systems, and have designed a $LiNbO_3$ modulator driver using 0.15-μm double-doped AlGaAs/InGaAs/AlGaAs pseudomorphic HEMTs. We obtained a 15-dB gain and 54-GHz bandwidth with GCPW lines, and a fairly clear 6-$V_{p\text{-}p}$ output with rms-jitter of 1.1 ps at 40 Gb/s. We also developed preamplifiers for both NRZ and RZ applications. Our preamplifier for the NRZ-data format has a 52-dBΩ gain and a 49-GHz bandwidth. It consists of a lumped element circuit and a distributed amplifier with InP-based HEMTs. We confirmed that a clear NRZ-output waveform was produced by a module that consisted of the preamplifier connected to a photodiode at 40-Gb/s. The preamplifier for the RZ-data format has an 11-dB gain and a 57-GHz bandwidth. We also developed a PIN/preamp module that has a 44-dBΩ gain and a 45-GHz bandwidth. The module showed a superior phase characteristic and produced a clear RZ eye-pattern. Furthermore, we fabricated an ultra-broadband amplifier for over-40-Gb/s applications. From this, we obtained a 9-dB gain and 70-GHz bandwidth. These results demonstrate that a distributed amplifier is the best candidate for over-40-Gb/s fiber optic communication systems. We expect further extension of the limits of distributed amplifiers to continue for the foreseeable future.

Acknowledgements

We thank Messrs. K. Makiyama, H. Hirano, and Tateno for the device processing, and thank Drs. M. Takigawa and Y. Watanabe for their encouragement and support.

References

1. Y. Yano, T. Ono, K. Fukuchi, T. Ito, H. Yamazaki, M. Yamaguchi, and K. Emura, "2.6-Terabit/s WDM transmission experiment using optical duobinary coding," ECOC'96 Postdeadline Papers. 1996, ThB. 3.1.
2. S. Kuwano, N. Takachio, K. Iwashita, T. Otsuji, Y. Imai, T. Enoki, K. Yoshino, and K. Wakita, "160-Gbit/s (4-ch x 40-Gbit/s electrically multiplexed data) WDM transmission over 320-km dispersion-shifted fiber," OFC'96 Postdeadline Papers, 1996, PD25.
3. M. Wurzer, J. Bock, H. Knapp, W. Zirwas, F. Schumann, and A. Felder, "A 40-Gb/s Integrated Clock and Data Recovery Circuit in 50-GHz f_T Silicon Bipolar Technology," IEEE Journal of Solid-State Circuits, vol.34, No.9, pp.1320, September 1999.
4. T. Otsuji, Y. Imai, E. Sano, S. Kimura, S. Yamaguchi, M. Yoneyama, T. Enoki, and Y. Umeda, "40-Gb/s IC's for Future Lightwave Communications Systems," IEEE Journal of Solid-State Circuits, vol.32, No.9, pp.1363, September 1997.
5. J. Godin, M. Riet, S. Blayac, Ph. Berdaguer, J.-L. Benchimol, A. Konczykowska, A. Kasbari, Ph. Andre, and N. Kauffmann, "Improved InGaAs/InP DHBT Technology for 40 Gbit/s Optical Communication Circuits," IEEE GaAs IC Symp., pp. 77-80, 2000.
6. T. Masuda, K. Ohhata, F. Arakawa, N. Shiramizu, E. Ohue, K. Oda, R. Hayami, M. Tanabe, H. Shimamoto, M. Kondo, T. Harada, and K. Washio, "45-GHz Transimpedance-, 32-dB Limiting Amplifier, and 40-Gb/s 1:4 High-Sensitivity Demultiplexer with Decision Circuit using SiGe HBTs for 40-Gb/s Optical Receiver," ISSCC, pp. 60, 2000.
7. M. Mokhatari, T. Swahn, R. H. Walden, W. E. Stanchina, M. Kardos, T. Juhola, G. Schuppener, H. Tenhunen, and T. Lewin, "InP-HBT-Chip-set for 40 Gbit/s fiber optical communication systems operational at 3 V," IEEE J. Solid-State Circuits, vol. 32, pp. 1371-1382, Sept. 1997.
8. H. Masuda, K. Ouchi, A. Terano, H. Suzuki, K. Watanabe, T. Oka, H. Matsubara, and T. Tanoue, "High performance InP/InGaAs HBTs for 40 Gb/s optical transmission ICs,"

IEICE Trans. Electron., vol. E82-C, no.3, pp.419-427, Mar. 1999.

9. M. Sokolich, C. H. Fields, S. Thomas, III, B. Shi, Y. K. Boegeman, M. Montes, R. Martinez, A. R. Kramer, and M. Madhav, "A Low-Power 72.8-GHz Static Frequency Divider in AlInAs/InGaAs HBT Technology," IEEE J. Solid-State Circuits, vol.36, pp.1328-1334, Sept. 2001.
10. T. Suzuki, H. Kano, Y. Nakasya, T. Takahashi, K. Imanishi, H. Ohnishi, and Y. Watanabe, "40-Gbit/s D-type Flip-Flop and Multiplexer Circuits Using InP HEMT", 2001 IEEE MTT-S Digest.
11. H. Suzuki, K. Watanabe, K. Ishikawa, H. Masuda, K. Ouchi, T. Tanoue, and R. Takeyari, "Very-High-Speed InP/InGaAs HBT IC's for Optical Transmission Systems," IEEE J. of Solid-State Circuits, vol.33, No.9, pp.1313, September 1998.
12. Y. Ayasli, R. L. Mozzi, J. L. Vorhaus, L. D. Reynolds, R. A. Pucel, "A monolithic GaAs 1-13 GHz traveling-wave amplifier," IEEE Trans. Microwave Theory Tech., vol. MTT-30, pp. 976-981, July 1982.
13. W. Kennan, T. Andrade, and C.C. Huang, "A 2-18 GHz Monolithic Distributed Amplifier Using Dual-Gate GaAs FETs," IEEE Trans. Microwave Theory Tech., vol. MTT-32, pp. 1693-1697, Dec. 1984.
14. Y. Yamashita, A. Endo, K. Shinohara, K. Hikosaka, T. Matsui, S. Hiyamizuz, and T. Mimura, "Pseudomorphic $In_{0.52}Al_{0.48}As/In_{0.7}Ga_{0.3}As$ HEMT with an Ultrahigh ft of 562GHz," to be published in IEEE Electorn Device Letters.
15. H. Shigematsu, M. Sato, T. Suzuki, T. Takahashi, K. Imanishi, N. Hara, H. Ohnishi, and Y. Watanabe, "49-GHz Preamplifier with a Transimpedance Gain of 52 dBΩ Using InP HEMTs," IEEE J. Solid-State Circuits, vol.36, pp.1309-1313, Sept. 2001.
16. H. Shigematsu, M. Sato, T. Hirose, and Y. Watanabe, "45-GHz Distributed Amplifier with a Linear 6-Vpp Output for a 40-Gb/s $LiNbO_3$ Modulator Driver Circuit," ," IEEE GaAs IC Symp., pp.137-140, 2001.
17. T. Takahashi, M. Nihei, K. Makiyama, M. Nishi, T. Suzuki and N. Hara, "Stable and Uniform InAlAs/InGaAs HEMT ICs for 40-Gbit/s Optical Communication Systems," Proc. 13th Int. Conf. InP and Related Materials, pp.614-617, 2001.
18. T. Enoki, H. Yokoyama Y. Umeda and T. Otsuji, "Ultrahigh-speed integrated circuits using InP-based HEMTs," Jpn. J. Appl. Phys. Vol.37, Part 1, No.3B, pp.1359-1364, March 1998.
19. Y. Nakasha, T. Suzuki, H. Kano, A. Ohya, K. Sawada, K. Makiyama, T. Takahashi, M. Nishi, T. Hirose, M. Takikawa and Y. Watanabe, "A 43-Gbit/s Full-Rate-Clock 4:1 Multiplexer in InP-based HEMT Technology," 2002 IEEE ISSCC Tech. Dig., pp.190-191
20. H. Kano, T. Suzuki, S. Yamaura, Y. Nakasha, K. Sawada, T. Takahashi, K. Makiyama, T. Hirose, and Y. Watanabe, "A 50-Gbit/s 1:4 Demultiplexer IC in InP-based HEMT Technology," IEEE IMS-2002, pp.75-78.
21. T. Suzuki, Y. Nakasha, T. Takahashi, K. Makiyama, K. Imanishi, T. Hirose, and Y. Watanabe, "A 90-Gbit/s 2:1 Multiplexer IC in InP-based HEMT Technology," 2002 IEEE ISSCC Tech. Dig., pp.192-193
22. M. Takechi, K. Araki, Y. Furukawa, K. Sato, T. Tokumitsu, M. Kobayashi, H. Ohnishi, A. K. Dutta, H. Shigematsu, "PIN/Preamp receiver modules for 40 Gb/s optical communication systems," ITCom Techinical Digest, 2002.

23. M. Sato, T. Hirose, and Y. Watanabe, "A 70-GHz Bandwidth and 9-dB Gain Travelling Wave Amplifier Using 0.15-μm Gate InGaP/InGaAs HEMTs with Coplanar Transmission Line Technology," EuMC Technical Digest 2000.
24. J. B. Beyer, S. N. Prasad, R. C. Becker, J. E. Nordman, and G. K. Hohenwarter, "MESFET Distributed Amplifier Design Guidelines," IEEE Trans. Microwave Theory Tech., vol.32, pp.268-261, Mar. 1984.
25. J. B. Beyer, S. N. Prasad, J. E. Nordman, R. C. Becker, and G. K. Hohenwarter, "Wideband monolithic microwave amplifier study," ONR Rep. NR243-033, Sept. 1983.
26. S. Kimura, Y. Imai, Y. Umeda, T. Enoki, "Loss-Compensated Distributed Baseband Amplifier IC's for Optical Transmission Systems," IEEE Trans. Microwave Theory Tech., vol.44, pp.1688-1693, Nov. 1996.
27. M. Sugiyama, M. Doi, S. Taniguchi, T. Nakazawa, and H. Onaka, "Driver-less 40 Gb/s LiNbO3 Modulator with Sub-1 V Drive Voltage", Tech. Dig. OFC2002 Post-deadline paper FB6, 2002
28. S. Masuda, T. Hirose, T. Takahashi, M. Nishi, S. Yokokawa, S. Iijima, K. Ono, N. Hara, and K. Joshin, "An Over-110-GHz InP HEMT Flip-chip Distributed Baseband Amplifier with Inverted Microstrip Line Structure for Optical Transmission Systems" to be published in IEEE GaAs IC Symp., 2003.

International Journal of High Speed Electronics and Systems
Vol. 13, No. 1 (2003) 141–173

World Scientific
www.worldscientific.com

The InP-HEMT IC Technology for 40-Gbit/s Optical Communications

YASURO YAMANE*AND KOICHI MURATA

NTT Photonics Laboratories. 3-1, Morinosato-Wakamiya,
Atsugi, Kanagawa Pref., 243-0198, Japan
*(*Present Affiliation: NTT Electronics Corporation)*

We present the outline of the InP HEMT IC technology. This technology realizes InP HEMT digital ICs for 40-Gbit/s optical fiber communication systems through the integration of 0.1-μm-gate-length HEMTs, vertical diodes, capacitors, and WSiN resistors with two level interconnections. This paper describes the high-speed digital IC circuit design and fabrication in InP HEMT technology for 40-Gbit/s/channel optical communication systems. Some results on InP HEMTs' reliability are also covered. Basic circuit design techniques utilizing SCFL topology and fundamental circuit elements of the selector and D-type flip-flop are discussed in detail. The basic digital ICs of MUX, D-FF, and DEMUX ICs fabricated with 0.1-μm-gate InP HEMTs successfully operated up to 50 Gbit/s in the packaged modules. These IC modules offer large speed margins for the 43-Gbit/s OTU-3 data rate. In order to develop cost-effective optical transmitters and receivers, we designed a PLL-based CDR with a full-rate architecture. The fully monolithic integrated CDR exhibited error-free operation for 2^{31}-1 PRBS data signal at the OTU-3 bit rate of 43.0184 Gbit/s. Four-bit MUX and DEMUX ICs are other key components, and could be implemented by using InP HEMT technology. Additionally, we describe InP-IC fabrication technology with two-level inter-connection. This is already fully matured for 40-Gbit/s SSI fabrication. The uniform FET characteristics and high-yield passive component fabrication technologies support this degree of maturity. InP HEMT lifetime is 10^7 hours at 100°C. These results prove the InP HEMT IC fabrication technology presented here, to be highly reliable. These investigations show that robust performance and yield when realizing SSI and MSI 40-Gbit/s functions.

Keywords: HEMT; InP; D-FF ; T-FF; MUX; DEMUX; SSI.

1. Introduction

High-speed integrated circuit (IC) technology has greatly contributed to the development of economical high-speed large-capacity optical communication systems. Due to the rapid growth of data traffic, telecommunication systems now require enhanced network flexibility, more capacity and reduced cost at the same time. Wavelength division multiplexing (WDM) systems based on high-speed electrical time division multiplexing (ETDM) are attractive for meeting these requirements. Recently, 40-Gbit/s/channel WDM system experiments based on a 40-Gbit/s ETDM have been carried out, and Tera-bit class transmission capacities have been achieved.[1, 2]

For 1-Tbit/s transmission, a 40-Gbit/s-based system has several advantages over systems based on 2.5 Gbit/s or 10 Gbit/s, such as simplicity of the network management and the reduced node cost. On the other hand, 40-Gbit/s signal transmission through existing optical fiber needs the error rate (BER) improvement. To satisfy this need, the optical channel transport unit (OTU) frame format specified by International Telecommunication Union-Telecommunication Standardization Sector (ITU-T) includes the out-of-band forward error correction code (FEC) at the expense of at least 7 % line rate increase. For making 40-Gbit/s transmission technology practical, physical transmission at a bit rate of 43 Gbit/s or higher is more suitable.

In the pursuit of high-speed devices, various device technologies, such as GaAs HBT, InP HBT, SiGe HBT and InP HEMT are under development.[3-51] InP HEMT technologies have been used in 40-Gbit/s transmission experiments[52] and achieved one of the best performances among all high-speed device technologies.

An InP HEMT utilizes the two-dimensional electron gas formed at the interface between $In_{0.52}Al_{0.48}As$ and $In_{0.53}Ga_{0.47}As$ lattice matched to an InP substrate as the carrier. The mobility of InP HEMTs is as large as 10000 $cm^2/V*sec$ at room temperature, which is 1.5 times that of GaAs HEMTs. Additionally, the saturation velocity of InP HEMTs is as high as 2.7×10^7 cm/sec. The ΔEc of InAlAs/InGaAs is 0.5 eV, while the ΔEc of AlGaAs/GaAs is only 0.3 eV. The large ΔEc of InAlAs/InGaAs contributes to a better electron confinement and increases device design freedom for the Ns. The highest f_T (about 400 GHz) ever reported for any electron device is measured on an InP HEMT.[53] As for FET-type devices, the required performance is a 180-GHz f_T and 1.2-S/mm g_m for 40-Gbit/s operating digital ICs.[54]. Uniformity and reproducibility of the FET characteristics are also indispensable. Because the HEMT fabrication is based on recess technology, Vth control is more difficult than in other devices based on the ion implantation. The uniformity data, which satisfy the requirement for high-speed SSI fabrication, are presented in this paper.

This paper describes design, fabrication, and reliability of high-speed InP HEMT digital ICs for 40-Gbit/s class optical fiber communication systems. The next section outlines optical communication ICs comprising an optical transmitter and receiver, and the basic circuit configurations of digital ICs. Section 3 describes the basic circuit design of a Source-Coupled FET Logic (SCFL) inverter circuit, which is widely used as the basic circuitry for ultrahigh-speed logic circuits. Section 4 discusses high-speed circuit design technology for the basic InP HEMT digital ICs, such as the multiplexer (MUX) IC and D-type flip-flop (D-FF). These ICs are primitive key components for optical transmitters and receivers. We detail the circuit designs of buffer, interface, and digital core circuits using InP HEMT device. Section 5 presents the circuit performance of the packaged ICs with the measurement setup. Section 6 discusses a fully monolithic clock and data recovery IC as recent progress in IC technology for the development of cost-effective transmitters and receivers. Section 7 outlines the IC structure and its components. The section 8 discusses FET performance and its reliability, and section 9 covers IC yield.

2. Optical Fiber Communication ICs

Figure 1 shows the basic configuration of an optical transmitter and receiver. The transmission data rate is 43 Gbit/s, because we assume a system based on the OTU-3 frame format.[55] The transmitter consists of a SONET/SDH framer, a multiplexer (MUX), a clock

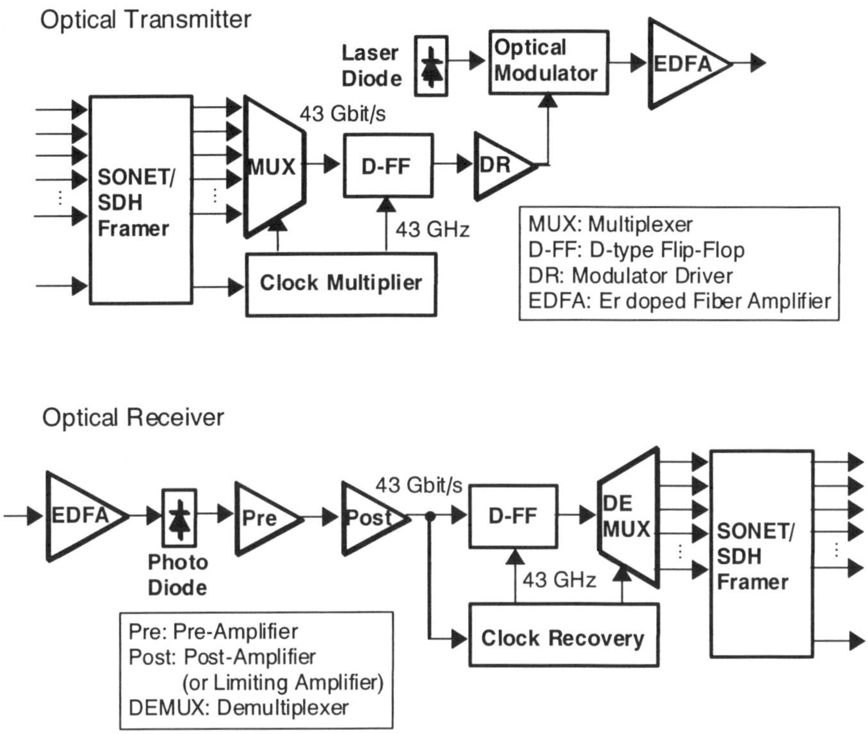

Fig. 1. Basic configuration of an optical transmitter and receiver.

signal multiplier, a D-type flip-flop (D-FF), a modulator driver (DR), a laser diode, and an optical modulator. The electrical interface between the framer and the MUX is now being discussed in the Optical Internetworking Forum (OIF), and the SFI-5 implementation agreement will be announced in the near future. The MUX multiplexes lower rate parallel signal into 43-Gbit/s serial. The clock multiplier generates the full-rate 43-GHz clock signal from a lower frequency clock signal. The D-FF reshapes the output waveforms of the MUX. The driver IC amplifies the data signal and drives the optical modulator. An output voltage swing of over 5 volts is required for a lithium niobate modulator, and one of over 2 volts is required for an electro-absorption modulator.

The receiver should provide reshaping, retiming, and regeneration functions. It consists of a photodiode, pre- and post-amplifiers, a clock recovery circuit, a D-FF, a demultiplexer (DEMUX), and a framer LSI. The photodiode converts the optical signal into an electrical signal, which is then amplified by the pre- and post-amplifiers. The clock recovery circuit extracts the full-rate clock signal from the data signal, and the D-FF regenerates the input data signal. The demultiplexer, or DEMUX, and the framer LSI recover the original low-speed electrical signals from the 43-Gbit/s bit-serial data stream.

Erbium-doped optical fiber amplifiers (EDFA) are used as a booster amplifier in the transmitter and a preamplifier in the receiver. In this receiver configuration, receiver sensitivity is determined by the EDFA noise, and the gain requirement for the post amplifier is

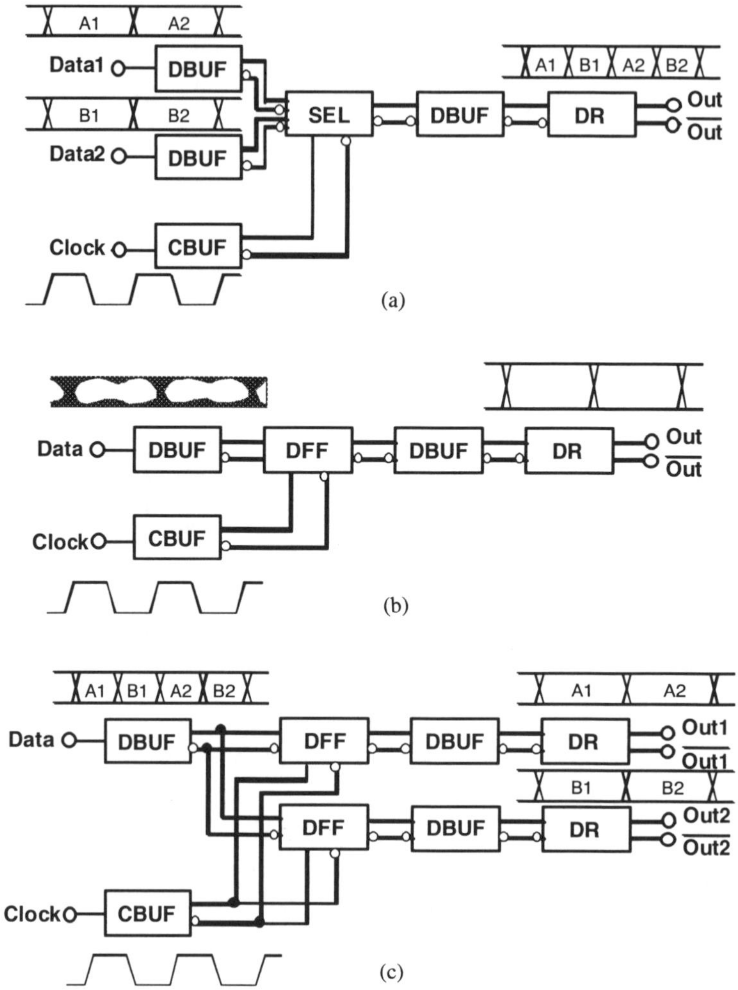

Fig.2. Block diagram of the basic digital ICs.

(a) Multiplexer (MUX). (b) D-FF. (c) Demultiplexer (DEMUX).

relaxed through the use of EDFA. All ICs except for the clock multiplier and clock recovery ICs require broadband operation from near DC to the maximum bit rate with good eye openings. Among the digital ICs, the D-FF limits the speed of the system because of its 43-GHz full-rate clock operation.

In terms of digital ICs, the MUX, D-FF, and DEMUX are the key components for multiplexing, regeneration, and demultiplexing functions. Figure 2 shows block diagrams of these ICs. All the ICs consist of data input and output buffers (DBUF), a clock input buffer (CBUF), a driver (DR), and logic core circuits. A selector circuit (SEL) is used as the core circuit of the MUX, while D-FFs are used for the D-FF IC and DEMUX. The input data and clock buffer circuits are configured with differential amplifiers, and generate a differential signal from input single-ended signal, which is needed for the internal circuit operations.

The output buffer and driver are inserted to amplify the output of the core circuit to the voltage level of around 1 Vp-p. The core circuit should operate at the transmission data rate, while the data buffer circuits should cover the bandwidth of approximately 70 % of the data rate. The clock buffer requires a sufficient gain at the clock frequency. All of the circuits are designed based on Source-Coupled FET Logic (SCFL), which is widely used in basic digital ICs for the purpose of high-speed operation.[56-58] The circuit design of the SCFL will be detailed in the following section.

3. Basic Circuit Design of Source-Coupled FET Logic Inverter

Basic circuit design of a Source-Coupled FET Logic (SCFL) inverter is described in this section. The circuit diagram of the SCFL inverter is shown in Fig. 3. The inverter consists of a source-coupled differential FET pair (*xf1*, *xf2*), current source FET (*xf3*), load resistor (*RL1*, *RL2*), and source follower circuits (*xf4* ~ *xf7*). The bias conditions of all FETs are designed to be in the saturation region, in which FETs are biased in the high-transconductance low-gate-drain capacitance region. The SCFL inverter operates with differential signals, and has a differential input port (*DT*, *DC*) and output port (*QT*, *QC*). Combinational logic gates have a sereies-gated configuration based on SCFL inverters. In addition, the sequential logic function in D-FF and toggle flip-flop (T-FF) is configured with a feedback loop along the data signal path. In the circuit design of the SCFL inverter, three major circuit parameters should be determined: the operating current of the differential FET pair, the internal logic swing, and the operating bias point of the differential pair.

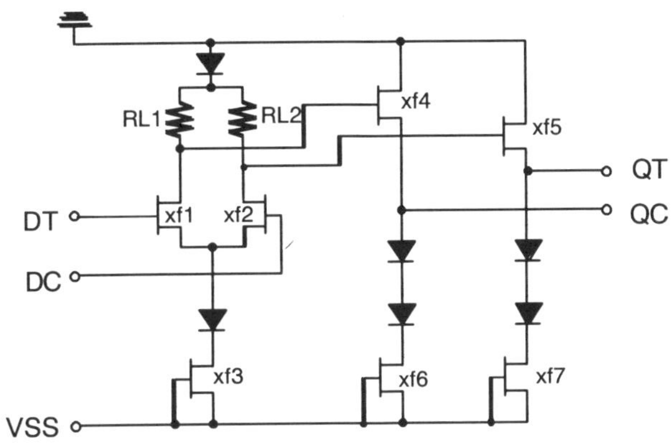

Fig. 3. Circuit diagram of an SCFL inverter.

3-A Design of the gate width of FET

In the following discussions, we assume that the FET is a depletion mode (D-mode) FET having a threshold voltage of around -0.5 V. In addition, all FETs in the inverter are assumed to have the same gate width. The operating current of the differential pair is determined by the gate width of the current source FET in the SCFL inverter, which corresponds to the gate width of the FETs. This parameter should be determined first. Gate delay time of the inverter is an important measure in designing the gate width of the FETs. Figure 4 plots gate delay times and operation currents for a single-stage SCFL inverter with the gate width of FET as a parameter. These data were determined from measured oscillation frequencies of a 19-stage ring oscillator. The product of the gate width and load resistance was designed to be 2724 μm•Ω, which corresponds to the internal voltage of around 800 mVp-p. The SPICE simulation results are also plotted by solid and dashed lines. We took the fringe capacitance and parasitic capacitance of the FET into account in our FET model for the SPICE simulations. As a result, the simulated gate delay times show good agreement with the measured values in the gate width range form 5 to 20 μm. The gate delay time of the inverter is almost constant for the gate width of over 10 μm, while it drastically increases for the gate width of less than 5 μm. The main cause of this is that the external capacitance associated with the gate electrode and gate fringe becomes large compared with the intrinsic gate capacitance in such small gate width FETs. To achieve faster operation with lower power consumption, the gate width should be set in the range from 10 to 20 μm. In our circuit design, we basically set the gate width to be 20 μm for all of the circuit components to achieve 40-Gbit/s circuit operation. In this case, the gate delay time is around 6 ps and the operating current for each inverter stage (inclusive the source followers) is around 28.5 mA.

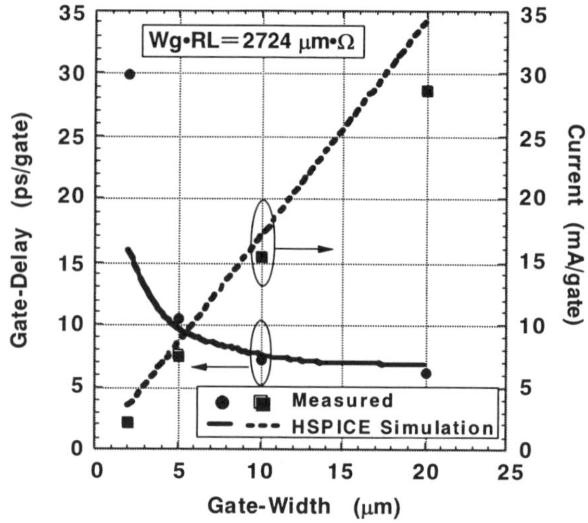

Fig. 4. Gate delay times and operating currents of an SCFL inverter.

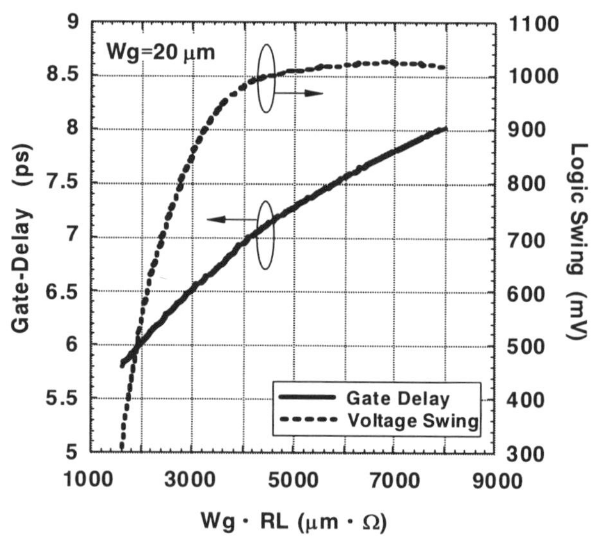

Fig. 5. Simulated gate delay times and internal logic swings.

3-B Design of internal logic swing

The next parameter to be determined is the internal logic swing, which corresponds to the product of gate width and load resistance (*Wg•RL*). Figure 5 shows the simulated gate delay times and the logic swings, where the horizontal axis is the product *Wg•RL*. Here, the gate width was set to 20 μm. The simulated results clearly indicate that shorter gate delay time could be achieved with smaller logic swing. However, the DC gain of the inverter is another important parameter. In the logic circuit design, each inverter stage should provide a DC gain of over 1, which is different from an analog amplifier design, because a limiting operation is indispensable for ensuring the internal logic voltage level. It is desirable that the DC gain is approximately 2 considering the variation of the threshold voltage of FET. Figure 6 shows simulated DC transfer curves of the inverter with the product *Wg•RL* as a parameter. Here, the reference voltage fed to the reference voltage terminal *DC* is -2.65 V. Figure 6 indicates that the value of the *Wg•RL* should be over 3000 μm•Ω to achieve a DC gain of 2. Considering both the simulated gate delay times and DC transfer curves, we basically set the *Wg•RL* to 3000 μm•Ω (*Wg*=20 μm, *RL*=150 Ω) in our inverter circuit design, which results in the expected gate delay time of 6.5 ps.

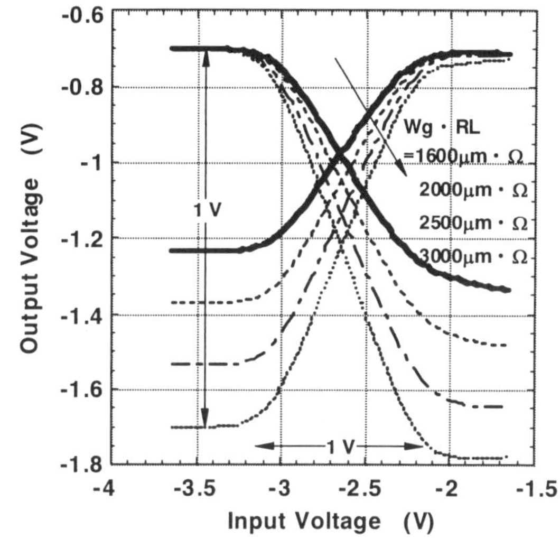

Fig. 6. Simulated DC transfer curves of an SCFL inverter.

3-C Optimization of the operating bias point of the differential FET pair

We discuss the optimization of the operating bias point of the differential FET pairs in this section. Figure 7 shows transconductance and drain current of our InP HFMT with the gate-source bias voltage (*Vgs*) as a parameter. Here, the gate width of the FET was 20 μm and drain-source voltage (*Vds*) was the constant of 1 V. The transconductance offers the maximum value at the *Vgs* of -70 mV. For high-speed circuit

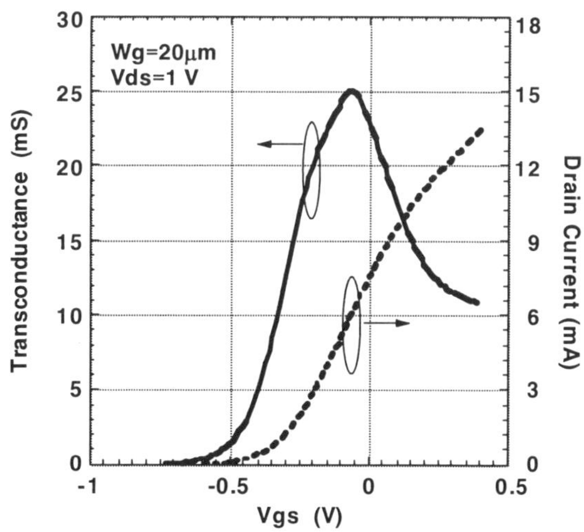

Fig. 7. Transconductance and drain current of the InP HEMT.

operation, the bias condition should be set in the higher transconductance region. The operating bias point can be designed by setting the ratio of the gate width of the differential pair FET (*Wg-sw*) and that of the current source FET (*Wg-cs*). Figure 8 shows simulated bias conditions in *Vgs* with the gate-width ratios of *Wg-sw/Wg-cs*. Here, the dashed lines are for the maximum and minimum values of *Vgs*, which indicates *Vgs* changes between these values in the switching operation. The solid line is for the center bias point in *Vgs*. For example, *Vgs* varies from -0.45 V to 0.0 V when the gate width ratio is 1. The operating region can be designed in the higher gate bias region by setting the ratio to less than 1. The center bias value for the gate ratio of 0.5 is as high as -70 mV, which seems to be suitable for high-speed operation because the transconductance exhibits the maximum value at the bias level. However, the DC gain should again satisfy the criteria mentioned above. Figure 9 shows the DC transfer curves with the gate width ratio *Wg-sw/Wg-cs* as a parameter. The DC transfer gain gradually degrades at the ratio less than 0.75. The ratio of the *Wg-sw/Wg-cs* should be optimized considering the trade-off relationship. In our circuit design, we set the ratio at 1 for inverters in data and clock buffer circuits, because we regard the DC gain as important. For core circuits such as selectors and D-FFs, we set the ratio to 0.8, considering circuit operation speed.

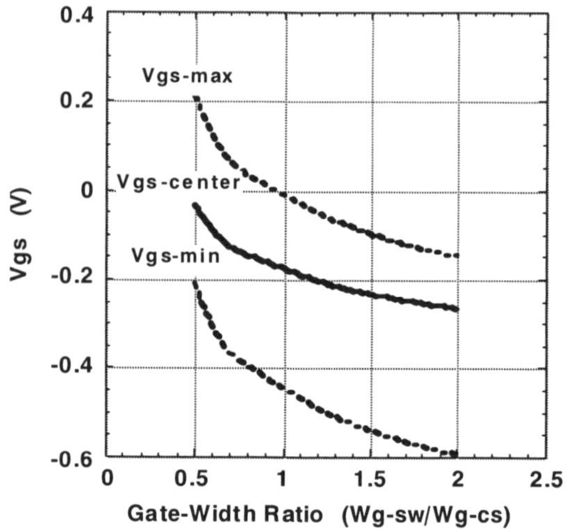

Fig. 8. Bias voltage of the differential pair FET.

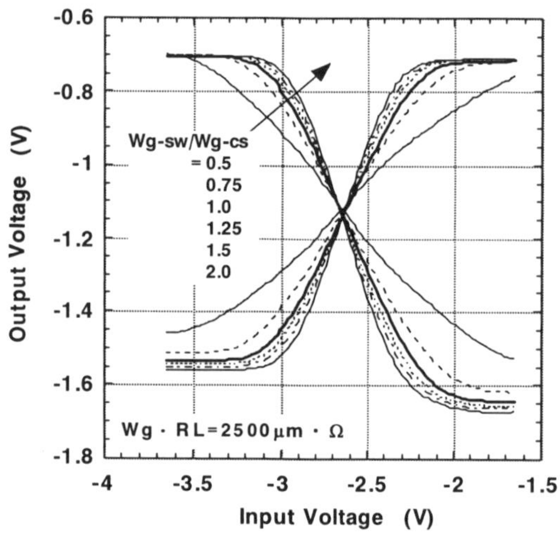

Fig. 9. Simulated DC transfer curves.

4. High-Speed Circuit Design for Basic Communication ICs

In this section, we describe the circuit design of the basic digital ICs, such as MUX and dynamic and static D-FF ICs, for 43-Gbit/s OTU-3 optical fiber communication systems.

4-A Buffer Circuits

For non-return-to-zero (NRZ) signals, the data buffer circuits should cover the bandwidth of approximately 70 % of the data rate, while the clock buffer should provide a sufficient gain at the clock frequency of 43 GHz. A capacitive feedback differential amplifier[59] is employed as the data buffer circuit. The circuit diagram is shown in Fig. 10. The capacitors inserted between the input terminal and the source terminal of the output source followers compensate the gate-drain capacitance of the differential pair FETs (*xf1*, *xf2*), resulting in improvement in the bandwidth of the differential amplifier. Connecting the capacitors to the source followers in parallel improves the bandwidth of the source followers in high-frequency region.[60] Three stage amplifiers were used as the input data buffers in the digital ICs. SPICE simulation indicates the buffer circuit has DC gain of 12 dB and 3-dB bandwidth of 27 GHz, which almost completely covers the bandwidth requirements for 40-Gbit/s NRZ signals.

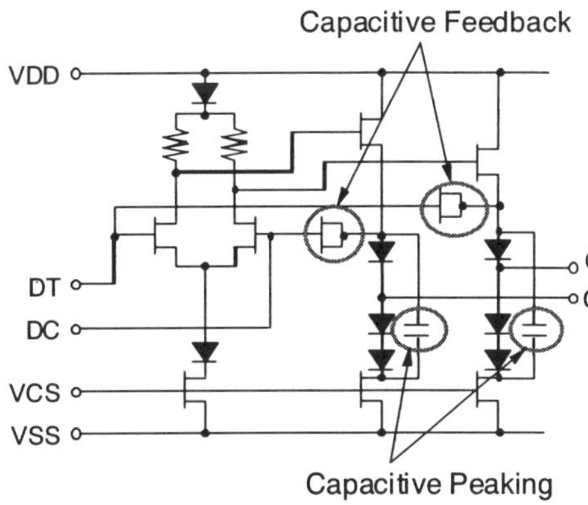

Fig. 10. Circuit diagram of the data buffer amplifier.

Figure 11 shows the circuit configuration of the clock buffer. The inductor peaking amplifier is used as the clock buffer.[60] We used AC coupled configuration for the input and the intermediate stage. SPICE simulation indicates that the lower cut-off frequency is around 1 GHz and the gain at 43 GHz is around 7 dB.

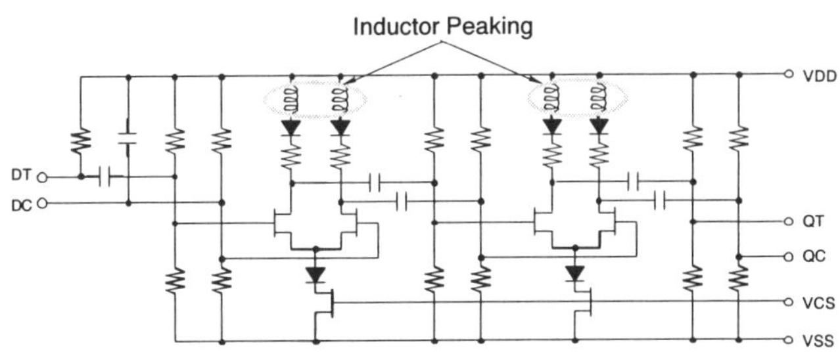

Fig.11. Circuit diagram of the clock buffer.

4-B Interface Circuit

An SCFL interface[61] was employed as the input and output interface of the InP HEMT ICs. The high-level of the output signal is the ground level, while the low-level is -0.9 V. The interface is suitable for high-bit rate transmissions because the high-level is the ground level. Figure 12 shows the circuit configurations of the output stage of a sender IC and the input stage of a receiver IC for the SCFL interface.

The output stage consists of an open-drain-type differential amplifier with impedance matched on-chip resistors (*RTS*). The value of the *RTS* should be 50 Ω for the ideal case,

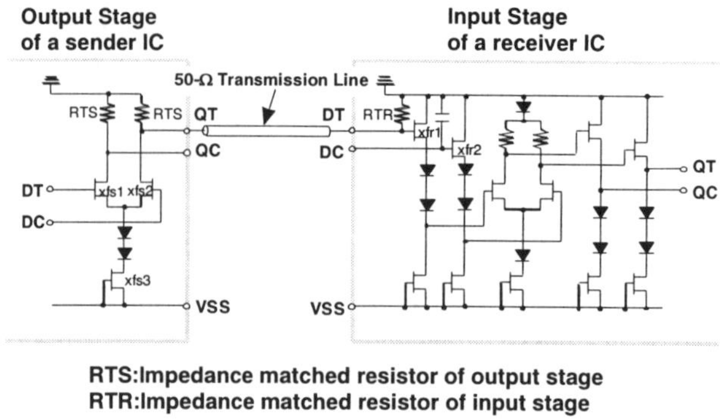

Fig. 12. Circuit configurations for the SCFL interface.

which means that the output impedance of the open-drain-type differential amplifier is infinite. However, the drain conductance of the driver FET (*xfs1, xfs2*) decreases the output impedance. In addition, the operating current of over 36 mA (100-μm gate width for the differential amplifier) is needed for the 50-Ω termination resistance to output the voltage swing of 0.9 Vp-p, which increases power consumption. To cope with this, we optimized the termination resistor so as to maintain the return loss of the output stage at less than -10 dB up to 40 GHz. The small-signal simulation showed that 67-Ω termination resistor satisfies the requirement in the case of a 70-μm gate width FET for the differential output driver.

The input stage consists of an impedance matched on-chip 50-Ω resistor (*RTR*) and source follower circuits that act as level shifters for the following differential amplifier. Since the transmitted signal is generally single ended, a reference voltage level input is needed for the receiver IC. This interface can offer DC-coupled signal transmission. The interface could also be easily enhanced to a differential scheme by adding a 50-Ω transmission line between the output terminal *QC* of sender IC and the input terminal *DC* of receiver IC. In that case, the capacitor between ground and the *DC* terminal should be eliminated.

4-C MUX IC and Selector Core Circuit

The circuit configuration of the MUX IC is basically the same as that reported in *Ref. 62*. The circuit design technique of the MUX is precisely described in the reference paper. In this paper, we focus on the circuit design of the core selector shown in Fig. 13. The circuit is a simple selector circuitry based on SCFL. The CR time constant at the drain node of the data input FET is a major factor limiting the circuit speed of the selector core. In the present circuit, the gate widths of the data input FETs (indicated in Fig. 13 by asterisks) are set smaller than those of the clock input FETs, which is different from the

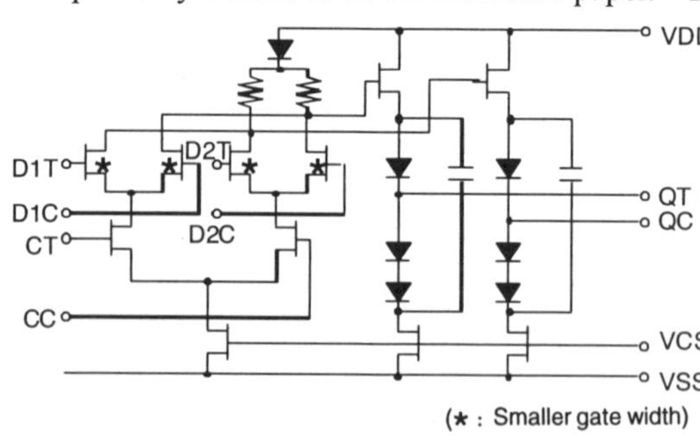

Fig. 13. Circuit diagrams of the selector core.

previous work. As described in the previous section, this gate width combination ensures that the first level differential pair is biased so as to operate in the higher transconduction region, which results in shorter switching time. To confirm this design concept, we simulated the output voltage swing of the selector core for a 2^7-1 pseudo random bit sequence (PRBS) at 80 Gbit/s. Here, the output voltage swing is defined as the eye height, i.e., the vertical opening of the eye diagram. Figure 14 shows the simulation results with the ratio of the gate-width of the data input FET to that of the clock input FET. This figure clearly indicates that decreasing the data-input FETs gate-width increases eye height. This result yields improvement of switching time because the peak-to-peak output voltage maintains an almost constant value regardless of the gate-width ratio. In the present circuit design, the gate-widths of the data input FETs of the core selector are designed to be 20 % smaller than those of clock input FETs. The speed of the multiplexer IC is expected to be improved by approximately 10 % compared to the conventional circuit design if the maximum operating speed is defined as the bit-rate at which the eye opening of the IC exceeds 700 mV.

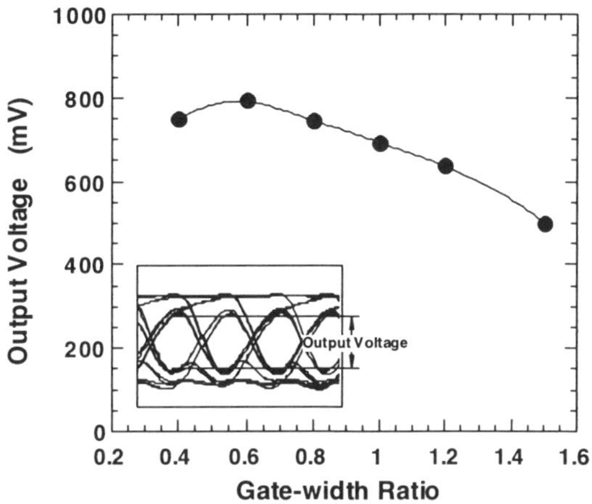

Fig. 14. Simulated output voltage of the selector core.

4-D Super-dynamic D-FF

A D-FF is a fundamental circuit element for a decision IC and a DEMUX IC, which limits the speed performance of the optical senders and receivers. Much effort is under way to boost circuit speed of the D-FF, several new circuit topologies, such as a high-speed latching operation flip-flop (HLO-FF)[63] and a super dynamic flip-flop,[64] have been reported. The dynamic operation principle is introduced into these flip-flops to boost their circuit speed, and the improvement in the circuit speed was confirmed by using GaAs MESFET[64, 65] and InP HEMT technologies[11]. The super-dynamic D-FF operates twice as fast as the conventional static master-slave flip-flop (MS-FF).[64, 65]

A super-dynamic D-FF was employed as the D-FF core for the reshaping and/or regenerating D-FF ICs, because of its 43-GHz full-rate clock signal operation. In this section, we focus on improvement of the jitter performance of the super-dynamic D-FF. Figure 15 shows the circuit diagram of the core D-FF. The circuit configuration of the D-FF IC is basically the same as those reported in *Ref. 60* except for the gate-width of the source couple negative feedback pair (SCNFP) in the latching circuit of the super-dynamic D-FF. As described in *Ref. 64*, the SCNFP instantaneously forces the complementary output level of the D-FF to equalize at the middle of the logic swing at every transition between latching and reading operations. Circuit operation can be a cause of significant pattern jitter if the SCNFP

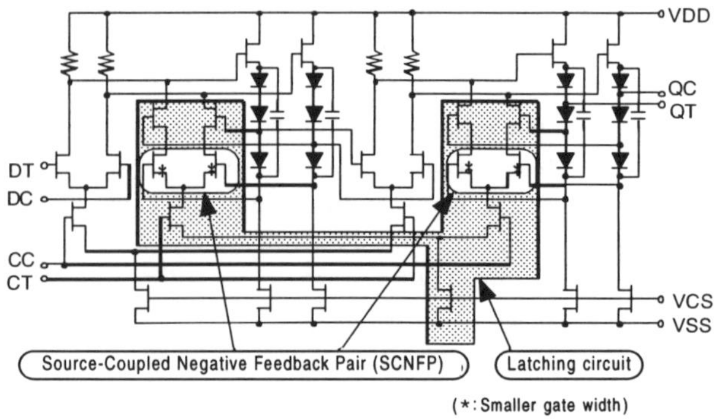

Fig. 15. Circuit diagram of the super-dynamic D-FF.

has a large gate width. Therefore, in the present circuit, the gate widths of the SCNFP (indicated in Fig. 15 by asterisk) are designed to be smaller than those of the latching circuit to reduce pattern jitter. Here, both the SCNFP and the latching circuit have the same gate width in the previous circuit design. Figure 16 shows the simulated pattern jitter of the D-FF for a 2^7-1 PRBS sequence at 50 Gbit/s. Here, the horizontal axis plots the ratio of the gate width of the SCNFP to that of the latching circuit, and the pattern jitter is defined as the time width (peak to peak) at the cross point of the D-FF output eye pattern. The simulated pattern jitter originates from the circuit response to the input data-signal patterns because no other jitter sources were considered in the simulations. As shown in Fig. 16, the pattern jitter decreases as the gate-width of the SCNFP decreases. In the present circuit, the gate-widths of the SCNFPs were designed to be half those of the latching circuit considering the minimum manufacturable gate width. The simulation indicates that the pattern jitter is expected to be reduced by 50 % compared to the previous circuit.

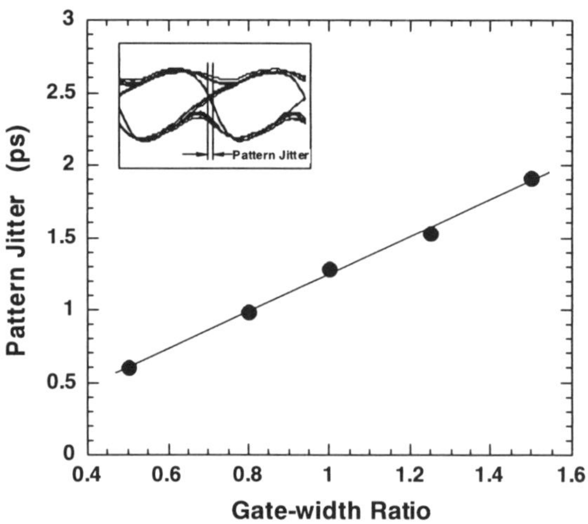

Fig. 16. Simulated pattern jitter of the D-FF.

4-E Static D-FF

The super dynamic D-FF is superior to the MS-FF in terms of speed. The D-FF, however, has a minimum operating frequency for the input clock signal, and the low and high levels of the D-FF output are periodically forced to the center level at each transition timing around its minimum operating frequency.[65] The MS-FF has wide operating clock frequency range, which is especially important for the application to demultiplexers and/or measurement instruments. The static D-FF IC based on the MS-FF was designed to confirm the feasibility of application to the 1:2 DEMUX.

Figure 17 shows the circuit diagram of the MS-DFF. The circuit is based on SCFL two-level series-gated master-slave circuitry. To boost the core circuit speed, we devised a circuit configuration in which we firstly decreased the gate widths of the first-level differential pairs by 20 % compared to those of the second-level differential pairs. This allows us to set the bias condition of the first-level differential pair in the higher transconductance region as described previously. Secondly, we set the voltage level input to the first-level differential pair through two-stage level shift diodes, which are encircled in the figure. This ensures lower gate-drain capacitance (*Cgd*) of the first-level differential pair FETs. Lastly, we incorporated capacitive peaking into the source followers of the D-FF, which compensates for loss at high frequencies.

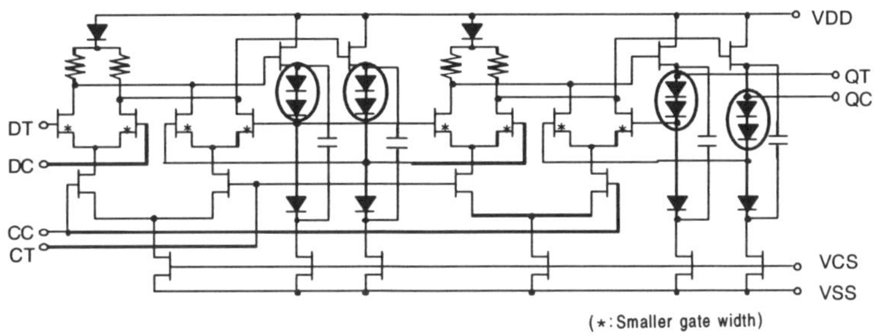

Fig. 17. Circuit diagram of the master-slave D-FF.

5. Experimental Results for Basic Communication ICs

All of the ICs were fabricated with 0.1-μm gate-length InAlAs/InGaAs/InP HEMTs.[66] After on-wafer testing, the ICs were mounted on our in-house metal package. The design concept and detailed structure of the package are described in *Ref. 67* and *Ref. 68.* In the following section, we describe the measurement setup and the circuit performance of the packaged IC modules.

5-A Measurement Setup

Waveform observation on digitizing sampling oscilloscope is necessary for characterizing circuit operation, but is not sufficient because the 40-Gbit/s NRZ signal has frequency components that are outside the bandwidth of available 50-GHz digitizing sampling oscilloscopes. The measurement of bit-error performance using the pseudo-random-bit stream (PRBS) signal is indispensable for confirming logic operation of the optical communication ICs, and is especially important for the measurement of the phase margin of the D-FF ICs and DEMUX ICs.

To cope with this, we construct the measurement setup shown in Fig. 18. This is an example setup for the device under test (DUT) of a D-FF IC. The data signal source so-called pulse pattern generator (PPG) configured with a 4-channel PPG (Anritsu: MP1778A), 4:2 multiplexer (MUX) unit (Anritsu: MP1779A), and our in-house MUX unit that contains an InP HEMT selector IC.[62]. The unit can output a complementary signal.

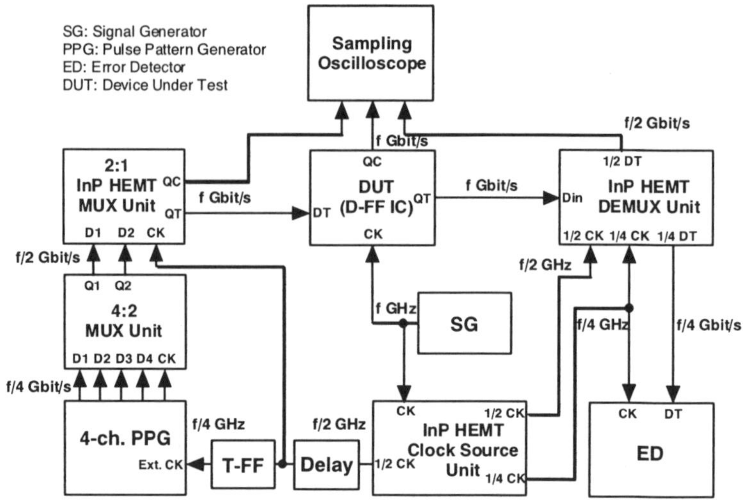

Fig. 18. Measurement setup for the D-FF IC.

The 4-channel PPG outputs four parallel PRBS signals while maintaining the delay of the signals relative to each other to one quarter of the data period. So, the four-times bit-multiplexed signal provides a true PRBS at the bit rate of the multiplexed signal. The delay line (Delay) allows us to set the phase between the multiplexed data signal and the clock signal input to the DUT, which is useful in the measurement of the phase margins of the D-FF IC.

The output signal of the DUT is demultiplexed two times by using an in-house DEMUX unit and then input to the error detector (ED). The DEMUX unit consists of an InP HEMT D-FF IC[60] and a Si-bipolar D-FF IC. The bit-error rate can be measured for all four original data signals by changing the clock phases input to the DEMUX unit. The maximum available data rate is 50 Gbit/s, which is limited by that of the 4-channel PPG, 4:2 MUX unit and ED.

All the measurement instruments are synchronized with a single signal generator (SG). Half-rate and quarter-rate clock signals, which are needed for each measurement instrument, are generated from the full rate clock signal of the SG. This clock distribution scheme is suitable for reducing unexpected system jitter. To generate lower clock signals, we constructed a clock source unit, which includes an InP HEMT T-FF IC[69] and a GaAs MESFET T-FF IC.[64]

5-B Circuit performance of the basic packaged digital IC

5-B-1 MUX IC

The MUX IC module was tested at 70 Gbit/s. Since the bit-rate of the input data signal to the IC module is 35 Gbit/s, the fundamental bit-rate of the 4-ch. PPG was set to 8.75 Gbit/s. The 2^{23}-1 PRBS signal was used for the fundamental data signal. A 35-GHz clock signal and the complementary data signals of 35 Gbit/s were appropriately delayed relative to each other before being input to the IC module. The output signals were monitored by a

50-GHz bandwidth digitizing scope connected by a 50-cm-long 50-GHz bandwidth coaxial cable.

Figure 19 shows operating waveforms of the MUX IC module at 70 Gbit/s.[70] Although the insufficient bandwidth of the measurement instruments degraded the transition speed of the waveform measurements, eye openings were still clear with 810 mVp-p output voltage swing. The power dissipation of the MUX module was 2.1 W.

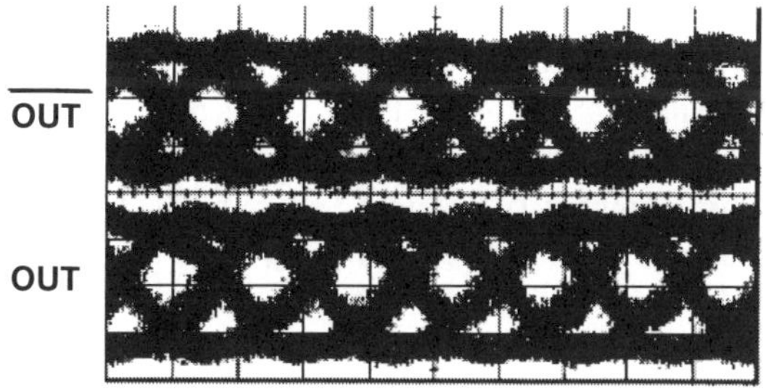

250 mV/div., 10 ps/div.
Fig. 19. Operating waveforms of the multiplexer IC module at 70 Gbit/s.

5-B-2 Dynamic D-FF IC

The dynamic D-FF IC module was tested at 50 Gbit/s using the measurement setup shown in Fig. 18. The bit rate of the 4-ch. PPG was set to 12.5 Gbit/s, which results in the final bit rate of 50 Gbit/s. Both the 50-Gbit/s data signal and the 50-GHz clock signal were input to the IC module.

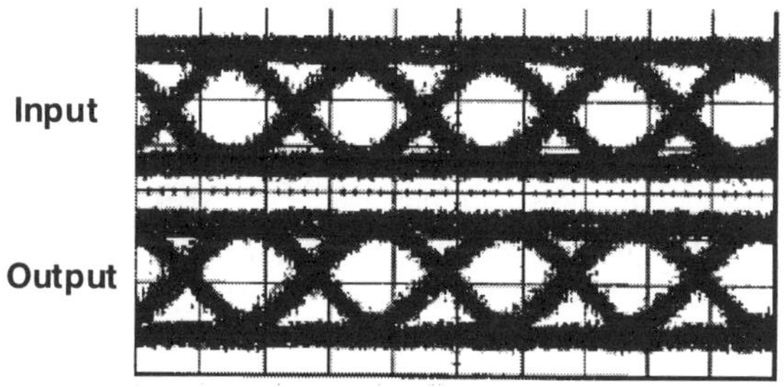

250 mV/div., 10 ps/div.
Fig. 20. Operating waveforms of the dynamic D-FF IC module at 50 Gbit/s.

Figure 20 shows operating waveforms of the IC module at 50 Gbit/s when error-free operation was confirmed for 2^{23}-1 PRBS input.[70] Good eye openings were obtained with 760-mVp-p output voltage swing. The retiming capability of the IC module was also confirmed by shifting the timing of the input data signal while maintaining the clock timing. The phase margin of 101 degrees was obtained at 50 Gbit/s. The power dissipation of the D-FF IC module was 1.7 W.

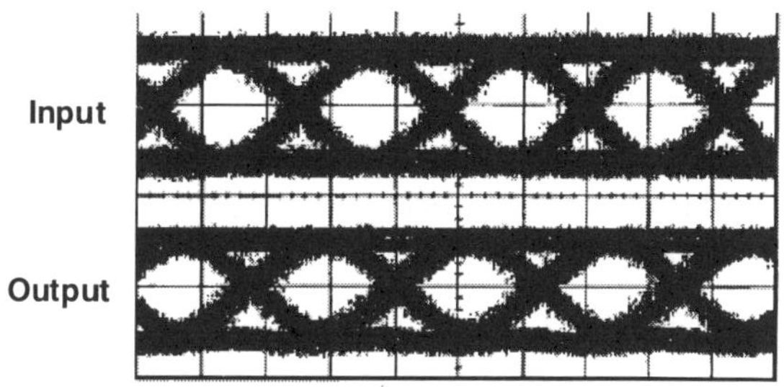

250 mV/div., 10 ps/div.
Fig. 21. Operating waveforms of the static D-FF IC module at 45 Gbit/s.

5-B-3 Static D-FF IC

The static D-FF IC module was tested at up to 45 Gbit/s. Figure 21 shows the operating waveforms of the IC module at the maximum operating bit-rate of 45 Gbit/s.[71] Good eye openings were obtained with 690 mVp-p output voltage swing. Figure 22 shows the measured phase margin and input sensitivity of the module. Error-free operation of the module was confirmed over a wide operating range from 1 Gbit/s to 45 Gbit/s for the 2^{23}-1 PRBS data input. A phase margin of 101 degrees and input sensitivity of 300 mV were obtained at 40 Gbit/s.

Error-free demultiplexing operation from 45 Gbit/s into 22.5 Gbit/s was also confirmed. In the demultiplexing operation, the demultiplexed 22.5-Gbit/s signal exhibited good eye-opening and low and high levels remained constant, as shown in Fig. 23. The phase margin of 158 degrees was obtained in the demultiplexing operation from 40 Gbit/s into 20 Gbit/s. The power dissipation of the IC module was 1.7 W.

These results indicate that the static D-FF core can be used as the core circuit in DEMUX ICs. A 1:2 DEMUX IC based on the circuit design techniques was reported in *Ref. 72*. The packaged IC exhibited error-free operation at up to 50 Gbit/s for 2^{23}-1 PRBS input. The measured phase margin was 170 degrees at 43 Gbit/s.

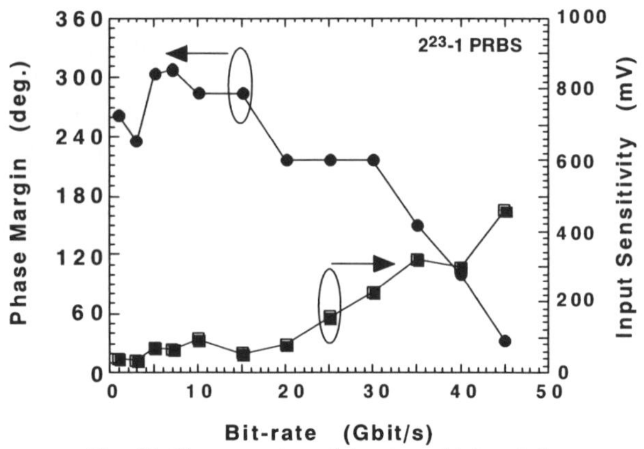

Fig. 22. Phase margin and input sensitivity of the static D-FF IC module.

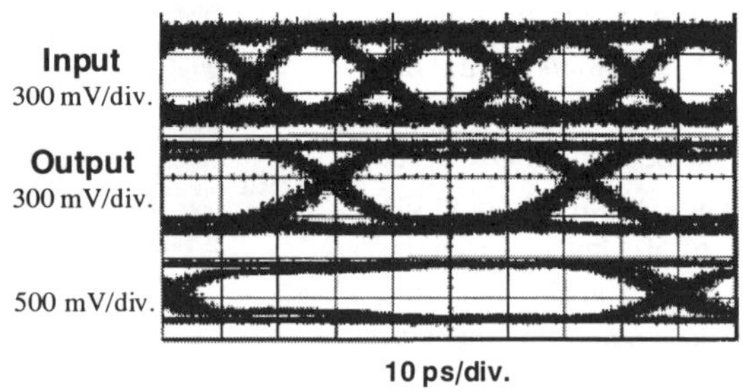

Upper: Input (45 Gbit/s)
Middle: Demultiplexed output (22.5 Gbit/s)
Lower: Demultiplexed data signal input to the error detecotor (11.25-Gbit/s)

Fig. 23. Operating waveforms of the static D-FF IC module in demultiplexing operation.

6. Highly functional InP HEMT IC

Table 1 lists the performance of the basic packaged digital ICs. All of them operate at up to 50 Gbit/s and have an output voltage swing of over 800 mVp-p. These packaged ICs were used for various 40-Gbit/s/channel WDM transmission experiments in the laboratory[73,74] and/or in field experiments.[75] The InP HEMT ICs are promising devices for building 43-Gbit/s OTU-3 optical

Table 1. Performances of InP-HEMT basic digital ICs.

Circuit	Operating Speed	Output	Power
2:1 Multiplexer	70 Gbit/s	1.0 Vpp	2.1 W
1:2 Demultiplexer	1 - 50 Gbit/s	0.8 Vpp	2.8 W
Decision	20 - 50 Gbit/s	0.9 Vpp	1.7 W
Frequency divider	8 - 50 GHz	0.8 Vpp	1.1 W

communication systems because of their large speed margins, large output voltage swings, and the excellent quality of their output waveforms. For the development of cost-effective optical transmitters and receivers, their integration density and functionality must be improved. A single-chip clock and data recovery circuit (CDR) is a key component that greatly reduces the size and power consumption of the transmitter and receiver. Figure 24 shows a tentative goal for the configuration of a transmitter and receiver made with InP HEMT technology. The bit rate of the lower-speed interface of the InP HEMT IC is 10.75 Gbit/s, because mature GaAs, Si, or SiGe IC technologies can cover that rate. In this configuration, the single-chip CDR is used for the clock multiplier in the transmitter and for the retiming and regeneration functions in the receiver. Up to now, 40-Gbit/s-class monolithic integrated

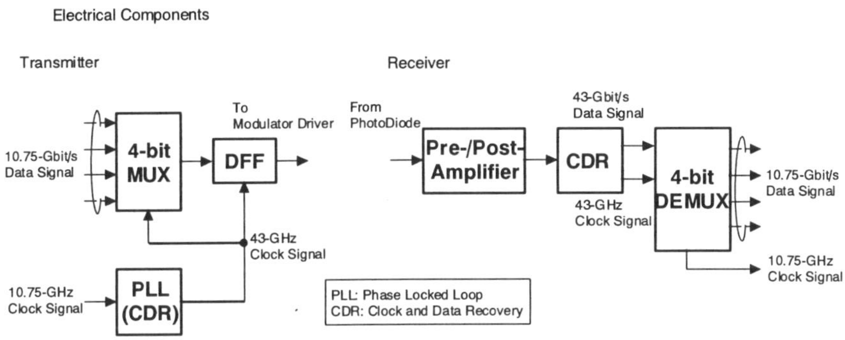

Fig. 24. Tentative goal for configuration of optical transmitter and receiver made with InP HEMT technology.

CDRs have been demonstrated by using Si-Ge bipolar[76] and GaAs HEMT[77] technologies. These CDRs use a circuit architecture that regenerates the input data signal with the recovered half-rate clock signal because of the difficulty in achieving high-speed circuit operation. For robust transmission systems, it's highly desirable that the CDR operates with the full-rate 43-GHz clock signal.

6-A Circuit configuration of the CDR

The circuit block diagram of the CDR is shown in Fig. 25.[78] The circuit architecture of the CDR is consistent with that in *Ref. 79*. The circuit is based on an analog PLL that consists of a D-type flip-flop (D-FF), a phase comparator, a 90-degree delay, a voltage-controlled oscillator (VCO), and a low pass filter (LPF). Incorporating the reference 90-degree delay into a conventional PLL enables both the recovered clock frequency and the phase timing to be adjusted automatically for the input data. In addition, a phase comparator circuit configuration that combines two multiplier circuits improves the tolerance for the data signal mark ratio variation.[79, 80]. The circuit was designed in the source-coupled FET logic, and the internal circuit operation is differential type. The CDR has a single-ended data input, a single-ended clock signal output, and a differential data output. The D-FF regenerates the input data signal using the output of the VCO.

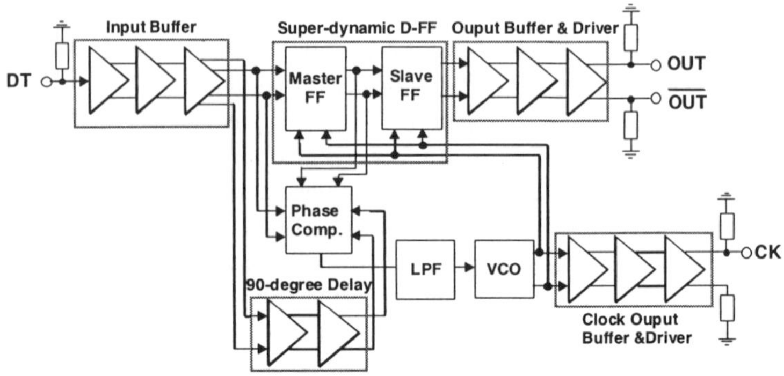

Fig. 25. Circuit block diagram of the CDR.

Except for the phase comparator, the key components in the CDR have different circuit configurations from those in *Ref. 79*. A super-dynamic D-FF[64] was used as a core D-FF to achieve 43-Gbit/s operation with a sufficient speed margin. The 90-degree delay circuit is composed of two-stage differential amplifiers. The simulated gate-delay time is around 12 ps, which corresponds to the half-bit of the 43-Gbit/s NRZ input data signal. In order to construct a compact, low-noise oscillator, a differential amplifier with tuned tank circuits was adopted as the VCO[81]. The VCO circuit configuration is shown in Fig. 26. The inductors are meander-type and formed with the second metal layer. Reverse-biased drain-source-shortened FETs were used for the varactors. The control voltage is input to the common gate of the FETs. A lag-lead-type circuit consisting of an on-chip MIM capacitor and two on-chip metal resistors was used as the LPF.

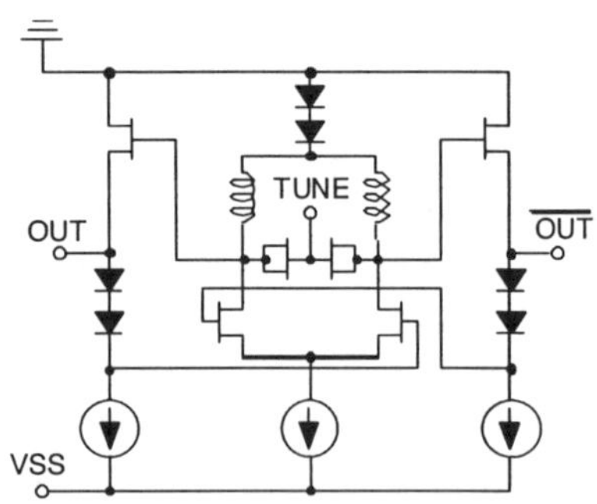

Fig. 26. Circuit configuration of the VCO.

6-B Experimental Results of the CDR

The CDR was fabricated with 0.1-μm gate-length InAlAs/InGaAs/InP HEMTs.[66] The averaged threshold voltage was -543 mV with a standard deviation of 27 mV in a 3-inch wafer. The averaged transconductance was 1.23 S/mm, and the current gain cut-off frequency was 173 GHz. A photograph of the fabricated IC is shown in Fig. 27. The chip size is 3 mm x 2 mm.

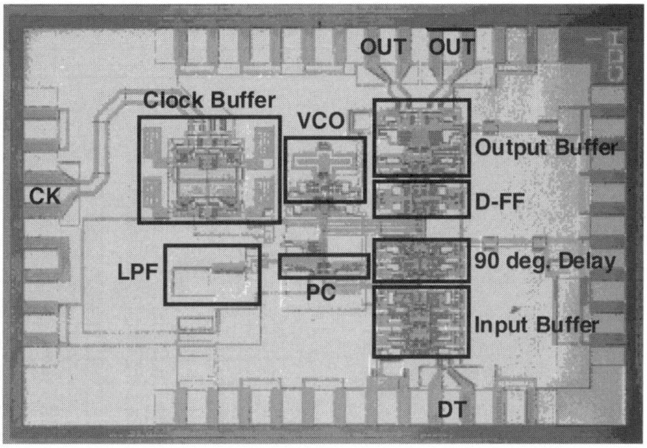

Fig. 27. Chip photograph of the CDR IC.

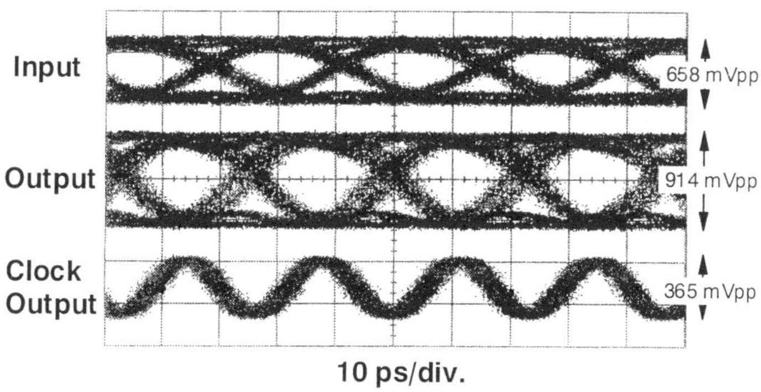

Fig. 28. Operating waveforms of the CDR IC at 43.0184 Gbit/s.

The IC was tested in on-wafer measurements. The measurement setup for the CDR is the same as that for D-FF IC except that the clock signal input to the DUT was eliminated. The input data rate was set to 43.0184 Gbit/s. The recovered clock signal was monitored by a spectrum analyzer and/or a digitizing sampling oscilloscope synthesized with the PPG. Figure 28 shows the operating waveforms of the CDR at 43.0184 Gbit/s.[78] Clear eye opening of the recovered data signal was observed with a 914-mVp-p voltage swing. The recovered clock signal is also clear, and its output voltage swing was 365 mVp-p (inclusive of 4-dB loss in the measurement setup). Error-free operation was confirmed for the 2^{31}-1 PRBS data input. Figures 29 (a) and (b) show the spectrum and the single side-band phase noise of the recovered clock signal, respectively, when the error-free operation was confirmed. The recovered clock signal has a high signal-to-noise ratio of around 40 dB. The measured phase noise was -102 dBc/Hz at the offset frequency of 1 MHz. The RMS jitter of the clock signal measured by the oscilloscope was 1.2 ps. The power dissipation of the IC was 2.79 W.

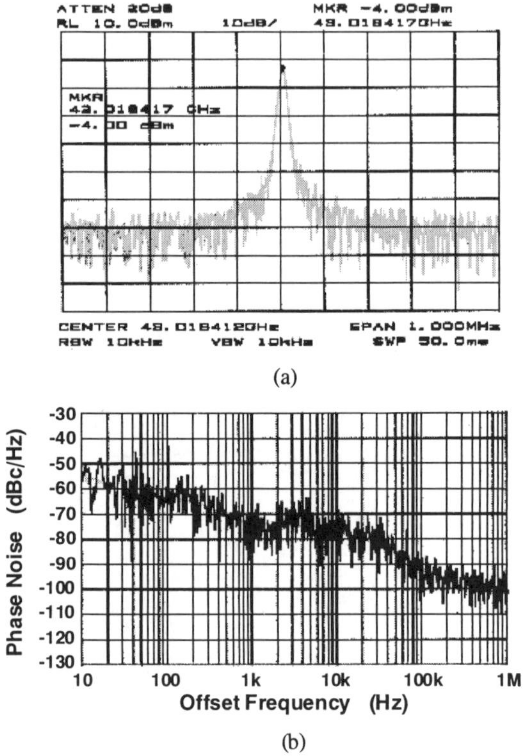

Fig. 29. Characteristics of the recovered clock signal.

(a) Spectrum of the recovered 43.0184 GHz clock signal.

(b) Single side-band phase noise of the recovered clock signal.

7. The InP HEMT IC structure

InP-HEMT ICs are formed on a 3-inch semi-insulating InP-substrate. MOCVD technology is used for crystal growth of HEMT and diode layers. The vertical Schottky diode layers are grown on the HEMT epitaxial layers. An InP etch-stop layer divides these layers. Figure 30 shows the cross section of an InP HEMT IC and Fig. 31 is a SEM photograph of a fabricated IC. The InP HEMT technology used here has two levels of interconnect. The first is made of evaporated Au, and the second is selectively grown by electroplating technology. The inter-layer film is 2-μm-thick benzocyclobutene (BCB). These interconnection layers form an inverted micro-strip line (IMSL) for matching line impedance and the impedance of the SCFL cell.[82] Additionally, this IC fabrication technology can integrate the capacitor, resistor, and the vertical Schottky diode. Table 2 summarizes the components integrated on the InP HEMT IC chip and their specifications.

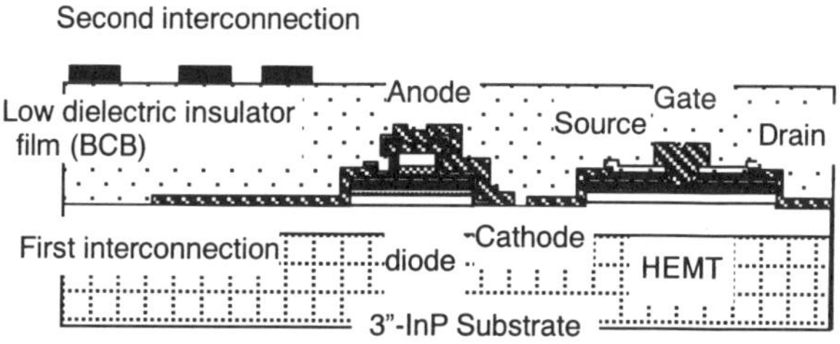

Fig. 30. Cross section of the InP-HEMT IC.

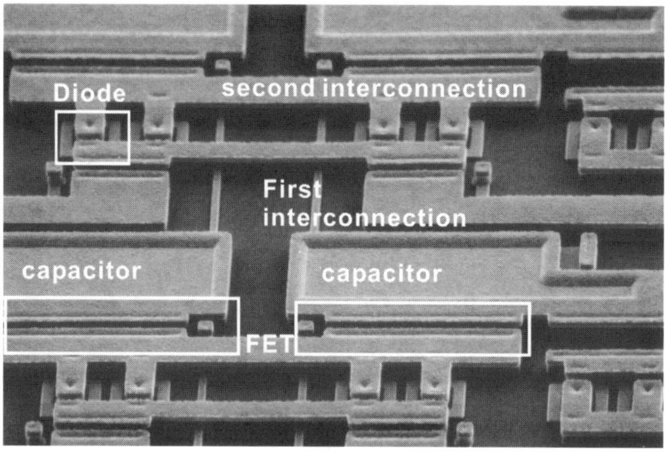

Fig. 31. InP-HEMT IC overview.

The vertical diode is used as a level-shift diode. Although the integration vertical Schottky diode complicates the epi-structure and process, it has lower series resistance than a lateral diode. The source resistance of a vertical diode is 20 Ω on average in a 3-inch wafer, whereas the FET source resistance is 50-60 Ω /10 μm. And the scattering of Rs is as low as below 0.2 Ω.

The 200-nm-thick p-CVD SiN film is used as a capacitor. The SiN film covers all the entire wafer surface including all FETs and all diodes, acting as passivation film. The resistor is made of sputtered WSiN. The sheet resistance is 150 Ω and the dispersion in 3-inch wafer is 2.2 Ω.

8. FET fabrication technology

This session describes the FET characteristics and reliability. FET performance is summarized in Table 2. Peak gm, f_T and f_{max} at the best Vg and 1-V Vd are 1 S/mm, 180 GHz and 300 GHz, on average for 42 FETs on a 3-inch wafer. The Vth is designed to be -500 mV, the standard deviation is below 20-mV. This Vth dispersion is low enough to fabricate SSIs, such as T-FF and D-FF. Figure 32 is the cross section of a 0.1-μm-gate-length HEMT. The 0.1-μm gate with the Au over-layer is formed in the 0.3-μm-recess region

Table 2. Components and its specifications

Component	Outline	Characteristics
FET	Lg:0.1μm, EB-lithography	Vth = -0.5 V, σVth = 20 mV on 3" wafer, g_m = 1.0 S/mm, f_T = 180 GHz, f_{max} = 300 GHz
Diode	Vertical type, 5x5 μm	Rs = 20ohm, Vbi = 0.55 V
resistor	sputterd WSiN	150 ohm/sq., σRsh = 2.2 ohm
capacitor	p-CVD 200nm SiN	0.3 $fF/\mu m^2$
1st-layer	evaporated 0.5 μm	30 ohm/μm
2nd-layer	electroplated 2 μm	7.5 ohm/μm
inter layer film	2.0 μm thick BCB	ε_r = 2.8
contact hole	2x2 μm	< 0.1 ohm

successfully. The source and drain electrodes are seen at either side. The fabrication process is outlined in *Ref. 54*. The gate patterning is performed by EB lithography. The other patterns are formed with an i-line stepper. There are two merits in applying of EB lithography to gate patterning. The fist is, of course, the capability of the fine pattern delineation, and the second is accuracy in pattern size. Accurate pattern size, particularly in gate formation, is required for reduction of FET characteristic dispersion. The yield and the performance of digital ICs suffer from the FET characteristics dispersion easily.

Fig. 32. TEM photograph of the 0.1-μm-gate-length InP-HEMT.

Figure 33 shows the EB-written gate pattern size distribution along with that for an i-line stepper. The EB-written pattern has 0.107-μm length on average and the standard deviation (std) is only 0.0029 μm, while the average size of the pattern formed with the i-line stepper is 0.357 μm with 0.0112-μm std, which is 3 times larger.

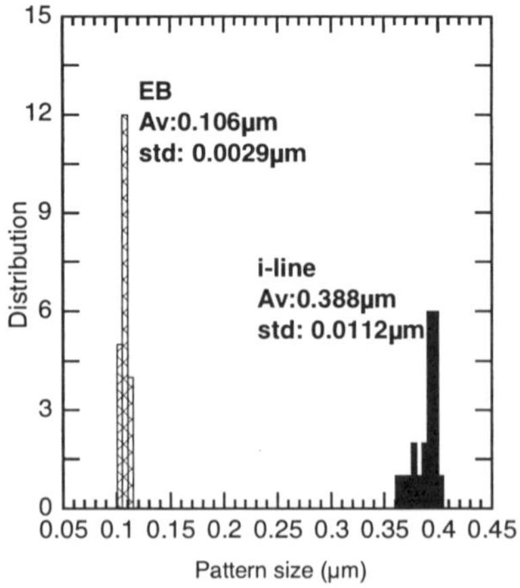

Fig. 33. Pattern size distribution of EB and i-line stepper.

Figure 34 shows FET Lg (gate length) dependence of Vth (threshold voltage). Close circles are the measured values and the line indicates the result of two-dimensional simulation. There are good coincidences between the two data, particularly in the region of over 0.1 μm. From these data, the effect of the gate length distribution on Vth can be estimated. The slope at 0.1 μm Lg is 1.33 V/mm. The above-mentioned gate pattern size scattering of 0.0029 μm results in the 4.0 mV Vth std. As presented above, the actual Vth distribution is about 20 mV. The main cause of the Vth distribution is therefore thought to be inaccuracy in the vertical structure. In other words, the distribution due to the recess process or the scattering of the doping concentration still causes Vth scattering even though this HEMT has an InP etch stop layer.[83] The accomplished value of 20 mV as the Vth std is enough for SSI or MSI fabrication, as shown later.

Since the latter part of the 90's, the thermal-stability or reliability-related issues have become as important as high-speed circuit operation. In the later part of this chapter, some results on the reliability of InP-HEMTs are described.

References 84-86 pointed out that the donor passivation takes place in only lattice-matched n-InAlAs. Here, we discuss the effect of substitutiting InAlP for the InAlAs carrier supply layer on the FET reliability. As InAlP is the strained material to an InP substrate, the thickness of the strained layer is limited to 2 nm. Additionally, the Al content is 0.25, which results in 0 eV ΔEc between InAlAs and InAlP. The change of carrier supply layer doesn't affect the initial FET characteristics. Typical FET degradation is shown in Fig. 35. The solid line indicates the initial characteristics and the dotted line those after a 1000-h high-temperature bias stress test. The stress is Vds = 1.0 V and Vgs = 0.0 V. The test results show several important points:

1. Vth shift is very small.

2. The g_m and I_{ds} decrease is more serious in the forward biased region.

3. The slope at low drain bias decreases, which is due to drain resistance increase.

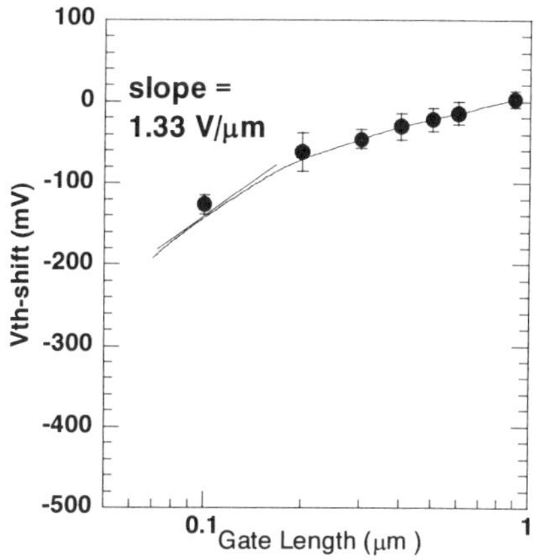

Fig. 34. Vth dependence on Lg.

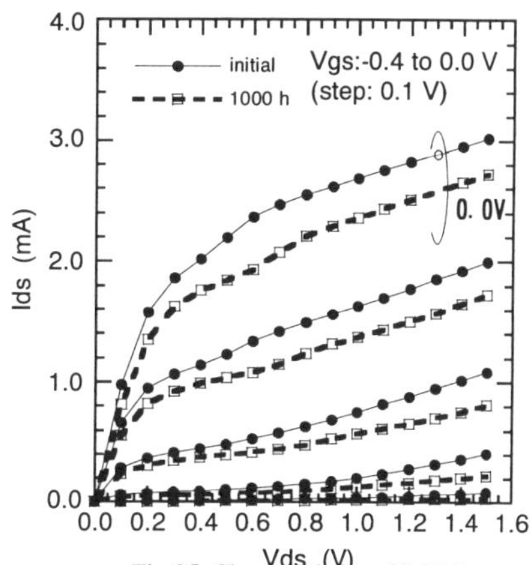

Fig 35. Characteristics of initial and bias-stressed FET.

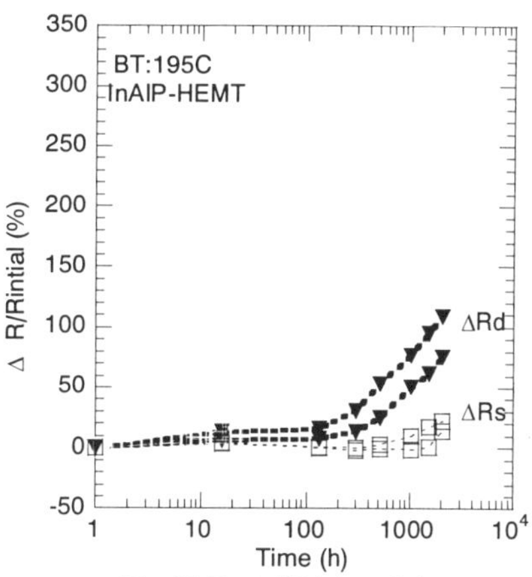

Fig. 36. Rs and Rd degradation.

These phenomena are also reported in *Ref. 87* and *Ref. 88*. This type of degradation is considered to be very common in InP HEMTs.

The drain resistance (Rd) increase is shown in Fig. 36 in comparison with the source resistance (Rs) increase. Both parasitic resistances increase with storage time. In this case, the HEMTs were biased during high-temperature storage. The bias condition was Vds = 1.0 V and Vgs = 0.0 V. The storage temperature was 195°C. The drain resistance increase is much larger than the source resistance increase. This asymmetric change means the electron current is not the main origin of the degradation in the parasitic resistance.

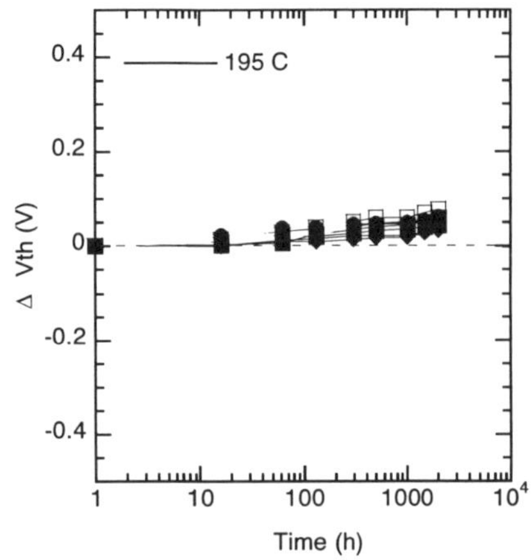

Fig. 37. Vth shift in BT test.

The Vth shift has been reported very often. The Vth shift at 195°C is shown in Fig. 37. The bias stress condition in Fig. 37 is the same as that in Fig. 36. The Vth of all HEMTs goes in the + direction. This is consistent with the assumption that the ionized dopants in the channel are neutralized by impurities or defects. The median life is over 1000 h at 195°C, under the failure criterion of a 50-mV Vth shift.

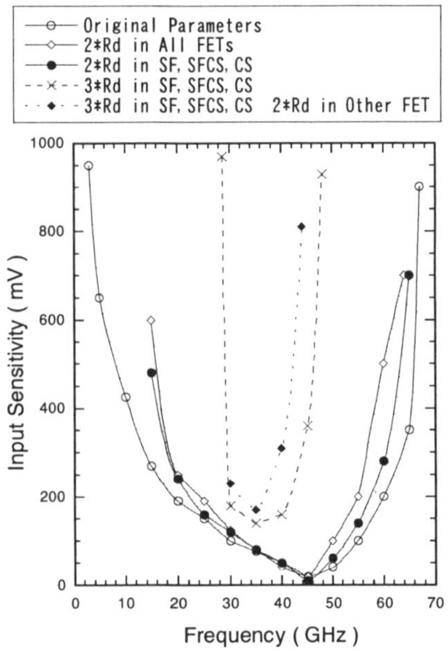

Fig. 38. Sensitivity degradation caused by Rd increase.

Thus, Rd increase is a notable feature in InP HEMT degradation. As the Rd increase dose not have a large effect on Ids or gm, its effect on circuit operation is not known. The failure criterion for Rd increase is necessary.

We checked the operation of SCFL-type T-FF circuits under several conditions. The FETs in SCFL are classified into four types by bias condition: differential-pair FETs (DCFET), current-source FET of differential-pair FETs (CSFET), source-follower FET (SFFET), and current-source FET of source follower FET (SFCSFET). The SFCS-FET is biased most severely. The input sensitivity of T-FF circuit was simulated by adding linear resistance to Rd. The simulation results are shown in Fig. 38. This graph shows the input sensitivity dependence of T-FF on the operating frequency under various Rd conditions. At first, the T-FF composed of SF-, CS-, or SFCS-FETs with doubled Rd and the one composed of all FETs with doubled Rd do not exhibit serious degradation compared to the T-FF composed of the original FET. On the other hand, a T-FF composed of the SF-, CS-, or SFCS-FETs with tripled Rd shows narrower sensitivity characteristics than the initial T-FF. Thus, it is reasonable that the failure criterion for Rd should be a 100% increase. The temperature dependence of t_{50} is shown in Fig. 39. The t_{50} is the time it takes for 50% of the FETs to break down. The x-axis is the inverse of the storage temperature. The self-heating effect is not taken into account. The activation energy of Rs and Ids is as large as 1.9 eV. The lifetime at 100°C is determined by the Rd degradation, as the Rd activation energy is as low as 1.35 eV. Additionally, in the high temperature storage test with no bias, lifetime was very long, as shown in Fig. 39. HEMTs with InAlP as a carrier-supply layer have t_{50} of 10^7 hours at 100°C as shown in Fig. 40. An InAlP CSL HEMT has 100 times longer lifetime than an InAlAs CSL FET. This result points to the excellent potential of utilizing InP-HEMT ICs.

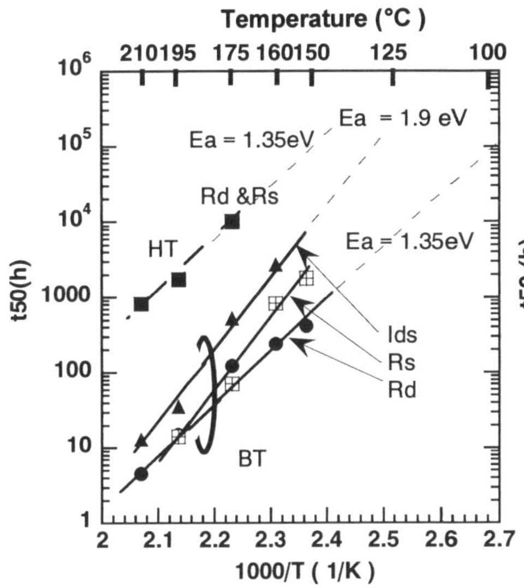

Fig. 39. Activation energy of various parameters.

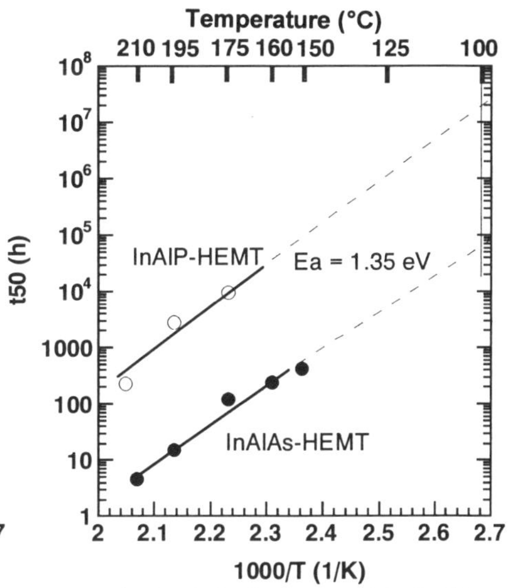

Fig. 40. Activation energy of the InAlP CSL and InAlAs-CSL FET.

9. InP-HEMT IC characteristics

Table 1 lists the developed digital ICs for 40-Gbit/s optical communications. Figure 41 is the photograph of the HLO-type T-FF. This T-FF is used as a process TEG for checking FET performance or passive component yield. It operates from 8 to 50 GHz with over 90 % yield. Figure 42 shows the yields of various scale circuits. The number of devices includes FETs and diodes. Practical yield is expected for an IC

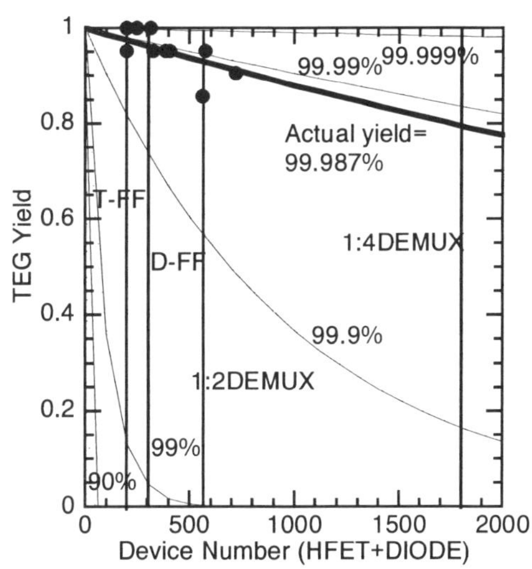

Fig. 41. 40 Gbit/s IC process.

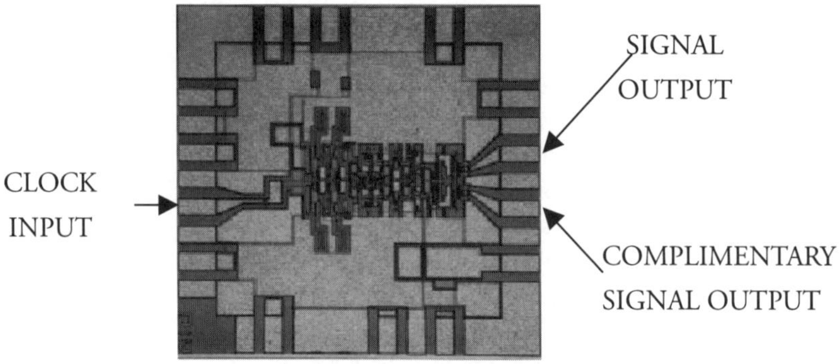

Fig. 42. An HLO Type T-FF.

Operating frequency is 8-50 GHz, output Vpp = 0.9 V.

composed of 1800 devices. This is equivalent to the number of devices for 1:16 DEMUX. The 40G SSI and MSI fabrication process is already matured enough from the standpoint of IC performance and fabrication yield. The next steps are

1. Verification of IC reliability and clarification of the relation between IC and FET reliability.
2. Achievement of much higher speeds toward 100 Gbit/s by shortening gate length.
3. Construction of OEICs by integration with PDs.

10. Conclusions

We described high-speed digital IC circuit design based on InP HEMT technology for 40-Gbit/s/channel optical communication systems. We have discussed in detail the basic circuit design techniques for the SCFL circuit and fundamental circuit elements of the selector and D-type flip-flop. The basic digital ICs, the MUX, D-FF, and DEMUX ICs, fabricated with 0.1-μm gate InP HEMTs successfully operate at up to 50 Gbit/s as packaged modules. These IC modules offer large speed margins for the 43-Gbit/s OTU-3 data rate. With respect to the development of cost-effective optical transmitters and receivers, we have designed a PLL-based CDR with a full-rate architecture. The fully monolithic integrated CDR exhibits error-free operation for 2^{31}-1 PRBS data signal at the OTU-3 bit rate of 43.0184 Gbit/s. Four-bit MUX and DEMUX ICs are other key components, and could be implemented by using InP HEMT technology.

Additionally, we presented InP IC fabrication technology with two-level interconnection. This technology is already fully matured for 40-Gbit/s SSI fabrication. The uniform FET characteristics and high yield passive component fabrication technologies support this degree of maturity. InP-HEMT lifetime reaches 10^7 hours at 100°C, which proves the InP-HEMT IC fabrication technology presented here is highly reliable.

InP-HEMT IC technology is a promising way to construct 40-Gbit/s/channel optical transport network systems.

Acknowledgement

The authors thank Masahiro Muraguchi, Hiromu Toba, Yasunobu Ishii, Eiichi Sano, Takatomo Enoki, and Takashi Kobayashi for their direction and encouragement throughout this work, and Taiichi Otsuji, Kimikazu Sano, Suehiro Sugitani, Hiroto Kitabayashi, Haruki Yokoyama, Mikio Yoneyama, and Yutaka Miyamoto for their fruitful discussions. They also thank Takashi Maruyama, Takashi Kusumoto for their support in FET and IC fabrication.

References

1. Y. Miyamoto, K. Yonenaga, S. Kuwahara, M. Tomizawa, A. Hirano, K. Murata, Y. Tada, Y. Umeda, and H. Miyazawa, "1.2 Tbit/s (30 x 42.7-Gbit/s ETDM optical channel) WDM transmission over 376 km with 125-km spacing using forward error correction and carrier-suppressed RZ format," in OFC'2000 PD-26, Feb., 2000.
2. K. Fukuchi, T. Kasamatsu, M. Morie, R. Ohhira, T. Ito, K. Sekiya, D. Ogasahara, and T. Ono, "10.92-Tb/s (273 x 40-Gb/s) triple-band/ultra-dense WDM optical-repeatered transmission experiment," in OFC2001, PD24-1, 2001.
3. Tohru Oka, Koji Hirata, Kiyoshi Ouchi, Hiroyuki Uchiyama, Takafumi Taniguchi, Kazuhiro Mochizuki, Tohru Nakamura, "Advanced Performance of Small-Scaled InGaP/GaAs HBT's with f_T over 150 GHz f_{max} over 250 GHz," in IEDM Tech. Dig., 1998, pp. 653-656.
4. Tohru Oka, Koji Hirata, Kiyoshi Ouchi, Hiroyuki Uchiyama, Kazuhiro Mochizuki, Tohru Nakamura, "InGaP/GaAs HBT's with High-Speed and Low-Current Operation Fabricated Using WSi/Ti as the Base Electrode and Burying SiO_2 in the Extrinsic Collector," in IEDM Tech. Dig., 1997, pp. 739-732.
5. H. Masuda, K. Ouchi, A. Terano, H. Suzuki, K. Watanabe, T. Oka, H. Matsubara, T.Tanoue, "Device Technology of InP/InGaAs HBTs for 40-Gb/s Optical Transmission Application," in GaAs IC Symp. Tech. Dig., 1997, pp. 139-142.
6. K. Runge, P. J. Zampardi, R. L. Pierson, P. B. Thomas, S. M. Beccue, R. Yu, K. C. Wang, "High Speed AlGaAs/GaAs HBT Circuits For Up To 40-Gb/s Optical Communication," in GaAs IC Symp. Tech. Dig., 1997, pp. 211-214.
7. Yasushi Amamiya, Takaki Niwa, Noboru Nagano, Masayuki Mamada, Yasuyuki Suzuki, Hidenori Shimawaki, "40-GHz Frequency Dividers with reduced Power Dissipation fabricated using High-speed small-Emitter-Area AlGaAs/InGaAs HBTs," in GaAs IC Symp. Tech. Dig., 1998, pp. 121-124.
8. T. Mathew, H.-J. Kim, D. Scott, S. Jaganathan, S. Krishnan, Y. Wei, M. Urteaga, S. Long, M. Rodwell, "75 GHz ECL Static Frequency Divider using InAlAs/InGaAs HBTs," Electron. Lett., vol. 37, no. 11, pp. 667-668, 2001.
9. Hiroki Nakajima, Eiichi Sano, Minoru Ida, Shoji Yamahata, "80-GHz 4:1 Frequency Divider IC using non-self-aligned InP/InGaAs Heterostructure Bipolar Transistor," Electron. Lett., vol. 36, no. 1, pp. 34-35, 2000.
10. Eiichi Sano, Hiroki Nakajima, N. Watanabe, Shoji Yamahata, Y. Ishii, "40-Gbit/s 1:4 demultiplexer IC using InP-based heterojunction bipolar transistors," Electron. Lett., vol. 35, no. 24, pp. 2116-2117, 1999.
11. Eiichi Sano, Hiroki Nakajima, N. Watanabe, Shoji Yamahata, "40-Gbit/s decision IC using

InP/InGaAs Composite-Collector Heterojunction Bipolar Transistors," Electron. Lett., vol. 35, no. 14, pp. 1194-1195, 1999.

12. R. Pullela, D. Mensa, Q. Lee, B. Agarwal, J. Guthrie, S. Jaganathan, M. Rodwell, "48-GHz static Frequency Dividers in Transferred-Substrate HBT Technology," Electron. Lett., vol. 34, no. 16, pp. 1580-1581, 1998.
13. D. Mensa, R. Pullela, Q. Lee, J. Guthrie, S. C. Nartin, R. P.Sith, S. Jaganathan, T. Mathew, B. Agarwal, S. I. Long, M. Rodwell, "48-GHz Digital IC's and 85-GHz Baseband Amplifier Using Transferred-Substrate HBT's," IEEE J. Solid-State Circuits, vol. 34, no. 9, pp. 1197-1203, 1999.
14. H. Suzuki, K. Watanabe, K. Ishikawa, H. Masuda, K. Ouchi, T. Tanoue, R. Takeyari, "Very-High-Speed InP/InGaAs HBT IC's for Optical Transmission Systems," IEEE J. Solid-State Circuits, vol. 33, no. 9, pp. 1313-1320, 1998.
15. P. Andre, J-L Benchimol, P. Desrousseaux, A-M. Duchenois, J. Godin, A. Konczykowska, M. Meghelli, M. Riet, A. Scavennec, "InP DHBT Technology and Design Methodology for High -Bit-Rate Optical Communications Circuits," IEEE J. Solid-State Circuits, vol. 33, no. 9, pp. 1328-1335, 1998.
16. M. Mokhtari, T. Swahn, R. H. Walden, W. E. Stanchina, M. Kardos, T. Juhola, G. Schuppener, H. Tenhunen, T. Lewin, "InP-HBT Chip-set for 40-Gb/s Fiber Optical Communication Systems Operational at 3 V," IEEE J. Solid-State Circuits, vol. 32, no. 9, pp. 1371-1383, 1997.
17. Hiroki Nakajima, Tadao Ishibashi, Eiichi Sano, Minoru Ida, Shoji Yamahata, Yasunobu Ishii, "InP-Based High-Speed Electronics," in IEDM Tech. Dig., 1999, pp. 771-774.
18. A. Gutierrez-Aitken, E. Kaneshiro, B. Tang, J. Notthoff, P. Chin, D. Streit, A. Oki, "69 GHz Frequency Divider with Cantilevered Base InP DHBT," in IEDM Tech. Dig , 1999, pp. 779-782.
19. H. Suzuki, K. Watanabe, K. Ishikawa, H. Masuda, K. Ouchi, T. Tanoue, R. Takeyari, "InP/InGaAs HBT ICs for 40-Gbit/s Optical Transmission Systems," in GaAs IC Symp. Tech. Dig., 1997, pp. 215-218.
20. J. Godin, P. Andre, JL. Benchimol, P. Desrousseaux, AM. Duchenois, A. Konczykowska, P. Launay, M. Meghelli, M. Riet, "A InP DHBT technology for High Bit-rate Optical Communications Circuits," in GaAs IC Symp. Tech. Dig., 1997, pp. 219-222.
21. G. Georgiou, Y. Baeyens, Y. -K. Chen, C. Groepper, P. Pashke, R. Puellela, M. Reinhold, C. Dorschky, J. P. Mattia, T. W. von Mohrenfels, C. Schulien, "Clock and Data Recovery IC for 40 Gb/s Fiber-Optic Receiver", in GaAs IC Symp. Tech. Dig., 2001, pp. 93-96.
22. Y. Baeyens, G. Georgiou, J. Weiner, V. Houtsma, P. Paschke, Q. Lee, A. Leven, R. Kopf, J. Frackoviak, C. Chen, C. T. Liu, Y. -K. Chen, "InP-D-HBT IC's for 40-Gb/s and Higher Bit rate Light wave Transceivers," in GaAs IC Symp.Tech. Dig., 2001, pp. 125-128.
23. M. Rodwell, Q. Lee, D. Mensa, J. Guthrie, S. Jaganathan, T. Mathew, S. Long, "48-GHz Digital ICs using Transferred Substrate HBT," in GaAs IC Symp.Tech. Dig., 1998, pp. 113-117.
24. Q. Lee, D. Mensa, S. Jaganathan, T. Mathew, Y. Betser, S. Krishnan, S. Crean, M. Rodwell, "66-GHz Static Frequency Divider in Transferred substrate HBT Technology," in Proc. IEEE MTT-S radio Frequency Integrated Circuits Symposium, 1999, pp. 87-90.
25. J. Godin, P. Andre, JL Benchimol, P. Berdaguer, S. Blayac, JR. Burie, P. Desrousseaux,

AM. Duchenois, N. Kauffmann, A. Konczykowska, M. Riet, "40-Gbit/s Optical Communications: InP DHBT Technology, Circuits and System Experiments," in GaAs IC Symp. Tech. Dig., 1999, pp. 185-188.

26. J. Godin, M. Riet, S. Blayac, Ph. Berdaguer, JL. Benchimol, A. Konczykowska, A. Kasbari, P. Andre, N. Kauffmann, "Improved InGaAs/InP DHBT Technology for 40-Gbit/s Optical Communication Circuits," in GaAs IC Symp.Tech. Dig., 2000, pp. 77-80.

27. M. Sokolich, C. Fields, B. Shi, Y. K. Brown, M. Montes, R. Martinez, A. R. Kramer, S. Thomas III, M. Madhav, "A Low Power 72.8-GHz Static Frequency Divider Implemented in AlInAs/InGaAs HBT IC Technology," in GaAs IC Symp.Tech. Dig., 2000, pp. 81-84.

28. Satoshi Tsunashima, Hiroki Nakajima, Eiichi Sano, Minoru Ida, Kenji Kurishima, Noriyuki Watanabe, Takatomo Enoki, Hirohiko Sugahara, "90-GHz Operation of a Novel Dynamic Frequency Divider using InP/InGaAs HBTs," in Proc. of the 14th Int. Conf. On InP and Rel. Mat., 2002, pp. 43-46.

29. S. Blayac, M. Riet, JL. Benchimol, F. Alexandre, P. Berdaguer, M. Kahn, A. Pinquier, E. Dutisseil, J. Moulu, A .Kasbari, A. Konczykowska, J. Godin, "MSI InP/InGaAs DHBT technology: beyond 40 Gbit/s Circuits," in Proc. of the 14th Int. Conf. On InP and Rel. Mat. , 2002, pp. 51-54.

30. W. E. Stanchina, M. Sokolich, K. R. Elliott, "Process and Integration Technologies for InP ICs," in Proc. of the 13th Int. Conf. On InP and Rel. Mat., 2001, pp. 489-492.

31. Shoji Yamahata, Hiroki Nakajima, Minoru Ida, Hajime Niiyama, Noriyuki Watanabe, Eiichi Sano, Yasunobu Ishii, "Reliable carbon-doped InP/InGaAs HBTs technology for Low Power 40-GHz Static frequency Divider," in Extended Abstracts Int. Conf. Solid State Devices and Materials, 1999, pp. 570-571.

32. Toshihide Suzuki, Yasuhiro Nakasha, Tsuyishi Takanashi, Kouzou Makiyama, Kenji Imanishi, Tatsuya Hirose, Yuu Watanabe, "A 90 Gb/s 2:1 Multiplexer IC in InP-based HEMT Technology," in Dig. Tech. Papers IEEE Int. Solid-State Circuit Conf., 2002, pp. 150-151.

33. Yasuhiro Nakasha, Toshihide Suzuki, Hideki Kano, Akio Ohya, Ken Sawada, Kozo Makiyama, Tsuyishi Takanashi, Nasahiro Nishi, Tatsuya Hirose, Masahiko Takikawa, Yuu Watanabe, "A 43-Gb/s Full-Rate-Clock 4:1 Multiplexer in InP-based HEMT Technology," in Dig. Tech .Papers IEEE Int. Solid-State Circuit Conf., 2002, pp. 148-149.

34. Taiichi Otsuji, Koichi Murata, Takatomo Enoki, Yohtaro Umeda, "An 80-Gbit/s Multiplexer IC Using InAlAs/InGaAs/InP HEMTs," in GaAs IC Symp. Tech. Dig., 1997, pp. 183-186.

35. T. Takahashi, M. Nihei, K. Makiyama, M. Nishi, T. Suzuki, N. Hara, "Stable and Uniform InAlAs/InGaAs HEMT ICs for 40-Gbit/s Optical Communication Systems," in Proc. of the 13th Int. Conf. on InP and Rel. Mat., 2001, pp. 614-617.

36. J. Mullrich, W. Klein, R. Khlifi, H.-M. Rein, "SiGe Regenerative Frequency Divider Operating up to 63-GHz," Ellectron. Lett., vol. 35, no. 20, pp. 1730-1731, 1999.

37. A. Felder, M. Moeller, M. Wurzer, M. Rest, T. F. Meister, H. -M. Rein, "60-Gbit/s Regenerating Demultiplexer in SiGe Bipolar Technology," Ellectron. Lett., vol. 33, no. 23, pp. 1984-1985, 1997.

38. M. Moeller, H. -M. Rein, A. Felder, T. F. Meister, "60-Gbit/s Time-Division Multiplexer in SiGe-Bipolar Technology with Special Regard to Mounting and Measuring technique,"

Ellectron. Lett., vol. 33, no. 8, pp. 679-680, 1997.

39. Mounir Meghelli, Alexander V. Rylyskov, Lei Shan, "50-Gb/s SiGe BiCMOS 4:1 Multiplexer and 1:4 Demultiplexer for Serial-Communication Systems," in Dig. Tech .Papers IEEE Int. Solid-State Circuit Conf., 2002, pp. 210-211.
40. A. Joseph, D. Coolbaugh, D. Freeman, S. Subbanna, M. Doherty, J. Dunn, C. Dickey, D. Greenberg, R. Groves, M. Meghelli, A. Rylyakov, M. Sorna, O. Shreiber, D. Herman, "0.13-μm 210-GHz f_T SiGe HBTs-Expanding the Horizons of SiGe BiCMOS," in Dig. Tech .Papers IEEE Int. Solid-State Circuit Conf., 2002, pp. 138-139.
41. Mario Reinhold, Claus Dorschky, Rajasekhar Pullera, Eduard Rose, Peter Mayer, Peter Paschke, Yves Baeyens, John-Paul Mattia, Frank Kuntz, "A Fully-Integrated 40-Gb/s Clock and Data Recovery / 1:4 DEMUX IC in SiGe Technology," in Dig. Tech .Papers IEEE Int. Solid-State Circuit Conf., 2001, pp. 84-85.
42. Herbert Knapp, Thomas F. Meister, Martin Wurzer, Dietmar Zoeschg, Klaus Aufinger, Ludwig Treitinger, "A 79-GHz Dynamic Frequency Divider in SiGe Bipolar Technology," in Dig. Tech .Papers IEEE Int. Solid-State Circuit Conf., 2000, pp. 208-209.
43. Katsuyoshi Washio, Eiji Ohue, Katsuya Oda, Reiko Hayami, Masamichi Tanabe, Hiromi Shimamoto, Takahashi Harada, Masao Kondo, "82-GHz Dynamic Frequency Divider in 5.5-ps ECL SiGe HBTs," in Dig. Tech .Papers IEEE Int. Solid-State Circuit Conf., 2000, pp. 210-211.
44. Toru Masuda, Ken-ichi Ohhata, Fumihiko Arakawa, Nobuuhiro Shiramizu, Eiji Ohue, Katsuya Oda, Reiko Hayami, Masamichi Tanabe, Hiromi Shimamoto, Masao Kondo, Takahashi Harada, Katsuyoshi Washio, "45-GHz Transimpedance 32dB Limiting Amplifier and 40-Gb/s 1:4 High-Sensitivity Demultiplexer with Decision Circuit using SiGe HBTs for 40-Gb/s Optical Receiver," in Dig. Tech .Papers IEEE Int. Solid-State Circuit Conf., 2000, pp. 60-61.
45. M. Wurzer, T. F. Meister, H. Schaefer, H. Knapp, J. Boeck, R. Stengl, K. Aufinger, M. Franosch, M. Rest, M. Moeller, H. -M. Rein, A. Felder, "42-GHz Static Frequency Divider in a Si/SiGe Bipolar Technology," in Dig. Tech .Papers IEEE Int. Solid-State Circuit Conf. ,1997, pp. 122-123.
46. John-Paul Mattia, Rajasekhar Pullera, Yves Baeyens, Young-Kai Chen, Huan-Shang Tsai, George Georgiou, Timo Winkler von Mohrenfels, Mario Reinhold, Carsten Groepper, Claus Dorschky, Christof Schulien, "A 1:4 Demultiplexer for 40-Gb/s Fiber-Optic Application," in Dig. Tech .Papers IEEE Int. Solid-State Circuit Conf., 2000, pp. 64-65.
47. Martin Wurzer, Thomas F. Meister, Herbert Knapp, Klaus Aufinger, Renate Schreiter, Sabine Boguth, Ludwig Treitinger, "53 GHz Static Frequency Divider in a Si/SiGe Bipolar Technology," in Dig. Tech .Papers IEEE Int. Solid-State Circuit Conf., 2000, pp. 206-207.
48. Katsuyoshi Washio, Eiji Ohue, Katsuya Oda, Masamichi Tanabe, Hiromi Shimamoto, T. Onai, "96-GHz f_T Self-Aligned Selective epitaxial SiGe HBT with SMI Electrode," in Dig. Tech .Papers IEEE Int. Solid-State Circuit Conf., 1998, pp. 312-313.
49. Toru Masuda, Ken-ichi Ohhata, Eiji Ohue, Katsuya Oda, Masamichi Tanabe, Hiromi Shimamoto, T. Onai, Katsuyoshi Washio, "40-Gb/s Analog IC Chipset for Optical Receiver using SiGe HBTs," in Dig. Tech .Papers IEEE Int. Solid-State Circuit Conf, 1998, pp. 314-315.
50. G. Freeman, Y. Kwark, M. Meghelli, S. Zier, A. Rylyakov, M. Sorna, T. Tnji, O. Schreiver,

K. Walter, J. Rieh, B. Jagannathan, A. Joseph, S. Subbanna, "40-Gbit/sec Circuits Built from a 120-GHz f_T SiGe Technology," in GaAs IC Symp. Tech. Dig., 2001, pp. 89-92.

51. K. Washio, R. Hayami, E. Ohue, K. Oda, M. Tanabe, H. Shimamoto, M. Kondo, "67 GHz Static Frequency Divider using 0.2μm Self-aligned SiGe HBTs," in Proc. IEEE MTT-S Radio Frequency Integrated Circuits Symposium, 2000, pp. 31-34.
52. S. Kuwano, N. Takachio, K. Iwasita, T. Otsuji, Y. Imai, T. Enoki, K. Yoshino, K. Wakita, "160 Gbit/s(4-ch x 40-Gbit/s Electrical Multiplexed Data) WDM Transmission over 320-km Dispersion-shift-Fiber", OFC'96, post deadline paper PD25,1996.
53. YoshimiYamashita, Keisuke Shinohara, Kohki Hikosaka, Toshiaki Matui, Satoshi Hiyamizu, Takashi Mimura, "Pseudomorphic $In_{0.52}Al_{0.48}As/In_{0.7}Ga_{0.3}As$ HEMTs With an Ultrahigh f_T of 562 GHz," IEEE Eletron Dvice Lett., vol. 23, No. 10, pp. 573 - 575, 2002.
54. T.Enoki, H.Yokoyama, Y.Umeda, T.Otsuji, "Ultrahigh-speed Integrated Circuits using InP-Based HEMTs", Jpn.J.Appl.Phys., vol. 37, pp. 1359-1364, 1998.
55. " Interface for Optical Transport Network (OTN)," ITU-T recommendation G.709, 2001.
56. T. Otsuji, E. Sano, Y. Imai, and T. Enoki," 40-Gbit/s ICs for Future Light wave Communications Systems," IEEE J. of Solid-State Circuits, vol. 32, no. 9, pp. 1363 - 1370, 1997.
57. N. Ohta and T. Takada, " High speed GaAs SCFL monolithic integrated decision circuit for Gbit/s optical repeaters," IEE Electron. Lett., vol. 19, no. 23, pp. 983 - 985, 1983.
58. M. Togashi, M. Ohhata, K. Murata, H. Kikuchi, M. Ino, M. Suzuki, and Y. Yamane, " 10 Gb/s GaAs MESFET IC's for Ultra High-Speed Transmission Systems, " in IEEE GaAs IC Symp. Tech. Dig., 1990, pp. 49 - 52.
59. M. Vadipour, "Capacitive Feedback Technique for Wide-Band Amplifiers," IEEE J. Solid-State Circuits, vol. 28, no.1, pp. 90 - 92, 1993.
60. M. Yoneyama, A. Sano, K. Hagimoto, T. Otsuji, K. Murata, Y. Imai, S. Yamaguchi, T. Enoki, and E. Sano, "Optical Repeater Circuit Design Based on InAlAs/InGaAs HEMT Digital IC Technology," IEEE Trans. Microwave Theory Tech., vol. 45, no. 12, pp. 2274-2282, 1997.
61. T. Takada and M. Ohhata, "A new interfacing method 'SCFL-interfacing' for ultra high-speed logic IC's," in Tech. Dig. IEEE GaAs IC Symp., pp. 211 - 214, 1990.
62. T. Otsuji, K. Murata, T. Enoki, and Y. Umeda, "An 80 Gbit/s Multiplexer IC Using InAlAs/InGaAs/InP HEMTs," IEEE J. Solid-State Circuits, vol. 33, no. 9, pp. 1321 - 1327, 1998.
63. K. Murata, T. Otsuji, M. Ohhata, M. Togashi, E. Sano, and M. Suzuki, "A Novel High-speed Latching Operation Flip-Flop (HLO-FF) Circuit and Its Application to a 19 Gb/s Decision Circuit Using 0.2 μm GaAs MESFET," IEEE J. Solid-State Circuits, vol. 30, no. 10, pp.1101 - 1108, 1995.
64. T. Otsuji, M. Yoneyama, K. Murata, and E. Sano, "A Super-Dynamic Flip-Flop Circuit for Broad-Band Applications up to 24 Gb/s Utilizing Production-Level 0.2-μm GaAs MESFET's," IEEE J. Solid-State Circuits, vol. 32, no. 9, pp. 1357-1362, 1997.
65. K. Murata, T. Otsuji, M. Yoneyama, and M. Tokumitsu, "A 40-Gbit/s Super-Dynamic Decision IC Fabricated with 0.12-μm GaAs MESFET's," IEEE J. Solid-State Circuits, vol. 33, no. 10, pp. 1527-1535, 1998.
66. T. Enoki, H. Ito, K. Ikuta, and Y. Ishii, "0.1 μm InAlAs/InGaAs HEMT with an InP-

Recess-Etch Stopper Grown by MOCVD," Int'l Conference on Indium Phosphide and related Materials. Conf., pp. 81-88, 1995.

67. T. Shibata, S. Kimura, H. Kimura, Y. Imai, Y. Umeda, and Y. Akazawa, "A design technique for a 60 GHz bandwidth distributed baseband amplifier IC module," IEEE J. Solid-State Circuits, vol. 29, pp. 1537-1543, 1994.
68. S. Yamaguchi, Y. Imai, S. Kimura, and H. Tsunetsugu, "New module structure using flip-chip technology for high-speed optical communication IC's," in IEEE MTT-S Int. Microwave Symp. Tech. Dig., pp. 243-246, 1996.
69. T. Otsuji, M. Yoneyama, K. Murata, Y. Imai, T. Enoki and Y. Umeda," 2 – 46.5 GHz quasi-static 2:1 frequency divider IC using InAlAs/InGaAs/InP HEMTs," IEE Electron. Lett., vol. 33, no. 16, pp. 1376-1377, 1997.
70. K. Murata, T. Otsuji, E. Sano, S. Kimura and Y. Yamane," 70-Gbit/s multiplexer and 50-Gbit/s decision IC modules using InAlAs/InGaAs/InP HEMTs," IEICE Trans. on Electron., vol. E83-C, no. 7, pp. 1166–1169, 2000.
71. K. Murata, T. Otsuji and Y. Yamane, "45 Gbit/s decision IC using InAlAs/InGaAs/InP HEMTs," IEE Electron. Lett., vol. 35, no. 16, pp. 1379-1380, 1999.
72. K. Sano, K. Murata, and Y. Yamane, "50-Gbit/s Demultiplexer IC Module Using InAlAs/InGaAs/InP HEMTs," IEICE Trans. on Electron., vol. E83-C, no. 11, pp. 1788-1790, 2000.
73. Y. Miyamoto, A. Horano, K. Yonenaga, A. Sano, H. Toba, K. Murata and O. Mitomi, "320 Gbit/s (8 x 40 Gbit/s) WDM transmission over 367 km with 120 km repeater spacing using carrier-suppressed return-to-zero format," IEE Electron. Lett., vol. 35, no. 23, pp. 2041–2042, 1999.
74. Y. Miyamoto, K. Yonenaga, S. Kuwahara, M. Tomizawa, A. Horano, H. Toba, K. Murata and Y. Tada, "1.2 Tbit/s (30 x 42.7 Gbit/s ETDM channel) WDM transmission over 3 x 125 km with forward error correction," IEE Electron. Lett., vol. 36, no. 9, pp. 812–813, 2000.
75. Y. Yonenaga, Y. Miyamoto, A. Hirano, A. Sano, S. Kuwahara, H. Kawakami, H. Toba, K. Murata, M. Fukutoku, Y. Yamane, K. Noguchi, T. Ishibashi and K. Nakajima, "320 Gbit/s WDM field experiment using 40 Gbit/s ETDM channels over 176 km dispersion-shifted fiber with nonlinearity-tolerant signal format," IEE Electron. Lett., vol. 36, no. 2, pp. 153–155, 2000.
76. M. Reinhold, C. Dorschky, R. Pullela, E. Rose, P. Mayer, P. Paschke, Y. Baeyens, J.-P. Mattia, F. Kunz, "A Fully-Integrated 40 Gbit/s Clock and Data Recovery / 1:4 DEMUX IC in SiGe Technology, " in IEEE ISSCC Tech. Dig., pp. 84–85, 2001.
77. M. Lang, Z. Wang, A. Thiede, H. Lienhart, T. Jakobus, W. Bronner, J. Hornung, and A. Hülsmann, "A Complete GaAs HEMT Single Chip Data Receiver for 40 Gbit/s Data Rates," in IEEE GaAs Symp. Tech. Dig., pp. 55 - 58, 1998.
78. K. Murata, K. Sano, E. Sano, S. Sugitani, and T. Enoki, "Fully monolithic integrated 43 Gbit/s clock and data recovery circuit in InP HEMT technology, " IEE Electron. Lett., vol. 37, no. 20, pp. 1235-1237, 2001.
79. N. Ishihara, S. Fujita, M. Togashi, S. Hino, Y. Arai, N. Tanaka, Y. Kobayashi, and Y. Akazawa, "3.5-Gb/s x 4-Ch Si bipolar LSI's for Optical Interconnections," IEEE Journal of Solid-State Circuits, vol. 30, pp.1493–1501, 1995.

80. N. Ishihara and Y. Akazawa, "A Monolithic 156 Mb/s Clock and Data Recovery PLL Circuit Using the Sample-and-Hold Technique," IEEE Journal of Solid-State Circuits, vol. 29, pp.1566–1571, 1994.
81. S. Nati and I. Kyles, "A Monolithic Gallium Arsenide Interval Timer IC with Integrated PLL Clock Synthesis Having 500-ps Single Shot Resolution," IEEE Journal of Solid-State Circuits, vol. 32, no. 9, pp. 1350–1356, 1997.
82. T. Otsuji, K. Murata, T. Enoki, Y. Umeda, "An 80-Gbit/s Multiplexer IC using InAlAs/InGaAs/InP HEMTs", in GaAs IC Symposium Tech. Dig., pp. 183, 1997.
83. T. Enoki, H, Ito, K. Ikuta, Y. Ishii, "0.1-μm InAlAs/InGaAs HEMTs with an InP-Recess-Etch stopper Grown by MOCVD," in Proceedings of International Conference InP and Related Materials, pp. 88, 1995.
84. N. Hayafuji, Y. Yamamoto, T. Ishida, K. Sato, "Degradation Mechanism of the AlInAs/GaInAs High Electron Mobility Transistor due to Fluorine Incorporation", Appl. Phys. Lett, vol. 69, no. 26, pp. 4075-4077, 1996.
85. T. Sonoda, Y. Yamamoto, N. Hayafuji, H. Yoshida, H. Sasaki, T. Kitano, S. Takamiya, M. Otsubo, "Manufacturability and Reliability of InP HEMTs", Solid-State Electronics, vol. 41, no. 11, pp. 1621-1628, 1997.
86. N. Hayafuji, Y. Yamamoto, T. Ishida, K. Sato, "Reliability Improvement of AlInAs/GaInAs High electron Mobility Transistor by Fluorine Incorporation Control", J. Electrochem. Soc., vol. 145, no. 8, pp. 2951-2954, 1998.
87. A. S. Wakita, H. Rohdin, C. Y. Su, N. Moll, A. Nagy, V. M. Robbins, "Drain Resistance Degradation under High Fields in AlInAs/GaInAs MODFETs", in Proceedings of International Conference InP and Related Materials, pp. 376, 1997.
88. T. Enoki, H. Ito, Y. Ishii, "Reliability study on InAlAs/InGaAs HEMTs with an InP Recess-Etch stopper and Refractory Gate Matal," Solid-State Electronics, vol. 41, no. 11, pp. 1651-1656, 1997.

International Journal of High Speed Electronics and Systems
Vol. 13, No. 1 (2003) 175–219

INTEGRATED SiGe AND Si DEVICE CAPABILITIES AND TRENDS FOR MULTI-GIGAHERTZ APPLICATIONS

G. G. FREEMAN, B. JAGANNATHAN, N. ZAMDMER, R. GROVES
IBM Microelectronics, Semiconductor Research and Development Center, Hopewell Junction NY 12533, USA

R. SINGH, Y. TRETIAKOV, M. KUMAR, J. B. JOHNSON
IBM Microelectronics, Semiconductor Research and Development Center, Burlington VT 05452, USA

J. O. PLOUCHART, D. R. GREENBERG, S. J. KOESTER, J. D. SCHAUB
IBM T.J. Watson Research Center, Yorktown Heights NY 10598, USA

Silicon-based devices, including the increasingly available SiGe-based devices, are now demonstrating f_T and f_{MAX} values over 200 GHz. These recent advances open the door to a wide range of silicon-based very high frequency, low power and highly integrated solutions. Trends in silicon MOS, SiGe HBT, SiGe MODFET and SiGe strained silicon FETs are reported. Silicon inroads to device functions viewed as the sole realm of III-V technologies are also being demonstrated. Capability and trends of the integrated silicon photodiode are reported here as an example. Integration of these high-speed devices into a complex circuit requires on-chip passive device functionality at such high frequency. Key devices to enable integration are the inductor, varactor, and transmission line, and operation of these devices at high frequency is reported. Further, we discuss noise isolation issues and techniques, which may be used when minimizing cross-talk within a conductive silicon substrate.

Keywords: SiGe HBT; MODFET, strained silicon, photodiode, varactor, inductor, transmission line, substrate isolation.

1. Introduction

Integration and cost will continue to be a strong draw toward silicon-based solutions. With the ever-shrinking device and interconnect features, and the introduction of SiGe films into certain devices, raw device performance has increased to levels beyond most III-V devices and will continue to increase. Enabling complex analog-mixed signal circuits, designers have available FETs for digital and analog complementary device functions, as well as the option of a bipolar device for high gain, current drive and the highest frequency performance. In addition, a wide range of analog passives may be made with the variety of layers in such processes. Junction and FET-based varactors, resistors with a variety of sheet resistance and temperature coefficient, metal and FET-based capacitors, and on-chip inductors, combine to enable analog / mixed signal system-on-a-chip applications.

This broad variety of devices may be utilized for functions not frequently cited in typical high-performance demonstrations, yet which are critical for competitive products. Most apparent is the integration of digital and analog on the same chip (see for example Figure 1).[1] Such single chip integration minimizes power loss and improves signal

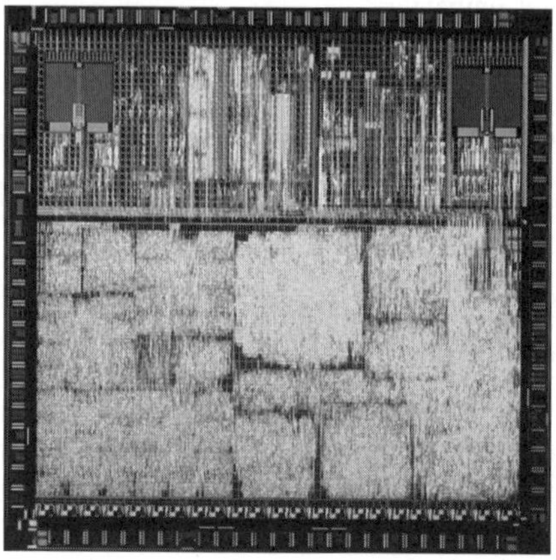

Figure 1: Chip micrograph of a highly integrated mixed-signal circuit.

integrity. Further, the digital logic can enable (for example) complex signal deskew and self-test capability on-chip. With the availability of complementary devices, analog elements such as op-amps, mixers, quadrature VCOs, and power management circuitry are simplified.

The trend toward high-speed silicon and silicon-germanium implementation is apparent in products recently introduced. One key market category in the regime above 10 GHz is optical networking. Network speeds of 10 Gb/s are becoming well-established, with multiple standards including Synchronous Optical Network (SONET) and Ethernet stemming from both the telecommunications and data networking sides. A typical block diagram for a 10 Gb/s Ethernet system is illustrated in Figure 2. Of the various building blocks of such a system, those handling the full-data-rate bit stream demand the highest level of performance. Such blocks include the transimpedance amplifier and laser/modulator driver to interface with the optical devices themselves, as well as the serializer/deserializer combined with the clock multiplication and clock/data recovery circuits to convert between the high-speed serial signals and slower, parallel paths to the switch fabric. Silicon began to address these building blocks soon after the first

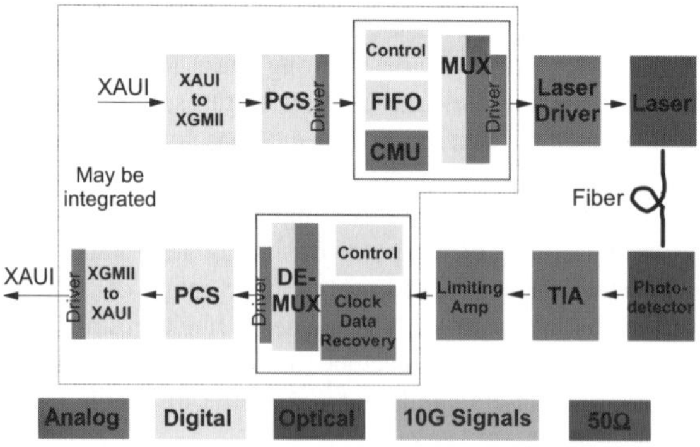

Figure 2: Typical 10 Gb/s Ethernet block diagram.

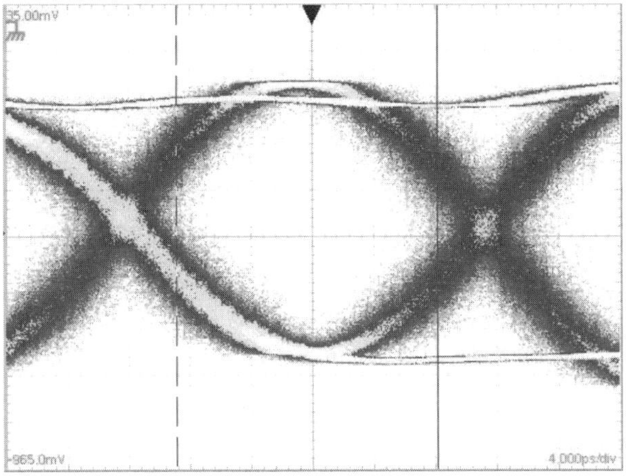

Figure 3: Eye diagram at 40 Gbit/s for multiplexer implemented by AMCC in f_T =120GHz SiGe BiCMOS.

commercial availability of 50 GHz SiGe BiCMOS technology, with companies marketing integrated transmitter (serializer plus clock multiplier unit) and receiver (deserializer plus clock/data recovery) chips as well as optoelectronic components.[2] Development of SiGe BiCMOS to performance levels in excess of 120 GHz is now enabling companies to demonstrate similar components for the emerging 40 Gb/s market[3,4] as illustrated by the open eye diagram from a fully-functional 40 Gb/s AMCC multiplexer in Figure 3. This achievement encroaches definitively into space previously considered the exclusive domain of InP.

With SiGe BiCMOS having served in this manner as the first-to-market cutting edge, establishing silicon's foot in the door, CMOS is now beginning to tackle the 10 Gb/s market, aiming to reduce both cost and power. A prime example is Broadcom's 10 Gb/s Ethernet transceiver, which combines all functions from the 4-lane 10 Gb Attachment Unit Interface (XAUI) to the full-data-rate serial Physical Medium Attach (PMA) interfaces.[5] The first generation of this part was developed and is being produced in 0.18 μm CMOS. This version has recently been followed up with a 0.13 μm CMOS design, highlighting the rapid drive toward greater ease of design, improved yield, reduced cost and lower power.

High-performance opportunities for silicon are not limited to wired applications. While the highest-volume wireless applications are presently concentrated below 5 GHz, the demand for available new spectrum and for the support of high data rates is spurring the development of new applications in the higher bands. Such applications include point-to-point communications such as Local Multipoint Distribution System (LMDS) at 28 GHz, as well as experimental LANs at 60 GHz for the distribution of multimedia content. In the automotive world, companies are exploring the 24 and 77 GHz bands for short-range radar aimed at adaptive cruise control and collision avoidance/preparation. Figure 4 illustrates the performance realized by a 19 GHz power amplifier developed for this market, designed to explore the output levels achievable by SiGe in this frequency regime.[6]

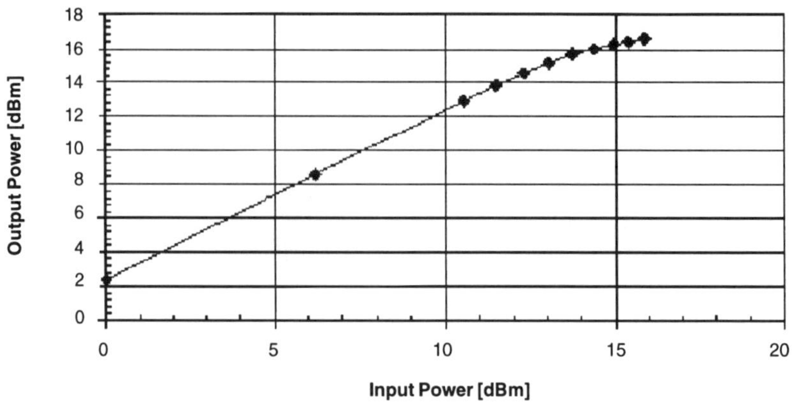

Figure 4: Output power vs. input power for a 19 GHz power amplifier implemented in SiGe BiCMOS.

To some, given the superior mobility and carrier velocity behavior of III-V materials, it seems that the laws of physics are being violated when silicon exceeds III-V device performance and demonstrates the behavior of such designs. In fact, it is in the details of the device construction and operation that silicon and silicon-germanium devices find advantages not available in III-V devices and technologies. From the basic material level to the device construction, advantages are found which translate to device performance and reliability. For one, the silicon and silicon-germanium crystal lattices are more robust, allowing higher current densities, and importantly, better heat conductivity compared to III-V materials. On the question of robustness, it is interesting to note that typical processing temperatures for silicon systems are approximately 1000°C with relatively well understood dopant diffusion properties at this temperature, whereas III-V systems cannot withstand temperatures greater than approximately 500°C without catastrophic effects. Such differences highlight the very different materials involved.

From the structure point of view, the ability to form thermal SiO_2, apply precise anisotropic etches, deposit high-temperature Chemical Vapor Deposition (CVD) films (> 700°C), form low-resistance self-aligned silicide films and temperature-tolerant low-contact resistance layers all contribute to substantial flexibility to create more complex structures in the silicon systems compared to the III-V systems.[7] Using these techniques, devices may be constructed with smaller geometries, and lower parasitic resistance and capacitance than a similar device in III-V systems. This goes beyond the active devices and extends into the metal system as well. For example, an additional, often-cited example of III-V benefits is the semi-insulating substrate, which enables low loss transmission lines. To counter the trend of thinner films for each silicon technology generation and enable comparable transmission lines, thick dielectric and metal layers are now built upon the standard tight-pitch thinner metal using planarization and damascene processes. An example of this is shown in Figure 5. These metal systems are found to have acceptable loss and relatively high packing density. [8, 9]

With the silicon material possessing a fraction of the thermal resistance of either GaAs or InP, concerns regarding self-heating and the consequent reliability are reduced. Complicating the comparison is that silicon systems operate at higher current densities, and may contain deep trench isolation which increases the effective thermal resistance.

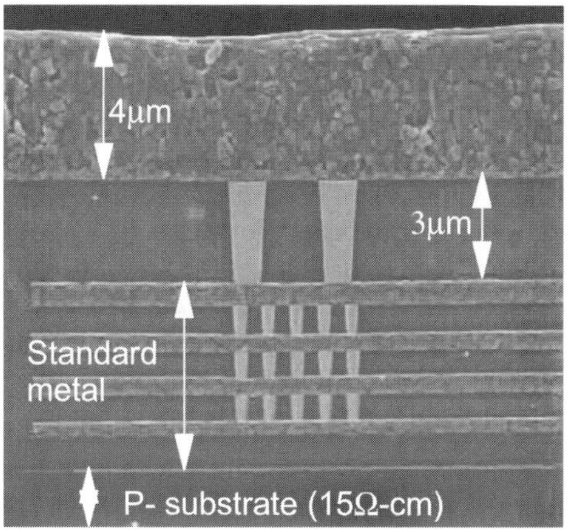

Figure 5: Planar metal system with standard digital CMOS levels and added thick "analog" metal.

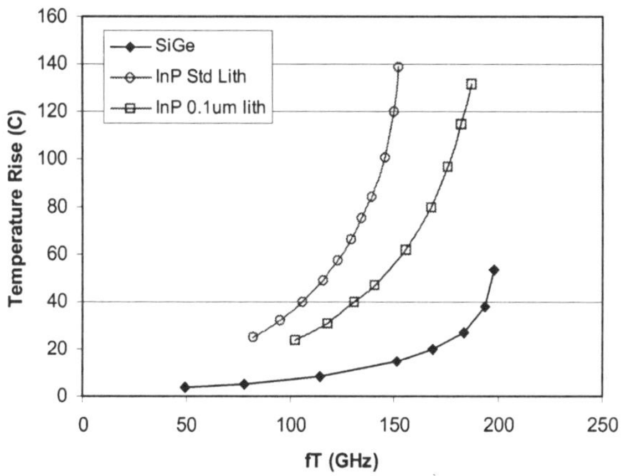

Figure 6: Device self-heating normalized to f_T. [10]

On the other hand, III-V devices may contain very high thermal resistance epitaxy layers, such as InGaAs, also increasing effective thermal resistance. A useful method of comparing self-heating of a device for required f_T performance is provided by Thomas et. al.[10] Shown in Figure 6 is this method utilized to compare two InP SHBT designs from Thomas et. al. [10] against a SiGe 200GHz HBT device. Despite the low current density in the InP HBT, the geometry is larger and the thermal resistance of the InGaAs collector is substantially higher, and thus the device temperature is higher. This is used to illustrate that the factors establishing device performance go beyond the mobility differences, and must also include parasitics, as well as thermal issues.

This paper covers silicon device performance improvement and trends to enable integrated high-speed electronics. The focus is on key enabling devices to achieve highly integrated designs operating over 10 GHz or 10 Gb/s. We initially cover the active devices, first describing trends in the workhorse silicon MOSFET, which appears to be establishing a solid foothold in 10 Gb/s systems today. We then cover the compound devices, with FETs containing SiGe to enhance mobility, followed by a discussion on the advances in the SiGe HBT device. We then attempt to bring in relatively non-standard devices, which may be future game-changers. The area is broad, and so we use in this paper the example of the integrated silicon trench photodiode, which is traditionally thought to be a device only suitably made in III-V systems. Many remaining parts of the integrated circuit are critically important to achieve the high speeds desired. We single out the inductor, varactor, and transmission-line to discuss as key enabling devices for high frequency design. Finally, the conducting substrate effects, including impact on isolation and on passive performance are discussed, with emphasis on layout and isolation solutions.

2. Active Devices

Although successful circuit design exploits a combination of both active and passive devices, the core functionality of a technology is influenced heavily by its active device library. Silicon processing enables a wide array of active device types and flavors, which may be combined in selected combinations to create technology offerings. This variety is intended to accommodate and attract a range of disparate customers, providing circuit designers with the freedom to implement desired applications with a desired level of performance and in the most time-efficient and cost-effective manner.

Silicon technologies fall broadly into three categories defined by the included mix of active devices. CMOS provides complementary NFET and PFET devices in several different variants. The evolution of CMOS has traditionally been driven by volume digital applications, with emphasis on gate delay and power consumption. Until recently, radio-frequency (RF) figures of merit, such as f_{MAX}, low-frequency and broadband noise and linearity have not been high priority. Bipolar-only technologies offer homojunction or SiGe heterojunction bipolar transistors as well as a variety of passives for RF/analog applications, but may not include a high performance CMOS component. Such technologies are generally specialized toward discrete or low-integration applications requiring high performance and low cost. Combining bipolar and CMOS devices results in BiCMOS technology. While some forms of BiCMOS employ a non-optimized or even parasitic bipolar in order to offer bipolar functionality for low performance or DC tasks, the form most suitable to modern high-speed communication applications features a high-performance bipolar module built upon an ASIC-compatible CMOS technology.

While BiCMOS provides the designer with the most flexibility due to its varied active device library, the additional processing for the bipolar device can add cost relative to CMOS at a given lithography generation. For the designer, choosing a technology thus comes down to determining what level of performance is good enough for the application at hand. With the market driving designers to achieve greater and greater functionality at lower cost, CMOS is becoming increasingly attractive, particularly at 0.25μm generation and beyond. This new market for CMOS has also impacted the development of CMOS itself, spawning new device variants that address the bottlenecks to RF/analog/mixed-signal performance, including gate resistance (impacting f_{MAX} and broadband noise), self-gain, matching and 1/f noise. These new FET variants, combined with RF-optimized passives and interconnects, form the backbone of a new family of RF CMOS technologies.

The designer tackling a given application can now choose from among several generations of both BiCMOS and RF CMOS. The choice creates both flexibility and confusion. To temper the confusion, technology selection can be broken down into two stages. The first stage considers the performance capabilities of the technology compared against the application specification, eliminating technologies which do not suffice. The second stage typically selects the least expensive option from among those sufficient for meeting specifications. For many applications, it is typically not necessary to exceed spec, except in areas that may have a marketing advantage, such as power consumption. In evaluating expense, raw wafer cost is only one consideration. For example, ease of design, related to quality of design kit, impacts the number of required prototype iterations and the ultimate time to market, both of which indirectly influence cost or sales. Integration level, such as single chip vs. multiple technologies on a module, may come into play as well.

In narrowing down the list of suitable technologies, a designer will typically compare figures of merit in a variety of categories. Bipolar devices enjoy an advantage in several categories. The parameter f_T in both a bipolar and FET device is given by a ratio between transconductance and input capacitance. Given a bipolar and FET of equal f_T, and thus equal ratio, the bipolar ratio is comprised of a larger transconductance divided by a larger capacitance. Its value of f_T is therefore less impacted by the addition of a fixed amount of additional capacitance, such as due to the loading from a driven stage. The degradation of CMOS f_T with loading can require twice the unloaded f_T compared with a bipolar technology in order to tackle a given application. The higher bipolar g_m also gives the bipolar emitter follower an advantage over the CMOS source follower, in both linearity and device size for given drive capability. This advantage is of particular interest in a circuit needing to drive a low impedance (e.g. 50 Ω) interface.

Since the active regions of a bipolar device are contained within silicon and away from contact with oxide interfaces, these devices also enjoy lower 1/f noise compared with CMOS. Finally bipolar devices also enjoy an advantage in matching. The emitter of a bipolar device need not be fabricated at difficult-to-control minimum feature sizes in order to achieve high f_T. Thus, current at a given drive voltage can exhibit less variation and better matching compared with CMOS, which requires aggressive gate dimensions for best performance. The resulting gate length variation impacts current through both transit time and threshold voltage, via the short channel effect.

CMOS enjoys a clear advantage over bipolar in wafer cost. Mixed-signal functionality requires the availability of CMOS, making bipolar devices available only as part of a more complex and thus more costly BiCMOS process. While CMOS wafer cost may be cheaper, there is a caveat. The lowest cost overall solution may still favor BiCMOS under certain circumstances. Although CMOS models continue to improve rapidly, bipolar models are currently more accurate for RF simulation, due to their historical head start in this application space. This presumably temporary advantage can translate into a greater chance of first pass success, and thus both lower development cost and a time-to-market advantage.

Regarding other figures of merit, such as broadband noise and linearity, bipolar and CMOS devices share greater parity. This is particularly true at lower frequencies, such as in the 1-2 GHz band containing mobile telephony. At higher frequencies above 5 GHz, however, the higher f_T and f_{MAX} of the most advanced bipolar technologies may confer a gain advantage, translating into lower noise figure and higher associated gain. Noise figure depends critically on bipolar base resistance and CMOS gate and channel resistance, and is thus a strong function of device layout as well.

2.1 SOI and bulk CMOS FETs

CMOS is becoming an attractive technology for high-frequency applications due to the high bandwidths enabled by technology scaling, its low cost, and its integrability with digital logic, SRAM and embedded DRAM. The MOS structure, through simple variation of gate dielectric process and well, extension and halo implants, can be optimized for the specific sub-functions of a system-on-chip, including analog functions.[11,12] The same structure also lends itself to high-Q, highly-tunable varactors.[13] The same many-level interconnect that is required by GHz microprocessors can be used for a variety of high-Q, high-density inductors and capacitors.[14,15]

FET scaling

CMOS is currently making inroads to circuitry in the 1 to 10 GHz frequency range, and scaling must be effective if CMOS is to be used at higher frequencies. Woerle et al.[16] simulated the RF figures of merit for bulk silicon FETs of future technology generations, and found continuous improvement in such parameters as f_T, maximum stable gain and high frequency noise out to the 50 nm technology node. To verify this prediction, we measured the small-signal properties of FETs at poly gate lengths typical of future technologies. The NFETs are from a 0.13-μm partially depleted SOI CMOS technology.[13,14] The results are promising: Figure 7 shows that f_T increases with decreasing channel length without saturation.

The expected improvement in RF performance with scaling is mainly driven by the predicted increase in g_m from generation to generation, since input and output capacitance will tend to be constant due to gate dielectric scaling. We verified that g_m increases with decreasing channel length. The FETs were biased at V_{DS} = 1.2 V, at the gate bias that maximized g_m. Figure 8 shows that g_m, like f_T, does not tend to saturate, indicating that the gate maintains control of the channel. The increase in g_m from the longest to the shortest channel length (about one technology generation's worth of scaling) is 21%, which is less than the 40% per generation predicted by Woerle et al. However, these authors' prediction assumes gate dielectric scaling; the 21% improvement that we observe is without such scaling, and is due solely to the increase in the effective velocity of electrons

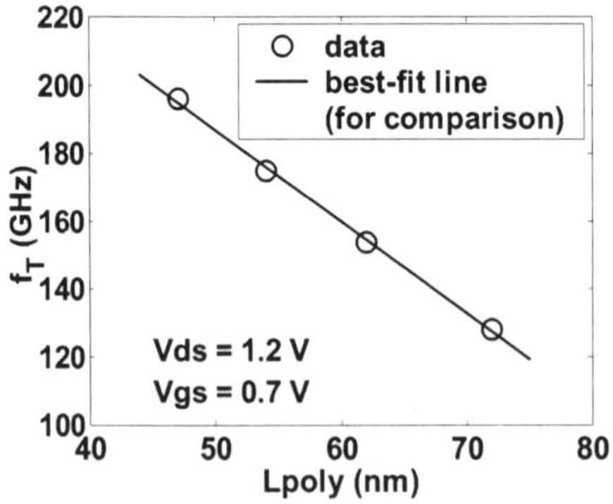

Figure 7. Unity current-gain frequency vs. polysilicon gate length.

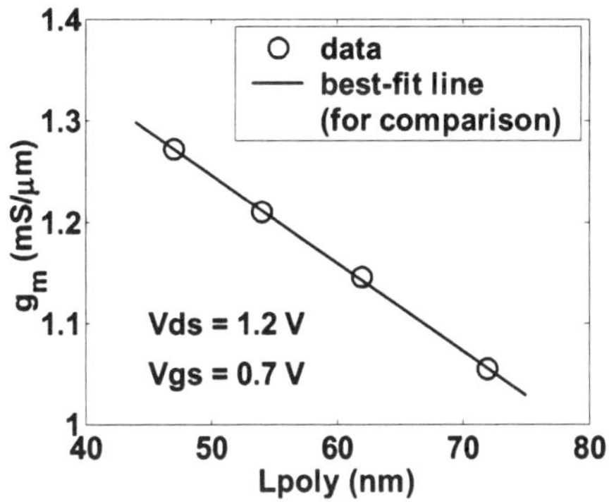

Figure 8. Transconductance vs. poly gate length.

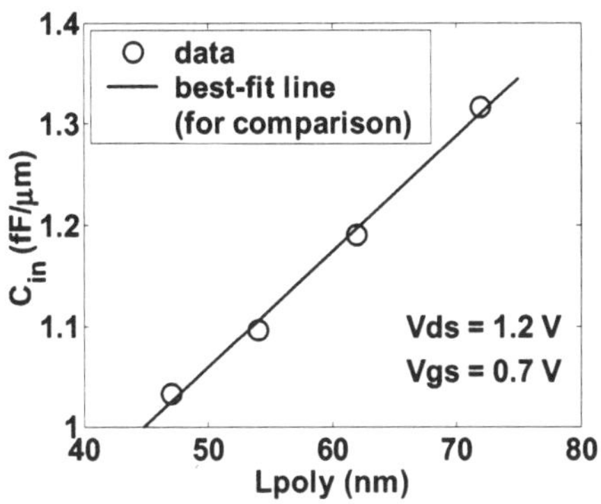

Figure 9. Input capacitance vs. poly gate length.

in the channel with increasing lateral electric field. This increase in effective velocity occurs even at high drain voltage.

Another possible obstacle to future FET performance improvement is parasitic capacitance, since this often does not decrease with scaling. A measure of parasitic capacitance is C_{IN}, the input capacitance derived from *S*-parameters, extrapolated to a gate length of zero. Figure 9 shows C_{IN} for the same devices described above. The value of C_{IN} extrapolated to zero gate length, normalized by the gate perimeter, is 0.24 fF/μm. This low value is comparable to the lowest FET Miller capacitances achieved today, and includes the parasitic capacitance from the diffusion wiring and contacts to the transistor gates. The parasitic component of C_{IN} is less than half the value of C_{IN} at a gate length of 47 nm, which shows that parasitic capacitance will not be a major limitation of performance for a few technology generations to come.

Another possible performance limitation is input resistance. As FET gate length shrinks, the sheet resistance of the silicided gate polysilicon tends to increase. Gate resistance limits the maximum available power gain, and a high-loss gate keeps designers from using wide FETs with lower variability and lower parasitics than narrow FETs. We

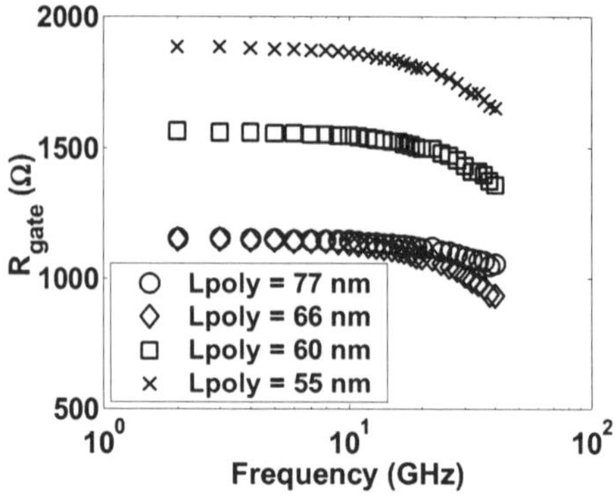

Figure 10. Gate resistance vs. frequency as measured on monitor structures (W = 14 μm).

measured the gate resistance on FETs with high, easily-measurable total gate resistance (~ 150 Ω) to investigate the gate-length dependence of silicide sheet resistance at gate lengths below 100 nm. The results are shown in Figure 10. The measured resistance is effectively constant below 20 GHz, as expected, reflecting accurate measurement. The rolldown in the gate resistance above 20 GHz is either due to parasitic inductance or inaccurate calibration. The sheet resistance is 6.3 Ω/square at a gate length of 77 nm and rises to 7.3 Ω/square at a gate length of 55 nm. This sheet resistance increase is moderate enough to allow increasing power gains at acceptable transistor widths for the next couple of generations, before metal gates become necessary to improve performance further still.

Targeting applications through integration of multiple FETs

Today's CMOS technologies targeted towards system-on-chip products combine a variety of gate oxides and well, extension and halo dopant implants in devices on the same wafer. Variation in these process steps has served a number of purposes. Threshold voltage variation through multiple well dopant implants has been used in digital logic to decrease power consumption. [11] Selective use of low-V_T FETs has been used to speed critical paths. Multiple oxide thicknesses have been used to integrate low-standby-power circuitry with other digital logic.[17] Similar variation has been made to optimize devices for specific analog and RF applications. For example, dual gate oxides can provide designers with FETs with reduced low-frequency noise.[14] Selective removal of the halo implant from an entire FET, or from only the drain side of an FET, can improve its self-gain (open-circuit voltage gain) and threshold voltage variability.[18] Figure 11 shows the observed improvement in self-gain (at DC) as a function of designed gate length for NFETs with the halo implant removed ("shadowed") from the entire device.[18] The improvement is significant in the 1 μm to 10 μm channel length range, a range often considered in the design of current sources, which require high self-gain.

Use of SOI technology gives the designer another set of FETs for targeted application: body-contacted FETs. Standard FETs on SOI have a body that is isolated from all other conductors by the buried oxide, the source/drain junctions, and by the gate. When isolated, the body is a high-impedance node with a variable, "floating" voltage. Ohmic contact can be made to the body, usually through localized implants placed at one end of an FET. Body-contacted SOI FETs have some advantages over floating-body FETs when

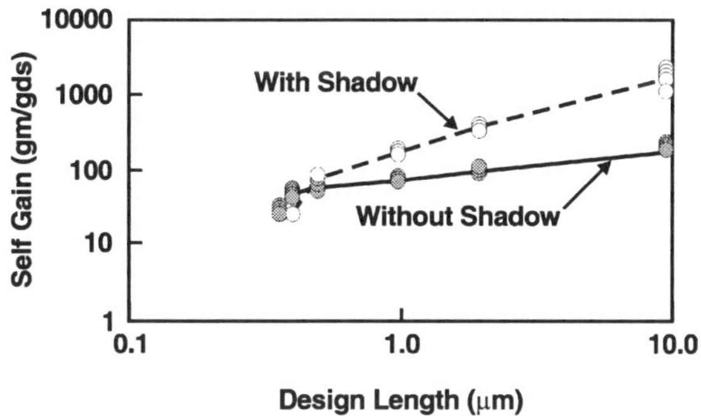

Figure 11. Self-gain vs. design gate length for NFETs with and without halo removal through "shadowing".

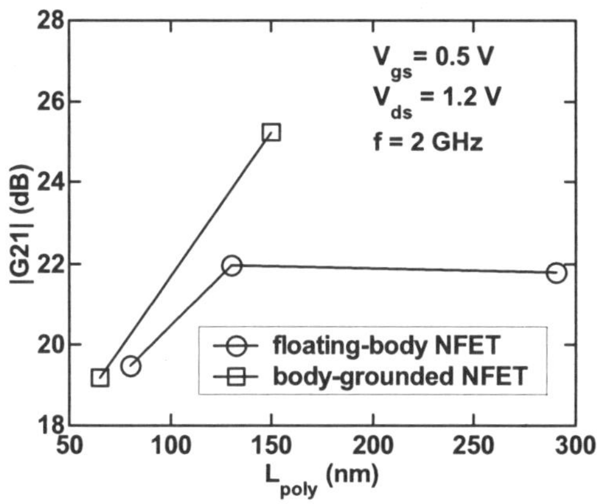

Figure 12. Self-gain (G21) vs. polysilicon gate length for floating-body and body-grounded SOI NFETs.

used in analog circuits. A body contact eliminates the contribution of the drain's diode current to low-frequency drain current noise.[14] The elimination of floating-body effects helps the body-grounded device achieve higher output resistance and therefore larger self-gain than the floating-body device. Figure 12 shows that the voltage gain at 2 GHz is higher for the body-grounded NFET device over all gate lengths.

2.2 *Si/SiGe strained-layer FETs*

SiGe-containing heterostructures have been the enabling factor in establishing Si-based technology as a viable contender in the RF communications marketplace. While SiGe HBTs are an established commercial technology, in recent years, tremendous progress has also been made on strained Si MOSFETs and modulation-doped FETs (MODFETs), and these devices hold the promise to further expand the capabilities of Si technology for high-speed communications. This section describes the basic concepts underlying Si/SiGe strained-layer FETs, and reviews the recent developments by IBM and other research groups in this area.

Effect of strain

Si/SiGe strained-layer FETs are based upon the concept of using strain to provide enhanced mobility to improve FET performance. The most common technique for creating these strained layers is to start with a low-defect-density relaxed $Si_{1-x}Ge_x$ buffer layer (grown on a Si substrate) that can then be used as a "template" for subsequent growth of a strained-layer heterostructure. For instance, as shown in Figure 13(a), a thin Si layer grown on relaxed SiGe will be under biaxial tensile strain. The strain splits the six-fold conduction-band degeneracy, reducing the in-plane electron effective mass as well as the inter-valley scattering rate. The tensile strain also splits the valence band degeneracy for light and heavy holes, leading to reduced scattering and increased mobility for holes. As shown in Figure 13(b), the strain splitting creates a staggered band alignment, with the formation of a potential well for electrons, and a potential barrier for holes. Similarly, a thin Ge layer grown on SiGe will be under biaxial compressive strain (Figure 13(c)). The strain also leads to band splitting, reduced scattering and improved mobility for holes. The band offset created by the strain produces a confining potential for holes, as shown in Figure 13(d). The compressive strain does not, however, provide mobility enhancement or confinement for electrons.

Mobility enhancement

The degree of mobility enhancement ctually observed in strained $Si_{1-x}Ge_x$ heterostructures depends upon the specific device implementation. A simple system consisting of a strained Si surface layer on relaxed SiGe is attractive for CMOS applications, because the strained layer is formed at the wafer surface, and enhancement in both the electron and

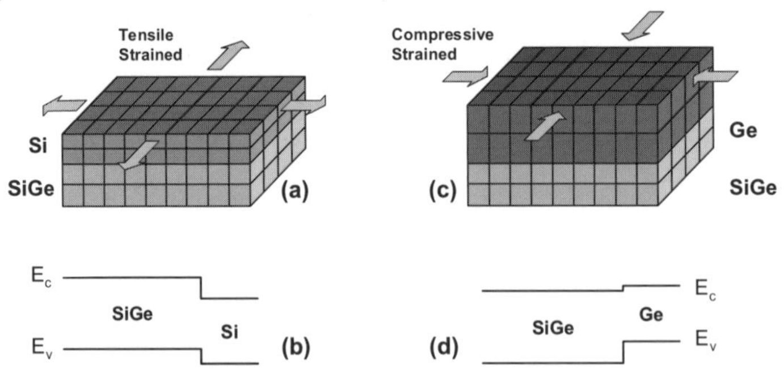

Figure 13. Schematic diagram of structure and band diagram of ((a)-(b)) strained Si and ((c)-(d)) strained Ge on relaxed SiGe.

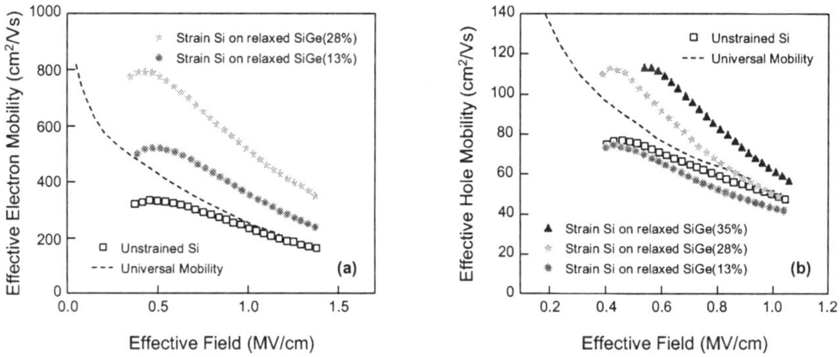

Figure 14. Effective mobility plotted vs. vertical effective field, for (a) strained Si n-MOSFETs and (b) p-MOSFETs with different Ge concentrations, as well as a bulk Si control. The universal mobility curve is shown for comparison.[21]

hole mobilities can be achieved.[19,20] A comparison of the electron and hole mobility in strained Si MOSFETs with bulk Si MOSFETs and the universal mobility curve is shown in Figure 14.[21] The strained Si n-MOSFETs show electron mobility enhancement factors, at high vertical fields (> 1 MV/cm), of about 60% (110 %) for $x = 0.13$ ($x = 0.28$), where x is the Ge mole fraction of the underlying $Si_{1-x}Ge_x$ layer. Strained Si p-MOSFETs require a higher Ge concentration in order for substantial mobility improvement to be observed. For these devices, at $x = 0.28$, the peak mobility enhancement is 45%, but at high vertical fields the enhancement virtually disappears. At $x = 0.35$, the high-field mobility enhancement is about 20%.

In order to obtain greater mobility enhancements compared to Si, buried-channel Si/SiGe/Ge heterostructures can be utilized. Since the band offsets created by the strain also provide carrier confinement, the well-known modulation-doping technique can also be used to populate the quantum well without inducing additional ionized impurity scattering. These layer structures also have improved mobility by eliminating the adverse effects of surface roughness scattering. Figure 15 shows the room-temperature electron mobility for a modulation-doped strained Si quantum well grown on relaxed $Si_{0.75}Ge_{0.25}$ compared with a bulk Si and a Si/SiO_2 inversion layer.[22] The plot shows that mobilities as

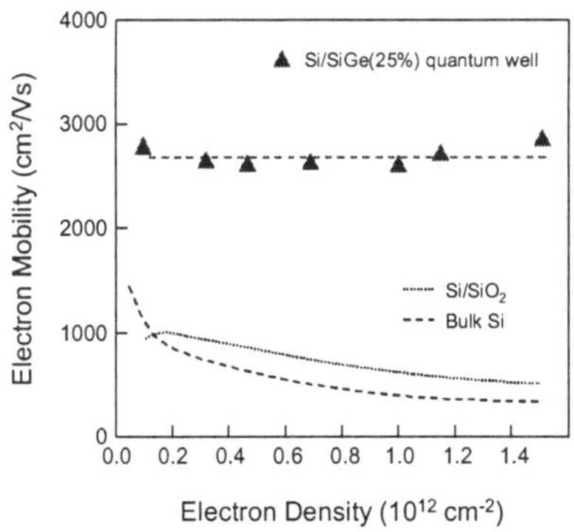

Figure 15. Plot of electron mobility vs. density for a modulation-doped $Si/Si_{0.75}Ge_{0.25}$ quantum well (symbols), bulk Si (solid line) and a Si/SiO_2 inversion layer (dashed line).[22]

high as 2800 cm²/Vs can been achieved in strained silicon; this represents an increase of 3-5 times compared to unstrained Si n-MOSFETs. The hole mobility of compressively-strained layers on SiGe shows even greater enhancement compared to unstrained Si. $Si_{0.2}Ge_{0.8}$ quantum wells grown on relaxed $Si_{0.7}Ge_{0.3}$ have produced mobilities over 1000 cm²/Vs[23], while Ge quantum wells on $Si_{0.4}Ge_{0.6}$ have yielded room-temperature hole mobilities around 1800 cm²/Vs.[24,25] The latter value is ~10 times higher than the mobility in Si p-MOSFETs for the same carrier density.

As FET scaling continues to into the deep sub-0.1 μm regime, there are several reasons why increased mobility is likely to still provide performance enhancement. First of all, as MOSFETs are scaled, the lateral electric field increases, but the channel mobility decreases. This is a consequence of the increased counter doping required to control short-channel effects, and the increased vertical electric fields that result from gate oxide scaling. Therefore, velocity saturation does not always dominate the device operation, particularly at low bias voltages, and low overdrive conditions.[26] Secondly, even at high fields, it has been predicted that strained Si should have increased velocity compared to bulk Si due to velocity overshoot effects. For instance, Monte Carlo simulations by Ferry *et al.*[27], predicted a transient velocity overshoot increase of 1.8 (1.4) for an applied field of 50 kV/cm (500 kV/cm) for strained Si compared to bulk Si. Device simulations also showed a steady-state peak velocity overshoot of 2.6 x 10^7 cm/s for a 0.18 μm-gate-length strained Si n-MODFET at room temperature. Finally, Lundstrom[28] has calculated that even in the quasi-ballistic transport regime, *i.e.*, where a significant fraction of electrons do not encounter any scattering events while traversing the channel, increased low-field mobility should increase the injection efficiency of ballistic electrons, thereby improving the current drive and transconductance. Experimental results by Ohba *et al.*[29] showed a universal correlation between low-field mobility and high-field carrier velocity, in agreement with the results of Lundstrom.

SiGe strained-layer FET performance

In recent years, a number of demonstrations have been made on the performance of short-channel strained Si n- and p-MOSFETs. For instance, Rim *et al.*[30] demonstrated

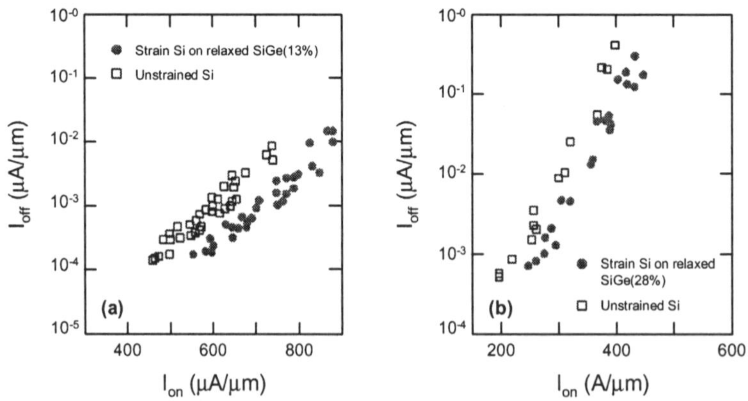

Figure 16. I_{on}-I_{off} characteristics for strained Si and bulk Si (a) n-MOSFETs and (b) p-MOSFETs. After removal of self-heating, the strained-Si devices exhibit 15% higher I_{on} for n-MOSFETs and 7-10% higher I_{on} for p-MOSFETs.[21]

60% intrinsic transconductance enhancement for strained Si n-MOSFETs with channel-lengths as small as 0.1 μm (L_{poly}). More recently, I_{on} enhancement for both p- and n-channel MOSFETs have been demonstrated (Figure 16). In both of these demonstrations, pulsed measurements were needed to remove the effects of self-heating in the strained Si devices. However, since strained Si MOSFETs have largely been targeted for logic applications, little work has been performed to determine the RF properties of these devices. Nevertheless, the fact that these devices exhibit enhanced I_{on} and g_m strongly suggests that improved RF and microwave performance compared to nominally identical bulk Si MOSFETs can be obtained.

A key advantage of the strained-Si surface channel MOSFETs is that they exhibit good thermal stability, and therefore can be fabricated using a process virtually identical to that of bulk Si MOSFETs. It has been shown that the strain is maintained for strained Si on relaxed SiGe even for annealing conditions of 1000 °C for 5 minutes.[31] Nevertheless, defects are a potential problem for these devices, and could cause yield and reliability problems, since relaxed SiGe layers grown on Si typically have threading dislocation densities in the range of 10^5 to 10^7 cm^{-2}. It is also important that the strained Si layer thickness remains below the critical thickness for thermodynamic stability, in order to prevent relaxation by misfit defect formation during thermal processing.[32]

Cross-sectional diagrams of typical n-MODFET and p-MODFET geometries are shown in Figure 17. A typical n-MODFET layer structure consists of a thin tensile-strained Si layer sandwiched between relaxed $Si_{1-x}Ge_x$ layers. Similarly, a p-MODFET can be realized by growing a thin $Si_{1-y}Ge_y$ layer on a relaxed $Si_{1-x}Ge_x$ substrate, where $y > x$. In the figure, n-MODFETs are shown with top-side modulation doping and p-MODFETs are shown with bottom-side doping, but, in general, the modulation doping can be on either or both sides of the quantum well. Because the buried channel separates the carriers from the surface, a Schottky gate is generally used to maximize the transconductance. A T-gate geometry is also generally used to minimize the gate resistance. In contrast to strained-Si MOSFETs, MODFETs have much more severe thermal budget restrictions because dopant diffusion from the modulation-doped supply layer into the conducting channel can degrade the high mobility. Nevertheless, provided that the process is modified to account

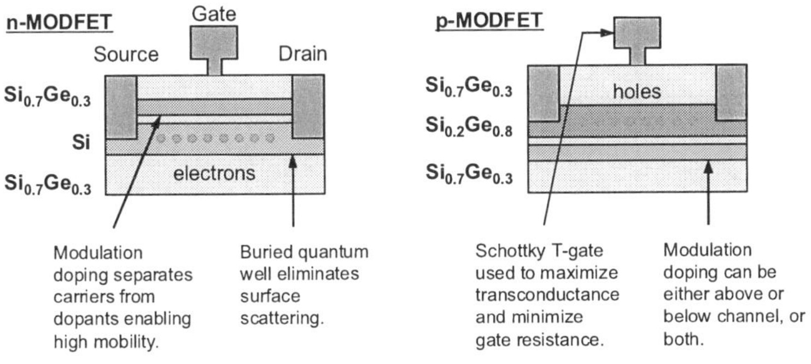

Figure 17. Schematic cross-sectional diagrams of Si/SiGe strained-layer n- and p-MODFETs, showing the salient features of the device geometry.

for the thermal budget restrictions of these devices, SiGe MODFETs can be fabricated in a standard Si CMOS manufacturing line.

Examination of previous results on long-channel devices indicates that Si/SiGe n-MODFETs have considerable performance enhancement compared to Si MOSFETs with similar gate dimensions. For instance, Ismail *et al.* fabricated 0.4 μm gate-length $Si/Si_{0.7}Ge_{0.3}$ n-MODFETs with f_T = 40 GHz and f_{max} = 56 GHz.[22] These devices also had extrinsic transconductance (g_m) of 420 mS/mm. More recent demonstrations produced n-MODFETs with gate length of 0.2 μm and source-to-drain spacing of 0.5 μm 33 with f_T = 62 GHz. Still more recently, Zeuner *et al.* fabricated 0.09 μm gate-length n-MODFETs with f_T = 90GHz.[34] Since the hole mobility enhancement in compressive-strained SiGe is even greater than for electrons in strained Si, SiGe p-MODFETs are capable of producing even greater performance enhancements compared to Si p-MOSFETs. For instance, p-MODFETs with 0.1 μm gate lengths on a $Si_{0.2}Ge_{0.8}/Si_{0.7}Ge_{0.3}$ layer structure showed f_T and f_{max} values of 50 GHz and 116 GHz, respectively.[35] Similar p-MODFETs with L_g = 0.1 μm have also produced f_T values as high as 70 GHz.[36] More recent work has focused on p-MODFETs with pure-Ge channels. Results from 0.1 μm gate-length p-MODFETs fabricated on a $Ge/Si_{0.4}Ge_{0.6}$ quantum well produced devices with peak f_{max} of 86 GHz and room-temperature transconductance of nearly 500 mS/mm.[24] Höck *et al.* obtained similar f_{max} results on 0.25 μm-gate-length p-MODFETs [37], and more recently, fabricated 0.1 μm Ge-channel p-MODFETs with f_{max} = 135 GHz.[38]

Due to the rapid progress in Si CMOS scaling, the high-frequency performance of Si MOSFETs has outpaced that of SiGe MODFETs, the technology for which is still relatively immature. Therefore, in order for the trend of SiGe MODFET improvement over Si MOSFETs to continue, improvements in the scalability of SiGe MODFETs will be needed. For instance, MODFETs have traditionally not utilized fully self-aligned geometries typical of Si MOSFETs. Also, for gate lengths less than ~0.1 μm, aggressive scaling of the vertical dimensions of the modulation-doped heterostructure will be needed in order to fully take advantage of the reduced gate lengths. However physical simulations show that with proper scaling of the lateral and vertical dimensions, f_T values of 300 GHz should be possible for MODFETs with gate lengths of 30 nm.[39]

The high mobility in strained-layer FETs produces the additional advantage that velocity saturation occurs at a lower electric field than in unstrained Si. This behavior

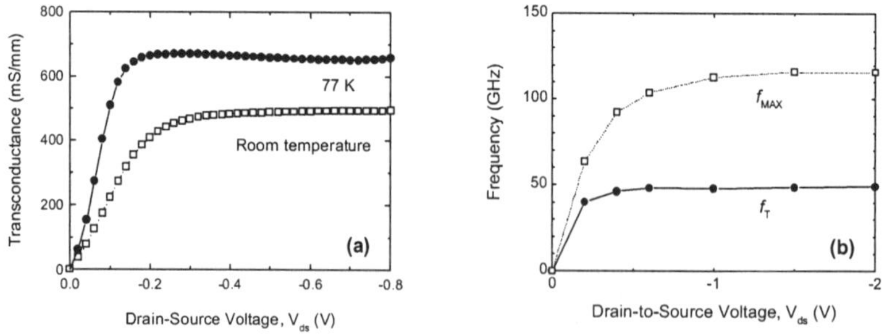

Figure 18. (a) Plot of extrinsic transconductance vs. drain-to-source voltage for a 0.1 μm gate-length Ge-channel p-MODFET. (b) Plot of f_T and f_{max} vs. drain-to- source voltage for a 0.1 μm gate-length $Si_{0.2}Ge_{0.8}$-channel p-MODFET.

allows the maximum g_m and f_T to be reached at a lower value of the drain-to-source bias voltage, leading to reduced on-state power consumption. These trends are shown below in Figure 18, where the drain-bias dependence of g_m, f_T and f_{max} are shown for a 0.1μm p-MODFET. Yet another advantage of SiGe MODFETs is the high mobility that occurs at low temperatures. Due to the extremely-low ionized impurity scattering in modulation-doped heterostructures, the mobility increases exponentially with decreasing temperature. Mobility values as high as 10,000 cm^2/Vs (40,000 cm^2/Vs) have been achieved at $T = 77$ K (20 K), for both n- and p-MODFETs. This behavior may enable extremely-high performance devices to be realized for use in cryogenic applications.

A final advantage of strained-layer MOSFETs and MODFETs are their potential for low noise operation. The simplest model for thermal noise in a FET is the well-known Fukui equation:[40]

$$F_{min} = 1 + KfC_{gs} \cdot \left(\frac{R_g + R_s}{g_m} \right)^{1/2} , \qquad (1)$$

where F_{min} is the minimum noise figure, g_m is the transconductance, C_{gs} is the gate capacitance, and R_g and R_s are the parasitic gate and source resistances, respectively. K is a proportionality constant determined by the device technology. A similar equation to (1) holds for bipolar transistors. According this equation, the ideal device for low noise operation, is one with low C_{gs} and high g_m. Increasing the mobility allows higher g_m for a given value of C_{gs}, thereby producing a lower noise figure.

In general, FETs should also have reduced noise figure compared to bipolar transistors. Assuming equal f_T values, the noise figure in a FET should be reduced since the input capacitance is much lower than in a bipolar transistor. Furthermore, the input resistance value (R_b for bipolar transistors, R_g for FETs) in FETs can be made much lower than in bipolar transistors. By using a metal gate or a metal-strapped gate, in combination with a multi-finger geometry, values of $R_g \sim 1\ \Omega$ can readily be achieved. Though high doping and silicide contacts can reduce the extrinsic and intrinsic base resistance, the base resistance of bipolar transistors remains over an order magnitude larger than in optimized FET geometries. The ultimate noise figure in HBTs is also limited by shot noise from the base current. For FETs, if an insulating gate is utilized, shot noise from gate leakage can be reduced to the extent that it does not influence the noise performance. For these reasons strained Si MOSFETs and SiGe MODFETs are a very promising technology for future low-noise communications applications, and may ultimately be the ideal devices for low-noise applications using a Si-based technology.

It has only been recently that noise parameter data on SiGe MODFETs has been reported. An example of these results is shown in Figure 19 for a 0.1μm gate-length $Si_{0.2}Ge_{0.8}/Si_{0.7}Ge_{0.3}$ p-MODFET. The figure shows the minimum noise figure, F_{min}, plotted vs. V_{gs} for two different frequencies of 3 and 20 GHz. At 3 GHz, F_{min} has a minimum value of 0.6 dB at $V_{gs} = 0$, while at 20 GHz, F_{min} has a minimum value of 2.5 dB at $V_{gs} = +0.3$ V.[35] The corresponding associated gain values are $G_a = 17.5$ dB at 3 GHz, and 7.5 dB at 20 GHz. These results are superior to IBM's early HBT technology, and still fairly comparable to the latest 0.18 μm HBT generations, despite the fact these MODFETs have f_T less than half of the bipolar devices. The high-frequency noise figure is also lower than reported values for bulk Si n-MOSFETs with similar gate lengths.[41] At 3 GHz, the F_{min} value in Figure 19 is limited by gate leakage current, I_g. With improvements in the device

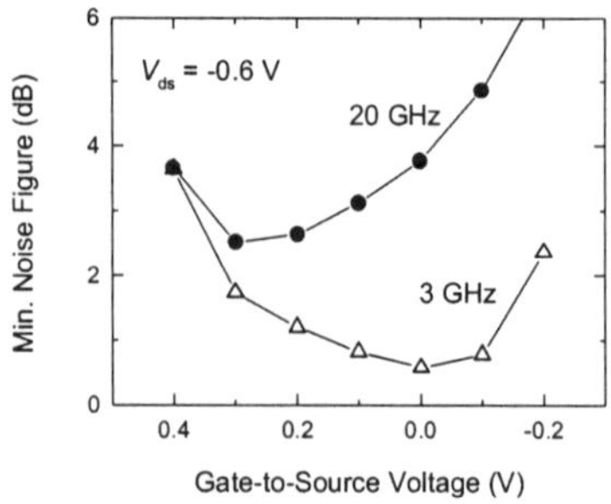

Figure 19. Plot of minimum noise figure vs. gate-to-source voltage for a 0.1 x 90μm^2 $Si_{0.2}Ge_{0.8}/Si_{0.7}Ge_{0.3}$ p-MODFET at 3 GHz and 20 GHz.

design and material technology to increase f_T and reduce I_g, considerable further reduction of the noise figure is expected.

Summary

In conclusion, Si/SiGe strained-layer field-effect transistors have performance that makes them well-suited to impact the high-speed communications marketplace. Compared to Si MOSFETs, these devices should have higher operating frequencies, lower on-state power dissipation, and reduced noise. Strained Si surface-channel MOSFETs are attractive because they are compatible with standard CMOS fabrication and scaling technology, making them excellent candidates for future high-performance digital and RF CMOS. SiGe MODFETs have promise for more substantial performance enhancement compared to Si MOSFETs, and the improved transport and resultant device characteristics are expected to provide performance competitive with GaAs FETs, with the advantage of reduced cost and improved integration capabilities. A host of challenges remain, however, including the ultimate scalability of the MODFETs as well as the effect of defects on yield and reliability.

2.3 *SiGe HBTs*

Bipolar devices continue to demonstrate higher f_T values compared to FETs, and performance continues to improve (Figure 20). Recently, SiGe HBT technology has demonstrated the highest f_{MAX} of 285 GHz[42] and the fastest operating circuits with ring oscillator delays of 4.2ps[43] for a production technology, including III-V technologies. Historically, bipolar devices have outperformed FET devices because, unlike FETs, bipolar devices depend more on vertical dimension reduction rather than lithographic dimension reduction to achieve the highest performance. Continuing performance improvements through vertical and lateral scaling, utilizing structural parasitic reduction techniques available to silicon technologies as discussed earlier, has maintained the bipolar f_T lead over the relentless silicon FET performance scaling.

The high f_T performance scaling in SiGe HBTs is achieved by reducing transit time and RC charging constants. To reduce transit time, film thicknesses are reduced as deposited, the germanium ramp is increased, and thermal cycles are reduced to minimize

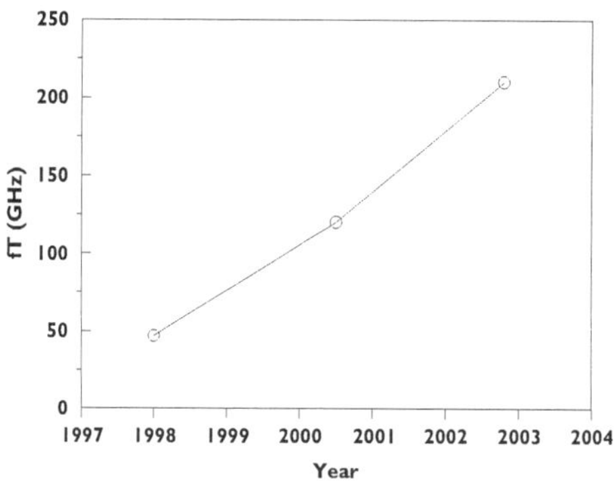

Figure 20: SiGe HBT device f_T trend vs. time.

spreading of the base dopant. Also significantly reducing the spread of base dopant during high temperature processing is the incorporation of carbon into the intrinsic base of the device. This has been shown to interact with the silicon interstitial defects, which would otherwise cause the boron to diffuse, diffusing the carbon instead.[44] Parasitic collector and emitter resistances are also minimized. Collector dopant concentration is increased, which reduces the collector-base space-charge width, reduces the collector resistance, as well as allows the transistor to minimize the kT/qI_C term of the charging delay by operating at a higher current density. At the same time, the emitter dimension is reduced, lowering thermal resistance, self-heating effects, and widening the metal for improved current handling. The result of such scaling is shown in Figure 21, where we show the f_T performance across several IBM generations of SiGe HBT. As noted in the figure, from one generation to the next, substantial power savings can be achieved since most (or perhaps all) transistors are not required to operate at peak f_T. The f_T can be traded for substantially lower current and thus lower power consumption.

Power gain cutoff frequency (f_{MAX}) improvements must also go hand-in-hand with the current gain cutoff frequency (f_T) improvements. Improving f_{MAX} is largely a matter of

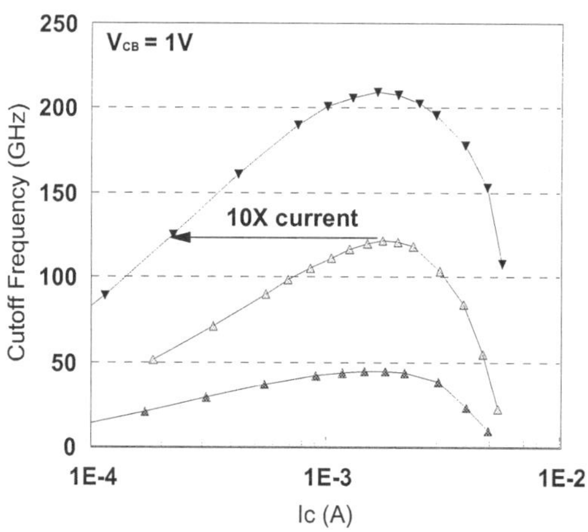

Figure 21: SiGe HBT scaling across device technology generations. While current densities increase, device dimensions are reduced with each generation.

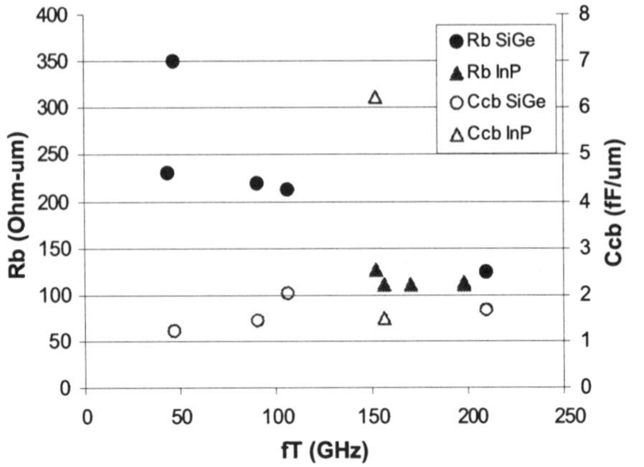

Figure 22: SiGe device parasitic trends compared to reported InP HBT parasitics.

reducing the parasitic R_B and C_{CB} components. This is where the structure innovations have the greatest effects. Shown in Figure 22 is a trend of R_B*L and C_{CB}/L vs. f_T, comparing different SiGe HBT devices to InP HBT devices.[4,45,46,47,48,49,50] With the increasing f_T, one can see that despite the increased collector doping concentration to achieve this f_T, the C_{CB} is kept relatively constant, and the R_B component is reduced. In the 200GHz f_T devices with 285GHz f_{MAX}, the parasitic component values are comparable to the best InP HBT devices. Despite the much higher pinched-base sheet resistance (to achieve high f_T), the SiGe devices are able to establish an overall low R_B value because of the availability of self-aligned silicide and spacer technologies, and creation of a low-resistance extrinsic layer adjacent to the intrinsic device. And despite the much smaller collector-base space-charge width and associated high unit area capacitance, the creation of a patterned collector region (also known as selectively implanted collector, SIC, or pedestal implant) minimizes the extent of this capacitance and reduces the overall C_{CB} value.

Key material and structural innovations make such high performance possible. As an illustration, we discuss here the raised-base structure which delivers the 200/285GHz f_T/f_{MAX} performance.[42] In the recent 10 years, the introduction of SiGe films into the bipolar device, and more recently the incorporation of carbon, have each provided a boost to device performance. These material advantages are now augmented with a raised extrinsic base to simultaneously reduce base resistance and collector base capacitance. Figure 23 contrasts this with the prior-generation implanted extrinsic base scheme. In Figure 23(a), we point to limitations in scaling the bipolar device with the implanted extrinsic base. On the one hand, the implanted base creates silicon interstitials, which at high temperature processing of 700°C or higher, cause the base dopants to diffuse. This limits the proximity of the extrinsic base to the intrinsic portion of the device and thus limits base resistance scaling. In addition, the deeper implanted junction intersects the pedestal implant at high concentrations, increasing the collector-base capacitance. The raised-base scheme minimizes both of these effects by putting the low-resistance extrinsic region above the intrinsic region, eliminating the implanted defects and minimizing the intersection of the collector and base dopants. The parasitic reduction results are shown

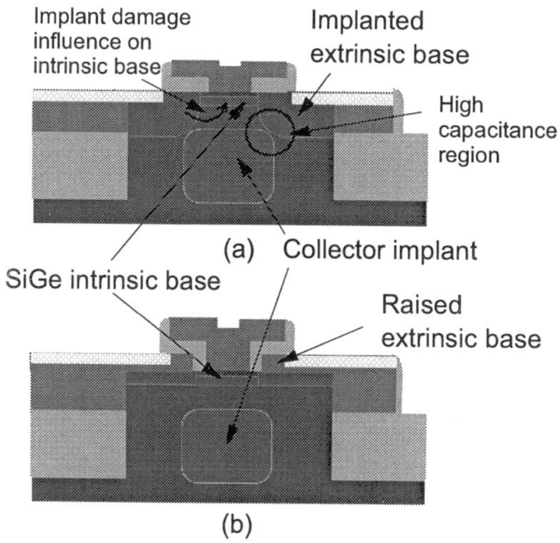

Figure 23: (a) implanted extrinsic base device structure and (b) raised extrinsic base structure.

in Figure 22. Further material and structural changes are expected to continue this trend to even higher performance.

3. Trench Photodiodes

Just as the silicon-based devices discussed so far have encroached upon functions previously believed to be the realm of III-V materials, the silicon photodiode also is making inroads into this space.

Applications that require high data rate transmission over relatively short distances, such as short-reach data communications and inside-the-box optical interconnects, are often implemented with widely available 850nm vertical cavity surface emitting lasers (VCSELs) and multimode fiber. For the optical receiver, recent work has demonstrated

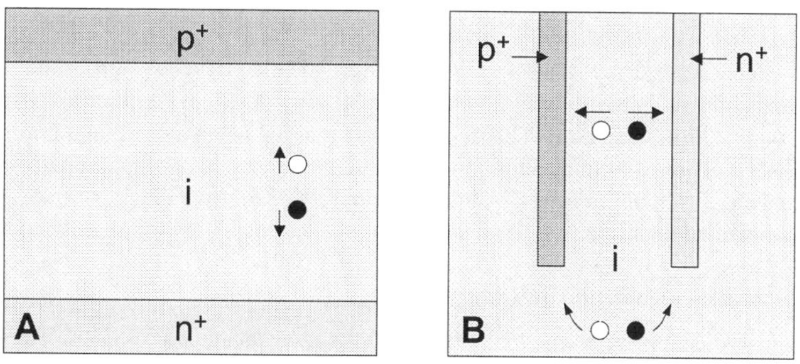

Figure 24: Cross-sectional diagram of the (A) vertical PIN photodiode and (B) lateral trench detector (LTD).

that it is possible to take advantage of the low cost and volume manufacturability of silicon technology, as opposed to the more traditional approach of using a GaAs-based photodetector to detect the optical signal. However, the absorption length at a wavelength of 850nm is 19μm in Si compared to 1.1μm for GaAs. As a result, silicon photodetectors have difficulty achieving both high bandwidth and high quantum efficiency at low operating voltages.

Silicon photodiodes that utilize a traditional vertical PIN structure can achieve high bandwidth (140GHz) but suffer from low quantum efficiency (0.9%).[51] Alternatively, lateral structures fabricated on an SOI substrate with electrodes at or near the top surface can achieve a lower bandwidth (2.1GHz) with better efficiency (21%).[52] Finally, photodiodes resonant cavity structures have been demonstrated with very high bandwidths (34GHz), low-voltage operation (5V) and good quantum efficiency at certain optical wavelengths (31% at λ=836nm).[53] However, the quantum efficiency is suppressed at non-resonant wavelengths, which limits their implementation in practical systems.

Recently, we have shown that this tradeoff between bandwidth, broad wavelength quantum efficiency and operating voltage can be dramatically improved by fabricating a lateral PIN structure based on deep trench EDRAM technology.[54,55,56] A cross sectional diagram of this lateral trench detector (LTD) structure is shown in Figure 24. Within the standard trench depth of 8μm, the length of the absorption region is decoupled from the carrier transit time. In addition, a uniform electric field is maintained between the trenches enabling low-voltage operation.

With the exception of SOI CMOS, nothing prevents implementation of an LTD-based optical receiver in the advanced silicon technologies discussed in this paper. While LTDs fabricated on thick SOI structures are useful, the ~0.1μm SOI CMOS is too thin for compatibility with the LTD. In the following sections, we will demonstrate the performance of large-area silicon LTDs and will compare the performance of the LTD and the GaAs PIN photodiode as the device dimensions shrink. We will show that the Si LTD is a promising alternative to GaAs as data rates increase.

3.1 Fabrication

The lateral trench photodetector (LTD) was fabricated on a p-type (100) silicon substrate with a resistivity of 11-16Ω·cm. A second LTD was fabricated on an 8μm thick silicon on insulator (SOI) substrate with similar resistivity. The 0.35μm wide and 8μm deep trenches were formed by reactive ion etching. These trenches were laid out in interdigitated finger patterns with a spacing of 3.3μm. The device featured a circular aperture with a 75μm diameter. The trenches were masked using a sacrificial borosilicate glass (BSG) deposition and alternatively filled with in-situ phosphorus-doped and boron-doped polysilicon, followed by annealing at 1000°C to drive the dopants away from silicon/polysilicon interface.

3.2 Performance on bulk and SOI substrates

The external quantum efficiency of the bulk substrate device was 68% at 850nm, as shown in Figure 25. Note, however, that only about 34% of the transmitted carriers are absorbed in the first 8μm of silicon. The remaining photons are absorbed below the trenches and do not contribute to a fast response. On the device fabricated on the SOI

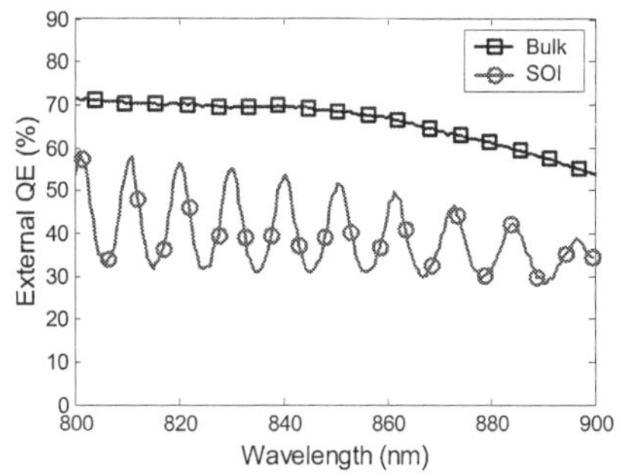

Figure 25: Measured external quantum efficiency as a function of wavelength for the silicon LTD on bulk and thick SOI substrates.

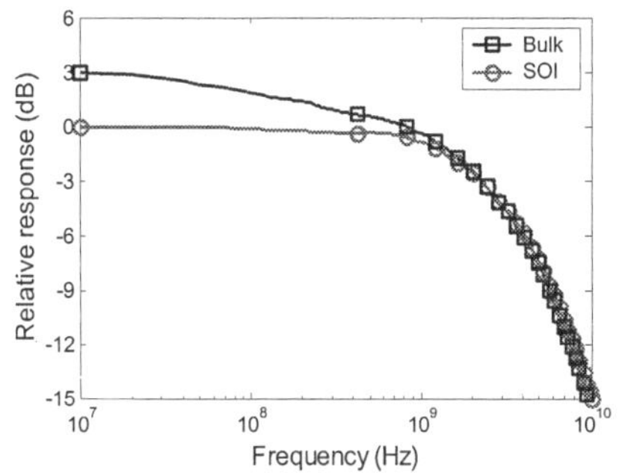

Figure 26: Measured frequency response of large area (75μm diameter) silicon LTDs on bulk and SOI substrates at a reverse bias of 5V.

substrate, a resonance was formed due to the reflection from the buried oxide. The peak quantum efficiency near 850nm was 51%.

The frequency response of the two devices is shown in Figure 26. On the bulk substrate, the LTD exhibited a low frequency tail that limited the bandwidth to <1GHz. The carriers generated below the bottom of the trenches experience a fringing electric field and drift slowly to the contacts, causing the tail in the response. No tail is observed from the LTD on the SOI substrate since the slow carriers are blocked by the buried oxide. The 3dB bandwidth of 2.2GHz is limited primarily by the capacitance of the large-area (75μm diameter) device.

3.3 GaAs PIN versus Si LTD

The primary advantage of the silicon LTD is the fact that it is fully compatible with standard silicon processing techniques and thus can be monolithically integrated with a silicon-based transimpedance amplifier (TIA) in volume manufacturing environments. In contrast, a hybrid optical receiver that uses a III-V-based photodetector requires the costly and complicated packaging steps of mounting and wirebonding the device in close proximity to the TIA. The advantage of the monolithic design is amplified in applications that require 1xN and NxN arrays.

Regardless of the packaging benefits, it is useful to compare the performance of a silicon LTD and a GaAs vertical PIN device, taking into account the various advantages of GaAs over Si. The higher absorption coefficient and velocity overshoot property of GaAs is somewhat offset by parasitic capacitance from the bonding pad and parasitic inductance from the wirebond. Independent of technology, the primary figure of merit of a photodetector (M) used in an optical receiver is the ratio of the quantum efficiency to the square root of the capacitance, as given in equation 2.[57]

$$M = \frac{QE}{\sqrt{Cap}} \quad . \tag{2}$$

The ratio of M_{GaAs} to M_{LTD} is a measure of the benefit of using a GaAs device over a Si LTD in an optical receiver. Figures 27 and 28 show a comparison between a silicon LTD, a GaAs PIN detector with no parasitic capacitance and a GaAs device with 60fF of parasitic capacitance. Note that 60fF of parasitic capacitance is actually quite low for a typical PIN device. These curves were generated by optimizing the n- and p-region

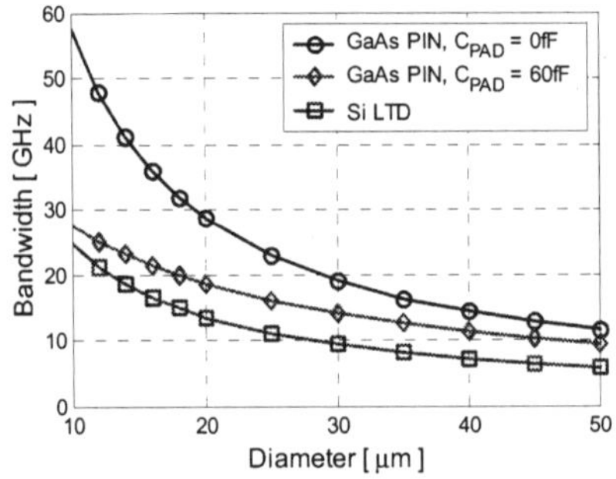

Figure 27: Bandwidth as a function of the device diameter for a silicon LTD and a GaAs PIN photodiode with and without pad capacitance at 5V. The spacing between the n- and p-electrodes has been optimized at each diameter to achieve the maximum bandwidth.

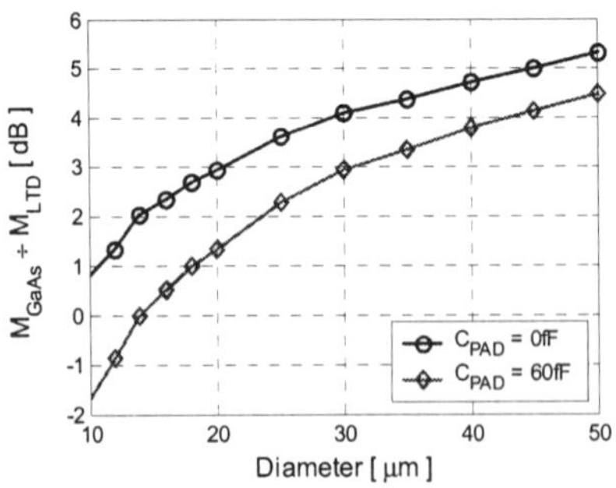

Figure 28: Ratio of the figure of merit for the GaAs vertical PIN device to the silicon LTD device as a function of the device diameter.

separation to optimize the total bandwidth (transit and RC) at a constant 5V bias. The transit time bandwidth takes into account the relationship of the carrier velocity to the electric field, including velocity overshoot in GaAs.[58] The RC-limited bandwidth assumes a standard 50Ω load, although the input impedance of the transimpedance amplifier is not limited to this value. The slow carriers generated below the trenches in the LTD are assumed to be cutoff from the response using SOI or other techniques.

Figure 27 shows calculation of the peak bandwidth as a function of the device diameter. As expected, the bandwidth of the nominal GaAs device is higher for all diameters due to (1) velocity overshoot in GaAs and (2) lower inherent capacitance of a vertical PIN structure compared to a deep trench LTD structure. If we include 60fF of capacitance for the GaAs device, the bandwidth is much lower, especially for small device diameters.

As shown in Figure 28, an optical receiver that utilizes a 50μm diameter GaAs photodiode can achieve a sensitivity over 5dB higher than a receiver with a Si LTD of the same diameter. However, this advantage decreases to 1dB at smaller dimensions. This trend is primarily due to the fact that quantum efficiency drops as the bandwidth of the GaAs device increases since the absorption length is coupled to the n- and p-region separation. In contrast, the absorption in the Si LTD remains constant as the diameter shrinks. If a parasitic capacitance of 60fF is included for the GaAs device, the performance of the LTD actually becomes more favorable at the smallest device dimensions.

Considering the inherent packaging benefits and the fact that optical power is not in short supply in short distance optical links, the Si LTD is a very promising alternative to the GaAs photodiode as data rates increase.

4. Key Passive Elements for High-Frequency Operation

4.1. Inductors

The explosion of wireless devices and the need for low cost solutions has promoted a great deal of research into design, fabrication and modeling of high Q passive inductor structures on silicon substrates. A fairly wide range of inductance values (0.1nH-10nH) are needed to enable diverse applications like filters, matching circuits, bandwidth enhancing circuits and LC oscillators at multi-gigahertz frequencies. The search for low power solutions necessitates need for high Q devices at these design frequencies. A number of references detail design and modeling approaches to realize high Q inductors on silicon.[59,60] The scaling of CMOS processes into the deep sub-micron regime has been beneficial for back-end-of-line (BEOL) passive devices like inductors because of an increase in the number of metal layers in CMOS flow and the transition from Al metallization to lower-resistivity Cu BEOL. Thick dielectric add-on modules with the option of stacking are now feasible because of these highly-planar Cu BEOL processes.

Typical on-chip spiral inductor are either octagonal or rectangular in geometry. More rectangular geometry is primarily seen because of the relative ease of modeling and layout of the structures. Depending on the inductance sought, the rectangular inductor can be implemented as a multi-turn spiral with an underpass. The device design parameters include metal width, number of turns, spacing between turns, and underpass metal width and location. At the technology level, the inductor characteristics are influenced by metal

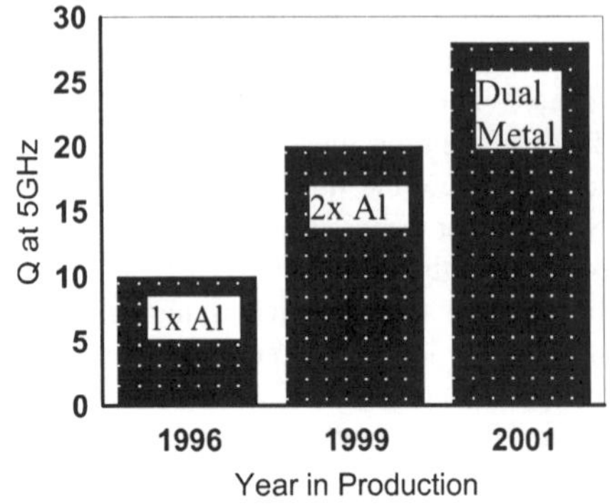

Figure 29. $\boldsymbol{Q}$ of ~1nH inductors in various generations of IBM production technology.

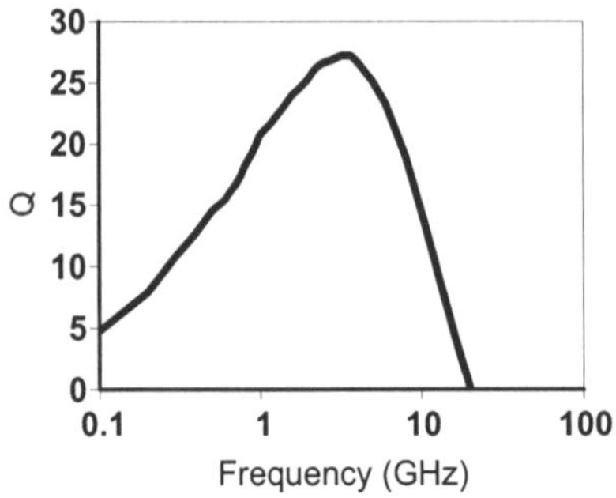

Figure 30. $\boldsymbol{Q}$ vs. Frequency for 1.1nH dual metal inductor

thickness, dielectric material and its spacing from the substrate, substrate resistivity and patterned ground-shield properties. The inductor losses on silicon substrates are primarily a result of series resistance in the inductor and losses originating from the coupling of the inductor to the silicon substrate underneath. Some relatively simple concepts coupled with process innovations are pushing the performance of these inductors. Traditionally the series resistance of spiral inductors has been minimized by increasing the thickness of the metal spiral and the underpass layer. Substrate losses are minimized by using high resistivity substrates, patterned ground planes and increasing the dielectric thickness. While typical substrate resitivities of 10-ohm-cm are commonly seen in bulk CMOS implementation, more latitude can be seen in technologies like SOI where higher resistivities can be supported. Taken together these incremental advancements can lead to inductors with high quality factors enabling integration of passives on chip.

Figure 29 shows the evolution of inductor $\boldsymbol{Q}$ at 5GHz in leading IBM production technologies over a period of time. The inductance is ~1nH in all cases. As can be seen, rapid improvements have been made by employing thicker metal with thicker dielectric separating the metal levels from each other as well as from the silicon substrate. Series stacking of inductors has also been adopted to increase area-utilization efficiency. A successful implementation of this approach can be seen in a dual metal inductor (Figure 30) where Q~28 at 3.5GHz (inductor optimized for this frequency) has been achieved for a 1.1nH inductor.[61] Here the inductance density was increased by 3.6x.

As application frequency increases, skin effect and current crowding in the spiral metal becomes very important. From a VCO design perspective with the increase in desired frequency the inductance value required also reduces. Skin effect limitation can be employed to an advantage by fabricating small inductors (area efficient) in the far BEOL with only single metal layer. The reduced parastics also have a beneficial effect of improving $\boldsymbol{Q}$.

4.2. Varactors

Varactors are another critical component in the tank design in an LC VCO. From the perspective of the circuit designer, the choice of the varactor affects the tank amplitude

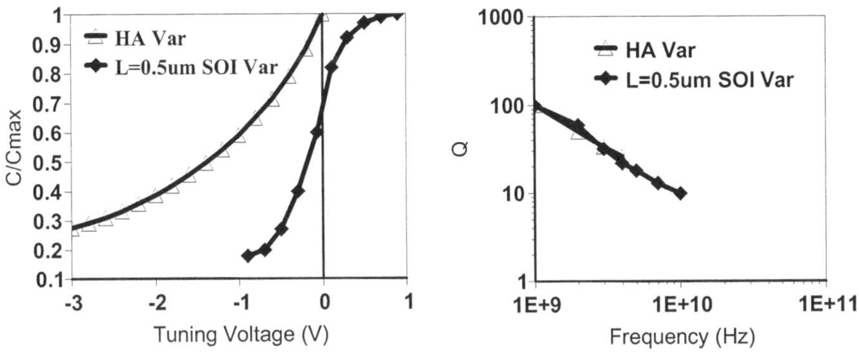

Figure 31 (a) Capacitance tuning and (b) Q characteristics for hyper abrupt and SOI NFET accumulation varactor.

and its losses which leads to degradation of VCO phase noise. The tank losses are proportional to the overall Q of the tank, which in turn are affected by the Q of the varactor. As application frequencies continue to rise beyond 10 GHz, the varactor Q becomes comparable to the on-chip inductor Q values and hence its design is of paramount importance in obtaining a low power VCO that can also meet the phase noise (jitter) specs. In addition to high Q characteristics, the varactor also has to provide sufficient tuning range to accommodate channel selection, process corners and temperature variation.

A typical CMOS or BiCMOS process flow provides a number of sequences that can be combined to design varactors. As implied these are free in the process, in that they do not require additional lithography or masking steps. Two typical varactor configurations that can be realized are the diode varactor and the MOS varactor. Junction varactors in a (Bi)CMOS process are implemented by the formation of p+/n- type diode, where is n-layer is contacted though a highly doped reach through. The Q in such devices is affected by the series resistance of the undepleted portion of the n- region. Maximum tuning can be obtained in the case of a hyper-abrupt (HA) doping profile. However, typical combinations of implants used in CMOS generally do not result in hyper-abrupt profiles. To obtain the tuning characteristics without sacrificing Q, dedicated implants are performed to fabricate such varactors. In such designs, non-scaling fixed capacitance must be minimized to obtain the full benefits of the improved tuning range. Since the varactor gain directly affects the phase noise characteristics of the VCO, the HA varactors can also be designed to minimize this gain. Figure 31 (a) shows the tuning range of the HA varactor and Figure 31 (b) shows Q as a function of frequency.

Recently much interest has also been focussed on the MOS varactors. In a MOS varactor, the capacitance variation results from the device transitioning from accumulation to depletion or inversion to depletion. Given the modes of device operation and the structures that can be constructed, a number of different varactor configurations are possible.[62] Generally speaking there are two broad categories of MOS varactor: one where the device is operated in an inversion-depletion mode and the other where it is operated in an accumulation-depletion mode. For inversion mode varactors, the capacitance change happens over a very small bias range (a few kTs) and also these devices have a smaller tuning range and suffer from lage parasitic channel resistance in weak inversion as compared to an accumulation mode varactor. As a consequence, NFET in Nwell accumulation mode varactor (where the poly silicon is also n-type doped) is a popular approach in realizing high Q MOS varactors. In these devices, the threshold voltage is shifted towards zero bias making this quite appealing to the designers. The Q of the varactor with the applied bias at a particular frequency is dependent upon a number of factors including the formation of parasitic resistance and capacitance. Typically, minimum gate length devices show the best Q characteristics at the expense of tuning range. The advantage of the MOS-type varactor is that a large tuning range can be easily achieved. The device also scales with each generation of the technology and is compatible with the power supply at that technology node. On the flip side the capacitance transitions over a smaller frequency range and as a result of this large gain, the phase noise of the VCO negatively impacted. However, some of these drawbacks can be compensated by innovative design techniques. As an example, a bank of varactors in an inband switching topology used to maintain tuning range while reducing the varactor gain has been shown to reduce phase noise.[63] This is an example of a quintessential CMOS solution wherein the control logic to select the various capacitors in the bank can be easily implemented without paying a penalty in the dissipated power. MOS varactors also benefit from other CMOS enhancement like the use of SOI substrates. Differential control topology has been proposed for improved noise rejection and suppression of common mode noise.[64] NFET in Nwell MOS varactors on SOI substrate have an additional advantage of symmetry of the gate with respect to the substrate, as result of which wide tuning range can still be sustained under differential operation. Figure 31(a) also shows tuning characteristics of NFET accumulation varactor with a gate length of 0.5μm fabricated on SOI substrates and Figure 31(b) shows its Q variation vs. frequency.

4.3. VCO performance

High performance VCOs have been designed and fabricated for use in 40Gb/s clock and data recovery (CDR) units in 0.18μm SiGe BiCMOS technology.[65] The VCO designs conform to strict specifications required in production CDR systems. Up to 27 GHz frequency oscillation has been measured with a phase noise better than -100 dBc/Hz at 1 MHz offset from the carrier. Coarse control in the PLL is performed using FET varactor bank and fine tuning is achieved by using junction varactors. Figure 32 shows the phase noise of the VCO across various coarse-control bit sequences and temperature of operation. A 40Gb/s VCO has also been demonstrated in 0.13μm SOI CMOS technology.[66] The tuning in this VCO is achieved by using accumulation mode NFET varactors. Because of low parasitic capacitance in the varactor, large varactors and small inductors could be used in the design, leading to high area-utilization efficiency.

4.4 Transmission line interconnects

Microstrip and strip transmission lines, coplanar strip lines and waveguides are often utilized in high frequency RF circuits, digital and mixed-signal applications to avoid signal integrity problems for critical nets. These interconnects can range from narrow digital clock nets over a wide common ground plane to wide (~100μm) analog transmission lines in power amplifiers.

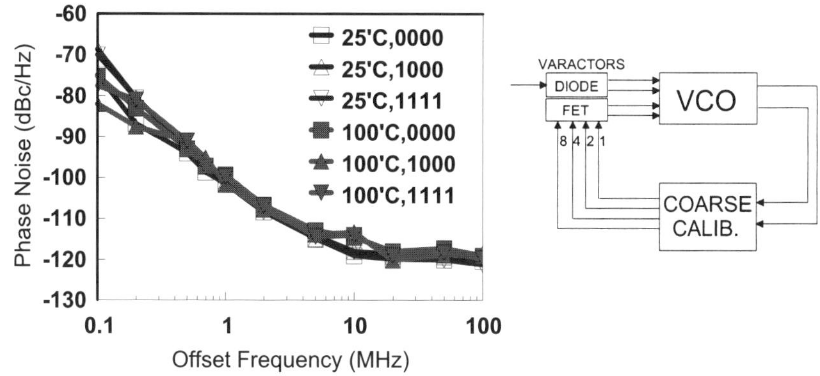

Figure 32 (a) 20 GHz VCO Phase noise as a function of temperature and 4-bit coarse frequency control. On the right is a block diagram of the VCO illustrating use of both FET and junction diode varactors.

Traditionally, transmission line interconnects are employed to control high frequency distributed impedance effects that become important as the wavelength approaches a fraction of the line length. Increasingly, transmission line structures are also employed to isolate critical interconnects from the lossy substrate either through reference ground plane shielding signal lines from below or field containment in coplanar-like structures.

Essential to transmission line capability is both the technology in which it is implemented, and also an accurate and flexible design system. The transmission line technology establishes structural constraints and thus defines the tradeoffs a designer faces such as between losses and impedance matching. The design system enables the designer to understand these tradeoffs and establish first-pass success.

Transmission line technology

From a technology viewpoint, the design of the metal interconnect stack takes into account many issues not obvious at first glance. Compatibility with high-performance CMOS, integration with low-cost metal-insulator-metal (MIM) capacitors, as well as low-cost wire-bond need to be taken into account.

The most simple transmission line structures to design and manufacture are microstrip lines depicted in Figure 33. These interconnects are effectively shielded from below with possible side shielding to eliminate the silicon substrate losses and coupling to other circuit elements. Optional side shielding are made of stacked vias. They work well as the side electromagnetic shield. Usually, top thick metal layers are used as signal wires with a ground shield on the lower copper layers. This enables the use of microstrip lines for the most critical interconnects including power lines in RF power amplifiers.

To enable high-performance CMOS on the same chip, the tight-pitch thin-wire metal system nearest the silicon must be established the same as the digital technology. On top of these thin metal layers, thick metal layers with thick inter-level dielectrics create low resistance films and low unit area capacitance. This combination is favorable for the low loss and high impedance required by designers. Figure 34 shows the comparison of two 50-Ohm single wire microstrip lines made of 4μm thick aluminum (width of 10.1μm and 9.25μm distance from a metal ground-plane) and a 1.25μm thick aluminum (width of 4.7μm and distance 4μm from a metal ground-plane). Both lines have the same length of

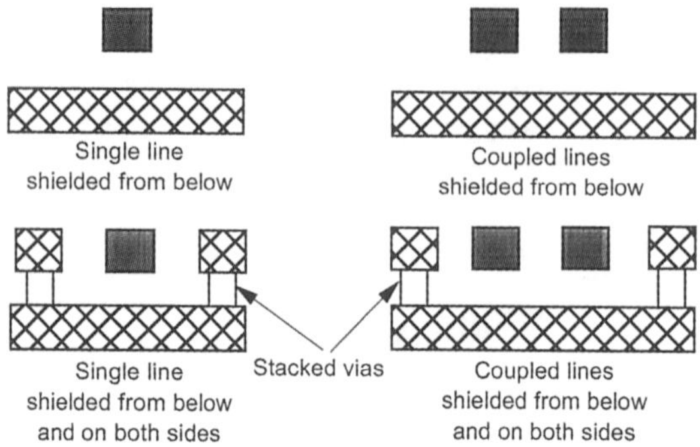

Figure 33: Microstrip transmission lines with optional side shielding.

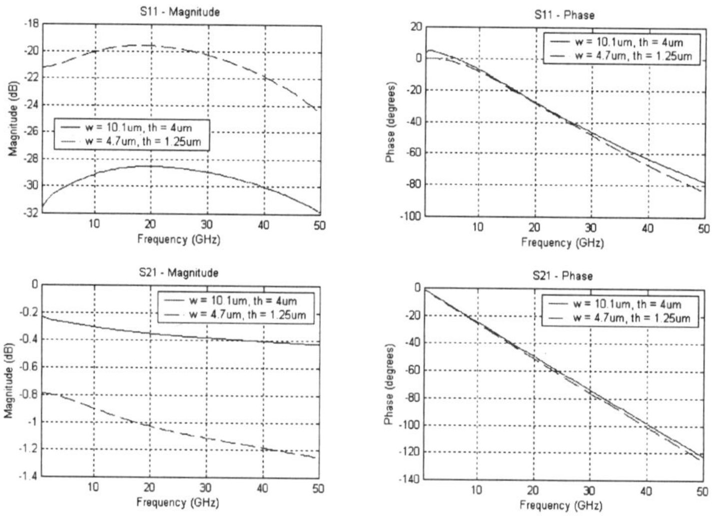

Figure 34: Modeled s-parameter comparison between two metal structures: thin 1.25μm metal lines (comparable to CMOS technology), and thicker 4.0μm metal lines with additional distance to the ground-plane. Both are 50Ω single lines with no side shielding.

1000μm. It can be clearly seen from Figure 34 that thicker interconnect has lowest losses. Figure 35 also shows extracted RLCZ parameters for the above lines. Again, thicker line has lower resistance and inductance. Its characteristic impedance is more close to 50 Ohm across the frequency range of interest.

Compared to a metal line over a 100μm semi-insulating GaAs or InP substrate with backside ground-plane, this ~10μm dielectric configuration exhibits somewhat greater loss. For highly integrated solutions however, one must consider packing density and signal cross-talk between lines. Comparison has been made between these metal configurations[67], and distances a factor of ten greater (i.e. comparable to the ground-plane separation difference) are required to maintain similar (e.g. < -10dB) coupling between lines. Thus there is a tradeoff in the loss and the attainable density, which favors the

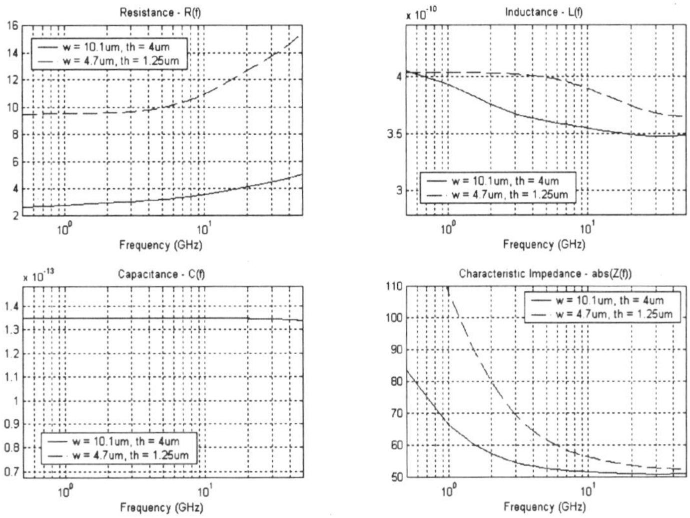

Figure 35: Modeled RLCZ parameter for same lines as represented in Figure 34.

topside groundplane approach over the backside approach for even moderate density designs.

Modeling aspects

A key element to design of transmission lines is the ability to rapidly optimize transmission line geometry. Device component equations are calculated in the component description form (CDF) as well as the model files, including compatible interconnect libraries for a broad range of industry-leading SPICE-compatible tools.

A suitable interconnect modeling methodology is through the development of SPICE-compatible lumped element models[68], preferably passive by construction. The models are parameterized frequency-dependent RLC lumped circuits that have been correlated using hardware and full-wave EM solver. The equivalent models consist of series RLC lumped segments in combination with ladder networks. Ladder networks are used to model the frequency-dependent skin effect. They consist of passive, frequency-independent elements and are relatively easy to netlist and simulate in commercial SPICE-like simulators. Examples of a ladder network can be found.[69] As far as RLC blocks are concerned, C and L parameters must be calculated carefully to take into account mutual electromagnetic coupling. The number of needed RLC cells is usually calculated in a way, such that there are at least 10 segments per shortest on-chip wavelength.[70,71]

The modeling technique, adopted for microstrip transmission line modeling, uses quasi-static transverse electromagnetic (TEM) approximation to calculate low frequency capacitance and inductance of an interconnect. The quasi-static approach has good accuracy because the cross section dimensions of the on-chip interconnects are very small in comparison with the shortest on-chip wavelength even for high frequencies of operation. After low frequency capacitance is estimated, it is used to calculate high-frequency inductance limit. The next step combines low and high-frequency inductance values to construct ladder networks. The final transmission line model has strong frequency dependencies of resistance and inductance. At the same time, capacitance is almost constant over a high-frequency bandwidth (see Figure 35).

Verification examples

As the first verification example, we simulated a single transmission line (singlewire) without side shielding (see Figure 33) using the adopted lumped element model in combination with Spectre simulation tool. Figure 36 represents *S*-parameter correlation results (model vs. EM solver) over a wide frequency range of 0.1-40 GHz. Other configurations show on both EM solver and model-hardware comparison predictability within 1 dB through 40 GHz.

We also compared the interconnect transmission model with measurement results. They can be found in Figure 37, and are accurate to within acceptable levels with S21 magnitude differing by less than 1 dB across the frequency range of interest.

These transmission-line characteristics, shown accurate up to 40 GHz, and the tools, at the center of which is an compact model methodology, enable the designers to design critical nets in complex metal systems of silicon technologies with great accuracy and with first-pass success.

5. Substrate Effects in RF and Microwave IC Designs

In the last decade, there has been a number of research papers published offering practical and usable approaches to model substrate coupling. This surge of research, in the areas of RF and mixed-signal designs, has been triggered by a realization in the chip design industry that substrate coupling is an important issue. This realization has in main arisen by the adoption of single IC solutions. Smaller gate lengths have lead to higher frequencies of circuit operations and to high levels of integration. A complete background to substrate modeling and coupling issues can be found in ref. 71.

Figure 38(a) shows a block level representation of a sensor system with a Digital Signal Processing (DSP) stage. This design is indicative of single IC solutions, now possible with recent technology improvements. Such designs can contain more than 1 million digitally switching transistors. The substrate coupling problem occurs when analog circuitry is integrated on to the same IC. However, it does not need such large designs for substrate coupling to be a problem. Substrate coupling can also degrade performance in small RF designs, as shown in Figure 38(b).

Given these trends, it is somewhat surprising that substrate modeling strategies are not yet common in commercial verification design flows. This is partly explained by the fact that the task of substrate modeling is not simple and inefficiencies can lead to impractical extraction and simulation time and memory requirements. There are many aspects to understanding and dealing effectively with substrate effects. In this paper, we study two key aspects:

(i) Use of technology CAD (TCAD) modeling to extend the work of test sites and utilize existing, calibrated, simulation of discrete devices and processes.

(ii) Measurement of test site structures to derive empirically fit equations for use in modeling substrate isolation between circuit blocks.

The results presented in this paper are the first part of a multi-stage project to coherently understand and then build equivalent models to predict the effects of the substrate. This

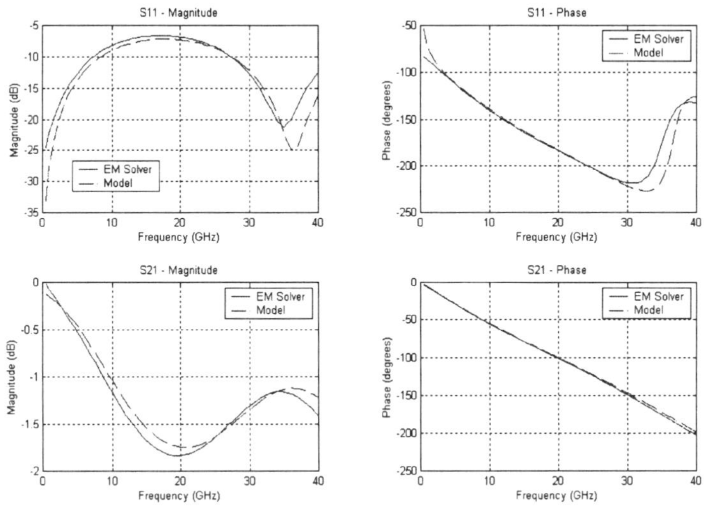

Figure 36: Model vs. EM Solver: S-parameters, single line, no side shielding.

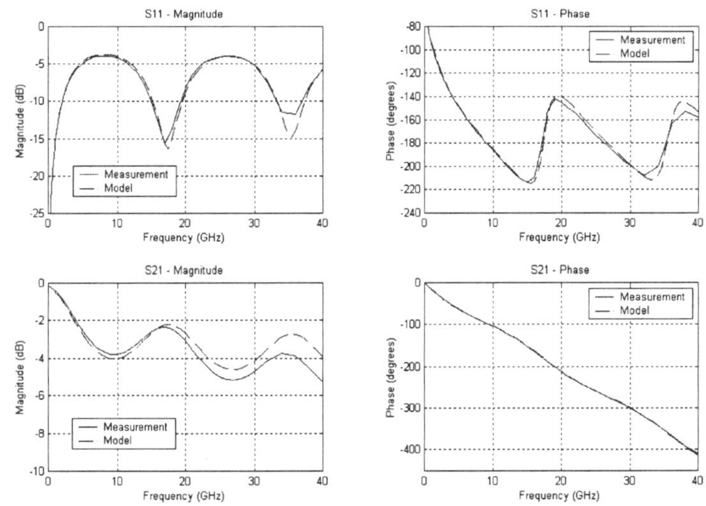

Figure 37: Model. vs. Hardware: S-parameters, single wire, no side shielding.

development work is currently focused on bulk silicon processes, and is planned to cover silicon-on-insulator substrates[72] in the future. It is also planned to use this work to understand the effects of substrate noise to and from spiral inductors.[73]

5.1 Substrate isolation evaluation with technology CAD

TCAD has been used for some time as a method for accurate modeling of active devices in the substrate, contributing significantly to both CMOS and bipolar technology scaling.

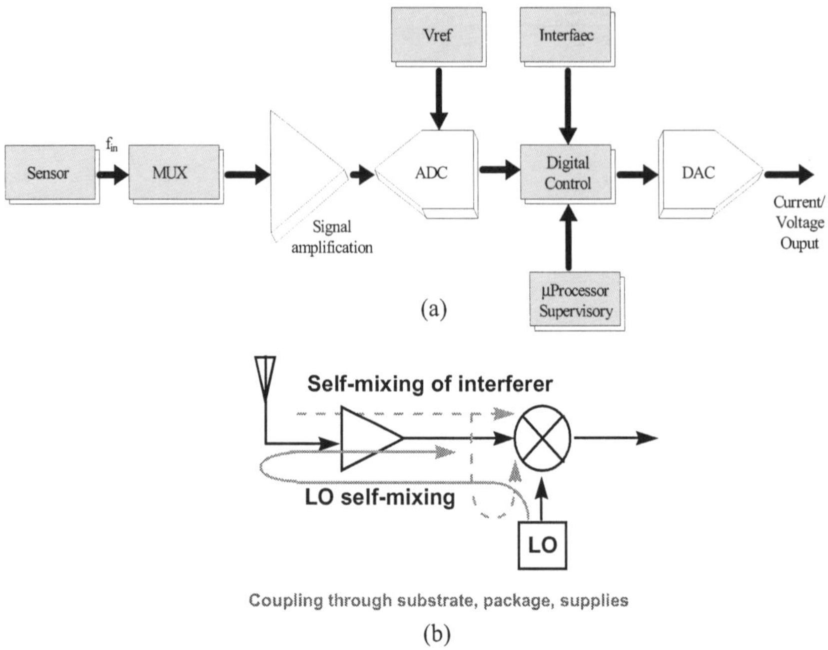

Figure 38: Mixed-signal and RF circuits that typically experience substrate noise coupling problems. (a) A high level representation of a DSP sensor system. (b) substrate coupling leading to LO feedthrough, in a direct-conversion receiver [74]

While its contribution to CMOS scaling is primarily concerned with simulation of DC parametrics such as threshold voltage and effective channel length, it has proven to be an invaluable tool for optimization of HBT AC characteristics such as f_T and f_{MAX} .[74] This approach is equally applicable to substrate isolation modeling .

Two approaches to applying TCAD to substrate signal propagation appear in the literature:
(i) Small-signal *S*-parameter simulations of large and fairly simple structures.[75,76,77,78]
(ii) Large-signal transient and DC simulations of more complex structures.[79,80,81,82,83]
Example of both of these two approaches to TCAD analysis of substrate isolation problems follow.

Large-signal simulations

Large-signal semiconductor device simulation consists simply of the standard DC semiconductor equations augmented by time-vary boundary conditions and addition of the partial derivatives of the carrier-concentrations with respect to time.[84] This technique, which has been observed to be numerically stable and robust, allows the simulation of arbitrarily complex transient waveforms with a resulting computer memory-storage burden equivalent to that of the DC problem.

As an example of the utility of the large-signal approach, an extensive set of two-dimensional large-signal transient simulations were performed on a receiver structure consisting of a nominal emitter-size DT-bounded f_T=47GHz SiGe HBT[1] biased as a simple common-emitter amplifier, at peak-f_T I_C, surrounded by a 1-µm wide oxide-lined

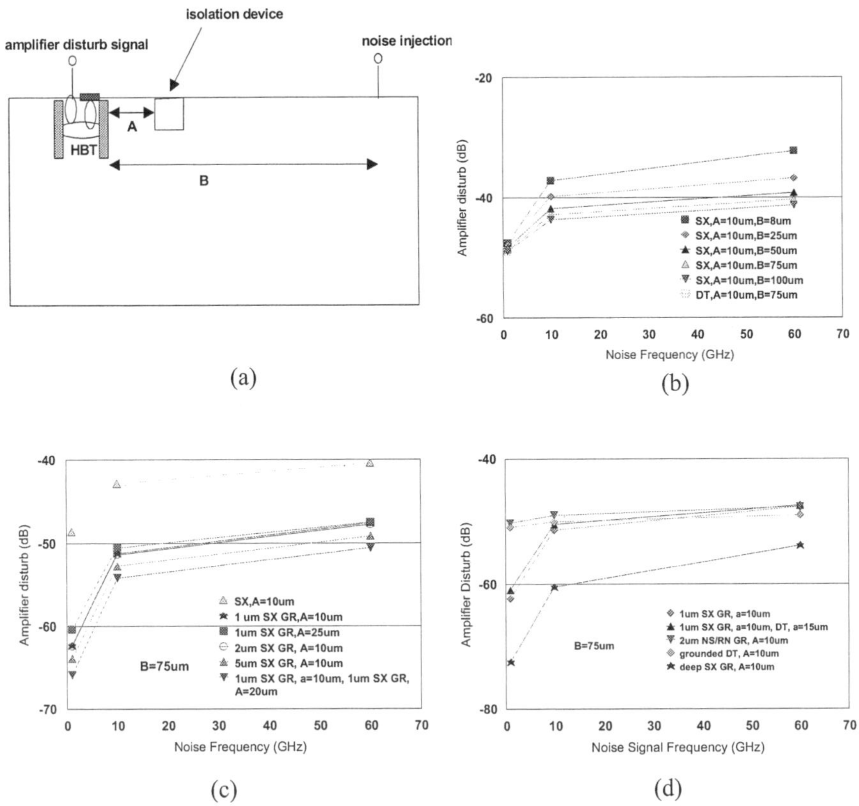

Figure 39: Large signal simulations (a) structure (b) p- substrate (SX) isolation comparison, without SX GR (c) grounded substrate contact (d) alternatives to the simple grounded substrate guard ring (SX GR).

and polysilicon-filled deep trench (DT). The structure was created by a detailed two-dimensional process simulation of the manufactured device, and includes a heavily-doped buried subcollector and implanted extrinsic base. A 100mV sinusoidal noise signal at three different frequencies, 1, 10 and 60GHz, was injected into the substrate at varying distances from the victim structure. In order to assess the most effective way to isolate the HBT from substrate noise, beyond the isolation provided by the DT, different isolation structures, also arising from detailed process simulation, were placed between the noise source and the HBT listener. In each of these cases, the substrate is 10 Ω-cm and the default p-well channel-stop implant is omitted except in the areas defined as a substrate contact. The general 2-D simulation layout is depicted in Figure 39(a). The resulting signal at the collector of the HBT was recorded for these noise inputs and is reported as the ratio of the maximum noise voltage to the maximum signal disturb at the collector of the listener HBT.

Figure 39(b) plots the simulation results for simply spacing the noise signal an increasing distance away from the HBT listener. In this case, the isolation structure is merely the lightly doped p- substrate. The advantage of increased spacing is observed to saturate for distances above 50μm. Augmenting the simple substrate isolation with an additional DT, spaced 10μm from the listener HBT, was found to provide no isolation benefit.

Figs. 39(c) and 39(d) plot the results of a second set of simulations run to evaluate the utility of alternative substrate isolation structures. The often-observed utility of a simple p+ guard-ring (SX GR), in this technology consisting of a PFET source/drain diffusion in a p-well, compared to the simple low-doped substrate, is apparent in Figure 39(c). Adding a DT between the SX GR and the noise source is observed to actually degrade the isolation, presumably because the DT isolates the SX GR from the noise signal. An additional SX GR is observed to add an additional 3.5 dB of isolation. Positioning the SX GR closer to the noise source marginally degrades the substrate isolation, while making the SX GR large marginally improves substrate isolation.

A further enhancement of the SX GR is to make it deeper via a sequence of implants in the MeV range. A process for this type of isolation structure was formulated and simulations of this structure, shown in Figure 39(d), indicate an additional 6-10 dB in noise isolation could be achieved by significantly extending the p+ SX GR depth. The utility of increased depth in the SX GR suggests the use of the HBT n-type buried subcollector, contacted by the HBT 'reachthrough' or 'sinker' diffusion as an isolation device, as it is very heavily doped and lies relatively deep in the substrate. However, the simulations of this device indicate that relative to the SX GR, it is an inferior isolation structure at lower noise frequencies, presumably due to the p-n junction it forms with the p- substrate, and has no advantage over the simple SX GR at higher frequencies. A final option considered, based on the observed utility of the deeper SX GR, is the polysilicon-filled deep trench. The polysilicon fill in the DT is floating in the normal process. However, straightforward process changes were devised to allow p-type doping and contacting (to ground) the DT polysilicon fill. When simulated, the isolation characteristics of this device were very similar to that of the HBT buried subcollector.

Small-signal simulations

When applied to the semiconductor device equations, the small-signal formalism[85] effectively duplicates standard AC measurement conditions and resulting device behaviour. Complex y-parameters are produced at the cost of significantly increased problem size resulting from simultaneous solution of the both the real and complex parts of the sinusoidal AC signal.

The utility of AC small-signal capability is demonstrated by simulation of the the structure in Figure 40(a). Two p+ diffusions in a p- substrate, without p-well channel-stop implant, are separated by large distances (50-150 μm) together with symmetrically placed isolation structures in-between. All contacts are at zero bias. In Figs. 40(b) and (c), isolation vs. frequency is shown for various isolation configurations. In each of these charts, note that the curves rise at highest frequencies, indicating that isolation is reduced at frequencies over 10GHz. This phenomenon will be discussed in the next section.

Figure 40(b) plots the results for the case where the isolation structures are simply the p- substrate. As in the large-signal case, for distances greater than 50 μm, the utility of increased injector/listener distance appears to be small, independent of frequency. Note that Figure 40(b) also indicates the negligible contribution of DT as a substrate isolation device. Figure 40(c) plots the results obtained from using grounded SX GR structures,

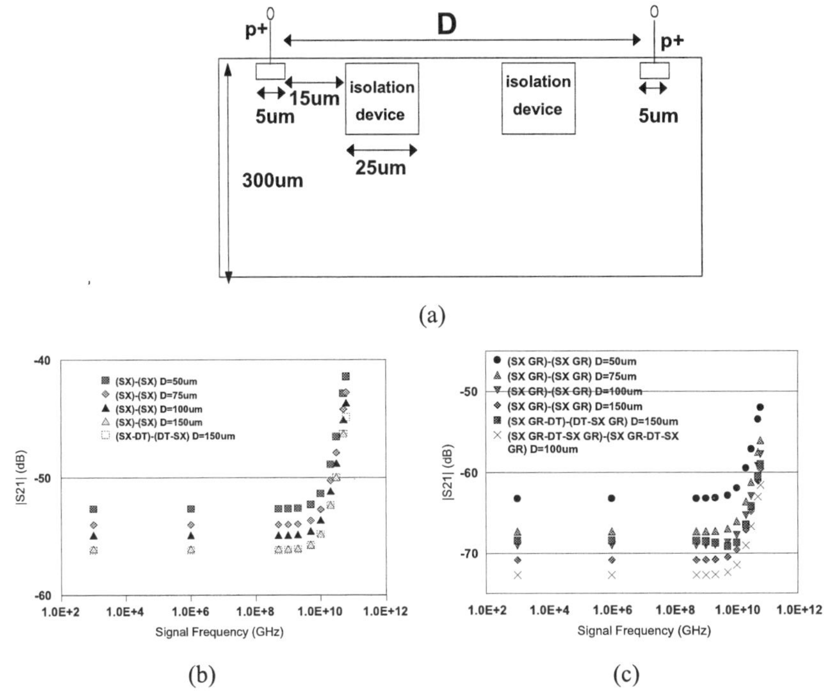

Figure 40: 2-D small-signal AC simulations. (a) structure where the two signal ports are assumed to be 5μm x 5μm and the substrate is doped p-; (b) isolation increasing distance without SX GR and (c) addition of SX GRs between signal and receiver ports.

showing improvements over the 50 μm separation case of approximately 10 dB for 75 μm separation and increasing to 15 dB of noise isolation for 150μm separation. Note that the addition of DT to the SX GR noticeably degrades the substrate noise isolation at very low frequencies, and that such structures actually are observed to show improved substrate isolation as noise-signal frequency increases between 1GHz and 10GHz, presumably as the DT becomes transparent to the electrical signal and no longer shields the SX GR from it. Additional SX GR isolation structures are observed to improve substrate noise isolation, but at a slower rate compared to the addition of the first one.

Summary

TCAD provides an analysis tool for substrate isolation issues that bridges the gap between hardware measurement and high-level circuit simulation. By basing the simulation on structures derived from detailed process simulation, insight into complex signal propagation phenomena in technologically-relevant structures can be gained without the time and expense of hardware build or the challenges of high frequency measurements. However, maximum leverage of TCAD for substrate isolation analysis, which will require

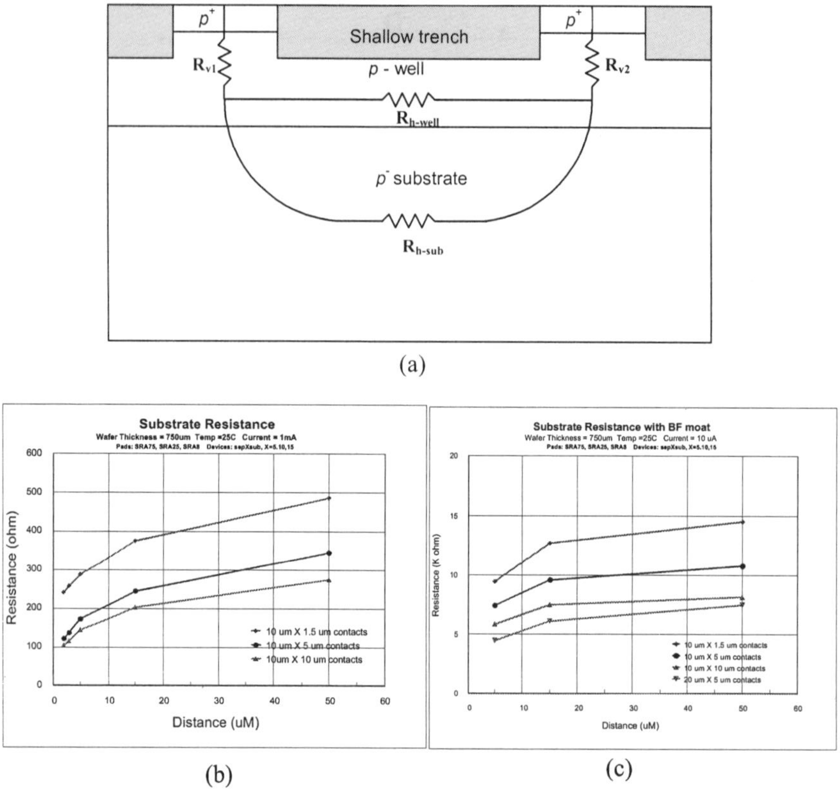

Figure 41: Substrate resistance measurement (a) structure and simple model representation (b) substrate resistance vs. separation distance for different contact sizes, with default pwell channel-stop under isolation (c) with p-well channel-stop implant omitted between contacts.

simulation of large-scale three-dimensional structures, represents a formidable computational challenge is just starting to be addressed.[75]

5.2 Hardware measurements

A series of systematic test-sites were developed, produced and measured for this work, using IBM's SiGe process. A p- substrate (10 Ω-cm) was implemented, with a p-well between the devices by default. These results are presented in this section as new important data for understanding the isolation effects of the substrate.

Based on measurements between identical substrate contacts of different sizes as a function of separation, it was concluded that the total resistance could be broken into horizontal and vertical components as shown in Figure 41(a). The vertical component depends on the substrate contact size and the horizontal component on the separation between the contacts. Due to relatively high doping near the surface in the p-well region, a majority of substrate current is expected to flow in the p-well region for short distances relative to the substrate thickness. Figure 41(b) shows the measurement results as a function of distance on a 750μm thick wafer. A logarithmic relationship was observed for the horizontal component of the substrate resistance. The empirical formula, shown in Equation (3), can be used to estimate substrate resistance up to about 50μm distance:

$$Rsub = 2*Rsv/wl + A*\log(d+0.4*w) \,||\, Rsh*(d + 0.4*w)/ l) \quad (3)$$

where Rsv = vertical resistance of the contact for unit area (= 1725 W-μm^2)

w = contact width (in μm)

l = contact length (in μm)

d = contact separation (in μm)

A = weak function of contact width

(205, 195 and 170 for 1.5 μm, 5 μm and 10 μm widths respectively)

Rsh = sheet resistance of the p-well

IBM's RF technologies offer layout options for increasing isolation between noisy and sensitive parts of the circuits. The default p-well channel-stop region can be omitted from selective regions of the layout, giving a lightly doped n-type region near the surface. Since the substrate current flows in the p- substrate in these isolated regions, higher substrate impedance can be achieved. Another isolation technique available in high performance BiCMOS technologies is the use of deep trench to force the current to flow deeper into the substrate (in the high resistive region). Figure 41(c) displays the substrate resistance in p-well omitted region. Comparing these results with Figure 40(b), isolation resistance can be increased by orders of magnitude by using such isolation.

IBM's SiGe processes use 11-16 ohm-cm substrates to minimize substrate losses. An interesting effect of using a lightly doped substrate is that the transition frequency Ft of the substrate (not to be confused with transistor current-gain cutoff frequency f_T) moves down to between 10 GHz and 15 GHz. For a substrate model, which consists of a network of resistors, the transition frequency is the frequency at which each of the resistors is shunted by a capacitive reactance, which is equal in magnitude to the value of each of the resistors. Equations (4) and (5) describe how to obtain values for F_t and C_{shunt}.[86]

$$F_t = 1/2\pi\rho\varepsilon \quad (4)$$

where:

F_t - substrate transition frequency (Hz)

ρ - substrate resistivity (Ohm-cm)

ε - substrate permitivity (Farads/cm)

$$C_{shunt} = \rho\varepsilon / R_{shunt} \quad (5)$$

where:

R_{shunt} - substrate resistive element (Ohms)

C_{shunt} - substrate capacitive element (Farads)

At higher frequencies, substrate currents may increase due to the ever-decreasing substrate impedance and substrate power dissipation may decrease since less current will be flowing through the resistive part of the substrate network. Parasitic capacitance of metal begins to be terminated into a series capacitor instead of a series resistor. Understanding such complex substrate effects, and providing guidance to the designers in the form of application notes for isolation optimization, and providing model capability will substantially improve the time to successful designs.

6. Summary

Lithography scaling, structure improvements, and material innovations are enabling silicon technologies to extend their reach well above 10 GHz. Device f_T values for SiGe bipolar devices are now exceeding 200GHz, with FET devices only slightly behind. For these devices, f_{MAX} values are also exceeding 200GHz. Other novel structures, such as the lateral trench photodetector, are further extending silicon capability into a realm of III-V devices. Also enabling these technologies are the passive elements, a variety of which may be fabricated on the same silicon chip. High quality inductors, varactors and transmission lines are now integrated with the high performance active devices. Further, substrate noise isolation has been explored and over 70 dB isolation is readily achieved between circuit blocks.

Acknowledgments

Acknowledgment is given to Sue Strang, Donald Jordan and Carl Dickey of the IBM Microelectronics (Burlington, Vermont, USA), David Goren, Rachel Gordin, and Michael Zelikson of the IBM Research and Development Laboratories in Haifa (Haifa, Israel) and Mike Longstreet of the IBM Microelectronics (Fishkill, New York, USA).

The authors would also like to acknowledge the support of Dennis Rogers and Min Yang for support and encouragement throughout the course of the trench photodetector work.

For the MODFET work presented here, the authors would like to acknowledge the efforts of numerous members of IBM research and microelectronics who contributed to the results presented in this paper. The authors would also like to acknowledge the partial support of SPAWAR Systems Center, SD, Office of Naval Research, under Contract No. N66001-99-C-6000, and DARPA under Contract No. N66001-00-C-8086.

References

1 S.A. St. Onge, D.L. Harame, J.S. Dunn, S. Subbanna, D. Ahlgren, G. Freeman, B. Jagannathan, S.J. Jeng, K. Schonenberg, K. Stein, R. Groves, D. Coolbaugh, N. Feilchenfeld, P. Geiss, M. Gordon, P. Gray, D. Hershberger, S. Kilpatrick, "A 0.24μm SiGe BiCMOS mixed-signal RF production technology featuring a 47 GHz f_T HBT and 0.18μm L_{EFF} CMOS," Proc. Bipolar/BiCMOS Circuits and Technology Meeting, 1999 pp. 117 -120

2 Y.M. Greshishchev, P. Schvan, J.L. Showell, Mu-Liang Xu. J.J. Ojha, J.E. Rogers, "A Fully Integrated SiGe Receiver IC for 10-Gb/s Data Rate" *IEEE Journal of Solid-State Circuits*, v. 35 no. 12 , Dec 2000 pp. 1949 -57

3 G. Freeman, M. Meghelli, Y. Kwark, S. Zier, A. Rylyakov, M. Sorna, T. Tanji, O. Schreiber, K. Walter, J.-S. Rieh, B. Jagannathan, A. Joseph, S. Subbanna, "40 Gb/s Circuits Built From a 120 GHz f_T SiGe Technology," to be published, *IEEE Journal of Solid-State Circuits*, Sept. 2002

4 AMCC press release, Sept. 10, 2001. Http://www.amcc.com/Compinfo/PressReleases/ S76803.htm, Part number AMCC S76803DI, reference http://www.amcc.com/Products/Sonet/ prod_322.htm

5 J. Cao, A. Momtaz, K. Vakilian, M. Green, D. Chung, K.-C. Jen, M. Caresosa, B. Tan, I. Fujimori and A. Hairapetian, "OC-192 receiver in standard 0.18μm CMOS," Tech. Dig. 2002 IEEE Int. Solid-State Circuits Conf. (San Francisco, CA), 250-251, 2002.

6 M. Thiel, H. Kuhnert, W. Heinrich "Active RF-SiGe Circuits on Standard IBM Foundry Process up to 20 GHz", Second Joint Symposium on Opto- and Microelectronic Devices and Circuits, pp. 31-37, March 2002

7 R. Williams, "Modern GaAs Processing Methods," Artech House, Inc, 1990

8 S. Subbanna, R. Groves, B. Jagannathan, D. Greenberg, G. Freeman, E. Eld, R. Volant, D. Ahlgren, B. Martin, K. Stein, D. Herman and B. Meyerson, "Using Silicon-Germanium Mainstream BICMOS Technology for X-Band and LMDS (25-30 GHz) Microwave Applications," *2002 IEEE MTT-S Proceedings*, p. 401

9 S. Voinigescu, P. Popescu, P. Banens, M. Copeland, G. Fortier, K. Howlett, M. Herod, D. Marchesan, J. Showell, S. Szilagyi, H. Tran, J. Weng, "Circuits and Technologies for Highly Integrated Optical Networking IC 's at 10 Gb/s to 40 Gb/s" 2001 Custom Integrated Circuits Conference

10 S. Thomas III, J.A. Foschaar, C.H. Fields, M.M. Madhav, M. Sokolich, R.D. Rajavel, B. Shi, "Effects of Device Design on InP-Based HBT Thermal Resistance," *IEEE Trans. On Device and Materials Reliability*, vol. 1, no. 4, 2001, pp. 185-189

11 S. V. Kosonocky, M. Immediato, P. Cottrell, T. Hook, R. Mann, J. Brown, "Enhanced Multi-Threshold (MTCMOS) Circuits Using Variable Well Bias", ISLPED (2001) 165.

12 D. Buss, "Device Issues in the Integration of Analog/RF Functions in Deep Submicron Digital CMOS", IEDM Tech. Dig. (1999) 423.

13 N. Zamdmer, A. Ray, J.-O. Plouchart, L. Wagner, N. Fong, K. A. Jenkins, W. Jin, P. Smeys, I. Yang, G. Shahidi, F. Assaderaghi, "A 0.13-μm SOI CMOS Technology for Low-power Digital and RF Applications", VLSI Tech. Symp. (2001) 85.

14 N. Zamdmer, J.-O. Plouchart, J. Kim, L.-H. Lu, S. Narasimha, P. A. O'Neil, A. Ray, M. Sherony, L. Wagner, "Suitability of Scaled SOI CMOS for High-Frequency Analog Circuits", to be published in Proc. ESSDERC (2002).

15 R. Aparicio and A. Hajimiri, "Capacity Limits and Matching Properties of Integrated Capacitors", IEEE J. Solid-State Circuits (1998) 2035.

16 P. H. Woerlee, M. J. Knitel, R. van Langevelde, D. B. M. Klaassen, L. F. Tiemeijer, A. J. Scholten and A. T. A Zegers-van Duijnhoven, "RF CMOS Performance Trends", IEEE Trans. on Elec. Dev. (2001) 1776.

17 G. C-F Yeap et al., "A 180 nm Copper/Low-k CMOS Technology with Dual Gate Oxide Optimized for Low Power and Low Cost Consumer Wireless Applications", VLSI Tech. Symp. (2000) 150.

18 T. B. Hook, J. Brown, M. Breitwisch, D. Hoyniak, R. Mann, "High-Performance Logic and High-Gain Analog CMOS Transistors Formed by a Shadow-Mask Technique with a Single Implant Step", submitted to IEEE Trans. on Elec. Dev.

19 J. Welser, J. L. Hoyt, S. Takagi, and J. F. Gibbons, "Strain Dependence of the Performance Enhancement in Strained-Si n-MOSFETs", Tech. Dig. Int. Electron Devices Meet. (1994) 373-376.

20 K. Rim, J. Welser, J. L. Hoyt, and J. F. Gibbons, "Enhanced Hole Mobilities in Surface-channel Strained-Si p-MOSFETs", Tech. Dig. Int. Electron Devices Meet. (1995) 517-520.

21 K. Rim, J. Chu, K. Jenkins, K. Lee, A. Mocuta, R. Roy, J. Newbury, H. Chen, H. Zhu, T. Kanarsky, D. Boyd, K. Petrarca, J. Ott, P. Mooney, D. Lacey, S. Koester, K. Chan, M. Ieong, and H.-S. Wong, "Characteristics and Device Design of Sub-100 nm Strained Si N- and PMOSFETs", 2002 Symposium on VLSI Technology, Honolulu, HI, Jun. 10-13, 2002.

22 K. Ismail, "Si/SiGe High-Speed Field-Effect Transistors", Tech. Dig. Int. Electron Devices

Meet. (1995) 509-512.

23 J. O. Chu, and R. Hammond, unpublished.

24 S. J. Koester, R. Hammond, J. O. Chu, "Extremely high transconductance $Ge/Si_{0.4}Ge_{0.6}$ pMODFET's grown by UHV-CVD", IEEE Elect. Dev. Lett. vol. 21 (2000) 110-112.

25 H. J. Herzog, T. Hackbarth, G. Höck, M. Zeuner, and U. König, "SiGe based FETs: buffer issues and device results", Thin Solid Films vol. 380 (2000) 36-41.

26 Y. Taur, and T. H. Ning, Fundamental of Modern VLSI Devices, Cambridge, 1998, p. 285.

27 D. K. Ferry, G. Formicone, and D. Vasileska., "Carrier Transport and Velocity Overshoot in Strained Si on SiGe Heterostructures", Mat. Res. Soc. Symp. Proc. vol. 533 (1988) 31-42.

28 M. Lundstrom, "Scattering Theory of the Short Channel MOSFET", Tech. Dig. Int. Electron Devices Meet. (1996) 387-390.

29 R. Ohba, and T. Mizuno, "Nonstationary Electron/Hole Transport in Sub-0.1 μm MOS Devices: Correlation with Mobility and Low-Power CMOS Application", IEEE Trans. Elect. Dev. vol. 48 (2001) 338-343.

30 K. Rim, J. L. Hoyt, and J. F. Gibbons, "Fabrication and analysis of deep-submicron strained-Si NMOSFET's", IEEE Trans. Elect. Dev. vol. 47 (2000) 1406-1415.

31 S. J. Koester, K. Rim, J. O. Chu, P. M. Mooney, J. A. Ott, and M. A. Hargrove, "The effect of thermal processing on strain relaxation and interdiffusion in Si/SiGe heterostructures studied using Raman spectroscopy", Appl. Phys. Lett. vol. 79 (2001) 2148-2150.

32 P. M. Mooney, S. J. Koester, J. A. Ott, J. L. Jordan-Sweet, J. O. Chu, and K. K. Chan, "Thermal stability of strained Si on relaxed $Si_{1-x}Ge_x$ buffer layers", Materials Research Society Meeting, Boston, MA, Nov. 26-30, 2001.

33 S. J. Koester, J. O. Chu, and R. A. Groves, "High f_T n-MODFETs fabricated on Si/SiGe heterostructures grown by UHV-CVD", Electron. Lett. vol. 35 (1999) 86-87.

34 M. Zeuner, A. Fox, T. Hackbarth, D. Behammer, and U. König, "90 GHz f_T HFET with fully optical self-aligned sub 100 nm gate", 60th Annual Device Research Conference, Santa Barbara, CA, June 24-26, 2002.

35 S. J. Koester, R. Hammond, J. O. Chu, P. M. Mooney, J. A. Ott, L. Perraud, K. A. Jenkins, C. S. Webster, I. Lagnado, and P. R. de la Houssaye, "SiGe pMODFETs on silicon-on-sapphire substrates with 116 GHz f_{max}", IEEE Elect. Dev. Lett. vol. 22 (2001) 92-94.

36 M. Arafa, K. Ismail, J. O. Chu, B. S. Meyerson, and I. Adesida, "A 70-GHz Low Operating Bias Self-Aligned p-Type SiGe MODFET", IEEE Elect. Dev. Lett. vol. 17 (1996) 586-588.

37 G. Höck, G. Hackbarth, U. Erben, E. Kohn, and U. König, "High performance 0.25 μm p type Ge/SiGe MODFETs", Electron. Lett. vol. 34, (1998) 1888-1889

38 G. Höck, T. Hackbarth, N. Käb, H.-J. Herzog, M. Enciso, F. Aniel, P. Crozat, R. Adde, E. Kohn, and U. König, "0.1 μm gate length p-type $Ge/Si_{0.4}Ge_{0.6}$ MODFET with 135 GHz f_{max}", Electron. Lett. vol. 36 (2000) 1428-1429.

39 Q. C. Ouyang, S. J. Koester, J. O. Chu, A. Grill, S. Subbanna, and D. A. Herman Jr., "A Comprehensive Simulation Study of Strained-Si/SiGe nMODFET Scaling for RF Applications", International Conference on Simulation of Semiconductor Processes and Devices, Kobe, Japan, Sep. 4-6, 2002.

40 H. Fukui, "Optimal noise figure of microwave GaAs MESFET's", IEEE Trans. Elect. Dev. vol. 26 (1979) 1032-1037.

41 J. Burghartz, M. Hargrove, C. S. Webster, R. A. Groves, M. Keene, K. A. Jenkins, R. Logan, and E. Nowak, "RF Potential of a 0.18-μm CMOS Logic Device Technology", IEEE Trans. Elect. Dev. vol. 47 (2000) 864-870.

42 B. Jagannathan, M. Khater, F. Pagette, J. –S. Rieh, D. Angell, H. Chen, J. Florkey, F. Golan, D. R. Greenberg, R. Groves, S. J. Jeng, J. Johnson, E. Mengistu, K. T. Schonenberg, C. M. Schnabel, P. Smith, A. Stricker, D. Ahlgren, G. Freeman, K. Stein, and S. Subbanna, "Self-Aligned SiGe NPN Transistors with 285GHz f_{MAX} and 207GHz f_T in a Manufacturable Technology," *IEEE Electron Device Letters* v.23 n.5 May 2002

43 B. Jagannathan, M. Meghelli, A. Rylyakov, R. Groves, A. Chinthakindi, C. Schnabel, D.

Ahlgren, G. Freeman, K. Stein, and S. Subbanna, "A 4.2ps ECL Ring-Oscillator with 250mV Voltage Swing in a 285GHz f_{MAX} SiGe Technology," Accepted for publication, *IEEE Electron Device Letters*

44 H. Rucker, B. Heinemann, D. Bolze, D. Knoll, D. Kruger, R. Kurps, H.J. Osten, P. Schley, B. Tillack, P. Zaumseil, "Dopant Diffusion in C-doped Si and SiGe: Physical Model and Experimental Verification," *IEDM Tech. Dig.*, 1999, pp. 345-348.

45 J. Bock, H. Schafer, H. Knapp, D. Zoschg, K. Aufinger, M. Wurzer, S. Boguth, R. Stengl, R. Schreiter, T.F. Meister, "High-speed SiGe:C bipolar technology," *IEDM Tech. Dig.*, 2001, pp. 344-347

46 M. Sokolich, C.H. Fields, S. Thomas III, B. Shi, Y.-K. Boegeman, M. Montes, R. Martinez, A.R. Kramer, M. Madhav, "A Low Power 72.8-GHz Static Frequency Divider in AlInAs/InGaAs HBT Technology," *IEEE Journal Solid State Circuits*, Vol. 36. no. 9, Sept 2001 pp. 1328-1334

47 K. Washio, M. Kondo, E. Ohue, K. Oda, R. Hayami, "A 0.2μm Self-Aligned SiGe HBT Featuring 107-GHz f_{MAX} and 6.7-ps ECL," *IEDM Tech. Dig* 1999, pp 557-560

48 T. Oka, K. Hirata, K. Ouchi, H. Uchiyama, T. Taniguchi, K. Mochizuki, T. Nakamura, "Advanced performance of small-scaled InGaP/GaAs HBT's with f_T over 150 GHz and f_{MAX} over 250 GHz," *IEDM Tech. Dig.*, 1998, pp. 653-656

49 D. Gloria, S. Gellida, G. Morin, "A New Extraction Method of High Frequency noise Parameters in the Temperature Range –55/150deg. For SiGe HBT in BiCMOS Process," *Proc. ICMTS* 2000, pp. 229-234

50 G. Freeman, D. Greenberg, K. Walter, S. Subbanna, "SiGe HBT Performance Improvements from Lateral Scaling," *Proceeding European Solid-State Device Research Conference, 1999*

51 M. Y. Liu, E. Chen, and S. Y. Chou, "140-GHz metal-semiconductor-metal photodetectors on silicon-on-insulator substrate with a scaled active layer", Appl. Phys. Lett. (1994) 887-888.

52 C. L. Schow, R. Li, J. D. Schaub, J. C. Campbell, "Design and Implementation of High-Speed Planar Si Photodiodes Fabricated on SOI Substrates," IEEE J. Quant. Elec. (1999), 1478-1482.

53 J. D. Schaub, R. Li, C. L. Schow, J. C. Campbell, G. W. Neudeck and J. Denton, "Resonant-Cavity-Enhanced High-Speed Si Photodiode Grown by Epitaxial Lateral Overgrowth," IEEE Photon. Technol. Lett. (1999) 1647-1649.

54 J. Y. L. Ho and K. S. Wong, "Bandwidth enhacement in Silicon Metal-Semiconductor-Metal Photodetector by Trench Formation," IEEE. Photon. Tech. Lett. (1996) 1064-1066.

55 M. Yang, K. Rim, D. Rogers, J, Schaub, J. Welser, D. Kuchta and D. Boyd, "A CMOS-Compatible High-Speed Silicon Lateral Trench Photodetector," 59th Dev. Res. Conf. (2001).

56 J. D. Schaub, D. M. Kuchta, D. L. Rogers, M. Yang, K. Rim, S. Zier and M, Sorna, "Multi Gb/s, high sensitivity all silicon 3.3V optical receiver using PIN lateral trench photodetector," OFC 2001 Postdeadline Papers, (2001) PD19-1.

57 D. L. Rogers, "Integrated Optical Receivers using MSM detectors," IEEE J. of Lightwave Technol., (1991) 1635-1638.

58 Sze, S. M., Physics of Semiconductor Devices, 2nd ed., John Wiley and Sons, 1981, pp. 45-46.

59 J. R. Long and M. L. Copeland, "The Modeling, Characterization, and Design of Monolithic Inductors for Silicon RF IC's," IEEE J. Solid-State Circuits, vol.32, 357-369, 1997.

60 J. N. Burghartz, D. C. Edelstein, M. Soyuer, H. A. Ainspan, and K. A. Jenkins, "RF Circuit Design Aspects of Spiral Inductors on Silicon," IEEE J. Solid-State Circuits, vol.33, 2028-2034, 1998.

61 D. Coolbaugh , E. Eshun, R. Groves , D. Harame, J. Johnson, M. Hammad , Z. He, V. Ramachandran , K. Stein , S. St Onge, S. Subbanna , D. Wang , R. Volant, X. Wang, and K. Watson, "Advanced Passive Devices For Enhanced Integrated RF Circuit Performance" Dig. 2002 IEEE RFIC Symp., pp. 341-344, 2002.

62 A-S. Poret, T. Melly, C. C. Enz, and E. A. Vittoz, "Design of High-Q Varactors for

Low-Power Wireless Applications using a Standard CMOS Process," IEEE J. Solid-State Circuits, vol. 35, pp. 337-345, 2000.

63 N.Fong, J-Plouchart, N. Zamdmer, D. Liu, L. Wagner, C. Plett and G. Tarr, "A Low-Voltage Multi-GHz VCO with 58% Tuning Range in SOI CMOS," Proc. Custom Integrated Circuits Conf., 2002.

64 D. J. Friedman, M. Meghelli, "Differential Control Topology for LC VCO," United States Patent, US 6,292,065 B1, September 2001.

65 G. Freeman, M. Meghelli, Y. Kwark, S. Zier, A. Rylyakov, M. A. Sorna, T Tanji, O. Schreiber, K. Walter, J-S. Rieh, B. Jagannathan, A. Joseph, S. Subbanna, "40 Gb/s Circuits Built From a 120 GHz f_T SiGe Technology," IEEE J. Solid-State Circuits, Accepted for Publication, 2002.

66 N. Fong, J-O. Plouchart, N. Zamdmer, D. Liu, L. Wagner, C. Plett and G. Tarr, "A 40 GHz VCO with 9 to 15% Tuning Range in 0.13μm SOI CMOS," IEEE Symp. VLSI Circuit, 2002.

67 S. Voinigescu, P. Popescu, P. Banens, M. Copeland, G. Fortier, K. Howlett, M. Herod, D. Marchesan, J. Showell, S. Szilagyi, H. Tran, J. Weng, "Circuits and Technologies for Highly Integrated Optical Networking IC 's at 10 Gb/s to 40 Gb/s," CICC 2001.

68 T. Dhaene and D. DeZutter, "Selection of lumped element models for coupled lossy transmission lines," IEEE Trans. Computer-Aided Design, vol. 11, no. 7, pp. 805-815, July 1992.

69 B.K. Sen and and R.L. Wheeler, "Skin effects models for transmission line structures using generic SPICE circuit simulators," EPEP'98-7th Topical Meeting on Electrical Performance of Electronic Packaging, pp. 128-131, Oct. 1998.

70 A. Deutsh, P.W. Coteus, G.V. Kopcsay, H.H. Smith, C.W. Surovic, B.L. Krauter, D.C. Edelstein, and P.J. Restle, "On-chip wiring design challenges for Gigahertz operation," IEEE Proc., vol. 89, no. 4, pp. 529-555, April 2001 .

71 R. Singh, *Signal Integrity Effects in Custom IC and ASIC Designs,* IEEE Press and Wiley & Sons, Nov. 2001.

72 K. Joardar, "Substrate Crosstalk in BiCMOS Mixed Mode Integrated Circuits," *Solid-State Elec.*, 39:4, 1996, pp. 511-516.

73 B. Razavi, *RF Microelectronics*, 1st ed., Prentice Hall, Inc., 1998.

74 J.B. Johnson, A. Stricker, A. Joseph, J.A. Slinkman, "A technology simulation methodology for AC-performance optimization of SiGe HBTs," *International Electron Devices Meeting*, 2001 pp. 21.4.1-4

75 K. Joardar, "Substrate Crosstalk in BiCMOS Mixed Mode Integrated Circuits," *Solid-State Elec.*, 39:4, 1996, pp. 511-516

76 S. Bharatan, P. Welch, K. H. to, R. Thoma and M. Huang, "3-D Simulation and Modeling of Signal Isolation in RF/IF Circuit," *Modeling and Simulation of Microsystems 2001*, pp. 470-473, 2001.

77 K. Joardar, "Signal Isolation in BiCMOS Mixed Mode Integrated Circuits," *Bipolar/BiCMOS Circuits and Technology Meeting*, pp. 178-181, 1995

78 K. Joardar, "Comparison of SOI and junction isolation for substrate crosstalk suppression in mixed mode integrated circuits," *Electronics Letters*, v. 31, pp. 1230-1231, 1995

79 D. K. Su, M. J. Loinaz, S. Masui and B. A. Wooley, "Experimental Results and Modeling Techniques for Substrate Noise in Mixed-Signal Integrated Circuits," *IEEE J. Solid-State Circuits*, v. 28, pp. 420-430, 1993.

80 J. M. Casalta, X. Aragones and A. Rubio, "Substrate Coupling Evaluation in BiCMOS Technology, " *IEEE J. Solid-State Circuits*, v. 32, pp. 598-603, 1997

81 S. Masui, "Simulation of Substrate Coupling in Mixed-Signal MOS circuits," *Symposium on VLSI Circuits Digest of Technical Papers*, pp. 42-43, 1992.

82 X. Aragones, F. Moll, M. Roca and A. Rubio, "Analysis and modelling of parasitic substrate coupling in CMOS circuits," *IEE Proc.-Circuits Devices Syst.*, v. 142, pp. 307-312, 1995.

83 K. Mayaram, "Substrate Noise Coupling Modeling and Applications to RF VCOs", presented

at IEEE SCV SSC November 2000 Meeting, 2000. Available at www.ewh.ieee.org/r6/scv/scv_sscnov2000.html

84 S. Selberherr, *Analysis and Simulation of Semiconductor Devices,* Springer-Verlag, New York, 1984, pp. 191-197

85 S. E. Laux, "Techniques for Small-Signal Analysis of Semiconductor Devices," *IEEE Transactions on Electron Devices,* 32:10, 1985, pp. 2028-2037

86 M. Pfost *et al*, "Modeling and Measurement of Substrate Coupling in Si-Bipolar IC's up to 40GHz", *IEEE J. Solid State Circuits*, Vol. 33, No. 4, April 1998

International Journal of High Speed Electronics and Systems
Vol. 13, No. 1 (2003) 221–237

World Scientific
www.worldscientific.com

HIGH-SPEED, LOW-POWER DIGITAL AND ANALOG CIRCUITS IMPLEMENTED IN IBM SiGe BiCMOS TECHNOLOGY

KARL E. FRITZ, BARBARA A. RANDALL, GREGG J. FOKKEN, MICHAEL J. DEGERSTROM, MICHAEL J. LORSUNG, JASON F. PRAIRIE, ERIC L. H. AMUNDSEN, SHAUN M. SCHREIBER, AND BARRY K. GILBERT
Special Purpose Processor Development Group, Mayo Foundation, Rochester MN, 55905, USA

DAVID R. GREENBERG
IBM East Fishkill Facility, Hopewell Junction, New York 12533, USA

ALVIN JOSEPH
IBM Burlington Facility, Essex Junction, Vermont, 05452, USA

Under the auspices of Defense Advanced Research Project Agency's Microsystems Technology Office (DARPA/MTO) Low Power Electronics Program, the Mayo Foundation Special Purpose Processor Development Group is exploring ways to reduce circuit power consumption, while maintaining or increasing functionality, for existing military systems. Applications presently being studied include all-digital radar receivers, electronic warfare receivers, and other types of digital signal processors. One of the integrated circuit technologies currently under investigation to support such military systems is the IBM Corporation silicon germanium (SiGe) BiCMOS process. In this paper, design methodology, simulations and test results from demonstration circuits developed for these applications and implemented in the IBM SiGe BiCMOS 5HP (50 GHz f_T HBTs with 0.5 um CMOS) and 7HP (120 GHz f_T HBTs with 0.18 um CMOS) technologies will be presented.

Key words: **Silicon Germanium (SiGe), BiCMOS, heterojunction bipolar transistor (HBT), CMOS, current mode logic (CML), low power, differential amplifier, multiplexer, demultiplexer, low voltage differential signalling (LVDS), radiation, single event upset (SEU), Smart Pixel Array (SPA)**

1. Introduction

An attractive feature of BiCMOS technologies is that high-speed bipolar circuits (>10 GHz clock rates for SiGe bipolar) can be processed on the same wafer as high-density low-power bulk CMOS circuits. This combination of devices can be very beneficial when high-speed signals must be captured, demultiplexed, and processed with digital signal processing using complex algorithms. This combination may also reduce the number of I/O by placing parallel streams of data into low power registers and transporting them on/off-chip serially. This strategy was used in the design of a polynomial multiplier described in this paper. The Special Purpose Processor Development Group (SPPDG) at the Mayo Foundation has designed numerous digital and analog test circuits to assess the viability of IBM's SiGe BiCMOS 5HP [1] and 7HP [2] technologies for these applications. The design methodology used, and measured results from fabricated circuits, are presented here.

2. Circuit design methodology

Through the course of four DARPA-sponsored "multi-user mask set" fabrication runs, a variety of digital circuits were designed to exploit the high-speed performance of IBM's SiGe heterojunction bipolar transistors (HBTs), as well as the low-power characteristics of the CMOS devices. High speed digital circuits, employing only HBTs, were designed, laid out, fabricated, and tested, including a divide-by-2-and-4 circuit, 4-bit linear feedback shift register, 4:1 multiplexer/1:4 demultiplexer, 32:4 multiplexer/4:32 demultiplexer, 32-bit single event upset (SEU) shift register, a 7-stage 2^7-1 pseudo-random pattern generator, and several standalone breakout macros used to test individual logic gate performance. Every one of the test circuits designed for the assessment of the SiGe BiCMOS technology and described in this paper was "borrowed" from larger integrated circuit or subsystem designs under development by or of interest to the military; only in this way could we validate the SiGe BiCMOS technology properly in the conceptual framework of future military system requirements and potential applications.

To achieve operating frequencies greater than 5 GHz for these HBT-only circuits, it was desired to employ some type of differential logic with coupled true and complement signals, to assure the rejection of common mode noise and increased signal integrity. These digital circuits were implemented using current mode logic (CML) technology, which consumes constant power over its entire clock rate range but generates very little switching noise. In addition to providing better signal quality, in many cases more functionality can be added with no increase in power due to the vertical "stacking" architecture of CML, as illustrated in Figure 1. One limitation of the "stacking" architecture is the requirement for logic "group levels" necessary to maintain the switching operation of the HBTs. As a result, some logic signal swings have a large DC offset with respect to other signal swings at different "group levels". To generate these different "group levels", emitter-follower "level shifting" circuits are employed. Level shifting translates a signal from one "group level" to a lower "group level". The collector-connected HBTs at the top level in the standard CML configuration are not able to drive a large amount of capacitive load caused by fanout and interconnect; therefore, emitter-follower level-shifters are also employed when it becomes necessary to drive a large capacitance. The V_{be} values of the HBTs and the selected power supply voltage determines the number of logic "group levels" allowed in a logic gate. The IBM SiGe HBTs, with a V_{be} of 0.9 V and a power supply voltage of -5 V were selected for these circuits, allowing logic gates to be designed with four "group levels", with each level exhibiting a 300 mV amplitude. Supply voltages can be decreased to be just sufficient to power the number of levels required in the stack. The less complex the logic gates need to be, the smaller the supply voltage can be, which in turn minimizes the gate power dissipation.

In addition to the HBTs, the NMOS and PMOS devices available in this technology were utilized in a 30-bit shift register to create a BiCMOS single-bit-operand polynomial multiplier. CML-to-CMOS and CMOS-to-CML translator circuits were designed to facilitate the communication between the SiGe HBTs and CMOS FETs. The FETs were used in the standard CMOS totem-pole configuration, with primary emphasis on reducing the power consumption.

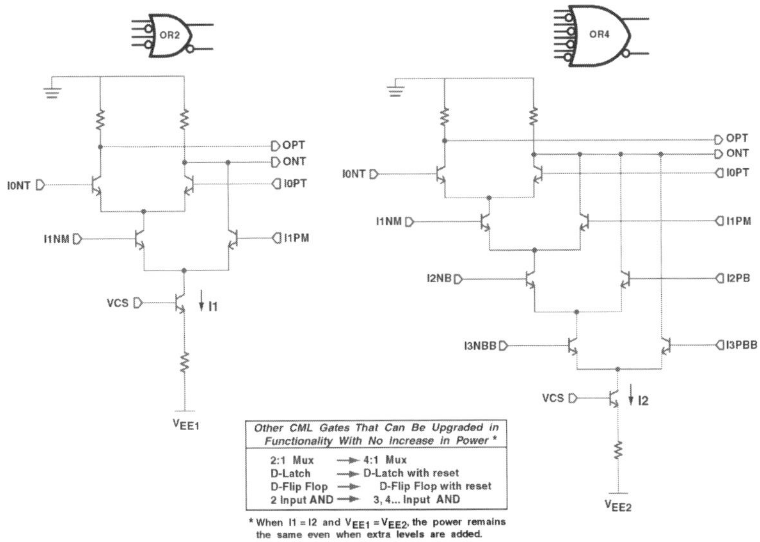

Figure 1: Illustration of the benefit of adding functionality to CML gates without a gain in power (16325)

PMOS devices were also used in the design of low voltage differential signaling (LVDS) input receivers and output drivers to reduce the power dissipation of the 32:4 multiplexer/4:32 demultiplexer circuit. LVDS I/O levels are specified to be 1.4 V to 1.0 V. Unlike the BiCMOS polynomial multiplier mentioned above, NMOS devices could not be used in these I/O circuits because body effect was a large problem in this design. The substrate is tied to -5.0 V for the CML circuits. Therefore, the p-well/substrate potential of the NMOS devices would be greater than 3.3 V with respect to the other terminals, which would either alter the performance of the circuit in ways that could lower the maximum speed, or in this case, possibly damage the NMOS devices. In the case of the BiCMOS single-bit-operand polynomial multiplier, both proof-of-concept and low power were considered priorities. Therefore, the body effect was accepted and compensated for as much as possible.

NMOS devices were used as current sources with bipolar circuits in the second IBM SiGe 7HP BiCMOS fabrication run. This design approach lowered the power dissipation compared to the previous circuits by eliminating a 0.9 V_{be} drop as well as a resistor with a 0.4 V drop across it, and replacing them with a simple NMOS device requiring a turn-on voltage of 0.4 V, thereby reducing the real estate as well as the power dissipation of the circuit.

3. Measured performance results from the first 5HP fabrication run

A simple 2-input CML OR-gate fabricated on the first DARPA-sponsored SiGe 5HP BiCMOS multi-user mask set was measured and found to function up to 15 GHz. Figure 2 shows the results at 1 GHz and 15 GHz. The average gate delay measured for the CML OR-gate was 17 ps with a power consumption of 19 mW when using a –5 V supply. Several breakout macros designed for DARPA/MTO's Advanced Digital Receiver Technology Program were also implemented on this fabrication run to evaluate the SiGe HBT technology for its application to all-digital radio frequency (RF) receivers for military applications. The measured results from two of these macros are presented in Figures 3 and 4. The 4-bit linear feedback shift register circuit operated successfully in our laboratory at clock rates of up to 9.5 GHz with 1.25 W power dissipation, while the divide-by-2-and-4 circuit operated up to 13 GHz with 1.0 W power dissipation. These circuits, as well as some of the others described in this paper, could have been designed to operate faster by inserting emitter followers between the master and slave portions of the flip-flops; however, the additional 25% increase in performance was not worth the approximate two times increase in power for our applications.

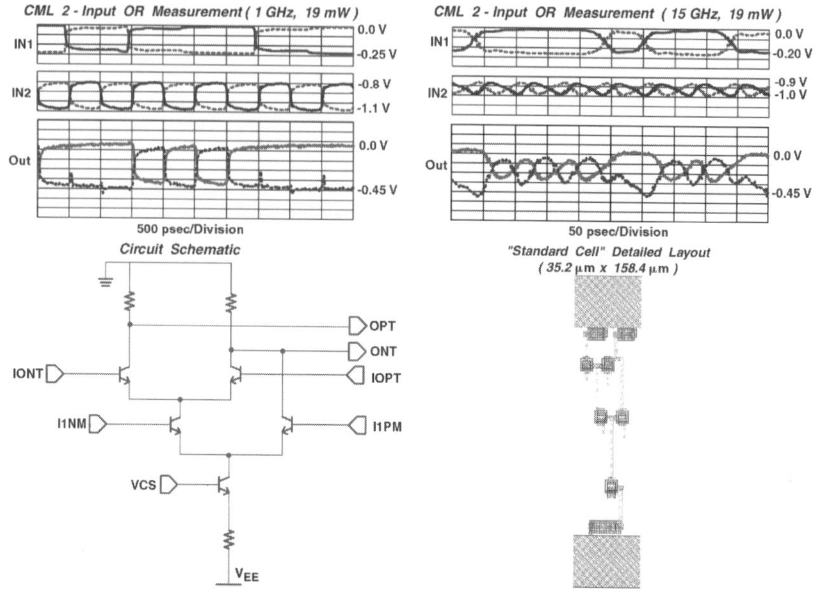

Figure 2: Measured results, schematic, and layout of Mayo-designed 2-input OR gate fabricated in IBM SiGe 5HP BiCMOS process on DARPA multi user mask set 1 (16307)

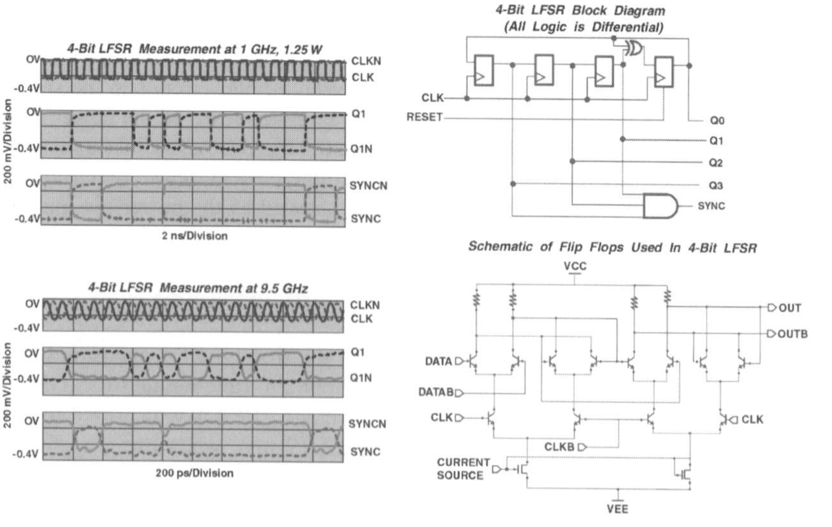

Figure 3: IBM SiGe 5HP BiCMOS DARPA multi user mask set 1 chip test/Mayo bipolar 4-bit linear feedback shift register (LFSR) circuit (15833)

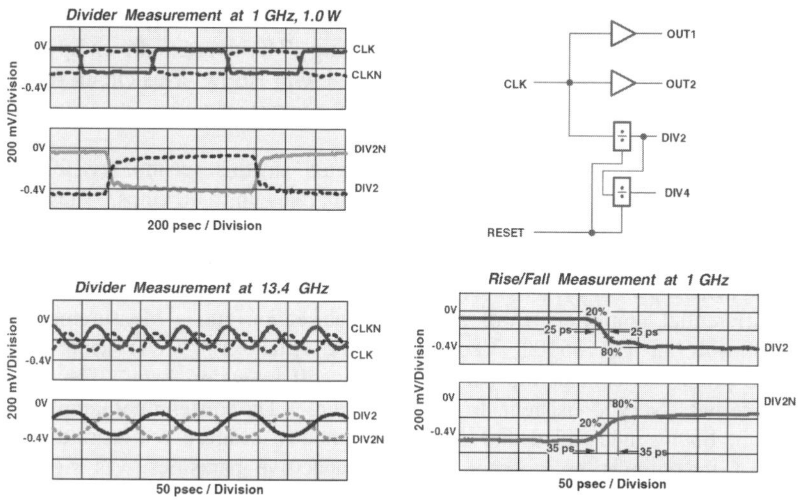

Figure 4: IBM SiGe 5HP BiCMOS DARPA multi user mask set 1 chip test/Mayo divider circuit (15765

The digital SiGe BiCMOS polynomial multiplier circuit fabricated on this mask set contains a 30-bit shift register implemented with CMOS devices, which serially shifts data into and out of the high speed multiplier implemented with HBT devices. The shift register portion of this circuit was successfully tested up to clock rates of 550 MHz, with 205 mW power dissipation. Simulations predicted that the multiplier portion of this circuit should operate in a stand alone mode at clock rates up to 8 GHz; however, the performance of the multiplier is limited to six times the maximum frequency of the CMOS register file that is loading and acquiring the data, or 3.3 GHz. Figure 5 presents the low speed measured data from the polynomial multiplier, including notations explaining the different stages of operation. In this figure the correct output has been placed in the Result 3 register and serially shifted out.

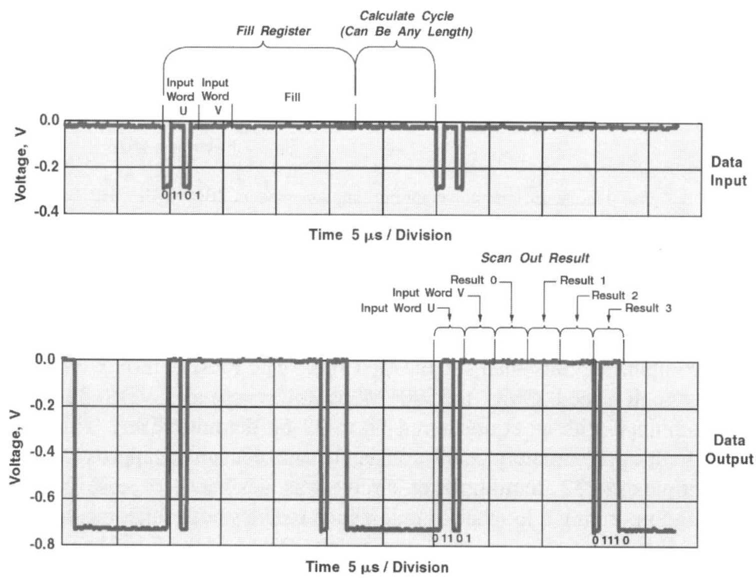

Figure 5: Measured results for Mayo 5-bit polynomial multiplier fabricated in IBM SiGe 5HP BiCMOS technology (17274)

One of the higher device count BiCMOS digital circuits implemented on this fabrication run was a CMOS inverter chain containing 1740 inverters driven by CML I/O buffers and CML-to-CMOS translators implemented with HBTs. This circuit functioned correctly, with an average delay of 98 ps per CMOS inverter gate.

Several analog circuits were also implemented in this technology for use in optoelectronic drivers and receivers operating at data rates up to 10 Gbps. These circuits included a differential amplifier, a laser driver, a transimpedance amplifier, and an automatic gain control amplifier. Figure 6 presents simulated S-parameters, with and without capacitive parasitics included, overlaid on top of measured S-parameters from several differential amplifiers from this SiGe fabrication run. Measurements were performed with an HP 8510C Vector Network Analyzer (VNA) using coplanar microwave probes to contact the bare die while still in wafer form. The measured performance was very consistent from die to die. The differential amplifier was measured in both single-ended and differential modes. Figure 6 illustrates that the accuracy of the simulation incorporating the capacitive parasitics was significantly greater than the simulation without the parasitics.

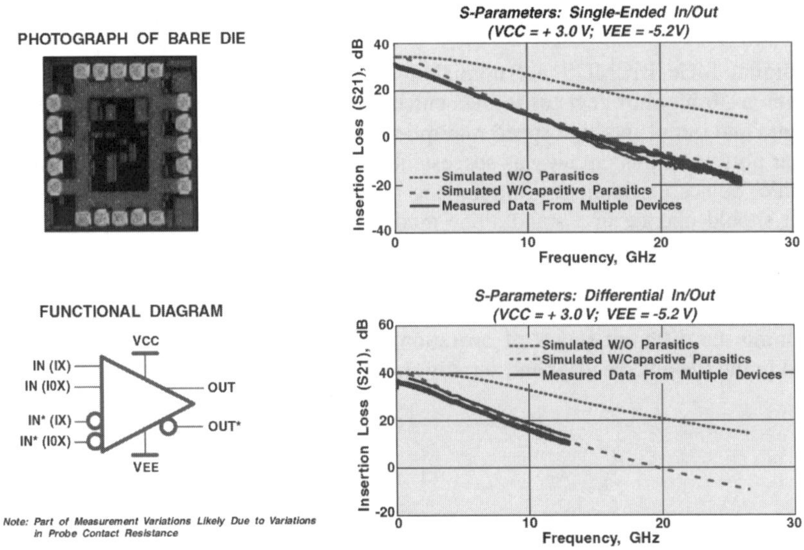

Figure 6: Mayo designed bipolar differential amplifier implemented in IBM SiGe 5HP BiCMOS process (17327)

4. Measured performance results from the second 5HP fabrication run

For the second DARPA-sponsored SiGe BiCMOS 5HP fabrication run, we prepared larger, more complex circuits than for the first run. The most complex circuit was a 4 GHz, 4 Gbps high speed CML to 500 Mbps lower speed LVDS 32-bit to 4-bit multiplexer combined with an equivalent 4-bit to 32-bit demultiplexer. This circuit was implemented with approximately 3600 transistors and consumed approximately 11 W. The 32:4 multiplexer/4:32 demultiplexer circuit was designed for ease of test in the laboratory by incorporating a loopback mode which enables both functions to be tested at the same time. Figure 7 depicts the measured results of one of the four channels of the demultiplexer-to-multiplexer loopback circuit tested at 1 GHz. This circuit can also perform the multiplexer-to-demultiplexer loopback function, as well as separate multiplexing and demultiplexing functions. These measurements were performed on-wafer with a low frequency probe card, which is limited to 1 GHz operation. At the time of this writing, a high speed test fixture is being fabricated to allow us to assess the maximum performance of this circuit. The use of LVDS output buffers on this circuit greatly reduced its power consumption. By way of comparison, a circuit of equivalent design was implemented with ECL I/O in place of the LVDS I/O for another program, in which the ECL output buffers alone consumed more power than this entire SiGe circuit.

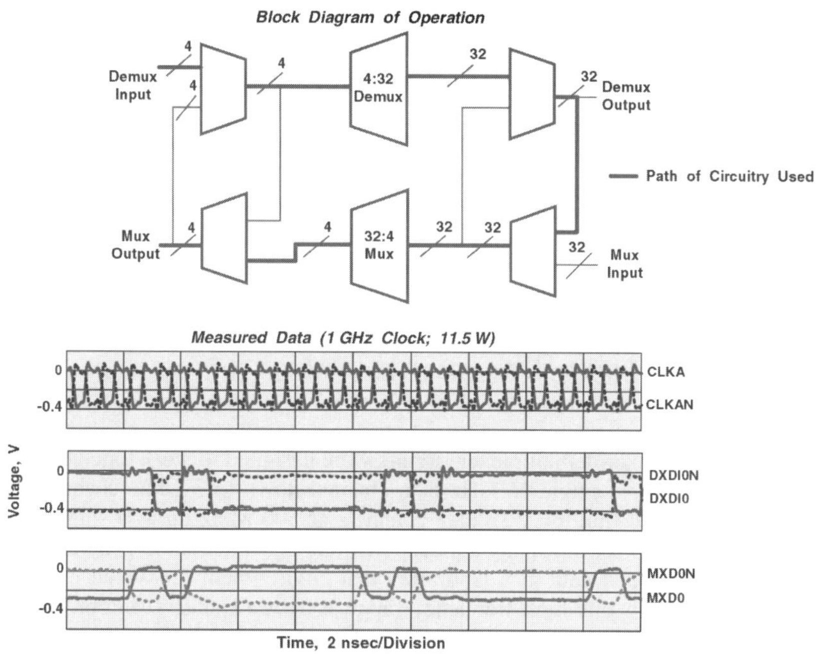

Figure 7: Measured results of Mayo BiCMOS 32:4 multiplexer/4:32 demultiplexer circuit fabricated in IBM SiGe 5HP BiCMOS technology on DARPA multi user mask set 2 (16524)

A smaller, higher speed multiplexer/demultiplexer was also placed on this second SiGe 5HP BiCMOS fabrication run, with the goal of achieving a 16 GHz, 16 Gbps high speed CML, to 4 Gbps lower-speed CML 4:1 multiplexer/1:4 demultiplexer so that it would be compatible with the larger chip described above. This circuit contained approximately 660 transistors and consumed 2.65 W; it incorporated the same ease-of-test features as the 32:4 multiplexer/4:32 demultiplexer circuit, such that the entire circuit could be tested using the loopback functions. The maximum simulated speed of the 4:1 multiplexer/1:4 demultiplexer (including parasitic capacitance) was 11 GHz at a power dissipation of 2.4 W, so this technology did not quite meet our goal. The maximum measured speed of the 4:1 multiplexer/1:4 demultiplexer was 8.7 GHz, with these measurements performed on-wafer with a probe card. At the time of this writing a high speed test fixture is being fabricated for further analysis of this multiplexer/demultiplexer circuit, and the measured performance is expected to improve in the resulting improved electrical test environment. The simulated and measured results of the serial data in the demultiplexer-to-multiplexer loopback mode may be observed in Figure 8.

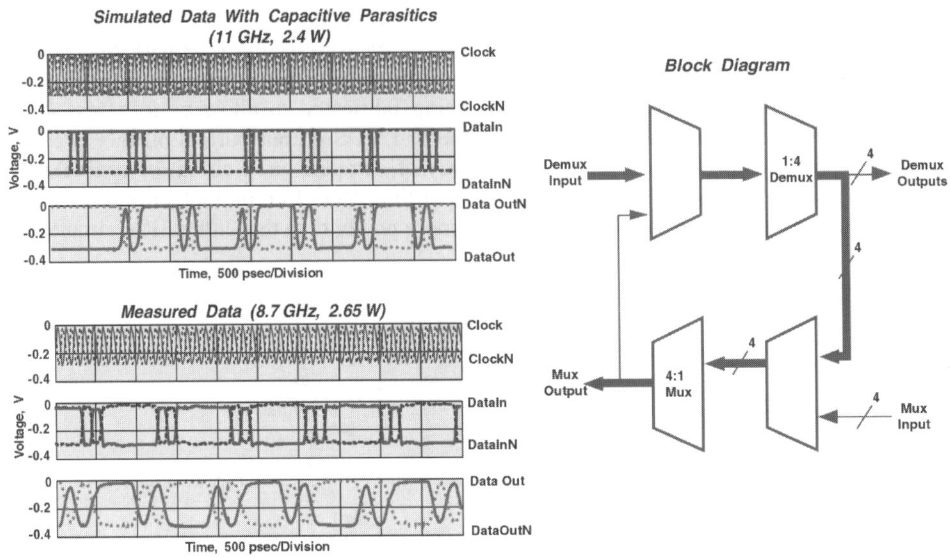

Figure 8: Simulated versus measured results of Mayo 4:1 multiplexer/1:4 demultiplexer circuit fabricated in IBM SiGe 5HP BiCMOS technology (17324)

These two multiplexer/demultiplexer circuits were designed to work together, with the goal of demultiplexing a single 16 GHz data channel down to 32 data channels at 500 MHz; or alternately, to multiplex 32 data channels at 500 MHz up to a single 16 GHz data stream. This chip set has many applications, one of which is the generation and/or acquisition of high speed data in a laboratory test environment for which existing test equipment operating in this high frequency range is very expensive and/or not widely available. Another application for these chips is interfacing very high speed analog-to-digital converter output data to a lower speed CMOS signal processor circuit.

A third circuit implemented on the second fabrication run was a 32-bit shift register designed to evaluate this technology for single event upset (SEU) effects in a radiation environment. This circuit was designed using a "current-shared hardening" approach developed and patented by Motorola[3]. Each component in the circuit was implemented with current mode logic, as illustrated in Figure 1; however, for each single transistor shown, five transistors are used in its place. This approach enables a majority current summing of the devices, so that if a single transistor experiences a radiation-induced upset, the remaining devices will collectively still draw enough current to overwhelm the affected device and produce the correct output. The test circuit was designed so that from three to five of the devices could be turned on, to evaluate the minimum number required for SEU hardness. The performance of the shift register with the different numbers of devices turned on was verified at Mayo, the results of which are shown in Figure 9 The 32-bit single event upset shift registers in their test fixtures were then delivered to Naval Research Laboratories, who performed a radiation sensitivity analysis. The results of this analysis have been reported elsewhere.[4]

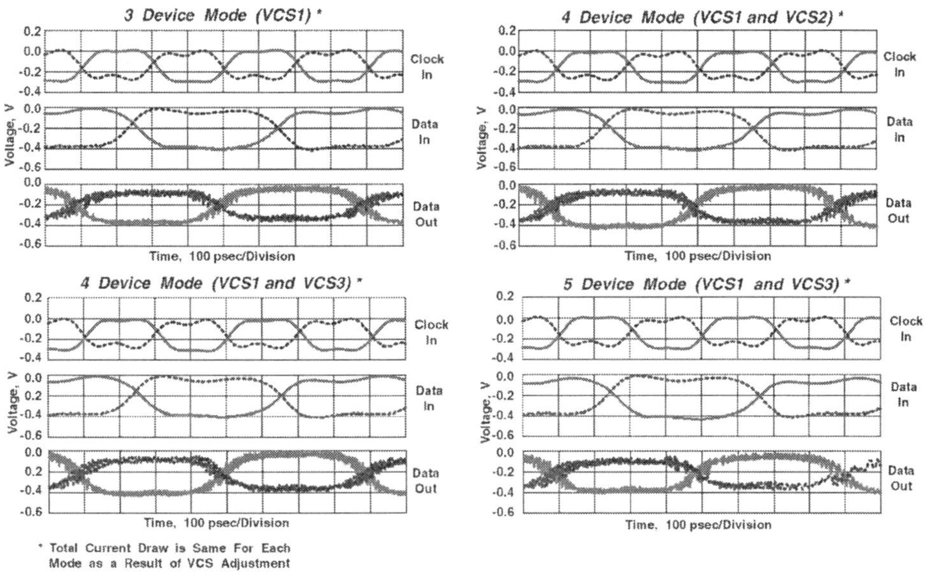

Figure 9: Measured results of Mayo 32-Bit SEU shift register fabricated in IBM SiGe 5HP BiCMOS technology on DARPA multi user mask set 2 (16557)

Breakout macros, smaller circuits used for performance analysis and debugging the larger circuits, were also placed on the second SiGe 5HP BiCMOS fabrication run. The LVDS drivers were of particular interest since this type of I/O is becoming a standard communications protocol between high speed CMOS circuits, or between CMOS circuits and gallium arsenide or indium phosphide chips. Figure 10 displays the measured and simulated eye diagram results of the Mayo-designed SiGe LVDS output buffer. Some LVDS buffer designs implement the required 100 Ohm termination at the source, while others place it at the far end destination. The Mayo version was source-terminated but will operate with a far end termination as well, with an output swing of half the normal amplitude. The maximum measured performance of the LVDS output buffer was 5 Gbps. This measurement exceeded the maximum simulation speed by a factor of two. Power consumption of the LVDS output buffer was 46 mW, which was 19 mW less than the CML output buffer. The maximum LVDS output buffer performance is difficult to ascertain because most very high speed test equipment provides a 50 Ohm environment, while the LVDS output buffer typically operates in a high impedance environment. To obtain the LVDS buffer measurements of Figure 10, an alternate approach was taken, as depicted in Figure 11.

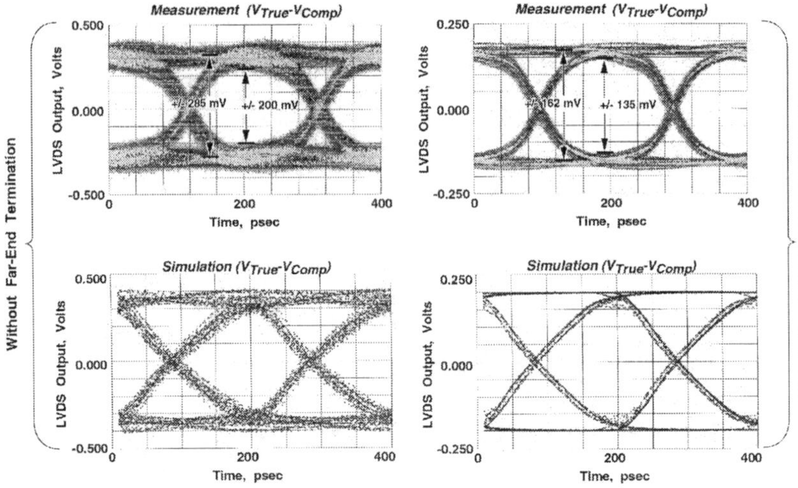

Figure 10: Measured vs. simulated eye diagrams of IBM 5HP SiGe source-terminated LVDS output buffer on DARPA multi-user mask set 2 (16544)

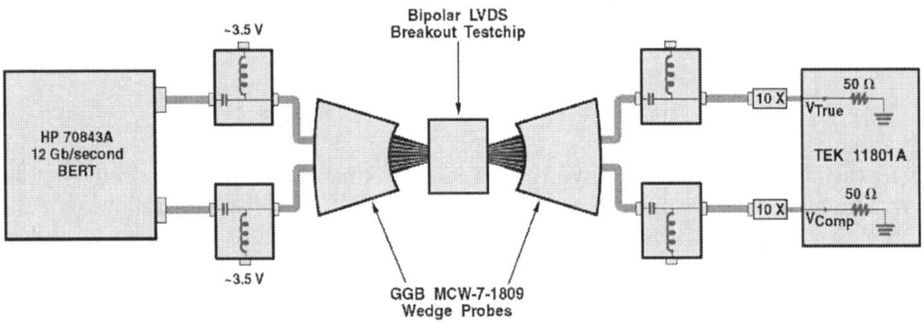

Figure 11: Block diagram of procedure to test bipolar implementation of low voltage differential signaling (LVDS) output buffer implemented in IBM SiGe BiCMOS process on DARPA multi-user mask set 2 (17305)

Blocking capacitors installed in series with the inputs to the high performance oscilloscope provided a far end termination and a half amplitude waveform. A 450 Ohm resistor placed immediately following the capacitor provided the high impedance environment necessary to achieve a full swing signal, as may be observed in the eye diagrams in Figure 10.

5. Measured and simulated performance results on the first 7HP fabrication run

IBM has continued to scale their SiGe BiCMOS technology [5] with much success, and the next two DARPA-funded multi user mask sets were fabricated in the 7HP technology. IBM's initial goals for the 7HP generation of SiGe BiCMOS were a 90 GHz f_T HBT coupled with the 0.18 um CMOS. IBM discovered that a minor change to the HBT could improve the device's f_T to 120 GHz, and all fabrication runs except this first one incorporated the higher performance HBT. Circuits for this first 7HP fabrication run were designed exclusively with HBT devices, as the early version of the design kit from IBM did not completely support using the FET devices. Circuits from prior 5HP runs, such as the divide-by-2-and-4 circuit and the 4:1 multiplexer/1:4 demultiplexer, were mapped into the 7HP technology. Simple circuits, such as ring oscillators to compare basic propagation delays between the two processes, as well as discrete HBTs to measure the device characteristics were also included.

The 19-stage ring oscillator in the 7HP technology demonstrated a basic buffer propagation delay of 11 ps, compared to 16 ps in the 5HP process. The divide-by-2-and-4 circuit in Figure 12 also showed considerable improvement, measuring 20.7 GHz for the divide-by-4 function and dissipating 885 mW. Simulated results including capacitive parasitics for the 4:1 multiplexer/1:4 demultiplexer predict 15 GHz operation and 2.02 W of power. Note that both the performance and the power for these circuits have improved compared to the 5HP technology. Recently, the 4:1 multiplexer/1:4 demultiplexer circuit was redesigned with the 120 GHz f_T devices, and this circuit simulates to operate up to 25 GHz, with a slight increase in power to 2.39 W.

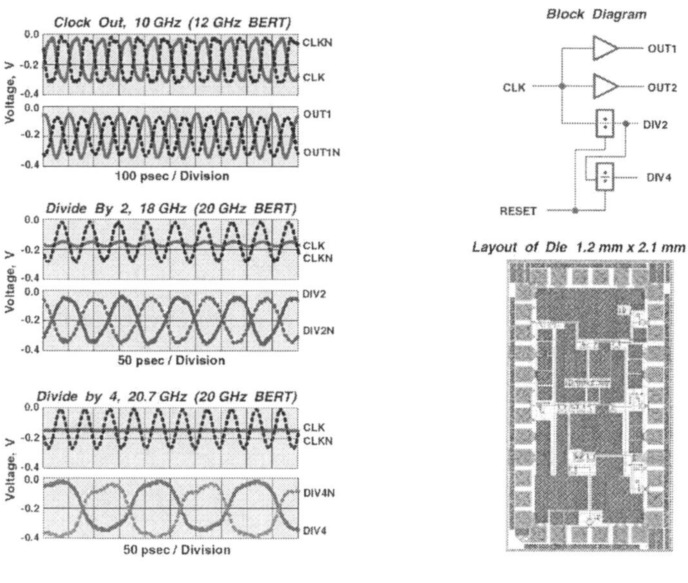

Figure 12: IBM SiGe 7HP BiCMOS (90 GHz f_T HBTs) multi-project wafer (MPW1) test results of Mayo divide by 2 and 4 circuit implemented with HBTs (17582)

The new circuit implemented on this fabrication run, a breakout macro that the Mayo group developed for another DARPA program, is a 7-stage 2^7-1 pseudo-random pattern generator (PRPG) designed to provide on-chip test patterns. Simulations including parasitic capacitance predicted the maximum frequency of operation to be 14 GHz. Due to the 12 GHz limit of the bit error rate test equipment, the circuit has only been tested up to 12 GHz. The power dissipation was measured to be 640 mW.

6. Circuits placed on second 7HP fabrication run

The second DARPA-sponsored fabrication run in the 7HP process featured the 120 GHz f_T HBTs. The measurements of this 7HP device appear in Figure 13 and can be compared to the measurements of the equivalent 5HP device shown in Figure 14.

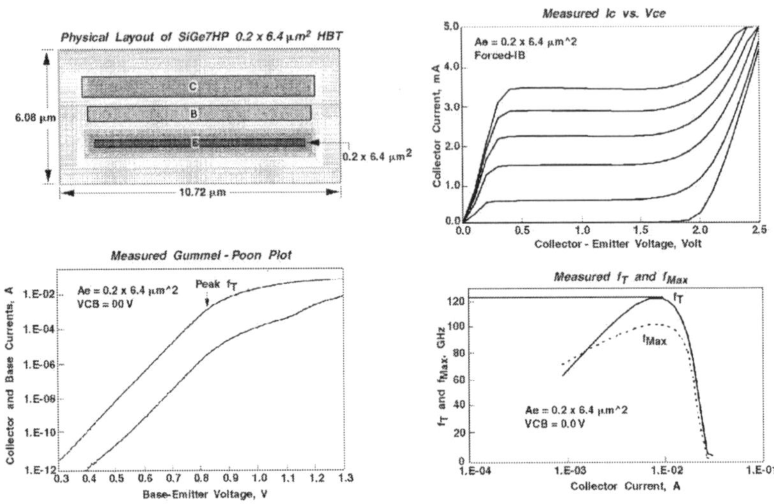

Figure 13: Characterization of 0.2 x 6.4 μm^2 IBM silicon germanium (SiGe) NPN heterojunction bipolar transistors (HBTs) fabricated on SiGe 7HP BiCMOS SPAWAR MPW2 Run (18102)

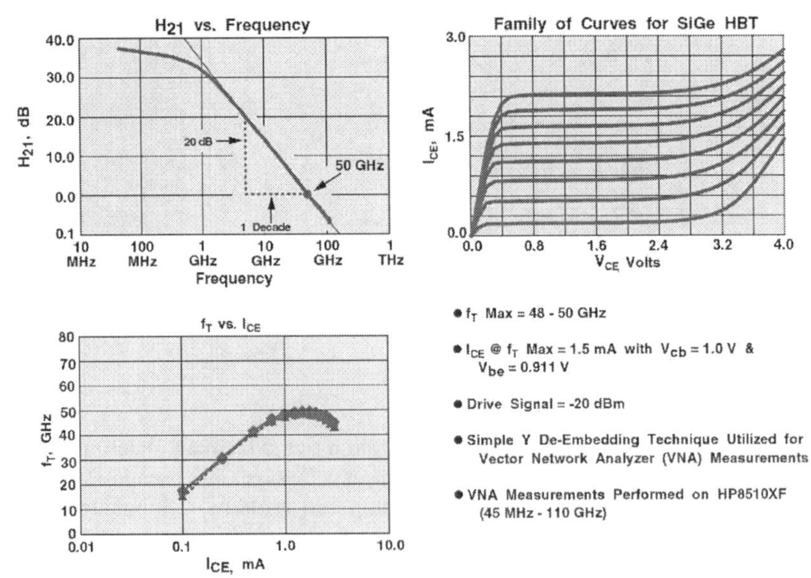

Figure 14: Characterization of 0.5 μm x 2.5 μm Silicon Germanium (SiGe) NPN heterojunction bipolar transistors (HBTs) fabricated on SiGe 7HP BiCMOS DARPA multi user mask set (16590)

The circuits implemented by Mayo on this fabrication run incorporated NFETs for current sources instead of the HBT-resistor combination used in previous circuits. This design approach decreased both the area of the circuits and their power dissipation, in comparison to the previous circuits, by eliminating a 0.9 V_{be} drop as well as a resistor with a 0.4 V drop across it, and replacing them with a simple NMOS device requiring a turn-on voltage of 0.4 V. This altered implementation enabled these logic gates to operate with a 3.3 V power supply instead of the 5 V supply typically employed. This methodology was utilized in the design of a high-speed smart-pixel array (SPA) for DARPA's VLSI Photonics Program. The SPA circuit was designed for a 10 Gb/s/channel free space optical chip-to-chip communication demonstration and is based on the FAST-Net optical interconnect scheme reported elsewhere.[6] In addition to the use of CMOS devices to decrease the power in the current sources, CMOS circuits were also used to serially load control signals, and then issue them in parallel, saving pad real estate as well as power. This circuit is currently being heterogeneously integrated with a GaAs vertical cavity surface emitting laser (VCSEL) and photodetector array, as illustrated in Figure 15.

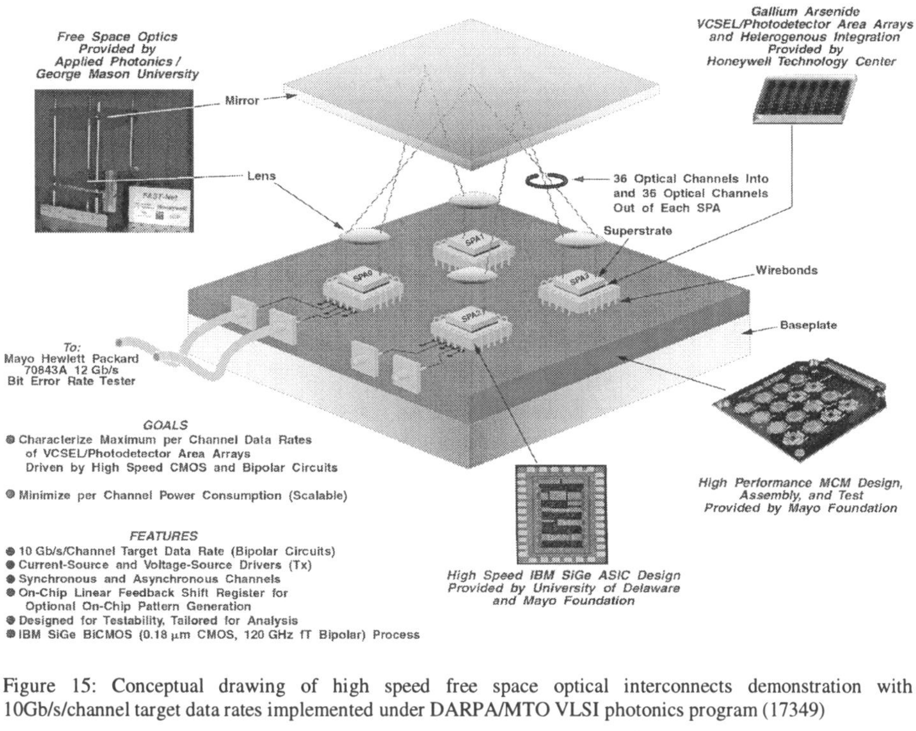

Figure 15: Conceptual drawing of high speed free space optical interconnects demonstration with 10Gb/s/channel target data rates implemented under DARPA/MTO VLSI photonics program (17349)

A bipolar differential amplifier was also included on this fabrication run to compare the bandwidth and gain to the differential amplifier of Figure 6. Results of a single-ended analog simulation including capacitive parasitics appear in Figure 16 which predicted a maximum gain of 36 dB and a maximum bandwidth of 38 GHz for this circuit. Measurements presented in this same figure predict a maximum bandwidth of 30 GHz through extrapolation. The single-ended gain of this amplifier is similar to the 5HP version, and both exhibit a maximum gain of about 34 dB. However, the 30 GHz bandwidth of the 7HP version is a factor of two improvement over the 15 GHz maximum bandwidth of the 5HP version.

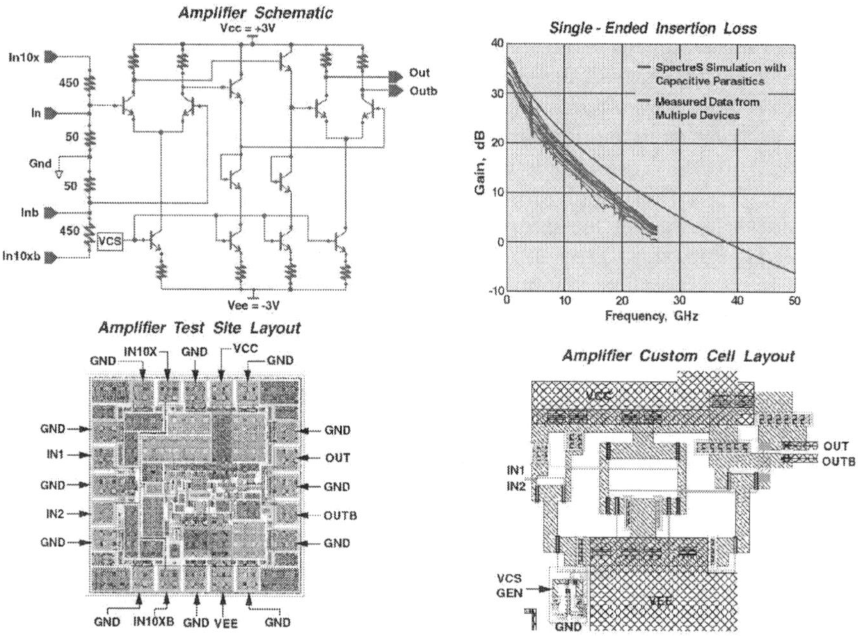

Figure 16: Mayo designed HBT differential amplifier implemented in IBM SiGe 7HP BiCMOS process (120 GHz f_T HBTs) (17800v1)

7. SiGe BiCMOS applications

The IBM SiGe BiCMOS 5HP & 7HP technologies exhibit higher performance than bulk silicon and consume less power than GaAs, making them well suited for next-generation applications operating in the multi-GHz frequency range.[7] With V_{be} values only 0.2 V greater than bulk silicon, and f_T values up to 120 GHz for the 7HP process, systems currently implemented with III-V compound semiconductor technologies could be implemented with SiGe BiCMOS. In addition, the SiGe BiCMOS technology is well suited for applications requiring high performance processing combined with dense, low power memory or circuitry. Several of the digital HBT circuits described in this paper, such as the divide-by-2-and-4 circuit, have maximum operating frequencies exceeding 10 GHz. Two examples of communications applications that require this bandwidth are OC-192 (10 Gbps) data rates, and the upcoming specification for 10 Gbps Ethernet. These performance results also demonstrate that SiGe BiCMOS is well suited for RF applications, such as wireless phones and pagers.

Applications such as all-digital receivers for radar and spread-spectrum communications for both civilian and military applications could be significantly improved in size, weight, and power by implementing them with SiGe BiCMOS technology. Using this technology, a wide bandwidth receiver, a serial-to-parallel demultiplexer, and a lower speed signal processor could all be fabricated on a single chip.

An additional benefit of the SiGe BiCMOS technology is that it can be fabricated in an existing bulk CMOS silicon manufacturing facility, thereby enabling the SiGe BiCMOS technology to leverage the benefits of a mature process with high yield and 8 inch wafers. Most III-V HBT integrated circuits are currently being manufactured on 3 inch or 4 inch wafers.

Thus we view this presently existing SiGe BiCMOS technology as a potential candidate for a large number of military systems. Many such systems will have to exploit the signal processing complexity achievable with large numbers of CMOS transistors operating at clock rates in the hundreds of MHz, combined with the multi-GHz clock rates achievable by a much smaller number of bipolar transistors for specialized input/output, or for demultiplexing / multiplexing high bit rate data streams. Other such military applications will require large amounts of digital logic combined with high-drive, very high speed transistors for complex electro-optic subsystems, such as those described in [8] which are infeasible or too cumbersome to construct with conventional components. Still other such systems will use the CMOS for digital logic and the bipolar transistors for specialized analog functions.

The MOSIS Service is currently offering the IBM SiGe BiCMOS 5HP technology described above, as well as the 6HP technology (50 GHz f_T HBTs with 0.25 um CMOS), through their multi-project fabrication runs, and plans to offer the 7HP technology in the future. Thus, designers of these military systems now have access to the IBM technology for implementation of critical elements or prototype versions of their systems.

Finally, as recently as one year before this paper was prepared, it appeared unlikely that the SiGe bipolar transistors would be able to operate at the speeds demonstrated for InP bipolar transistors. As a result, it was predicted that the SiGe BiCMOS would serve as a natural matching technology to serve in the data rate downconversion/upconversion role between InP components and slower speed digital functions implemented in bulk CMOS or directly on the BiCMOS chip. However, IBM has now announced the development of an improved SiGe HBT with f_T values of 210 GHz, with the possibility of even higher f_T values in the future. [9 10] The new HBT is scheduled to be incorporated into the next-generation 8HP BiCMOS technology, perhaps allowing the SiGe BiCMOS process to capture much of InP's "market" for extremely high speed and/or high bandwidth integrated circuits, while enabling the co-located CMOS devices to conduct subsequent stages of processing on the same chip. A significant number of military systems would benefit from such a combination of devices, which could implement analog functions in the K_u band (12-18 GHz) or even at K_a band (27-40 GHz), and very large and complex digital processing sections at clock rates in the many hundreds of MHz, all on the same integrated circuit. Based on the Mayo group's collaboration with the Department of Defense and its contractors on prototype integrated circuit and system developments over more than two decades, we believe that it is not an overstatement that this continually evolving SiGe BiCMOS technology represents a "silver bullet" for many future compact, low power, extremely capable next-generation military systems, as well as for civilian systems needing equivalent functionality.

8. Summary

The ability to fabricate HBTs to execute algorithms requiring multi-GHz digital processing functions and/or wide bandwidth analog functions, on the same integrated circuit with low power CMOS for slower speed but gate-count-intensive digital signal processing algorithms, has been demonstrated. Analog and digital, HBT-only, CMOS-only, and BiCMOS circuits were implemented and tested successfully. This paper clearly demonstrates that this integrated circuit technology is capable of improving performance, while reducing power consumption, for a number of military, as well as consumer, systems. One example of the SiGe's capability of improving performance is the SiGe 7HP (90 GHz f_T) divide-by-2-and-4 circuit. This circuit measured a 40% increase in speed, and a 10% power reduction over the SiGe 5HP version. Current Mode Logic architecture has also helped achieve high speeds through its differential configuration, as well as lower power consumption through its vertical stacking configuration. Lower power consumption through BiCMOS design has also been demonstrated in the VLSI Photonics chip, where NMOS devices replaced the HBT and resistor combination, thereby allowing a lower voltage power supply to be used versus other Current Mode Logic circuits mentioned in this paper. Lastly, SiGe BiCMOS circuits can be manufactured in an existing silicon fabrication facility, providing performance comparable to III-V technologies, but with higher yield and quantity.

Acknowledgements

An abbreviated version of this manuscript describing only the 5HP technology was presented at the 1999 GaAs Integrated Circuit Symposium.

This work was supported in part by the Microsystems Technology Office of the Defense Advanced Research Projects Agency (DARPA/MTO), through Contracts N66001-94-C-0051 and N66001-99-C-8605 from SPAWAR Systems Center San Diego. The authors wish to thank Drs. D. Radack and J. Murphy, DARPA/MTO, and Dr. C. Hanson, SPAWAR Code 8505, for administrative support and many helpful discussions; and S. Richardson, D. Jensen, E. Doherty, T. Funk, and L. Stewart, Mayo Foundation, for assistance in preparation of text and figures.

References

1. Ahlgren, D.C., G. Freeman, S. Subbanna, R. Groves, D. Greenberg, J. Malinowski, D. Nguyen-Ngoc, S. J. Jeng, K. Stein, K. Schonenberg, D. Kiesling, B. Martin, S. Wu, D. L. Harame, and B. Meyerson: A SiGe HBT BiCMOS technology for mixed signal RF applications. Proceedings of 1997 IEEE Bipolar/BiCMOS Circuits and Technology Meeting, September 1997, pp. 195-197.

2. Freeman, G., D. Ahlgren, D.R. Greenberg, R. Groves, F. Huang, G. Hugo, B. Jagannathan, S.J. Jeng, J. Johnson, K. Schonenberg, K. Stein, R. Volant, S. Subbanna: A 0.18 um 90 GHz f_T SiGe HBT BiCMOS, ASIC-Compatible, Copper Interconnect Technology for RF and Microwave Applications. IEDM, 1999.

3. Motorola Corporation, U.S. Patent number 5,600,260.

4. Marshall, P., M. Carts, A. Campbell, D. McMorrow, S. Buchner, R. Stewart, B. Randall, B. Gilbert, and R. Reed: Single Event Upset Sensitivity of Circuit Hardened SiGe HBT Logic at Gigabit per Second Data Rates. IEEE Transactions on Nuclear Science, Vol. 47, No. 6, December, 2000.

5. Freeman, G., D.R. Greenberg, K. Walter, S. Subbanna: SiGe HBT Performance Improvements from Lateral Scaling. 29th ESSDERC, pp. 724-727, 1999.

6. Haney, M.W., M.P. Christensen, P. Milojkovic, G.J. Fokken, M. Vickberg, B.K. Gilbert, J. Rieve, J. Ekman, P. Chandramani, F. Kiamilev: Description and Evaluation of the FAST-Net Smart Pixel-Based Optical Interconnect Prototype. Proceedings of the IEEE, Vol. 88, No. 6, June 2000.

7. Subbanna, S., J. Johnson, G. Freeman, R. Volant, R. Groves, D. Herman, B. Meyerson: Prospects of Silicon-Germanium-Based Technology for Very High-Speed Circuits. IEEE MTT-S International Symposium, 2000.

8. Ekman, J., F. Kiamilev, G. Fokken, S. Sommerfeldt, B. Gilbert, M. Vickberg, Y. Liu, A. Cox, P. Gui, P. Chandramani, X. Wang, M. Haney, M. Christensen, P. Milojkovic, K. Driscoll, and B. Vanvoorst: System design and packaging for an optically interconnected MCM switch for parallel computing. Proceedings of InterPACK'01: The PACIFIC RIM/International, Intersociety, Electronic Packaging Technical/Business Conference, July 8-13, 2001, Kauai, Hawaii, USA.

9. Jeng, S.J., B. Jagannathan, J.-S Rieh, J. Johnson, K.T. Schonenberg, D.R. Greenberg, A Strickler, H. Chen, M. Khater, D. Ahlgren, G. Freeman, K. Stein, and S. Subbanna: A 210-GHz f_T SiGe HBT with a non-self-aligned structure. IEEE Electron Device Letters, Vol. 22, No. 11, November 2001

10. B. Jagannathan, M. Khater, F. Paguette, J.-S Rieh, D. Angell, H. Chen, J. Florkey, F. Golan, D.R. Greenberg, R. Groves, S.J. Jeng, J. Johnson, E. Mengistu, K.T. Schoenenberg, C.M. Schnabel, P. Smith, A. Strickler, D. Ahlgren, G. Freeman, K. Stein, and S. Subbana: Self-Aligned SiGe NPN Transistors with 285 GHz f_{MAX} and 207 GHz f_T in a manufacturable technology. Submitted to Electron Device Letters.

International Journal of High Speed Electronics and Systems
Vol. 13, No. 1 (2003) 239–263

A SiGe HBT IC CHIPSET for 40-Gb/s OPTICAL TRANSMISSION SYSTEMS

T. MASUDA, N. SHIRAMIZU, E. OHUE, K. ODA, R. HAYAMI,
M. KONDO, T. ONAI, and K. WASHIO
ULSI Research Department, Central Research Laboratory, Hitachi, Ltd.,
1-280 Higashi-koigakubo, Kokubunji, Tokyo 185-8601, Japan

K. OHHATA, F. ARAKAWA, M. TANABE, and H. SHIMAMOTO
Musashino Office, Hitachi Device Engineering, Co. Ltd.,
1-280 Higashi-koigakubo, Kokubunji, Tokyo 185-8601, Japan

T. HARADA
Device Development Center, Hitachi, Ltd.,
6-16-3 Shin-machi, Ome, Tokyo, 198-8512, Japan

Using a 0.2-μm self-aligned epitaxial-growth silicon-germanium (SiGe) heterojunction bipolar transistor (HBT) technology, we have developed a chipset for 40-Gb/s time-division multiplexing optical transmission systems. In this paper, we describe seven analog and digital ICs: a 45-GHz bandwidth transimpedance amplifier, a 48.7-GHz bandwidth automatic-gain-controllable amplifier, a 40-Gb/s decision circuit, a 40-Gb/s full-wave rectifier, a 40-Gb/s limiting amplifier with a 32-dB gain, a 45-Gb/s 1:4 demultiplexer, and a 45-Gb/s 4:1 multiplexer. To increase bandwidth of the transimpedance amplifier, a common-base input stage is introduced. In order to have high gain and wide bandwidth simultaneously, active load circuits composed of a differential transimpedance amplifier are used for the AGC amplifier, the limiting amplifier, and the decision circuit. Full-rate clocking is employed to reduce the influence caused by clock-duty variation in digital circuits such as the decision circuit, the demultiplexer, and the multiplexer. All ICs were characterized by using on-wafer probes, and some of them were built in brass-packages for bit-error rate measurement.

Keywords: SiGe HBT; transimpedance amplifier; automatic-gain-controllable amplifier; decision circuit; full-wave rectifier; limiting amplifier; demultiplexer; multiplexer.

1. Introduction

There is an urgent need to expand transmission capacity because of the rapid growth of multimedia services. Long-haul optical transmission systems have been rapidly developed and are now acknowledged as essential for high-speed telecommunication and data transmission. Therefore, Synchronous Optical Network (SONET) and Synchronous Digital Hierarchy (SDH) have been used for constructing backbone networks and the maximum data rate has reached 10 Gb/s in commercial use. Time-division multiplexing (TDM) systems and wavelength-division multiplexing (WDM) systems had been extensively researched with the aim of increasing overall transmission capacity. Because a WDM system is based on several TDM systems, the increase of the maximum data-rate in a TDM system is essential. At present, 40-Gb/s TDM systems are expected as the next generation in SONET/SDH, and the demand for components comprising optical transmission systems is rising rapidly.

A typical circuit block diagram of a 40-Gb/s optical transmission system is shown in Fig. 1. The transmitter consists a 4:1 multiplexer (MUX), a modulator driver, an external modulator, and a laser diode. The MUX serializes four channels of 10-Gb/s data to a 40-Gb/s data stream, and the modulator driver circuit amplifies such high-speed data to drive the external modulator at a sufficient voltage swing. Then the external modulator converts continuous light-wave from laser diode to modulated optical signal by the driver's output. The optical signal is transmitted to a receiver via an optical fiber. The signal path of the receiver is composed of a photo detector, a preamplifier, an automatic gain controllable (AGC) amplifier, a decision circuit, and a 1:4 demultiplexer (DEMUX). The photo detector transforms the optical signal to an electrical current, and the preamplifier amplifies the current signal to a voltage signal with a low noise characteristic. The AGC amplifier provides a constant voltage swing, which is independent of the input voltage swing. The decision circuit makes a threshold decision and regenerates the data signal by using a 40-GHz clock signal. The 1:4 DEMUX separates the 40-Gb/s data into four channels of 10-Gb/s data in parallel. To generate the clock signal for the decision circuit and the DEMUX, the AGC amplifier output is fed to a clock-extraction circuit composed of a full-wave rectifier, a band pass filter, and a limiting amplifier.

To meet the demand for 40-Gb/s optical transmission systems, high-speed analog and digital ICs based on several technologies had been reported. Compound semiconductor technologies originally developed key ICs for 40-Gb/s systems because of their high cutoff frequency (f_T) and maximum oscillation frequency (f_{max}). [1-5] Their high breakdown voltages are a promising feature for the design of driver circuits, which generate large voltage swing. And the semi-insulating substrates are useful to realize low-loss transmission lines. However, high cost of compound semiconductor technology remains a major problem preventing its use for growing commercial applications, which require mass-production. Si technology also obtained high-speed capability by utilizing the effect of the SiGe hetero-structure. In recent years, such SiGe HBTs had comparable f_T and f_{max} with those of transistors based on compound semiconductor technologies. [6-8] Since the device geometry of SiGe HBTs is generally smaller than those of compound-semiconductor-based transistors, it helps to reduce parasitic capacitances of both the device itself and the interconnection metal patterns in a circuit. Therefore, high-speed operation is possible with a highly functional circuit. The features derived from Si technology's nature leads to the good cost effectiveness and high reliability, which are indispensable factor for mass-production. [9,10] Additionally, a SiGe HBT driver IC, which is capable of operation at 48 Gb/s with a single-ended voltage swing of 3.5 Vpp, has been demonstrated.[11] Therefore we developed a 0.2-μm self-aligned selective-epitaxial-growth (SEG) SiGe HBT that has both high performance and high productivity. In this paper, we report on a SiGe HBT IC chipset based on the SiGe HBT technology for 40-Gb/s optical transmission systems. All electronic circuit blocks for a 40-Gb/s system shown in Fig. 1, except for a modulator driver, can be realized with this chipset.

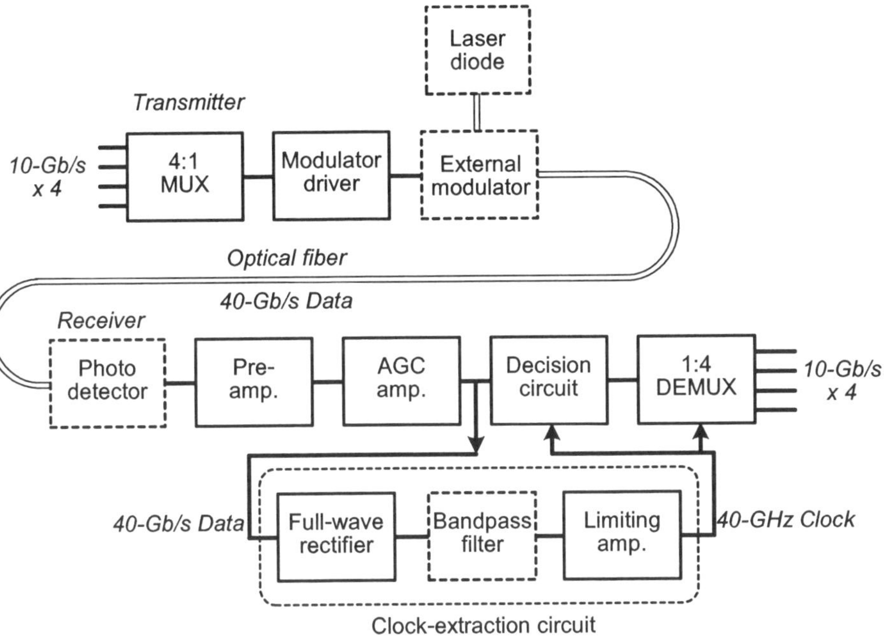

Fig. 1. Block diagram of a 40-Gb/s optical transmission system.

2. SiGe HBT Technology

The silicon-germanium hetero-system allows specific tailoring of the band structure, and can thus provide new generations of silicon-based transistors, which dominate the semiconductor market. Si-based SiGe-heterostructure devices can open up the way to the access to applications beyond the limits of Si. They can at least fill the performance gap in high-speed capability to the III/Vs compound semiconductor devices. Especially, SiGe HBTs have both high cutoff frequency and high maximum oscillation frequency at quite low cost. The cost effectiveness has been proven.[12] The total cost of SiGe is about \$0.12 /mm^2, which is almost same with that of Si, while the costs of GaAs and InP are as high as \$0.5 /mm^2 and \$1.2 /mm^2. The high-speed capability of SiGe HBTs was also presented.[7,8] A 71-GHz operation of static frequency divider is attributed to a cutoff frequency more than 100 GHz. The following is a brief description of a SiGe HBT technology used for fabricating a chipset for 40-Gb/s TDM optical transmission systems.

Figure 2 shows a schematic cross section of a 0.2-μm self-aligned selective-epitaxial-growth (SEG) SiGe HBT. The SiGe-base/Si-cap multi-layer was selectively grown by ultra-high-vacuum (UHV) chemical vapor deposition (CVD) to obtain high cutoff frequency. The as-grown SiGe-base layer containing boron at a concentration of 2 x 10^{19} cm^{-3} was 15 nm thick. A poly-Si assisted self-aligned SEG (PASS) structure was introduced to provide a good link between the intrinsic and extrinsic bases. The PASS structure ensures that the selectively grown 0.6-μm-wide Si-cap/SiGe-base multi-layer self-aligns to the 0.2-μm-wide emitter and the base poly-Si. This self-aligned structure of the active region produced both low collector capacitance and low base resistance. Furthermore, to reduce the parasitic capacitance of the collector and substrate, shallow-trench (0.4-μm deep) and dual 0.6-μm-wide deep-trench (3-μm deep) isolations were used. To reduce the parasitic resistance of all electrodes, Ti-salicide layers with a sheet resistance of 3 Ω/square were formed. As shown in the SEM cross sectional view in Fig. 3, the device structure mentioned above was successfully formed on an 8-inch wafer.

Typical parameters of a transistor with an emitter area of 0.2 x 2.0 μm are listed in Table 1. Current gain h_{FE} is 1,400, and collector-emitter breakdown voltage BV_{CER} is 3.0 V. The BV_{CER} was measured with a 2k-Ω termination-resistor at the base node. Collector parasitic capacitance C_{TC} is 3.6 fF and the collector-substrate capacitance C_{TS} is only 1.8 fF. Figure 4 shows the dependence of cut-off frequency and maximum oscillation frequency on collector current. The peak value of cut-off frequency is 122 GHz and that of the maximum oscillation frequency is 163 GHz when the collector current is 3.0 mA at a collector-emitter bias voltage of 2 V.

To find an optimal internal voltage-swing for digital circuits based on emitter-coupled-logic (ECL), we usually investigate the dependence of ECL propagation delay time (t_{pd}) on internal voltage swing (V_L). Regarding with this SiGe HBTs technology, the minimum delay time for a differential ECL circuit is 5.5 ps with an internal voltage swing of 250 mVpp. The internal voltage swing of 250 mVpp was used in the design for a delayed flip-flop circuit, which is the core circuit in digital circuits such as a MUX and a DEMUX.

This SiGe HBT technology has four aluminum metal layers for interconnection. A metal-insulator-metal (MIM) capacitor formed by the first and second-metal layers has a capacitance of 0.7 fF/μm^2, and it enables signal decoupling and stabilization of power-supply patterns. The fourth-metal layer (top layer) was mainly used for high-speed signal connection between the circuit blocks. A micro-strip transmission line, which formed by the fourth-metal layer for signal line and first-metal layer for a ground plane, is used for long-distance connection in a chip.

While we utilize a SiGe HBTs' features mentioned above, there are some issues that must be considered during circuit design. The influence of the Si substrate caused by imperfect insulation is one of the important issues.[13] Because of the conductivity of the Si substrate, transmission lines have relatively large power-loss compared to that on an insulated substrate used in compound semiconductor technology. Therefore, amplifiers based on Si technology have to attain further gain in order to compensate the loss on transmission lines.

Table 1. Typical transistor parameter for a 0.2-μm SiGe HBT.

A_E	0.2 x 2.0	μm
h_{FE}	1400	
BV_{CER}	3.0	V
R_B	90	Ω
C_{TC}	3.6	fF
C_{TS}	1.8	fF
f_T	122	GHz
f_{max}	163	GHz
t_{pd}^{min}	5.5 (@V_L=250 mV_{pp})	ps
Frequency Divider	67	GHz

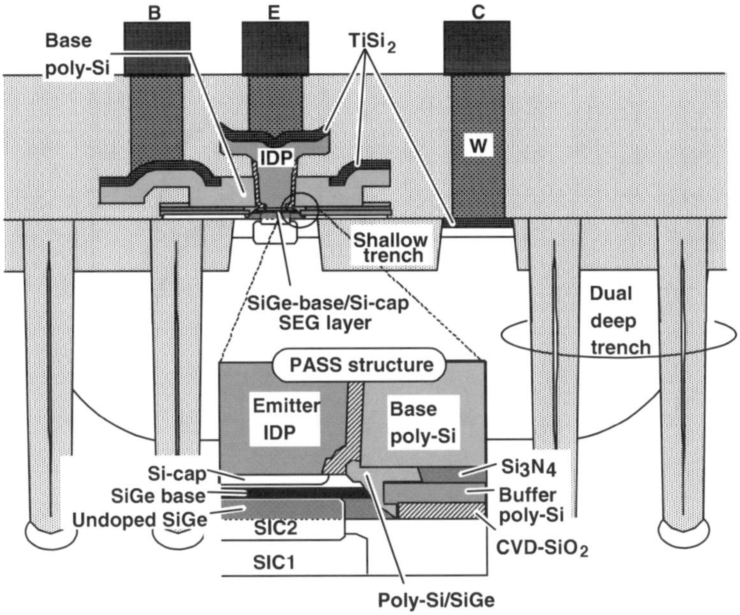

Fig. 2. Cross sectional schematic of a 0.2-μm SiGe HBT.

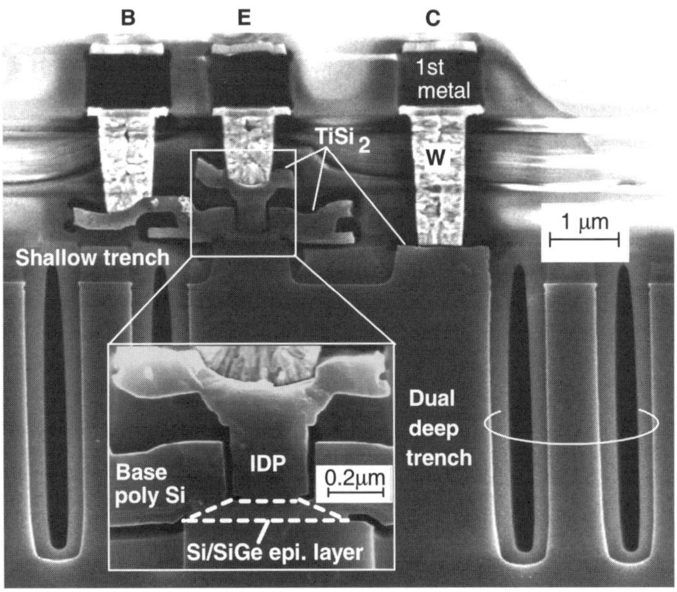

Fig. 3. SEM cross sectional view of a 0.2-μm SiGe HBT.

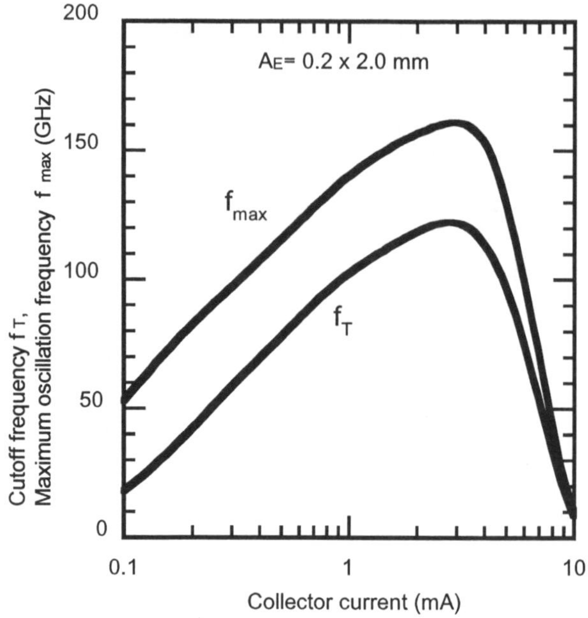

Fig. 4. Collector current dependence on cut-off frequency and maximum oscillation frequency for a 0.2-μm SiGe HBT at a collector-emitter voltage of 2 V.

3. High-Speed ICs for Optical Transmission Systems

In this chapter, we report our IC chipset for a 40-Gb/s optical transmission system. Almost of the circuit blocks in Fig. 1 had designed and verified the performance through the fabrication by using our 0.2-μm SiGe HBT technology, which is described in the previous chapter.

3.1. Receiver

3.1.1 Transimpedance amplifier.[14]

A preamplifier must amplify the current signal from a photo detector to ensure that there is a proper output voltage. It is desirable to achieve low noise, high transimpedance-gain, wide bandwidth characteristics, simultaneously. For use in 40-Gb/s optical receivers, a wide bandwidth of over 40 GHz is especially needed to ensure the total bandwidth is more than 30 GHz with the AGC amplifier and decision circuit that follows. Moreover, to reduce the receiver cost by improving the packaging yield, the preamplifier must provide the required bandwidth even when the input capacitance is as high as 100 fF and fluctuates.

There are three principle types of preamplifiers used in an optical receiver.[15] First is a low impedance preamplifier composed of a photo detector, which is terminated by a 50-Ω resistor as an input termination resistor, and an amplifier with 50-Ω input impedance. The photo detector is connected to input node of the amplifier through a transmission line. This configuration can be constructed easily, but it has the drawback of low sensitivity. Second is a high impedance preamplifier with a large input termination resistor for high sensitivity. However, an equalizer designed carefully is required to compensate the input pole for wide bandwidth. This approach is sometimes used but tends to have a limited dynamic range and is difficult to integrate due to the need for component matching. Third is a transimpedance-type preamplifier; it is the most frequently used because of its combination of high dynamic range, high sensitivity, and wide bandwidth. We therefore focused on developing a transimpedance-type amplifier.

Circuit Design

Figure 5 shows the design concept for improving the bandwidth of a preamplifier. In a conventional transimpedance preamplifier, input capacitance C_{in} and preamplifier input impedance Z_{in} are the dominant factors affecting frequency response. That is, the time constant of the preamplifier input node determines the dominant pole of the frequency response for the transimpedance gain Z_T. Capacitance C_{in} is the sum of photo diode capacitance C_{PD} and pad capacitance C_{pad}. The input capacitance of conventional transimpedance amplifier C_{in_TIA} equals C_{in}. Impedance Z_{in} is approximately given by $R_F/(A+1)$, where R_F is feedback resistance and A is the open-loop voltage-gain of the amplifier. In a frequency range above the upper bandwidth of the gain A, Z_{in} increases with frequency and total bandwidth decreases. To improve the bandwidth, both input capacitance and input impedance should be reduced. A small photo diode can be used to reduce input capacitance, but doing so increases the production cost because of the difficulty of mounting photo diode with high accuracy. Decreasing R_F for small Z_{in} decreases transimpedance gain and increases equivalent input noise current. In the case of a conventional transimpedance amplifier, it is thus difficult to improve the bandwidth without the depredating the other performance factors.

To overcome this difficulty, we placed a common-base input stage (CBIS) in front of the conventional transimpedance amplifier. The CBIS lowered the input impedance below that of the transimpedance amplifier up to high frequency. The input impedance when the CBIS is inserted is given by V_T/I_{CB}, where V_T is thermal voltage and Ic is the bias current of the common base transistor Q_{CB}. The time constant of preamplifier input (C_{in} x V_T/I_{CB}) can be designed lower than that of conventional transimpedance amplifier when the I_{CB} is set to a proper current. CBIS also separates C_{in} from the transimpedance amplifier. Capacitance C_{in_TIA} is roughly given by the collector parasitic capacitance of Q_{CB} (C_{TC}) and parasitic capacitances caused by the metal patterns used for the connection. This increases the bandwidth of the transimpedance amplifier. Accordingly, the CBIS provides a wide-bandwidth preamplifier without reducing of C_{in} and allows us to use a high input capacitance to obtain a bandwidth of more than 40 GHz. It is effective to improve the packaging yield of optical receiver.

Figure 6 shows a schematic of a transimpedance amplifier with a CBIS. It consists of a CBIS, a transimpedance amplifier stage, and emitter-followers as an output buffer. The CBIS is composed of a common base transistor, Q_{CB}, and bias resistors R_1 and R_2. The transimpedance amplifier stage consists of a common emitter amplifier (Q_1 and R_L) and a shunt feedback loop through Q_2 and R_F. Emitter followers with a two-stage configuration are used to drive a 50-Ω load without reducing bandwidth. A resistor, R_o, was introduced to set the output impedance to 50 Ω. V_{RL} bias circuit is designed to supply a stable bias voltage for R_L, because the gain of a common emitter transistor is sensitive to the variation of V_{RL}. Especially, to reduce the influence of V_{CC} variation, we designed V_{RL} to be defined from the ground level (shown in Fig. 6 (b)).

To take full advantage of the CBIS, the bias current and the emitter size and of Q_{CB} must be optimized to obtain a wide bandwidth and a low input equivalent current noise. Because shot noise decreases with bias current I_{CB}, small I_{CB} reduces the input-equivalent current-noise. However, at the same time, input impedance Z_{in} increases and the bandwidth is strongly decreased. On the other hand, a large transistor has a small base parasitic-resistance and the noise contribution becomes small. Larger transistor has larger collector parasitic-capacitance, thus bandwidth becomes narrow. We therefore found optimum values of emitter size and bias current.

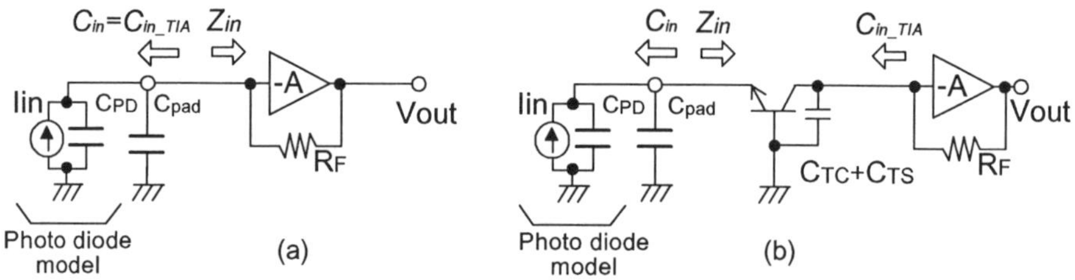

Fig. 5. Concept of a transimpedance amplifier with a common base input stage.

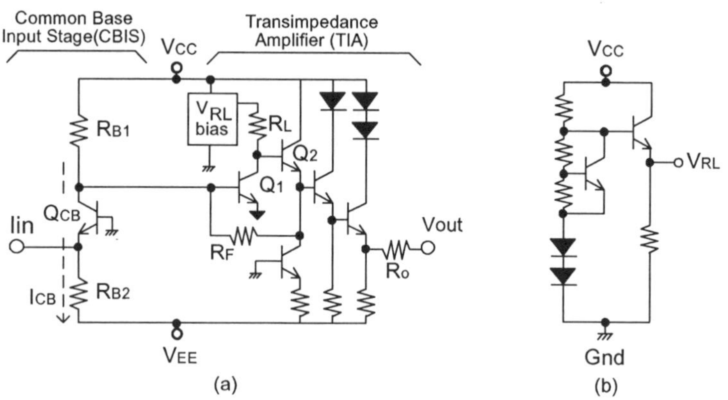

Fig. 6. (a) Schematic of the transimpedance amplifier with a common base input stage and (b) V_{RL} bias circuit.

Circuit Performance

A chip micrograph of transimpedance amplifier with CBIS is shown in Fig. 7. The chip size is 1.2 x 1.8 mm and the circuit core occupies 100 x 100 μm, and is located close to the input pad. A 900-μm-length micro-strip transmission line connects the circuit core to the output pad.

Figure 8(a) shows the frequency response of the transimpedance amplifier's input impedance when the CBIS is presented or not. Within a frequency range from DC to 40 GHz, the input impedance of a transimpedance amplifier with a CBIS is much lower than that of the preamplifier without a CBIS. We calculated transimpedance gain from measured scattering parameters, taking the input capacitance into account (Fig. 7 (b)). The bandwidth of the preamplifier without the CBIS was 26 GHz when Cin was 100 fF. On the other hand, the transimpedance amplifier with the CBIS produced a 45-GHz bandwidth under the same conditions; that is, a 70% increase in bandwidth. The transimpedance gain at low frequency is 50 dBΩ. Moreover, the bandwidth is more than 40 GHz even when input capacitance rose to 125 fF. We attribute these results to the low input impedance of the CBIS. These results demonstrate that the transimpedance amplifier can be used even if input capacitance rises to 125 fF because of deviation in the photo detector assembly, and use of CBIS is effective to reduce the cost of the 40 Gb/s optical receiver. The group-delay deviation of transimpedance gain is 8 ps within a frequency range up to 30 GHz. The loop gain in transimpedance stage should be optimized in order to reduce the group-delay deviation. The power supply voltages V_{CC} and V_{EE} are 5.0 V and -5.2 V, respectively, and the power consumption is 0.3 W. V_{EE} of –5.2 V is redundant but is used as a standard supply voltage for high-speed signal ICs. V_{EE} can be reduced to –2 V without any performance degradation.

To our knowledge, the bandwidth of transimpedance gain is the best yet attained by Si-based technology. Preamplifiers fabricated by using compound semiconductor technologies also had performance comparable to that of the SiGe transimpedance amplifier. A GaAs HBT transimpedance amplifier with 45-GHz bandwidth and 45-dBΩ gain and a GaAs metamorphic high-electron-mobility transistor (MHEMT) transimpedance amplifier with 47-GHz bandwidth and 48.4-dBΩ gain have been reported.[16,17] Moreover, based on InP HBT technology, a high-gain transimpedance amplifier with both linear and limiting modes has a high input sensitivity and can be operated at a high speed (i.e. a data rate of 48 Gb/s).[5] The combination of a transimpedance amplifier and a traveling-wave amplifier based on InP high-electron-mobility transistor (HEMT) technology was an effective way to produce 49-GHz bandwidth and 52-dBΩ gain.[18] Si-based preamplifiers have the advantage of being able to integrate with other functional circuit blocks at a low cost. On the other hand, a valuable feature of preamplifier fabricated by using compound semiconductor technologies is that these preamplifiers can be combined with optical ICs such as photo detectors. It is also cost effective for construction of a high-speed optical transmission system. At present, regarding preamplifiers for 40-Gb/s systems, both SiGe HBT technology and compound semiconductor technologies are used.

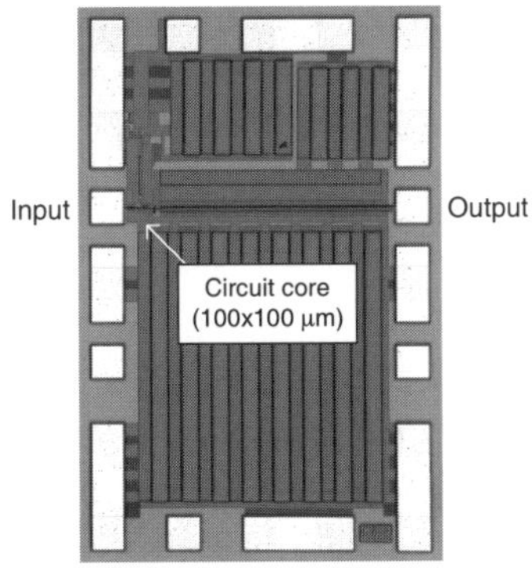

Fig. 7. Chip micrograph of the transimpedance amplifier with a common base input stage. 1.2 x 1.8 mm.

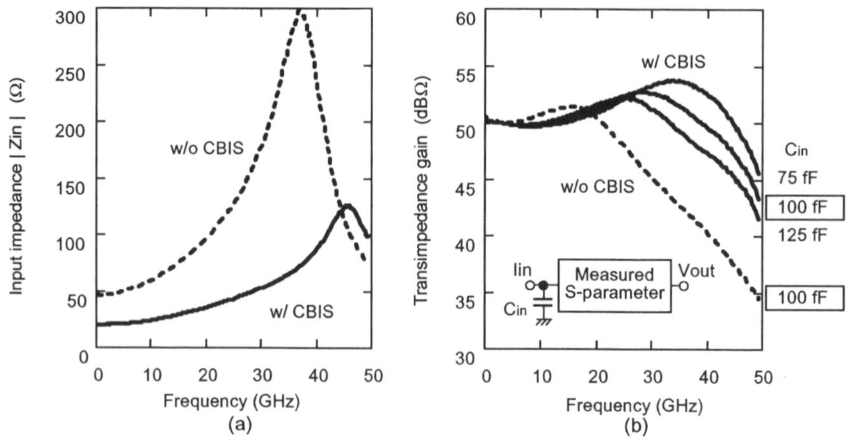

Fig. 8. Measured frequency response of the transimpedance amplifiers. (a) Input impedance, (b) transimpedance gain.

3.1.2 AGC amplifier[19]

In an optical receiver, an AGC amplifier is used to amplify the output voltage of the preamplifier to a value required by the succeeding decision circuit. While operating at high speed and high gain, an important demand on this circuit is to keep the output voltage swing constant even if the input voltage swing has a large dynamic range. Some limiting amplifiers can meet this demand. However, limiting may degrade sensitivity of optical receivers that used avalanche photo detectors. It is difficult to find an optimal threshold-decision for minimizing bit-error rate (BER) of the optical receiver. On the other hand, AGC amplifiers are able to amplify whole an input waveform and keep a shape. This is the reason why AGC amplifier is widely used for optical receivers. We therefore developed an AGC amplifier for 40Gb/s optical transmission systems. As in a previous work, a packaged AGC amplifier with a bandwidth of 26 GHz and a dynamic range of 6 dB, developed using AlGaAs/GaAs HBT technology, has been reported[20]; however, both its bandwidth and dynamic range were insufficient for 40-Gb/s systems. We also reported on a SiGe-HBT AGC amplifier having a wide bandwidth of 32.7 GHz and a dynamic range of 19 dB[21], but it needs to be further improved for practical use in 40-Gb/s systems.

Circuit Design

A block diagram of the AGC amplifier is shown in Fig. 9. In the signal path, there is an input buffer (IB), two variable-gain amplifier stages (VGA), and an output buffer (OB) with a constant gain. A peak detector (PD) outputs signals (PD0 and PD1) in proportion to the output amplitude of Output0 and Output1. Another output (PDM) indicates the average output voltage and is used as a reference voltage for automatic gain control. The PDM voltage is equal to PD0 and PD1 when no signal is input. According to the outputs of PD, a gain control circuit (GC) controls the gain of the VGAs to maintain a constant voltage swing of Output0 and Output1. The voltage difference between PD0 and PDM is utilized for offset voltage cancellation at the input nodes. As shown in Fig. 9, to compose AGC feedback loop, several external amplifiers and low pass filters are required.

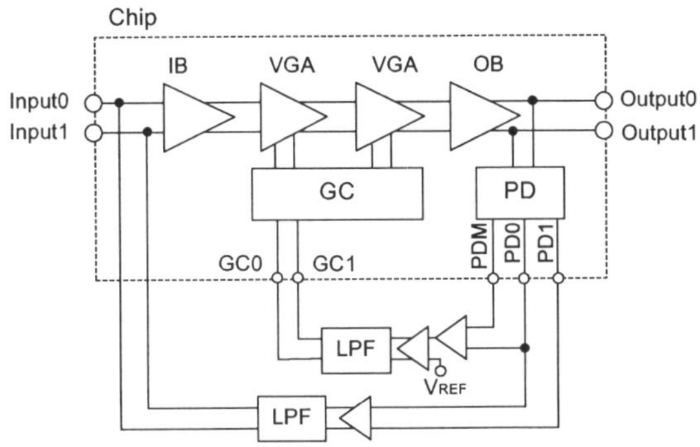

Fig. 9. Block diagram of the AGC amplifier.

The two VGAs have the same circuit configuration (Fig. 10). Each VGA consists of a Gilbert multiplier, a transimpedance load circuit, and three emitter followers in series. Signals Input0 and Inputt1 are input into the lower transistor pair (Q1 and Q2) and gain control signals VCNT0 and VCNT1 are input into the upper transistor pair (Q3-Q6) in the gain control stage. An AGC amplifier is needed to provide uniform frequency dependence and bandwidth within the tunable gain range. However, the bandwidth of a conventional gain-control stage, whose configuration does not have resistor R_E (Fig. 10), depends on gain-control voltage. In short, the bandwidth of the AGC amplifier decreases with the gain of the VGA.[21, 22] This is because the gain-control stage has a minimum bandwidth when the collector current of Q3 is the same as that of Q4. We therefore inserted bandwidth-compensating resistors (R_E) in emitter nodes of Q3-Q6. Resistance R_E makes the bandwidth of the gain-control stage insensitive to variation in collector current. It, therefore, provides uniform bandwidth of the AGC amplifier within a wide gain-tunable range.

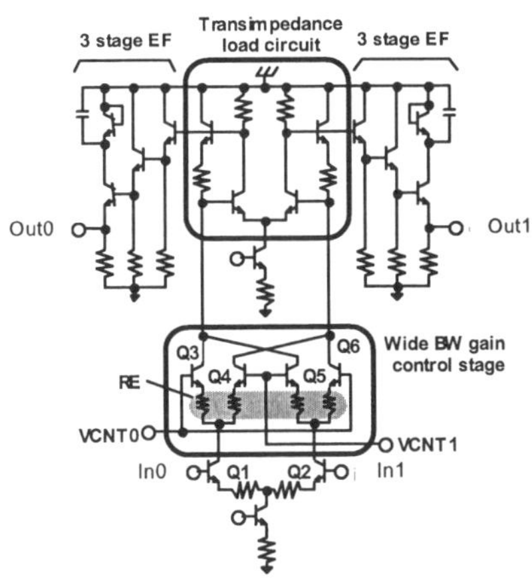

Fig. 10. Schematic of the variable gain amplifier (VGA) stage.

A peaking technique can also produce wide bandwidth. A transimpedance circuit is used as a load circuit.[23, 24] At the collector nodes of Q3 to Q6, the low input impedance of the active load reduces the time constant and increases the bandwidth. Due to the small time constant at the collector nodes, collector currents of Q3 to Q6 are converted to a voltage up to high frequency. Also, the transimpedance load circuit has a peaking effect due to its feedback loop and it helps to keep a gain at high frequency. However, peaking should be carefully controlled because excessive peaking degrades the flatness in the frequency response, resulting in waveform distortion.[21]

Circuit Performance

Figure 11 shows a chip micrograph of the AGC amplifier. The circuit core of the AGC amplifier (surrounded by dashed lines in Fig. 9) was only fabricated. But it is enough to ensure the high-speed capability of the circuit. The active area, 420 x 540 μm, is located in the center of the chip area: 1.2 x 1.8 mm. Regarding with chip layout, the circuit elements were arranged symmetrically, because each circuit block is based on differential amplifiers.

The gain-versus-frequency characteristics of the AGC amplifier measured by on-wafer RF probes are shown in Fig. 12. The bandwidth is 47.8 GHz at a maximum gain of 21.5 dB, and the bandwidth is 48.3 GHz at a minimum gain of 0.5 dB. The small fluctuation in the bandwidth when the gain was changed can be attributed to the bandwidth-compensating resistor in the VGA. The active load contributed to high gain of 21.5 dB with the wide bandwidth.

Figure 13 shows the operational waveform of the AGC amplifier module. The module is consisted of a ceramic substrate soldered on a CuW spacer and a brass block with co-axial connectors. AGC amplifier chip was glued onto the spacer and ultrasonically bonded to the co-planar line on the ceramic substrate at the same level. A 10-Gb/s pulse-pattern generator (PPG) and a 4:1 MUX module were used to generate 40-Gb/s pseudorandom bit streams.[21] The eye pattern is well-opened at a data rate of 40 Gb/s. The output voltage swing was 350 mVpp. V_{EE} supply voltage of -7.5 V is required because of a large voltage-drop in the transimpedance load circuit. The power consumption was 1.2 W with a V_{EE} of –7.5 V.

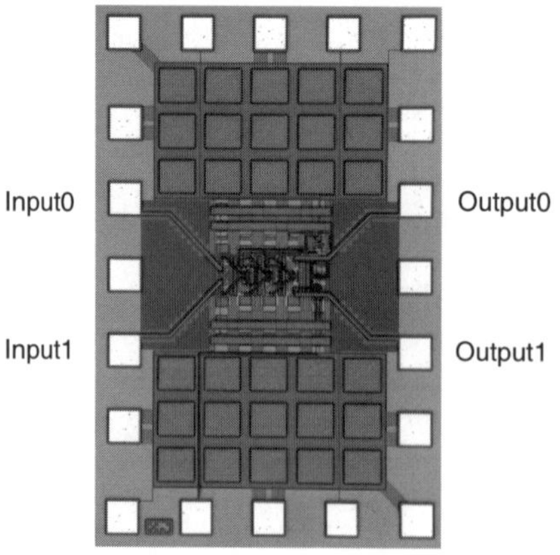

Fig. 11. Chip micrograph of the AGC amplifier. 1.2 x 1.8 mm.

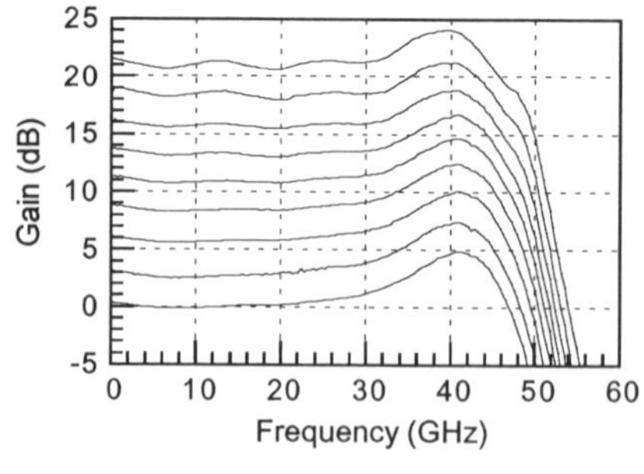

Fig. 12. Measured frequency response of gain.

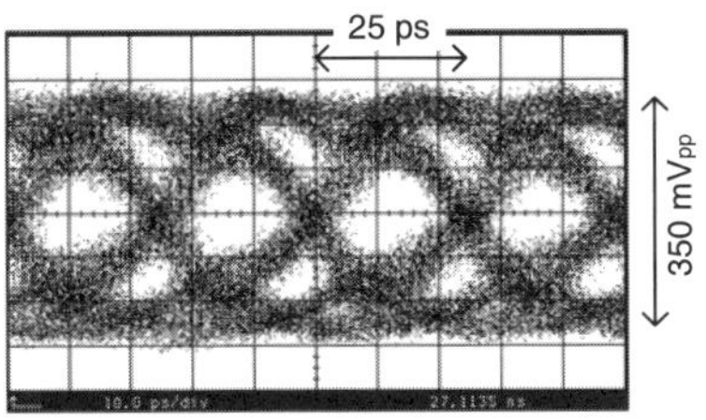

Fig. 13. Measured output waveform of the AGC amplifier module at a data rate of 40 Gb/s.

3.1.3 Decision circuit[19]

To regenerate a data stream that is reshaped and eliminated jitter, a decision circuit must be placed after the AGC amplifier. Input analog signals are converted to a binary data stream by a slice amplifier and retimed by a master-slave delayed-flip-flop (MS-DFF) with a clock signal. The slice amplifier has a threshold level for determining the data state. The threshold level can be adjusted to minimize the error rate of the receiver. Regarding works on decision circuits, some of the ICs fabricated by using compound semiconductor technologies have obtained excellent performance, namely, maximum operation speed is 45 Gb/s.[26] Recently, a decision circuit IC with InP/InGaAs DHBTs (double heterojunction bipolar transistor) showed good power efficiency[27]; at a data rate of 43 Gb/s with a power consumption of 0.7 W.

Circuit Design

Figure 14 shows a block diagram of decision circuit, which is composed of an input buffer (IB), a two-stage preamplifier (Pre1 and Pre2) as a slice amplifier, a MS-DFF, a pre-output buffer (PB), an output buffer (OB), and a clock buffer (CB). Triple-stage emitter followers (EFs) are used as input and clock buffers. The total gain of the two-stage preamplifier is 22 dB and the bandwidth is more than 40 GHz. A schematic of the preamplifier is shown in Fig. 15. As well as the AGC amplifier mentioned in section 3.1.2, a transimpedance load circuit is used in Pre1 to widen the bandwidth, and a cascode amplifier is used in Pre2 because the output DC level of Pre1 is too low to drive a MS-DFF. A delayed flip-flop (DFF), shown in Fig. 16, is a fundamental, speed-critical circuit for any digital circuit in an optical transmission system. In digital circuit designs such as a decision circuit, a MUX, a DEMUX, it is important whether a MS-DFF can be used with a full-rate clock of 40 GHz or not, because data retiming by full-rate clocking is preferred to reduce jitter caused by clock duty variation. In order to maximize operation speed of the MS-DFF, the internal voltage swing was set to be 250 mVpp. A static frequency divider composed of toggled flip-flops, which are similar with delayed flip-flops, can be operated at a frequency of more than 60 GHz.[7, 8] These result assure that the MS-DFF works well with a 40GHz full-rate clock.

Circuit Performance

A chip micrograph of the decision circuit (chip area: 1.2 mm x 1.8 mm) is shown in Fig. 17. Figure 18 shows the measured output waveforms of the decision circuit. On-wafer probes were used for the measuring the output waveform on an oscilloscope via a 0.5-m coaxial cable. The eye pattern is well-opened at the data rate of 40 Gb/s. As shown in the figure, the jitter in input signal is suppressed at the output node, and each period of the eye pattern is the same.

These results indicate the MS-DFF in the decision circuit can operate with a 40-GHz clock. The output voltage swing is 500 mVpp; the power consumption was 0.9 W with a supply voltage V_{EE} of -5.2 V.

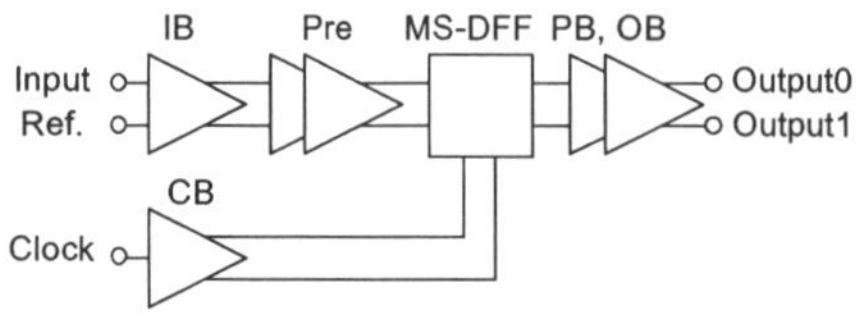

Fig. 14. Block diagram of the decision circuit.

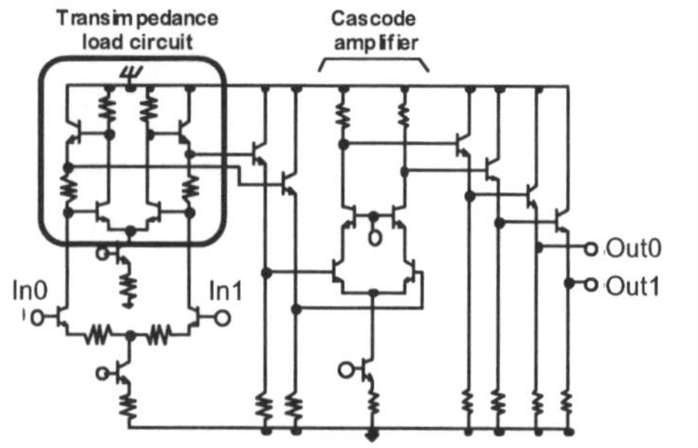

Fig. 15. Schematic of the preamplifier for the decision circuit.

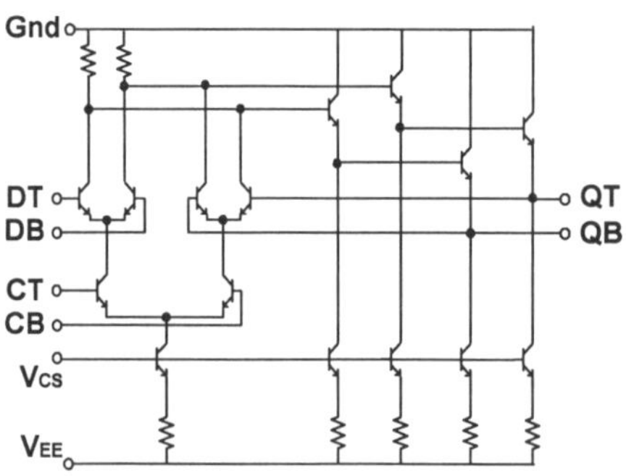

Fig. 16. Schematic of the delayed flip-flop.

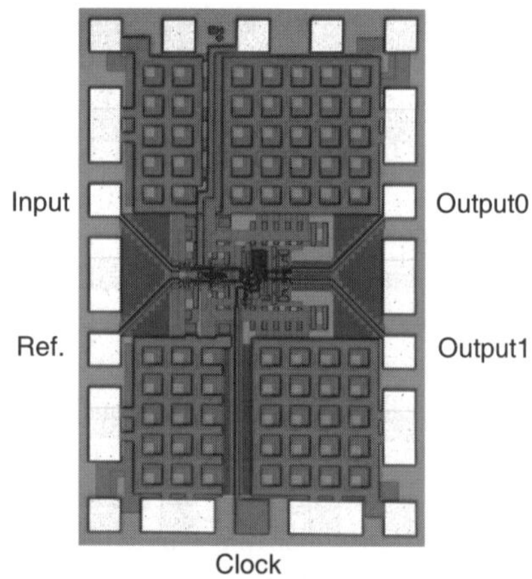

Fig. 17. Chip micrograph of the decision circuit. 1.2 x 1.8 mm.

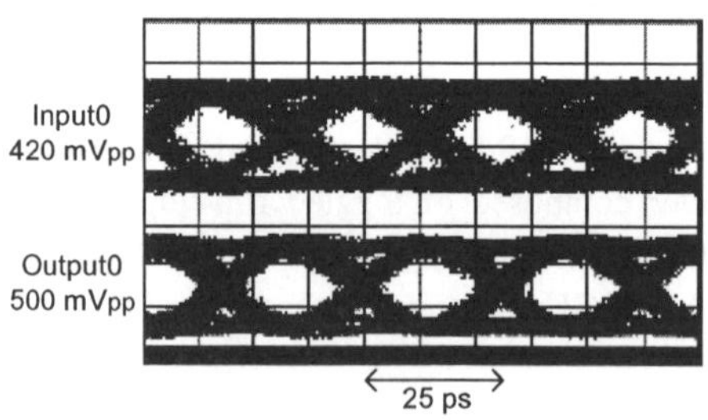

Fig. 18. Measured waveforms of the decision circuit at a data rate of 40 Gb/s.

3.1.4 DEMUX[25]

A DEMUX converts a high-speed data signal to a few channels of lower speed data signals with a reference clock. For a practical use in 40-Gb/s optical transmission systems, the

DEMUX is needed to handle at least four channels of 10-Gb/s data stream. In case that three 1:2 DEMUX ICs are combined in a multi-chip module, it is costly to adjust clock delay between the ICs. Therefore, a 1:4 DEMUX chip that fabricated by a low-cost technology is strongly required. To meet this requirement, a 40-Gb/s 1:4 AlInAs/InGaAs HBT DEMUX with a 20-GHz clock, a 50-Gb/s 1:4 InP-based HEMT DEMUX IC, and a 40-Gb/s clock-data-recovery (CDR) circuit IC with 1:4 DEMUX in SiGe technology have been developed.[28-30]

Circuit Design

Figure 19 shows a block diagram of the 1:4 DEMUX IC, which consists of a MS-DFF, three 1:2 DEMUXes, and a clock-distribution circuit (CDC). A MS-DFF for data retiming is used in front of the first 1:2 DEMUX. A 1:2 DEMUX is composed of a MS-DFF and master-slave-slave delayed-flip-flop (MSS-DFF) to maximize the phase margin. In the CDC, byte-synchronization is enabled by bit-rotation[31]; this synchronization is indispensable for practical use. The CDC is composed of phase-shifters (PS), static frequency dividers (FD), 2-bit counters, and exclusive ORs (EXOR1 and EXOR2). The CDC provides 1/2 clock (CKA) and 1/4 clock (CKB). The phase relationship between these clocks is controlled by the four states of the two EX-ORs' outputs, that is, two output-states of EXOR1 and two output-states of EXOR2. These outputs are determined by the bit-rotation signal via two 2-bit counters (B0 and B1). Therefore, only one bit-rotation signal is needed to change the data until the output bit pattern of the 1:4 DEMUX becomes properly aligned. The PS adjusts the CKA to the optimal clock phase for data latching in the first 1:2 DEMUX.

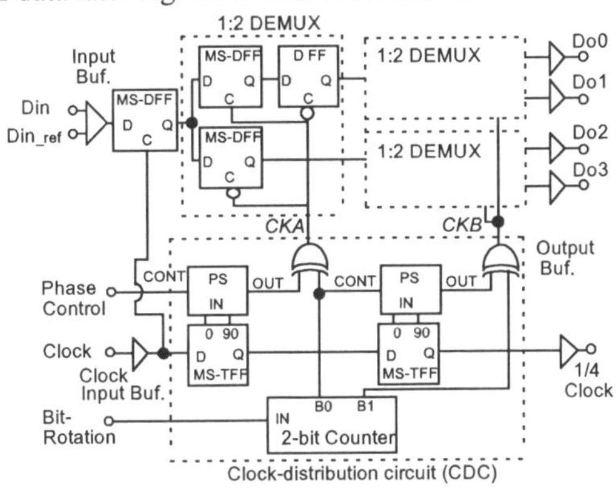

Fig. 19. Block diagram of the 1:4 demultiplexer.

Circuit Performance

A chip micrograph of the DEMUX is shown in Fig. 20. The 550 x 950-μm area circuit core was set at the center of the chip. A micro-strip transmission line, formed out of the fourth metal as a signal line and the first metal as a ground plane, connects the signal pad and the circuit core. MIM capacitors with a total capacitance of 700 pF around the circuit core eliminate interference among the power supply patterns.

The 1:4 DEMUX chip and 4:1 MUX (described in 3.3.1) chip are mounted on module-package as same as AGC amplifier module for the evaluation of BER. The measured input phase margin was 202 degrees and the input sensitivity was 115 mVpp at 40-Gb/s operations. At the back-to-back measurement with a DEMUX module and a MUX module, the maximum operation speed at 10^{-12} BER was a data rate of 45 Gb/s. For the input data, pseudo-random bit-sequences (PRBS) of 400-mVpp were used and the data length was 2^{31}-1. The maximum operation speed is determined by that of the DEMUX module. Figure 21 shows the output waveforms at 48 Gb/s measured by on-wafer RF probes. Well-opened demultiplexed data patterns (Do1-Do3) of 400 mVpp at 12 Gb/s were obtained. The phase differences between the output data are due to the length differences of the coaxial cables used in the measurements. The power supply voltage V_{EE} was -5.2 V, and the power consumption was 3.2 W for a 1.8 x 2.2-mm chip.

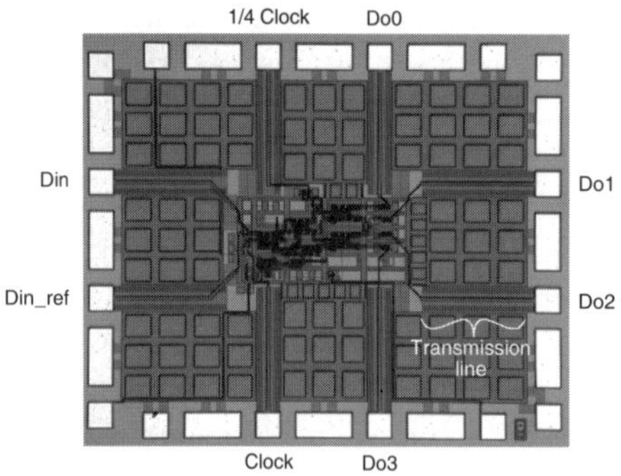

Fig. 20. Chip micrograph of the 1:4 demultiplexer. 2.2 x 1.8 mm.

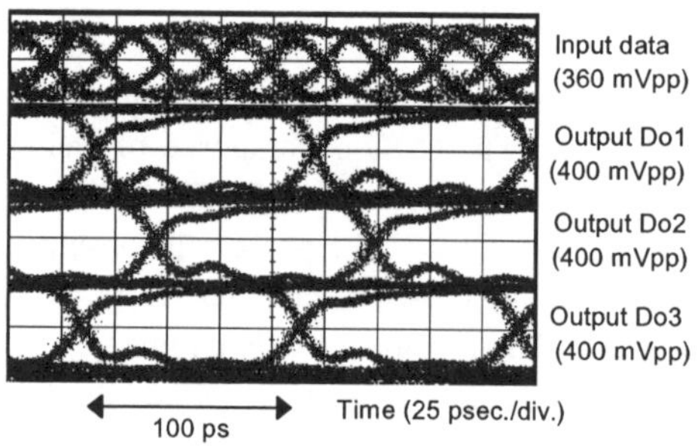

Fig. 21. Output waveform of the 1:4 demultiplexer at a data rate of 48 Gb/s.

Low power and low cost solution

For an optical receiver, single-chip integration of digital circuits, such as a decision circuit and a DEMUX, is an effective solution to minimize power consumption and cost. When a decision circuit IC and a DEMUX IC are each packaged as one module, the clock distribution for them must be carefully designed. Because a 40-GHz clock for data retiming has to be distributed to each module, clock delay-time deviation is caused by the packaging between the decision circuit and the DEMUX. To synchronize the data from the decision-circuit module and the clock for the DEMUX module, a highly accurate clock-timing adjustment using a delay line must be carried out. Thus, a decision circuit and a DEMUX are both needed on a chip to eliminate the need for this clock-timing adjustment. This setup also allows the total power consumption and the number of system components to be reduced because output buffer of the decision circuit and one of the data-retiming MS-DFFs can be eliminated. Consequently, we developed a highly integrated 1:4 DEMUX combined with a decision circuit and confirmed its data rate of 40 Gb/s. [14]

3.2 Clock recovery circuit

A receiver must first recover the clock signal to re-establish synchronous timing of the data. Some clock recovery circuits (CRU) are based on the use of a phase-locked loop (PLL), which requires a sensitive phase detector and a fairly linear voltage-controlled oscillator (VCO) with low phase noise. Due to the difficulty of designing a phase detector and a VCO, another primary approach is often used for clock recovery.[32] A clock-recovery circuit with a high quality-factor resonator is shown as a block diagram in Fig. 1. This approach is not so suitable for implementation on a chip, but has strong points such as the simplicity of the circuit configuration and faster re-acquisition period than that of a phase-locked-loop based CRU.[33] It is composed of a full-wave rectifier, a band-pass filter, and a limiting amplifier. The rectifier extracts a 40-GHz frequency component from a 40-Gb/s non return-to zero (NRZ) data. The band-pass filter, which consists of a high quality-factor resonator, eliminates the unwanted harmonic components caused by the rectifier. Because the output power of the band-pass filter depends on the input data pattern, a limiting amplifier is needed to provide a 40-GHz clock with a constant voltage swing.

3.2.1 Full-wave rectifier[19]

Circuit Design

Figure 22 shows a block diagram of a full-wave rectifier IC. The full-wave rectifier is composed of an input buffer (IB), a full-wave rectifier core (FR), an amplifier (AMP), and an output buffer (OB). The input signal is rectified by FR, and is amplified by AMP and OB. Both AMP and OB have a gain of 10 dB. A schematic of the FR is shown in Fig. 23. The FR is composed of a wide bandwidth differential amplifier and a wired-OR type full-wave rectifier. In the differential amplifier, a transimpedance load circuit is used to increase bandwidth.[23, 24] To rectify data and extract a 40-GHz clock component from a 40-Gb/s data stream, the wired-OR type full-wave rectifier is composed of differential emitter followers which outputs are coupled. The AMP converts a single-ended input voltage to a differential output voltages with a reference voltage Vref.

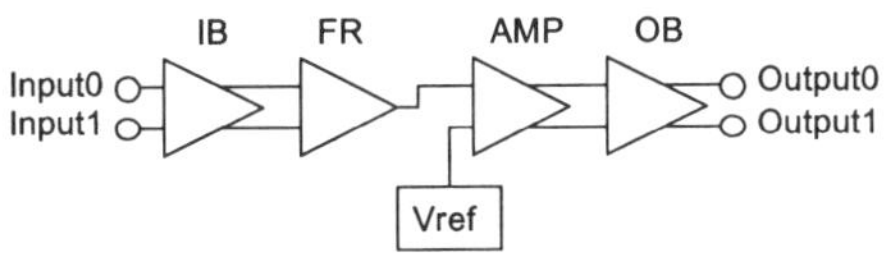

Fig. 22. Block diagram of the full-wave rectifier.

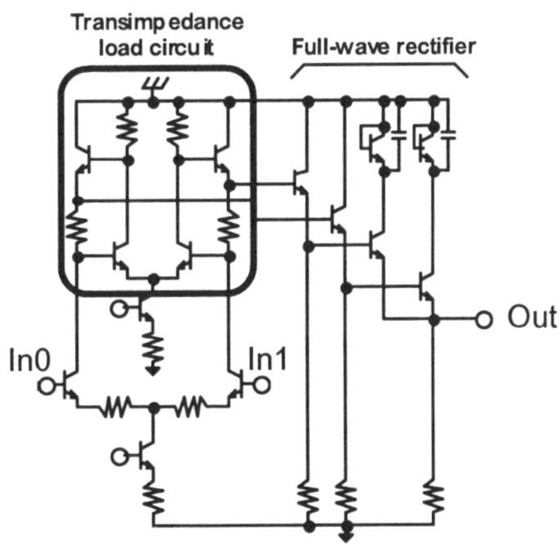

Fig. 23. Schematic of the full-wave rectifier core circuit.

Circuit Performance

Figure 24 shows a chip micrograph of a full-wave rectifier with a chip area of 1.2 x 1.8 mm. The operational waveform of the full-wave rectifier was measured by on-wafer RF probes (Fig. 25). The output waveform at 40 Gb/s is clear, and a 90-mVpp 40-GHz clock signal was obtained through a lossy band-pass filter. The power consumption was 1.6 W with a power supply voltage, V_{EE}, of –5.2 V.

Fig. 24. Chip micrograph of the full-wave rectifier. 1.2 x 1.8 mm.

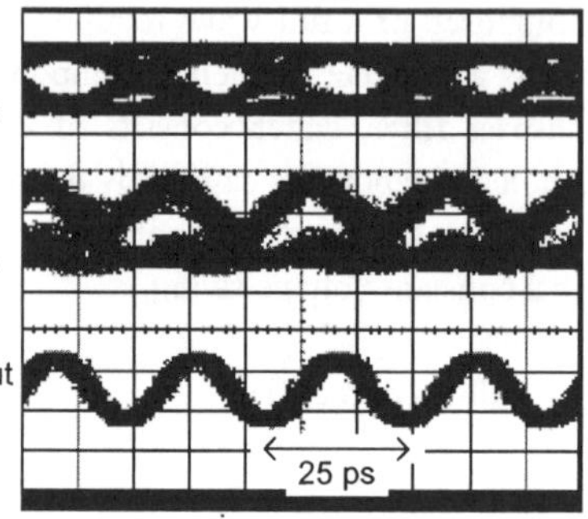

Fig. 25. Measured waveforms at a data rate of 40 Gb/s. (a) Input data waveform, (b) output waveform, (c) filtered waveform by a band-pass filter.

3.2.2 Limiting amplifier[14]

A limiting amplifier is used to provide a 40-GHz clock with constant amplitude to a decision circuit and a DEMUX. For this purpose, both high gain and large output swing are required. We set the design goal at 30-dB gain and 400-mVpp output-swing, simultaneously. In recent years, a limiting amplifier IC based on InP DHBT technology with a 40-GHz bandwidth and a 22-dB gain was developed. [34]

Circuit Design

Figure 26 shows a schematic of the limiting amplifier, which consists of an input buffer and three constant-gain stages in series. For a gain of more than 30 dB at 40 GHz, we introduced a transimpedance amplifier as an active load in the constant-gain stage. The constant-gain stage consists of a differential transistor pair (Q_1, Q_2, and R_E), and an active load (Q_3-Q_6, R_F, and R_L). This active load suppresses the Miller effect caused by parasitic capacitor of Q_1 and Q_2, and increases the bandwidth. In addition, we used two-stage emitter followers with damping resistors (Q_7, Q_8, and R_D) in the gain stage to create the effect of gain peaking at high frequencies.[23, 24] This also helped to achieve a high gain at 40 GHz. Two damping resistors, R_D, in stages A1 and A2 are used to avoid excessive peaking and oscillation. Because the stage A3 also operates as an output buffer, the damping resistors are set to 44 Ω to achieve 50-Ω output-matching.

To generate a large output voltage swing, the emitter node voltage of transistors Q_5 and Q_6 should be driven sufficiently. We therefore added a biasing resistor R_I at the emitter nodes of Q_5 and Q_6. When a constant-gain stage is operated in the saturation region, transistor Q_1 works in both the cut-off and active region. When Q_1 is in cut-off, the emitter voltage of transistor Q_5 must be simultaneously pulled down. However, bias current of Q_5 is only the base current of Q_3. Therefore, Q_5 cannot drive the following circuits with a sufficient voltage swing during the period. To overcome this problem, we added biasing resistors R_I to emitter node of Q_5 and Q_6. Resistor R_I then produces a current I_{RI} that is constantly much larger than the base current of transistor Q_3. Because of this constant bias current for Q_5 and Q_6, the output waveform is less distorted.

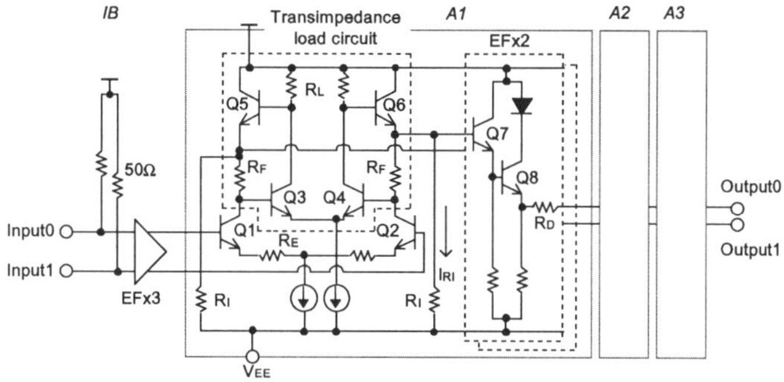

Fig. 26. Schematic of the limiting amplifier.

Circuit Performance
We measured the small-signal frequency response of the gain (S21) by using on-wafer probes. When the limiting amplifier had two-stage emitter-followers in each gain stage (as shown in Fig. 26), the gain at 40 GHz was 32 dB, while the gain was 23 dB with a single-stage emitter follower. We attributed increase of the gain to the peaking effect caused by two-stage emitter followers.

Figure 27 shows a chip micrograph of the limiting amplifier IC. Figure 28 shows output power and phase deviation dependence on input power at 40 GHz. The minimum saturation input power was -30 dBm. Within an input-power range from -30 to -10 dBm, the output power saturated at -2 dBm. The signal generator limited the maximum input power, i.e., -10 dBm, at 40 GHz. To have a stable operation of a decision circuit and a DEMUX, small phase deviation is required even when the input voltage swing is fluctuated. Phase deviation within an input power range from -30 to -10 dBm was less than 70 degrees. In the linear-operation region below -30 dBm, the gain, i.e., output power divided by input power, was about 25 dB. In Fig. 29, the waveform of a 500-mVpp output-voltage swing was obtained. The input power for operation in the saturation region was -20 dBm. The slight waveform distortion is attributed to the biasing resistors R_I. The power supply voltage V_{EE} is -7.5 V and power consumption is 1.4 W for a 1.2 x 1.8-mm chip.

3.3 Transmitter

3.3.1 4:1 MUX[25]

MUXes convert a few channels of lower speed data signals to a high-speed data signal. For a 40-Gb/s system, like the DEMUX described in section 3.1.4, four channels of 10-Gb/s data should be handled in a MUX for practical use. Recently, a 4:1 MUX using based on HEMT

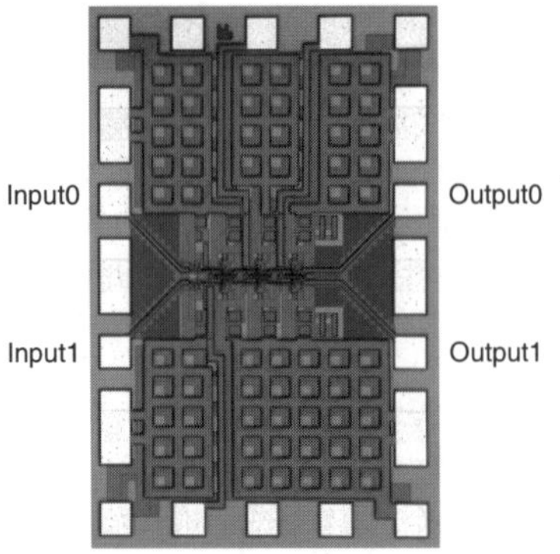

Fig. 27. Chip micrograph of the limiting amplifier. 1.2 x 1.8 mm.

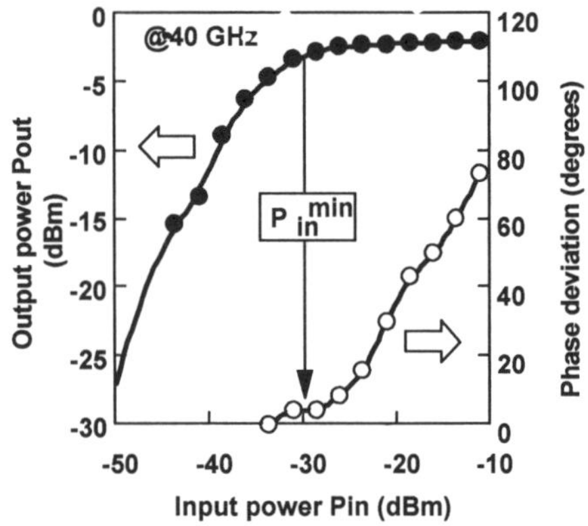

Fig. 28. Transfer characteristic of the limiting amplifier.

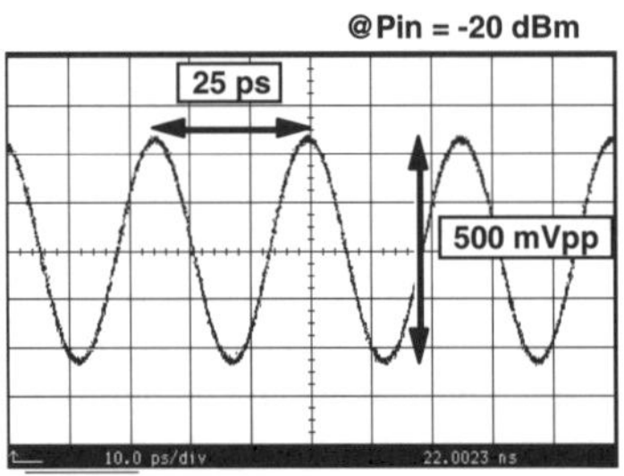

Fig. 29. Output waveform of the limiting amplifier at a frequency of 40 GHz.

technology has achieved operated up to 45 Gb/s at full-rate clock.[35] However, the power consumption of 7.9 W is two times larger than that of a SiGe HBT 4:1 full-rate MUX.[21] A half-rate clocking MUX can provide high-speed operation and low power consumption at the same time. A 56-Gb/s operation of 4:1 MUX using SiGe HBT technology obtained with a low power consumption of 1.5 W[36]. To get a wide tolerance on clock-duty variation, we focused on developing a full-rate clocking MUX.

Circuit Design

Figure 30 shows a block diagram of the 2:1 MUX, which is the core circuit of the 4:1 MUX. Both positive and negative edges of a clock signal are used for a selector; thus, a second bit (D2 in the selector) must be delayed one-half clock period to maximize the phase margin between the input data transitions and the select clock (S in selector). This requirement is fulfilled by using a MS-DFF for the first bit and a MSS-DFF for the second bit. Figure 31 shows a block diagram of the 4:1 MUX. It consists of three 2:1 MUXes connected in a tree structure, an output MS-DFF, an output buffer, and two master-slave toggled-flip-flops (MS-TFFs) in series. If the output MS-DFF is not used, the clock signal for the 2:1 MUX (M3) needs to have a precise 50% duty cycle. However, it is difficult to have a clock whose rising time (tr) and falling time (tf) are equal. Therefore, to allow for the duty variation of the clock and eliminate any output jitter of the selector in M3, an output MS-DFF is used for retiming using only the positive edge. To provide clock signals with proper voltage swing to 2:1 MUX and MS-DFFs, cascoded amplifiers are used in the internal clock buffers.

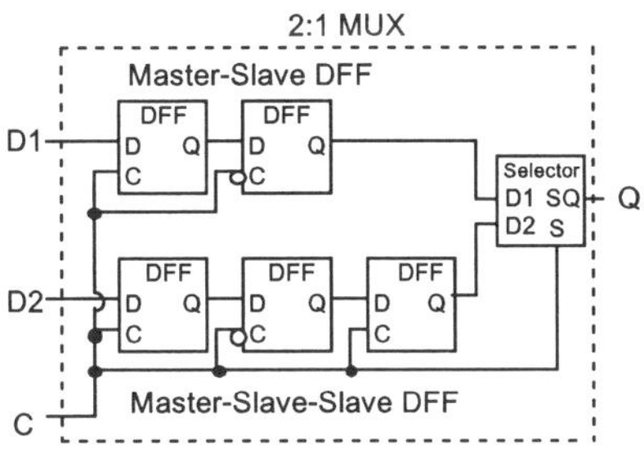

Fig. 30. Circuit block of the 2:1 multiplexer.

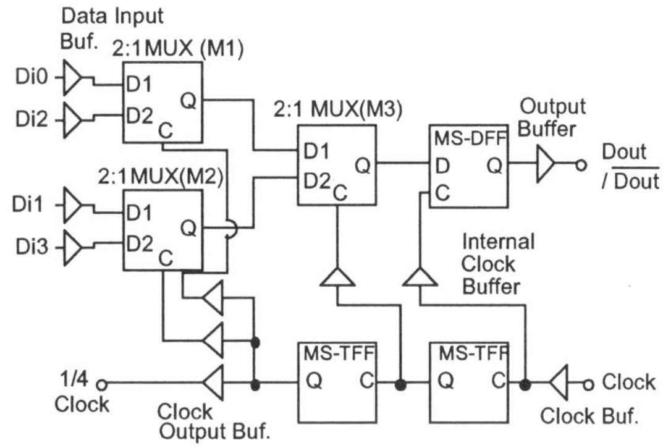

Fig. 31. Block diagram of the 4:1 multiplexer.

Circuit Performance

A chip micrograph of the 4:1 MUX is shown in Fig. 32. The chip is 1.8 x 2.2 mm. At BER evaluation with a DEMUX module, the 4:1 MUX module achieved error-free operation (< 10^{-12} BER) at a data rate of 45 GHz. The input voltage-level for data and clock are 400 mVpp and 600 mVpp. PRBS output waveform (2^{31}-1 length) is observed up to a data-rate of 48 Gb/s. The output voltage swing is 300 mVpp, rise and fall times (20-80%) are 15 and 17 ps, respectively.

Figure 33 shows the output eye diagrams of the 4:1 MUX IC for a 40-Gb/s and 50-Gb/s PRBS measured by on-wafer probes. Well-opened eye diagrams with an output swing of 400 mVpp are obtained. The upper measurement limit was 50 Gb/s, because of the 12.5-Gb/s maximum operation speed of the pulse pattern generator. The power consumption was 3.2 W with a power supply voltage, V_{EE}, of –5.2 V.

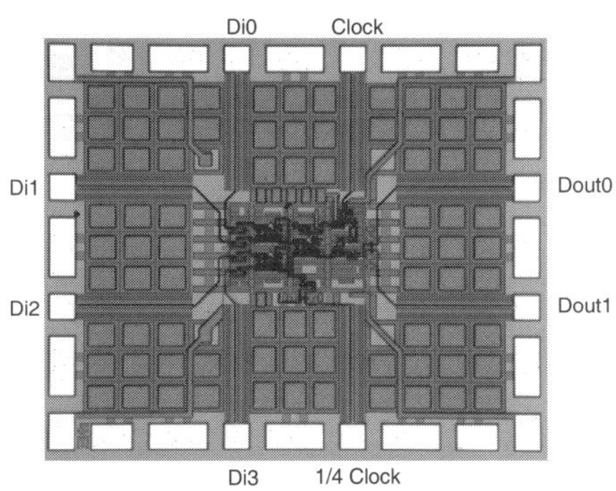

Fig. 32. Chip micrograph of the 4:1 multiplexer. 2.2 x 1.8 mm.

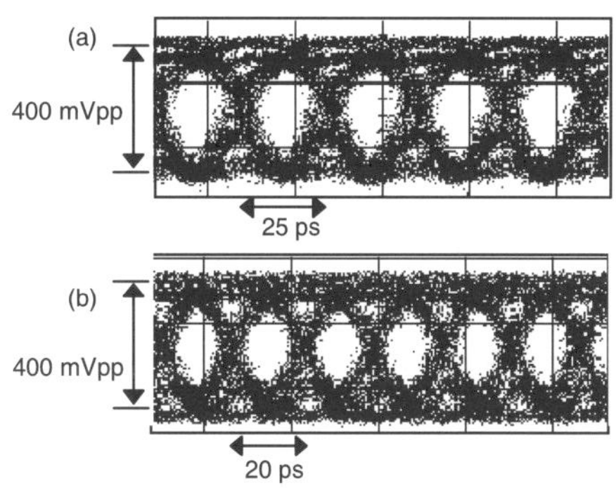

Fig. 33. Measured 4:1 multiplexer eye-pattern. (a) Operation at 40 Gb/s and (b) at 50 Gb/s.

4. Summary

In this paper, we have presented a 40-Gb/s optical-transmission IC chipset, fabricated in a 0.2-μm self-aligned epitaxial-growth SiGe HBT technology. The SiGe HBTs can provide high f_T, i.e. 122 GHz, and f_{max}, i.e. 163 GHz comparable to those devices produced by compound semiconductor technology. With the high-speed capability of the SiGe HBTs, circuit design techniques were used to ensure IC performance enough for operating at more than 40 Gb/s. To reduce the influence from large input parasitic capacitance caused by a photo detector, a common base transistor was introduced in transimpedance amplifier. To increase the gain-bandwidth product of amplifiers, active-load circuits were used and carefully optimized. For AGC amplifier, bandwidth-compensating resistors in gain control stage were used to equalize bandwidths within a variable gain range. For digital circuits, jitter in data stream was suppressed by a MS-DFF with full-rate clocking at more than 40 GHz. As a result, we attained a 40-Gb/s SiGe HBT IC chipset summarized in table 2; a transimpedance amplifier with a 45-GHz bandwidth, a AGC amplifier with a 47.8-Gb/s bandwidth, a 40-Gb/s decision circuit, a 45-Gb/s 1:4 DEMUX with full-rate clock, a 40-Gb/s full-wave rectifier, a limiting amplifier with a 32-dB gain at 40 GHz, and 45-Gb/s 4:1 MUX with full-rate clock. The static frequency dividers in these digital circuits can be operated at more than 60 Gb/s. Therefore, for digital circuits, the SiGe HBT technology will be applicable even when the operation speed must be increased 43 Gb/s or 50 Gb/s for forward error correction. In that case, the clock distribution path in the chips must be carefully designed. The sufficient bandwidths for analog circuits have also been demonstrated. However, the ability of analog ICs in practical 40-Gb/s systems must be discussed with other requirements.

Such high-speed capability, combined with cost effectiveness and high reliability brought by Si-based technology, makes the means that SiGe HBT technology is the most suitable candidate for high-speed optical transmission systems.

Table 2. Summary of the chipset for 40-Gb/s optical transmission systems using a 0.2-μm SiGe HBT technology.

Circuits	Bandwidth / Bit Rate	Gain	Output	Power consumption
Transimpedance amplifier [14]	45 GHz	50 dBΩ		0.3 W
AGC amplifer [19]	47.8 GHz	21.5 dB (max.)	350 mVpp	1.2 W
Decision circuit [19]	40 Gb/s		500 mVpp	0.9 W
1:4 demultiplexer [25]	45 Gb/s		400 mVpp	3.2 W
full-wave rectifier [19]	40 Gb/s		477 mVpp	1.6 W
limiting amplifier [14]	40 Gb/s	32 dB	500 mVpp	1.4 W
4:1 multiplexer [25]	45 Gb/s		400 mVpp	3.2 W
Static frequency divider [7]	67 GHz		250 mVpp	0.175 W

Acknowledgements

The authors greatly acknowledge the technical support of Dr. Y. Tamaki and the process staff members of Hitachi Device Development Center (HDDC), Tokyo, Japan, for their contribution to the HBT development. The authors also would like to thank Dr. Anzai, and Y.Hatta, HDDC, Tokyo, Japan, Dr. K. Seki, and K. Kimura, Central Research Laboratory, Hitachi, Ltd., Tokyo, Japan, for their encouragement. Finally the authors wish to thank R. Taketari for technical suggestions and support for the measurement.

References

1. T. Swahn, et al., "40 Gb/s, 3 volt InP HBT ICs for a fiber optic demonstrator system," IEEE GaAs IC Symposium, Digest of Technical Papers, 1996, pp. 125-128.
2. M. Mokhtari, et al., "InP-HBT chip-set for 40-Gb/s fiber optical communication systems operational at 3 V," IEEE, Journal of Solid-State Circuits, Vol. 32, No. 9, Sept. 1997, pp. 1371-1383.
3. M. Lang, et al., "20-40 Gb/s 0.2-μm GaAs HEMT chip set for optical data receiver," IEEE, Journal of Solid-State Circuits, Vol. 32, No. 9, Sept. 1997, pp. 1384-1393.
4. M. Yoneyama, et al., "Optical repeater circuit design based on InAlAs/InGaAs HEMT digital IC technology," IEEE Transactions on Microwave Theory and Techniques, Vol. 45, No. 9, Dec. 1997, pp. 2274-2282.
5. J. Müllrich, et al., "High-gain transimpedance amplifier in InP-based HBT technology for the receiver in 40 Gb/s optical-fiber TDM links," IEEE GaAs IC Symposium, Digest of Technical Papers, 1999, pp. 99-102.
6. K. Washio, et al., "A 0.2-μm 180-GHz-f_{max} 6.7-ps-ECL SOI/HRS self-aligned SEG SiGe HBT/CMOS technology for microwave and high-speed digital applications," IEEE Transaction on Electron Devices, Vol. 49, No. 2, Feb. 2002, pp. 271-278.
7. K. Washio, et al. "67-GHz static frequency divider using 0.2-μm self-aligned SiGe HBTs," IEEE Transactions on Microwave Theory and Techniques, Vol. 49, No. 1, Jan., 2001, pp. 3-8.
8. E. Ohue, et al. "5.3-ps ECL and 71-GHz static frequency divider in self-aligned SEG SiGe HBT", 2001 Bipolar/BiCMOS Circuits and Technology Meeting, 2001, pp. 26-29.
9. S. Ueno, et al., "A single-chip 10 Gb/s transceiver LSI using SiGe SOI/BiCMOS," 2001 IEEE International Solid-State Circuits Conference, 2001, Digest of Technical Papers, pp. 82-83.
10. T. Masuda, et al., "40 Gb/s 4:1 multiplexer and 1:4 demultiplexer IC module using SiGe HBTs," IEEE MTT-S International Microwave Symposium Digest, Vol. 3, 2001, pp. 1697-1700.
11. http://www.amcc.com/Compinfo/PressReleases/S76803.htm
12. U. Konig, et al., "High frequency SiGe heterostructure devices," IEEE/Cornell Conference on Advanced Concepts in High Speed Semiconductor Devices and Circuits, 1997, pp. 14-23.
13. M. Pfost, et al., "Modeling and measurement of substrate coupling in Si-bipolar IC's up to

40 GHz," IEEE, Journal of Solid-State Circuits, Vol. 33, No. 4, April, 1998, pp. 582-591.

14. T. Masuda, et al., "45 GHz transimpedance 32 dB limiting amplifier and 40 Gb/s 1:4 high sensitivity demultiplexer with decision circuit using SiGe HBT for 40 Gb/s optical receiver," 2000 IEEE International Solid-State Circuits Conference, 2000, Digest of Technical Papers, pp. 60-61.
15. K. Pedrotti, "High Speed Circuits for Lightwave Communications," International Journal of High Speed Electronics and Systems, Vol. 9, No. 2 , 1998, pp. 313-346.
16. Y. Amamiya, et al., "40-Gb/s Optical Receiver IC Chipset - including a Transimpedance Amplifier, a Differential Amplifier, and a Decision Circuit - using GaAs based HBT Technology," IEEE MTT-S International Microwave Symposium Digest, 2002, pp. 87-90.
17. C. F. Campbell, et al., "A 0.15-μm GaAs MHEMT Transimpedance Amplifier IC for 40-Gb/s Applications," IEEE MTT-S International Microwave Symposium Digest, 2002, pp. 79-82.
18. H. Shigematsu, et al., "A 49-GHz Preamplifier With a Transimpedance Gain of 52 dBΩ Using InP HEMTs," IEEE, Journal of Solid-State Circuits, Vol. 36, No. 9, April 2001, pp. 1309-1313.
19. K. Ohhata, et al. "40 Gb/s analog IC chipset for optical receivers - AGC amplifier, full-wave rectifier and decision circuit implemented using self-aligned SiGe HBTs," IEEE MTT-S International Microwave Symposium Digest, 2001, pp. 1701-1704.
20. R.Yu, et al., "A packaged broad-band monolithic variable gain amplifier implemented in AlGaAs/GaAs HBT technology," IEEE, Journal of Solid-State Circuits, Vol. 31, No. 10, 1996, pp. 1380-1387.
21. K. Ohhata, et al., "Design of a 32.7-GHz Bandwidth AGC Amplifier IC with Wide Dynamic Range Implemented in SiGe HBT," IEEE, Journal of Solid-State Circuits, Vol. 34, No. 9, 1999, pp. 1290-1297.
22. M. Möller, et al. "13 Gb/s Si-bipolar AGC amplifier IC with high gain and wide dynamic range for optical-fiber receivers," IEEE, Journal of Solid-State Circuits, Vol. 29, No. 7, 1995, pp. 815-822.
23. E. M. Cherry, et al., "The design of wide-band transistor feedback amplifiers," proc. IEE Vol.110, No. 2, 1963, pp. 375-389.
24. N. Ishihara, et al., "9 GHz bandwidth, 8-20 dB controllable-gain monolithic amplifiers using AlGaAs/GaAs HBT technology," Electron. Lett., Vol. 25, No. 19, pp.1317-1318, Sept. 1989
25. T. Masuda, et al., "40 Gb/s 4:1 multiplexer and 1:4 demultiplexer IC module using SiGe HBTs," IEEE MTT-S International Microwave Symposium Digest, 2001, pp. 1698-1701.
26. K. Murata, et al., "45 G/s decision IC module using InAlAs/InGaAs/InP HEMTs," Electronics Letters, 5th, August, 1999, vol. 35, No. 16, pp. 1379-1380.
27. K. Ishii, et al., "High-input-sensitivity, low-power 43 Gbit/s decision circuit using InP/InGaAs DHBTs," Electronics Letters, 6th, June, 2002, vol. 38, No. 12, pp. 557-558.
28. J. P. Mattia, et al., "A 1:4 Demultiplexer for 40Gb/s Fiber-Optic Applications," 2000 IEEE International Solid-State Circuits Conference, 2000, Digest of Technical Papers, pp. 64-65.
29. H. Kano, et al., "A 50-Gbit/s 1:4 Demultiplexer IC in InP-based HEMT Technology," IEEE MTT-S International Microwave Symposium Digest, 2002, pp. 75-78.

30. M. Reinhold, et al., "A Fully Integrated 40-Gb/s Clock and Data Recovery IC With 1:4 DEMUX in SiGe Technology," IEEE, Journal of Solid-State Circuits, Vol. 36, No. 12, 2001, pp. 1937-1945.
31. Z. Lao, et al., "Si Bipolar 14Gb/s 1:4 Demultiplexer IC for System Applications," IEEE, Journal of Solid-State Circuits, Vol.31, No.1, 1998, pp. 54-60.
32. R. Takeyari, et al., "Fully monolithically integrated 40-Gbit/s transmitter and receiver," OFC 2001, WO1.
33. S. Kunasani, "CRU Recovers 40-Gb/s OC-768 Clock Signals," MICROWAVES &RF, April 2002, pp. 94-98.
34. Y. Beayens, et al., "InP D-HBT IC'S FOR 40 Gb/s AND HIGHER BITRATE LIGHTWAVE TRANCEIVERS," 2001 IEEE GaAs IC Digest, pp. 125-128.
35. Y. Nakasha, et al., "A 43Gb/s Full-Rate-Clock 4:1 Multiplexer in InP-based HEMT Technology," 2002 IEEE International Solid-State Circuits Conference, 2002, pp. 190-191.
36. M. Meghelli, et al., "50Gb/s SiGe BiCMOS 4:1 Multiplexer and 1:4 Demultiplexer for Serial Communication Systems," 2002 IEEE International Solid-State Circuits Conference, 2002, pp. 260-261.

International Journal of High Speed Electronics and Systems
Vol. 13, No. 1 (2003) 265–275

MICROWAVE GaN-BASED POWER TRANSISTORS ON LARGE-SCALE SILICON WAFERS

S. MANOHAR and A. PHAM
University of California
Department of Electrical and Computer Engineering
Davis, CA 95616
pham@ece.ucdavis.edu

J. BROWN, R. BORGES, and K. LINTHICUM
Nitronex Corporation
Raleigh, NC 27606

This paper presents the development of microwave Gallium nitride (GaN) heterostructure field-effect transistors (HFETs) on silicon (Si). GaN-on-Si provides a low-cost manufacturable platform that could lead to the commercialization of GaN-based power devices for wireless applications. Small periphery GaN high electron mobility transistors (HEMTs) on Si exhibited a maximum drain current of 900mA/mm, a peak g_m of 300 mS/mm, and a microwave output power density of 1.5 W/mm at 2 GHz. Microwave characterization and device modeling of GaN HEMTs on Si are discussed.

Keywords: GaN, GaN-on-Si, and microwave power

1. Introduction

With the proliferation of wireless devices in recent years and market expansion for cellular and mobile communication networks, a variety of power amplifier technologies are vying for their market share. The material properties of wide-band gap semiconductors such as silicon carbide (SiC) and GaN are inherently superior to those of conventional silicon and Gallium arsenide (GaAs) counterparts for microwave power amplifier applications. GaN material has many properties that are near ideal for high power microwave and millimeter wave devices. The attractive physical properties of GaN include: a high critical breakdown field (3.3 MV/cm) [1,2,3], a high peak and saturated electron velocity ($1.5x10^7$ cm/s and $2.7x10^7$ cm/s), and good thermal conductivity. In addition, the large conduction band offset and the large piezoelectric and spontaneous polarization effects in this material system result in high channel sheet electron densities. These combined properties have enabled GaN-based devices to achieve high output power density in smaller devices that are easy to fabricate and require simpler matching circuitry. High voltage and high temperature operation also allows these devices to perform in harsh environments. The major obstacle in the development of GaN-based devices has been the lack of indigenous substrates on which quality expitaxial films are grown.

Most AlGaN/GaN high electron mobility transistors to date have been produced on Sapphire or SiC substrates due to limited availability of bulk GaN substrates. While GaN HEMTs have been demonstrated with remarkable performance on Sapphire (4.2 W/mm at

4 GHz [4]) and SiC (6.9 W/mm at 10 GHz) [5], the commercial viability of devices grown on these substrates is hampered by cost and high-volume manufacturing issues. However, the use of a mature silicon technology as the substrate material offers large-scale wafer fabrication and low-cost manufacturing. Significant progress has been made in the development of GaN-on-Si high power devices [6-11]. The use of silicon substrates also allows for the possibility of integrating silicon-based electronics on the same chip. In this paper, we present GaN-based power devices on Si and detailed DC, transient, small signal and large signal characterization and modeling at microwave frequencies.

2. Device Structure and Layout

GaN layers are grown on silicon wafers by a metal-organic chemical vapor deposition (MOCVD) using a transition layer scheme that addresses mismatches in the material system. The device structure is similar to that employed by other groups for fabricating AlGaN/GaN FETs. The HFET structure consists of an undoped AlGaN spacer layer of 50 Å, an AlGaN donor layer containing 2×10^{18} Si atoms, and undoped AlGaN followed by GaN cap layers. The aluminum mole fraction in all AlGaN layers was nominally 0.22. Figure 1 provides a schematic diagram of a device cross section. The layout was designed for high power operation at a frequency of 2 GHz. The nominal gate length was 1 μm, the source-gate spacing was 1 μm, and the gate-drain spacing was 3 μm for all devices reported in this work.

The HFETs were fabricated using contact lithography techniques along with a typical GaN liftoff based process. Source and drain ohmic contacts were fabricated by evaporating Ti/Al/Ni/Au followed by annealing for 30 seconds at 900°C. Dry etching of mesa structures in a reactive ion etching system employing a chlorine-based chemistry was used for device isolation. Ni/Au was used for the schottky gate metallization. Finally, the active devices were passivated with silicon nitride after the gate contacts were in place. Small periphery devices were characterized using an on-wafer probe station and ground-signal-ground microprobers.

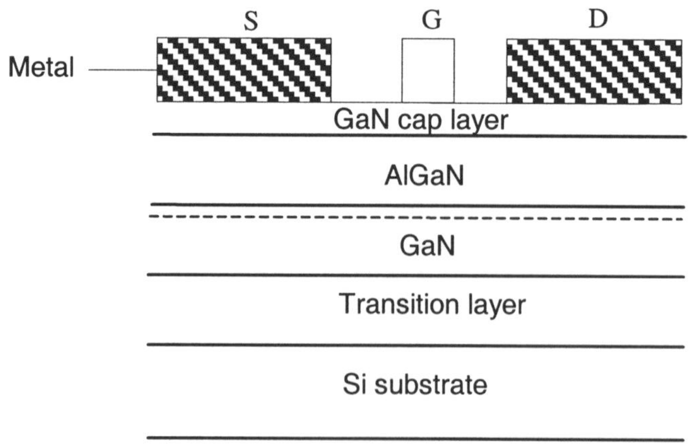

Figure 1. Cross section of an AlGaN/GaN HFET on Si.

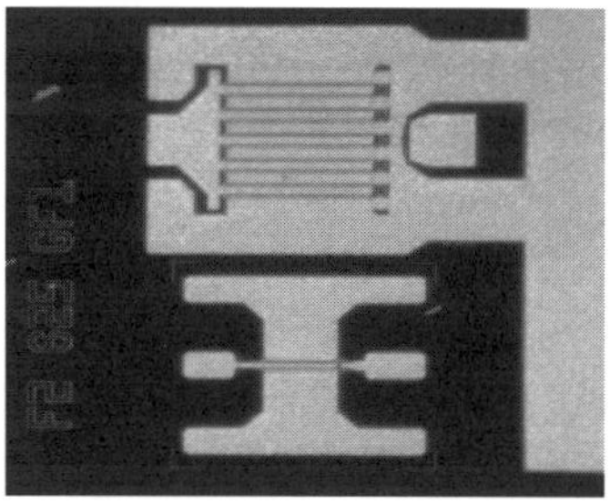

Figure 2. Photograph of the fabricated AlGaN/GaN HFET structures on silicon.

3. Electrical Results

Microwave characterization and device modeling of GaN HEMTs on silicon are described in subsequent sections. DC measurements show that our devices exhibit a maximum drain current of 900mA/mm and a peak transconductance of 300mA/mm. Gate lag measurements performed on these devices show near ideal responses after surface passivation indicating minimal trapping effects arising from trapping states. Drain lag measurements indicate an increase in the output conductance of the device with frequency. Load pull measurements indicate premature output power saturation that we attribute to substrate effects. Small signal models developed for these devices involved the use of extrinsic RC-networks that can be used to model substrate effects.

3.1. DC and Transient Analysis

Typical static characteristics of 200-μm gate width AlGaN/GaN HEMTs on Si are shown in Figure 3. The threshold voltage for the devices was found to be –3 V. The devices exhibited a saturation current of 662mA/mm and a maximum drain current of 887mA/mm as shown in Figure 3. The knee voltage for the devices was around 3 V. One of the advantages of using a silicon substrate is the superior heat dissipation handling compared to sapphire. However, some amount of self-heating is evident in the drop of drain current at high drain voltage shown in Figure 3.

The transfer characteristics of AlGaN/GaN HEMTs for a drain source voltage of 15 V are shown in Figure 4. The peak extrinsic trans-conductance was found to be 288ms/mm at a gate source voltage of –1.5 V (Figure 4). Although the static characteristics are comparable to previously reported results of AlGaN/GaN devices on sapphire and SiC, the microwave characteristics of these devices are affected by the conductive substrate in addition to the well-known DC-RF dispersion arising from trapping effects. In order to understand the influence of the respective effects, transient and small signal characteristics of these devices were investigated.

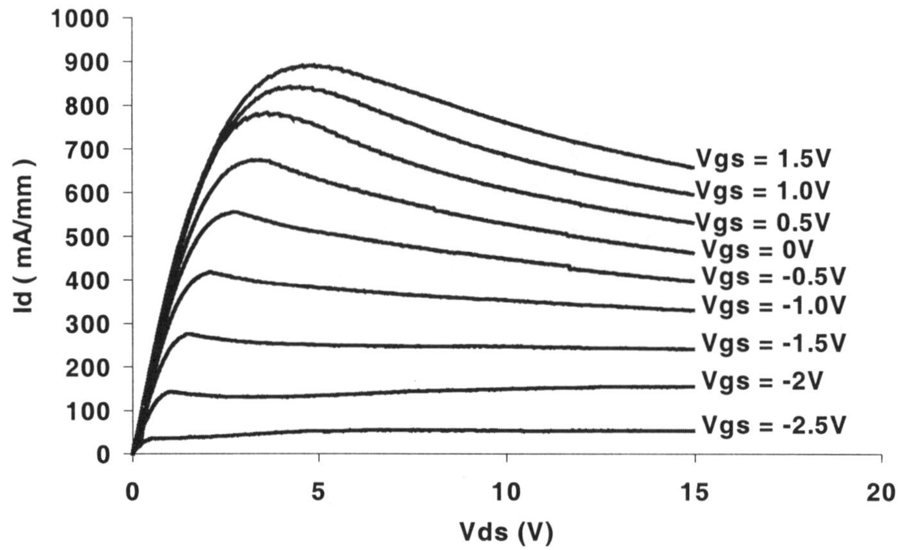

Figure 3. Measured I-V characteristics of the AlGaN/GaN HEMT on Si substrates for a gate width of 0.2 mm.

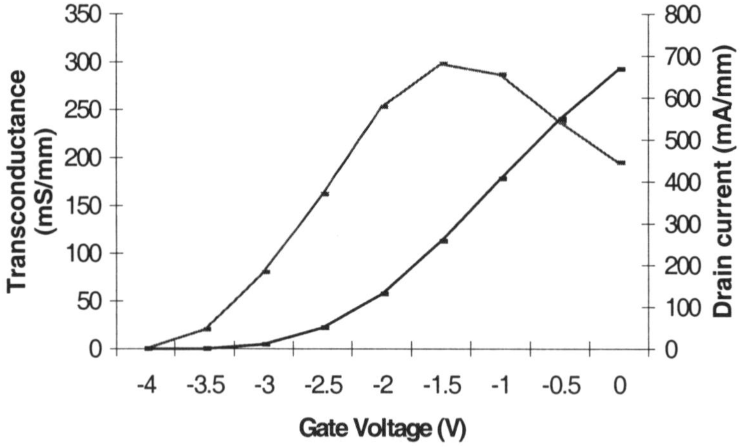

Figure 4. Transfer characteristics of the AlGaN/GaN HEMT at drain source voltage of 15 V. The peak extrinsic trans-conductance was found to be 300ms/mm.

In III-V device technologies, gate lag measurements are more sensitive to surface trapping effects and drain lag effects are more sensitive to buffer-layer trapping effects. Surface trapping leads to a 'lag' or a slower drain current response to a polarity reversing signal at the gate, hence restricting microwave output power. Buffer trapping leads to an increase in knee voltage and a reduction in maximum current, thus restricting output power [12]. Figure 5 shows the normalized drain current, which is the ratio of transient current to static current at the same bias drain voltage. The gate source voltage was pulsed from the threshold voltage to 0 V while the drain source voltage was held constant. It must be noted that the pulse width is short relative to the time constant of the gate lag. This is why one cannot observe the minimal 'lag' for the structure [13]. For a short pulse width, surface trapping is indicated as a drop in the transient drain current response. One can observe this from the normalized drain current ratio of less than 1. However, the normalized drain current close to 1 indicates a near-ideal gate lag response in the SiN-passivated devices and demonstrates minimal surface trapping [14].

Drain lag measurements were conducted on the same devices with drain step amplitudes of 8, 10 and 12 V for a gate-source voltage of 0 V. Ideally, in the absence of parasitic loading effects, the drain current must not increase with decrease in pulse width beyond a certain point (typically microseconds) as the trapped charges cannot respond to time-varying signals below this limit. However, we noticed an increase in the drain current as the pulse width is further reduced from the microsecond range indicating an increase in output conductance. Earlier work has indicated that there is a dispersion of output conductance due to the conductive substrate charge being coupled under small signal conditions [7]. As frequency dispersion of the output conductance corresponds to the small signal limit of the drain lag phenomena, we believe that a similar effect is responsible for the observed phenomenon.

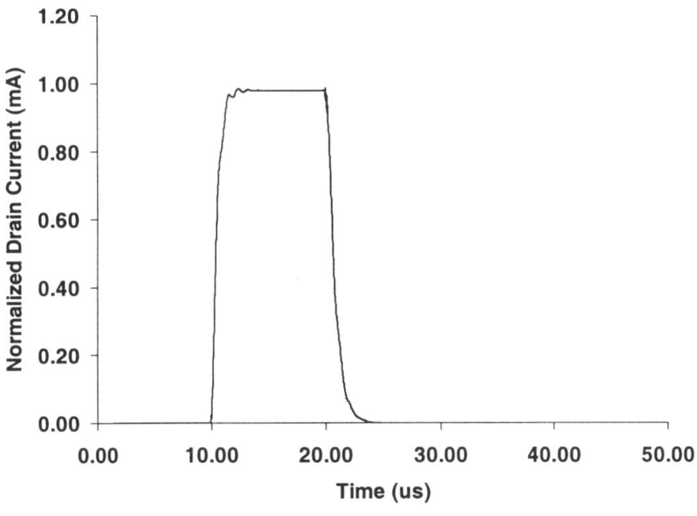

Figure 5. Gate lag response for the 200-μm AlGaN/GaN HEMT. The response was found to be near ideal after SiN passivation.

3.2. Large-signal Characterization

On-wafer load-pull measurements were conducted on 200 μm HEMTs in order to determine the continuous waveform output power. Focus Microwave computer controlled tuners were used in this measurement. The device was biased in Class A operation with a drain source voltage of 15 V and a gate source voltage of –1.5 V. A power density of 1.5 W/mm at 2 GHz was observed as shown in Figure 6. For a FET biased in class A, the output power can be estimated from the static I-V characteristics as

$$P_{out} = I_{max} \frac{V_{breakdown} - V_{Knee}}{8} \tag{1}$$

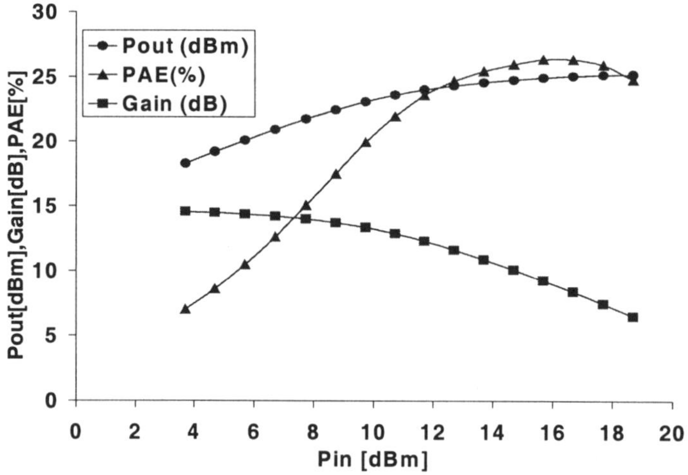

Figure 6. Class A load-pull measurements on a 200 μm HEMT. The measured power density is 1.5 W/mm at 2 GHz and a power added efficiency (PAE) of 26%.

The calculated theoretical output power density was found to be roughly 2.5 W/mm. However, the measured maximum output power density from these devices was 1.5 W/mm. As discussed earlier, transient analysis indicates that DC-RF dispersion arising from trapping effects in the surface and the bulk of the semiconductor are not enough to explain the reduction in power density. Since it has been postulated that the conductivity of the silicon substrate causes 'parasitic loading' [5] and premature output power saturation, it was decided that an equivalent circuit model of the HEMT be developed for our devices in order to gauge the extent of this effect.

3.3. Development of Small Signal Device Models

S-parameter measurements under hot-FET, pinched-off and cold-FET and forward bias cold-FET conditions were taken for 200 μm devices with the help of an 8510C network analyzer and on-wafer coplanar probes. Line-reflect-match calibration was done prior to S-parameter measurements in order to transfer the measurement reference planes to the probe tip. It was found that the development of the small signal models using traditional extraction methods [15] yielded frequency dependent intrinsic parameters. In particular, one saw a pronounced frequency dispersion of the output conductance. Previous work indicates that this effect was due to the influence of the conductive silicon substrate allowing the metal/GaN/Si substrate to act as a p-I-n diode in series with resistors that reflect the substrate conductivity and load the device. One difference between the devices used in [7] and ours was the use of an n-type Si substrate as opposed to the p-type substrate used in [7]. Looking at the measured S-parameters, one may notice that S_{22} does not follow the typical constant output conductance circle on the Smith chart. This is also true of S_{11}, albeit to a lesser extent. This is usually an indication of a non-constant output conductance and input resistance with frequency. As frequency dispersion of output conductance in traditional modeling efforts has been done employing extrinsic RC-networks, a similar approach was adopted. The topology of the equivalent circuit model is shown in Figure 7. The number of RC-networks in parallel indicates the strength of substrate effects.

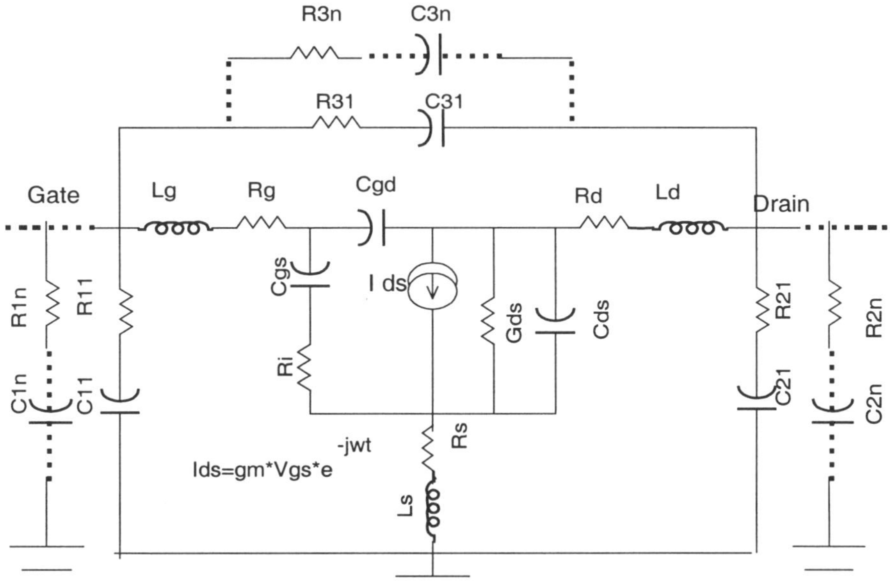

Figure 7. Equivalent circuit model topology of an AlGaN/GaN HEMT on silicon substrate.

Under forward bias conditions and low frequencies, the RC network was found to have minimum effect on the Z parameters. Conventional cold-FET techniques were used to extract the parasitic resistors and inductors. Pinched-off cold FET measurements were then taken. As the parasitics in the RC network are assumed to be independent of bias, determining their values by optimization under Cold-FET conditions is much simpler as the number of unknowns is reduced. Optimization was done using an Agilent Advanced Design System (ADS) tool. Once a good fit to the measured S-parameters under cold FET conditions was obtained, all the extrinsic parasitic elements were fixed and then de-embedded from the hot-FET measurements. The intrinsic FET parameters are determined from the de-embedded hot-FET S-parameters. The parameters of the HEMT equivalent circuit model are summarized in Tables 1-3. The bias point was V_{ds} = 15 V and V_{gs} = -1.5 V. The agreement between measured and modeled S-parameters up to 18 GHz is shown in Figure 8. Parasitic extraction of multi-finger GaN HEMTs of various sizes and geometries was also performed in order to understand the trade-offs involved in the optimization process. Based on our small signal analysis and load pull measurements conducted on multi-finger devices, we found that a trade-off between thermal resistance and substrate effects exists. Better thermal management from an increase in gate pitch causes higher currents to be supported. This in turn means a larger pad size and more significant substrate effects. Thus, the usual increase in power density with larger gate pitch is offset by the parasitic loading. The achievement of better power performance at high frequencies must involve the reduction of the parasitic loading effects.

Table 1. Extracted intrinsic element values for the AlGaN/GaN HEMT

g_m (mS)	τ (psec)	R_{ds} (Ω)	C_{gs} (pF)	C_{ds} (fF)	C_{gd} (fF)	R_i (Ω)
53.13	2.89	984	0.6	80	35	5.04

Table 2. Extrinsic element values

L_g (pH)	L_d (pH)	L_s (pH)	R_g (Ω)	R_d (Ω)	R_s (Ω)
71	103	12	2.35	2.89	0.966

Table 3. R-C element values

C_{11} (fF)	C_{21} (fF)	C_{31} (fF)	R_{11} (Ω)	R_{21} (Ω)	R_{31} (Ω)
230	120	10	208.82	155.39	53.89

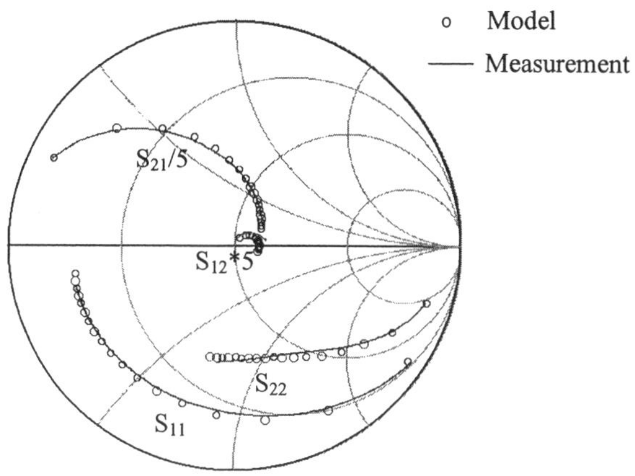

Measured vs. Modeled S-parameters (1-18 GHz)

Figure 8. Agreement between measured and modeled S-parameters up to 18 GHz (V_{ds} = 15 V and V_{gs} = -1.5 V)

4. Target Applications

Potential target applications of GaN-based devices are in the microwave power amplification for wireless communications. As far as RF power amplifiers are concerned, use of GaN on Si devices as power amplifiers in base station transmitters is a potential application. Typical base-station amplifier requirements are summarized in Table 4. As new applications and more stringent standards are applied to next generation wireless systems pushing conventional power amplifier performance beyond its limits, devices based on GaN based technologies are expected to deliver. The inherently higher linearity and efficiency of GaN based devices should give designers higher flexibility and room to meet their specifications.

Table 4. Typical base station amplifier requirements [17]

Frequency ranges	824 - 894 MHz, 1810 - 1870 MHz, and 1930 -1990 MHz
Output power	8 - 45 W
Gain	30 - 40 dB
Gain flatness	±0.5 dB
Input match	1.2 to 1
Adjacent channel power ratio (ACPR)	< 45 dBc @ ±885 kHz offset, < 60 dBc @ ±1.98 kHz and ± 2.25 kHz offsets
Group delay	4 ns ±2 ns

5. Conclusions

In this paper, we present an overview of GaN HFETs on Si for microwave power amplification. Detailed characterization and modeling of GaN HFETs on Si are discussed. The electrical performance of these devices reveals their viability in microwave power applications.

Acknowledgement

We acknowledge support in part by Nitronex Corporation and Clemson University.

References

1. J. C. Zolper, "Wide bandgap semiconductor microwave technologies: from promise to practice," *in IEEE IEDM Technical Digest*, 1999, pp. 389-392.
2. J. C. Zolper, and B. V. Shanabrook, "Wide bandgap semiconductor devices," *Proceedings of the IEEE* **90** (2002) 939-941.
3. U. K Mishra, P. Parikh, Yi-Feng Wu, "AlGaN HEMTs-an overview of device operation and applications," *Proceedings of the IEEE* **90** (2002) 1022-1031.
4. B. Green, K. K. Chu, E. M. Chumbes, J. A. Smart, J. R. Shealy, L. F. Eastman, "Role of surface passivation on the power optimization on AlGaN/GaN HEMTs," *IEEE Electron Device Lett.* **21** (2002) 268-270.
5. S. T. Sheppard, K. Doverspike, W. L. Pribble, S. T. Allen, J. W. Palmour, L. T. Kehias, T. J. Jenkins, "High-power microwave AlGaN/GaN HEMT's on semi-insulating silicon carbide substrates," *IEEE Electron Device Lett.* **20** (1999) 161-163.
6. R. Borges, E. Piner, A. Vescan, J. D. Brown, S. Singhal, and R. Therrien, "AlGaN/GaN HFETs fabricated on 100nm GaN on silicon (111) substrates," *in International Semiconductor Device Research Symposium Digest*, 2001, pp. 382-383.
7. E. M. Chumbes, A. T. Schremer, J. A. Smart, Y. Wang, N. C. Mac-Donald, D. Hogue, J. J. Komial, S. J. Lichwalla, R. E. Leoni, and J. R. Shealy, "AlGaN/GaN high electron mobility transistors on Si (111) substrates," *IEEE Transactions on electron devices* **48** (2001) 420-425.
8. M. Seon, T. Prokofyeva, M. Holtz, S. A. Nikshin, N. N. Faleev and H. Timkin, "Selective growth of high quality GaN on Si substrates", *Applied physics letters* **76** (2000) 1842-1844.
9. P. Javorka, A. Alam, M. Wolter, A. Fox, M. Marso, M. Heuken, H. Luth, and P. Kordos, "AlGaN/GaN HEMTs on (111) silicon substrates", *IEEE Electron Device Letters* **23** (2002) 4-6.
10. S. Manohar, A. Narayanan, A. Keerti, A. Pham, J. Brown, R. Borges, and K. Linthicum, "Characteristics of Microwave Power GaN HEMTs on 4-Inch Si Wafers" *in IEEE International Microwave Symposium Digest*, June 2002, pp. 1401-91404.
11. M. Marso, P. Javorka, A. Alam, M. Wolter, A. Fox, M. Heuken, P. Kordoš, H. Lüth, R. H. Jansen, H. Kalisch, H. Klockenhoff, A. Szymakowski, Y. Dikme, "AlGaN/GaN HEMTs on silicon substrates," *in IEEE IMS Wideband Gap Technologies: Applications, Devices and Circuit Issues Workshop*, June 2002.
12. S. C. Binari, K. Ikossi, J. A. Roussos, W. Kruppa, D. Park, H. B. Dietrich, D. D. Koleske, A. E. Wickenden, R. L. Henry, "Trapping effects and microwave power

performance in AlGaN/GaN HEMTs," *IEEE Trans. on Electron Devices* **48** (2001) 465-471.

13. M. Rocchi, "Status of the surface and bulk parasitic effects limiting the performances of GaAs IC's," *Physica* **129B** 119-137.
14. Y. F. Wu, D. Kapolnek, J. Ibbetson, N. Q. Zhang, P. Parikh, B. P. Keller, and U. K. Mishra, "High Al-content AlGaN/GaN HEMT's on SiC substrates with very high power performance", *in IEEE IEDM Technical Digest*, 1999, pp. 925-927.
15. G. D. Dambrine, A. Cappy, F. Helidore, and E. Playez, "A new method for determining the FET small-signal equivalent circuit," *IEEE Trans. on Microwave Theory Tech.* **36** (1988) 1151–1159.
16. Agilent Application Note, " Testing CDMA base station amplifiers," AN-1307.

International Journal of High Speed Electronics and Systems
Vol. 13, No. 1 (2003) 277–292

World Scientific
www.worldscientific.com

RADIATION EFFECTS IN HIGH SPEED III-V INTEGRATED CIRCUITS

TODD R. WEATHERFORD
Department of Electrical and Computer Engineering,
Naval Postgraduate School, Code EC/WT, Monterey, CA 93943-5121, USA
trweathe@nps.navy.mil

The types of applications affected by radiation effects in III-V devices have significantly changed over the last four decades. For most applications III-V ICs have provided sufficient radiation hardness. Some expectations for hardened soft error applications did not materialize until much later. Years of research defined that not only material properties, but device structures, layout practices and circuit design influenced how III-V devices were susceptible to certain radiation effects. The highest performance III-V ICs due to their low power-speed energy products will provide challenges in ionizing radiation environments from sea level to space.

Keywords: Radiation; GaAs, InP, GaN, SiC, compound semiconductors, total dose, displacement damage, single event effects

1. Introduction

The concept of radiation effects in electronics has changed significantly over the last half century. Initially from nuclear weapons and reactor applications to space effects and background terrestrial sources each application requires a special understanding of the radiation source and the delivery spatially and temporally to the target material. Once the incident energy enters the target material it transfers into the electrical circuit (ionizing radiation) or modifies the semiconductor crystal (displacement damage). This review explains how high-speed III-V integrated circuits are affected by radiation. Neither the incident particle types or the basic physics of the particle interaction has changed; only the size, materials and switching energy requirements of our circuits have changed dramatically. Today ICs switch on the order of 1,000s of electrons, instead of tens of millions electrons switched per bit in the 1970's. Due to this continual reduction of switching energy, more background particles can influence device operation. Compound semiconductor ICs has led this reduction of power-speed products (Joules/gate). Low power-speed product circuits are susceptible to more ionizing particles in the environment than with earlier generation electronic technologies.

Compound semiconductors were pursued in Department of Defense research to provide a substantial improvement over silicon-based ICs. In most cases, the first GaAs integrated circuits provided promising rad-hard results that supported the use of III-V devices for the military needs for high performance ICs.[1] The lack of oxides and wider band-gaps led many to believe that all III-V ICs would provide radiation-hardened ICs without substantial modifications. However, the substitution of material alone could not provide radiation-hardened devices at a time when feature sizes, geometries, new structures and circuit topologies contributed to sensitivities to fast local ionization, i.e. single event effects (SEE). The following review of radiation effects in the high-speed compound semiconductors outlines the past history and future directions. Before investigating the specifics, the cause for some misunderstanding lies in understanding the types of radiation effects. The following provides a brief summary of the most common radiation effects, background in our learning process on radiation effects in III-Vs, specific issues describing the advantages and disadvantages observed in compound semiconductor ICs, and the author's expectations of the future susceptibility of III-V ICs.

1.1. *Radiation effects categories*

Radiation can mean any source of energy that in influences the operation of electronic devices. Light and heat can be considered radiation. However, specifically when the term radiation effects are used this refers to the situation where concentrated energy has influence on electronic devices beyond the normal realm of applied voltage. Semiconductors predominantly interacts with electrons, photons and phonons on the order of a couple of eVs. Any particle more energetic than thermalized electrons could be considered radiation. An example of such particles would be gammas from a couple eV up to GeV heavy ions in space. These various radiation sources can cause reliability problems anywhere from temporary loss of data, circuit degradation, loss of functional operation to destruction of the semiconductor device, either by ionizing carrier densities to very high levels, or creating defects above background levels.

In the field of radiation effects in electronics we must divide the source characteristics and the delivery of these high-energy particles into subsets of effects regardless of the target material. The four most common categories are:

- total dose or total ionizing dose (TID),
- dose rate, gamma dot,
- single event effects (SEE), and
- displacement damage effects

The first three categories generally relate to ionizing radiation, where high-energy particles create electron and holes in pairs. In the last category, higher energy particles can not only ionize electrons but will introduce the atomic displacements in the semiconductor lattice. Such displacement introduces defects that alter the electrical properties of the semiconductor. Therefore depending on the environment, ionizing and displacement damage effects can occur, or only ionizing (charged particles, photons) or only displacement damage by uncharged particles (neutrons).

One common misperception is that the resilience of a technology against one of the several radiation effects implies that the technology is immune to all radiation effects. It is vital that the designers of devices, circuits and systems be aware of the high energy particles available in their operating environment be it, sea level, high altitude avionics, near heavy metals, space or strategic weapon environments. The device must be characterized for each aspect of the radiation environment.

1.1.1. *Total Dose Effects*

The total dose environment refers to the absorbed dose of energy in rads (ergs/g) that the semiconductor material has been exposed to. If the target material is silicon the unit of radiation is referred as Rad(Si). Total dose only accounts for absorbed energy for ionizing electrons. Other names for these effects are gamma, or total ionizing dose. Such environments would include:

- Gammas, electrons, protons in a space environment
- X-rays, gamma radiation from nuclear weapon detonations
- Gammas emitted from radioactive material

The most susceptible transistor materials to total dose effects are insulators. Ionized positive charges in an oxide move relatively slow (possibly years to move microns). Charges trapped in oxide-insulated field effect transistors (FET) shift threshold voltages. Charge trapped in gate or field oxides alters the device characteristics. Due to the lack of oxides in III-V devices compound semiconductor ICs have not shown serious degradation to total dose effects. III-V MESFETs or HEMTs are essentially immune in the harshest total dose environments.

1.1.2. *Dose Rate Effects*

Dose rate effects relate to the rate of absorbed ionizing dose is occurring, independent of the total dose. The effect is only discussed in the design of devices that must operate and/ or survive in the vicinity of a nuclear event. Such rates are described in units of Rad/sec absorbed, i.e. 10^9 Rad/sec. Even though a dose rate of 10^{11} Rad/sec appears high the source of radiation may only exist for several nanoseconds, thus the total absorbed dose is on the order of thousands of Rad(Si). High dose rates during the irradiation period convert a semi-insulating substrate into a highly conductive substrate for nanoseconds. Power and ground buses on the substrate become linked in temporary low impedance paths usually resulting in damage due to high currents. Hardening to dose rate effects requires considerable attention to circuit layout, packaging and power bus design.

Insulating or high resistivity substrates do provide the best protection to dose rate effects. Semiconductors materials with wide band-gaps (i.e. low intrinsic carrier densities) are the most desirable. Wide band-gap compound semiconductor devices produce lower photocurrents than silicon-based ICs.

In recent years the term "dose rate" has also been applied (to the opposite extreme) to testing of total dose effects using Co^{60} gamma sources. When radiation testing oxides the amount of oxide trapped charge is dependent on the experiment's dose rate. The dose rates of total dose experiments are operated from mRad/sec to 100's of Rad/sec much lower and much longer periods than dose rate effects related to nuclear events.

1.1.3. *Single Event Effects*

The two previous ionizing effects relate to absorbed dose over the complete IC. The situation of an individual energetic particle inducing errors or failure to an individual circuit node, or device is categorized as a Single Event Effects (SEE). Unlike Total dose and dose rate effects which have been examined since the 1950's, SEE only became known once electronic devices had miniaturized to require switching energies comparable to energy transferring cosmic rays in the space environment[2] and then alphas from packaging materials in 1978.[3] Wallmark in 1962 had predicted that eventually cosmic ray interaction would become a reliability issue.[4] In 1990's neutrons at aircraft altitudes were recognized to induce SEE.[5] In the late 1980's soft errors from solder emitting alpha particles forced IC manufacturers to obtain low alpha or alpha-free solder.[6] The environments that can induce SEE have and will increase with IC generations that continue to lower power-speed products.

GaAs SRAMs early on showed some susceptibility to SEE, which became its Achilles' heel for radiation applications requiring SEE immunity.[7] Early predictions expected SEE immunity would be based on the larger band-gaps to create less ionization than observed in a Si substrate. The high speed III-V circuits operated on much smaller switching energies due to low voltage swings, low capacitance and additionally had long diffusion lengths in semi-insulating substrates and non-insulating gates that amplified the SEE susceptibility. Today much more is understood about SEE in III-V integrated circuits, and the capabilities of new growth, materials and circuit techniques can provide designs hardened for SEE.

1.1.4. *Displacement Damage Effects*

Displacement damage introduces defects in semiconductors. The electron and proton Van Allen belts, neutrons from nuclear events, or implantation in manufacturing are all environments where displacement damage is observed. Low-doped or intrinsic structures such as solar cells and detectors are sensitive to displacement damage effects. Mid-gap defects increase trapping and recombination processes, analog devices such as bipolar transistors that are highly dependent on base recombination normally degrade due to displacement effects.

The displacement energy transferred to a crystal lattice is referred as non-ionizing energy loss (NIEL), and the electrical defects introduced is related to the binding energies of the atoms, the band structure, and defect transport in the semiconductor. Compound semiconductors like the Group 4 semiconductors can vary their response to displacement damage effects. Intrinsic and low-doped semiconductors are most vulnerable to defects introduced by displacement damage. Depending on the application, one or more of the previously mentioned radiation effects will require characterization. In the next section the chronologically of those applications are discussed.

2. Background

In Table 1 a chronological listing of initial events in the study of III-V radiation effects ICs is shown. Examining Table 1 by the types of studies performed these studies can be divided into three general periods or applications.

- Cold War radiation studies (1960's through 1980's) – Application of III-Vs ICs to military applications, mainly analog amplifiers and photodetectors. Radiation effects related to neutron, total dose and dose rate effects were important to these applications. Most of this work was on discrete devices.

- Space systems use of III-V devices (1980's – 1990's) – The prospect of replacing CMOS/SOS and Si technologies with digital GaAs ICs required radiation characterization in total dose and SEU.

- High-speed communications (1990's- 2000+) – The advent of high speed III-V ICs (GaAs, InP) for Internet and wireless applications require low bit error rates. Applications in both space (cosmic particles) and terrestrial (solder emitting alphas) are susceptible to SEE (i.e. soft errors). These new high speed ICs required understanding of radiation effects at high frequencies.

Future application of compound semiconductors will further require radiation effect studies in,

- High power III-V devices – The future use of GaN and SiC for microwave and low frequency power in military and commercial space applications will require knowledge of displacement damage and ionizing effects.

Table 1 - Timeline of Radiation Effects firsts related to III-V devices

Year	Event
1961	Electron radiation effects on unipolars[8]
1970	Neutron irradiation of GaAs JFETs[9]
1978	Total dose measurements? on GaAs MESFETs[10]
1979	First dose rate study of GaAs FETs[11]
	Dose rate studies on GaAs ICs[12]
1981	Dose rate effects on GaAs photodetectors[13]
1982	Dose rate latchup studies in GaAs ICs[14]
1983	First JFET SRAM SEU proton experiments[7]
	First JFET SRAM heavy ion SEU experiments[15]
1991	First GaAs MESFET SEU data from space experiments[16]
1995	First experiments on dynamic SEU testing of GaAs SEU[17]

Chaffin summarizes early work on radiation damage in GaAs in 1973.[18] A further review of total dose, dose rate and displacement damage (mainly neutrons) has been presented by Simons[1]. Zuleeg presented a chapter on radiation effects in 1985 that reviews the work on weapons effects and emerging space applications.[19] Srour et al provided a review of radiation effects in ICs including a section on GaAs effects.[20] Zuleeg provided an extensive review article in 1989.[21] Zuleeg's work provided a thorough discussion on dose rate and neutron damage effects and later discusses the SEU work on Complementary JFETs.

There was a large amount of open literature up to the mid-1980's investigating total dose and dose rate effects on GaAs. Much of this work showed that GaAs ICs were superior to silicon-based ICs in the weapon environments where dose rate, neutron displacement and total dose effects are important.[21] Most of the devices in these applications were discrete transistors and photodetectors.

In other applications such as space, where memory and logic ICs in satellite systems were required, digital GaAs technologies were just becoming available. Since these ICs were in their infancy in the early 1980's, little data was available on any radiation effect, particular SEE. Additionally the SEE field started in the early 1980s after May et al's work in Si DRAMs.[3] The only III-V ICs available at the time were SRAMs that showed excellent total dose hardness but few results were available in the 1980's on GaAs SEU.[19] Since little concrete SEU data was available and GaAs ICs had shown tolerance to total dose effects the general assumption was the SEU issue would not be important.

Despite some optimistic expectations, the SEE performance of digital circuitry based on compound semiconductors was essentially no better than that of non-hardened silicon CMOS. The CMOS/SOS technologies and CMOS/SOI SRAMs showed superior radiation hardness to the reported GaAs SEU results.[22] From the late 1980's to the mid-1990's investigations utilized circuit modeling similar to Si SEE radiation studies, but the problems required deeper understanding with device simulation in the early 1990's. The 2-D modeling codes were becoming available in the early 1990's that allowed researchers to study the SEE phenomena.[23] Additionally the experimental data into charge collection mechanisms was available by the early 1990's.[24] The reasons for the SEE sensitivity were several fold, small power speed products,[16] long diffusion lengths in semi-insulating substrate, backgating and parasitic bipolar transistor effects in the charge collection process and material ionization rates.[25] In 1995 McMorrow et al presented a review of these GaAs SEE mechanisms.[26] The advantage of GaAs's e-h pair ionization energy (4.8eV) over Si (3.6eV) was not sufficient to outweigh the other disadvantages. In the mean time the size of VLSI Si and CMOS/SOI far outpaced the performance of GaAs SRAMs and microprocessors in gate count and performance/cost ratio.

By 1995 solutions were available to make GaAs ICs immune to SEE[27], and for a short time Motorola was ramping up capability to manufacture rad-hard large high-speed GaAs ICs for the Celestri satellite constellation.[28] However market pressures eliminated Celestri, and Motorola joined with Teledesic, thus dropping the digital GaAs effort. During this time the Motorola's rad-hard digital GaAs technology was incorporated into a few satellite systems.

Today many high-speed communications circuits are being transitioned to other III-V technologies (i.e. InP-based) from GaAs FET ICs. Many of these applications require optoelectronic capability and high-speed requirements that for the most part are only available with direct-bandgap III-V ICs. A small subset of these applications is used in military and commercial space applications. But instead of SRAMs and microprocessors, which were the focus in the 1980's, researchers have investigated these new types of III-V ICs in such applications as crosspoint switches and Application Specific ICs (ASICs).

SEE is still the critical radiation effect for the majority of applications, but testing and studies have migrated to high speed dynamic testing where unlike a decade ago all SEE studies concentrated on soft errors in SRAM storage. Additionally the designs of consumer ICs in terrestrial applications are requiring radiation related reliability studies. The terrestrial sources are atmosphere neutrons; package or solder emitted alpha particles, or cosmic particles at sea level.

Looking ahead, the transfer of GaN and SiC-based devices to high power space applications will also require new studies on displacement damage, destructive SEE effects, and possible passivation issues with total dose effects. The extensive study by DoD laboratories in the 1970's and 1980's is less applicable to the new class of III-V ICs because new compounds, different device structures, geometries and fabrication methods exist today than those used two to three decades ago.

3. Specific Effects in III-V ICs

3.1. *Why are III-V ICs resilient to total dose and displacement damage?*

The absence of oxides in III-V transistors was a major advantage that III-V FETs and HBTs had over the MOSFETs for total dose effects. If Si MESFETs or JFETs were available for circuits, they also would provide higher thresholds to total dose effects over MOSFETs. The III-V substrate material was indirectly responsible for this advantage because gate-insulating oxides were not available. Co^{60} gamma testing provided many exceptional results for GaAs ICs, all above the Mrad(Si) level. MMICs are shown to operate above 10^9 rads before degradation.[29] GaAs JFET SRAMs showed operation to 100Mrad(Si).[30] These levels are magnitudes above those required for the most severe natural environments.

Displacement damage from neutrons showed levels between 10^{14} to 10^{16} n/cm^2.[31] In comparison studies in 1994 of neutron fluence damage on Si JFETs, Si MOSFETs, GaAs MESFETs, the GaAs devices were unaffected to 55Mrads.[32] For both total dose and displacement damage effects devices fabricated on epitaxial material normally performed better than ICs in ion-implanted processes.

The insensitivity of III-V devices to these radiation effects was well known by the early 1980's. Several defense contractors had positioned themselves to provide radhard

GaAs ICs for the DARPA's GaAs initiative, which developed high performance ICs for military systems. Several other factors also came together in the early 1980's. One was the personal computer flourished, increasing the overall demand for cheap ICs (i.e. bulk CMOS), thus increasing commercial demand that eventually attracted IC defense contractors to move into the commercial market. By the end of the 1980's almost all of the initial DoD GaAs foundries had migrated into commercial microwave or ASIC products. Second, the soft error phenomena (i.e. SEE) became important in the radiation effects field, not solely because of materials, but also due to reduced feature size. The GaAs devices further pressed lithography and lowered power-speed products to obtain the highest performance to keep ahead of Si benchmarks. A third factor to exacerbate the SEE was the semi-insulating substrates. The dose rate effects in GaAs benefited over Si substrates due to the higher resistivity but the high density of the GaAs material provided higher ionization of carriers per unit length than silicon substrates.

3.2. *Why is there a high sensitivity to soft errors in III-V ICs?*

The three factors previously mentioned, small feature size, high resistivity substrates and the higher ionization per unit length turned the III-V FETs and HBTs into efficient detectors. This was not recognized at first. The first SEU results in MESFET and JFET SRAMs did not provide the required soft error rates for space systems.[16] When poor results were observed it was not normally published.[33] By the late 1980's some GaAs IC manufacturers had to provide SEE solutions to continue to attract DoD support. The initial approaches to SEU hardening GaAs ICs examined circuit techniques that were similarly used in Si memories.[34] Resistive and capacitive circuit approaches were attempted on CJFET SRAMs;[22] others utilized redundancy,[35] while other investigated superlattices to isolate the substrate.[36]

To understand the SEE sensitivity we need to look at electron and hole carrier generation in the semiconductor, the carrier transport and how charge collected at the contacts influence the circuit. Photocurrent generation in a semiconductor is related to two parameters, creation energy for an electron-hole pair and the density of the material. The creation energy is proportional to the band-gap of the semiconductor as shown by Klein.[37] The band structure of the semiconductor requires conservation of momentum and energy to absorb the kinetic energy of the incident particles, thus the energy to move a valance electron to the conduction band requires higher energy to conserve energy and momentum with the band structure. The density of the material determines how much matter is available for interaction in a particular volume. Therefore the amount of free carriers released is calculated from,

$$LET\ (ehp/um) = LET\,\frac{(MeV * cm^2)}{mg} \times \varepsilon_i(\frac{ehp}{eV}) \times density\,(\frac{gm}{cm^3}) \tag{1}$$

where LET is the linear energy transfer of a particular incident ion and ε_i is the creation energy for an electron-hole pair. LETs for heavy ions in space range between 1 to 100 MeV/mg/cm^2. Table 2 includes a summary of electron-hole pair ionization and band gap for various compound semiconductors and Table 3 provides the charge ionized per unit length in the semiconductor.[38]

Table 2 - Electron-hole pair ionization energies for various semiconductors.

Semiconductor	E_g (eV)	ε_I (eV)
Ge	0.66	2.60
$In_{0.47}Ga_{0.53}As$	0.75	2.85
$Si_{0.9}Ge_{0.1}$	1.03	3.63
Si	1.12	3.64
InP	1.35	4.53
GaAs	1.43	4.75
$Al_{0.7}Ga_{0.3}As$	2.06	6.52
AlAs	2.17	6.83
SiC	2.86	8.76
GaN	3.4	10.3
AlN	6.1	17.8

Table 3 gives a direct comparison of various semiconductors for the ionization induced by an ion with a LET of 1 MeV/mg/cm^2. The fourth column provides the ionized charge (fC) in the conduction band per micron of material in the penetrating ion's path. Comparing GaAs to Si, if an ion penetrated two equivalent volumes of Si and GaAs, the GaAs material would have 70% more free carriers produced. As can be seen in this table, compound semiconductors with heavy masses such as the In-based materials provide even higher free carrier densities, and in contrast low Z compound semiconductors such as GaN and SiC are expected to produce lower amounts of carrier ionization. Back in the 1980's the incorrect assumption was that the higher band gap of GaAs (compared to Si) alone would provide the radiation hardness, the density was not addressed.

The other two issues for SEE susceptibility were the new feature sizes associated with the new circuit designs and the high resistivity substrates. Due to the nature of the JFET and MESFET logic gates, the power supplies were below 2 volts; the lower gate capacitances and the small area gates provided fairly low power-speed switching energies, a very beneficial aspect for high performance integrated circuits. The reduced switching energy increased SEE susceptibility. In some MESFET SRAM designs with impedance loads the circuit only required 5 fC of charge to induce a bit flip.[39] Figure 1 is a compilation of previous reported data[16] and the author's notes estimating geosynchronous orbit heavy ion soft error rates using the Peterson heavy ion equation. The progression of each technology generation is to move left on the plot to lower energy products. Hardening with substrate techniques reduces soft error rates. This is observed with both III-V approaches with low temperature grown (LTG) buffers or with insulating buried oxides (CMOS/SOI). Particular with III-V FETs, devices with improved gate isolation (i.e. HFETs) provide lower soft error rates and lower power speed products.

Table 3 - Charge per unit length of LET = 1 MeV/mg/cm^2 in various semiconductors.

Target Semiconductor	eV / eh-pair	Density (gm/cm^3)	fC/um for an LET=1 MeV/mg/cm^2	Divide LET by X for pC/um collected
Si	3.6	2.32	10.4	97
GaAs	4.8	5.32	17.8	56
InP	4.5	4.81	17.1	58
$In_{0.47}Ga_{0.53}As$	2.9	5.49	30.3	33
SiC	8.7	3.21	5.9	169
GaN	10.3	6.11	9.5	105

The additional issue was the high resistivity semi-insulating substrate. The semi-insulating substrate is ideal for isolation between transistors, and also eliminated additional steps for isolation. When carriers are ionized the region in the substrate acts

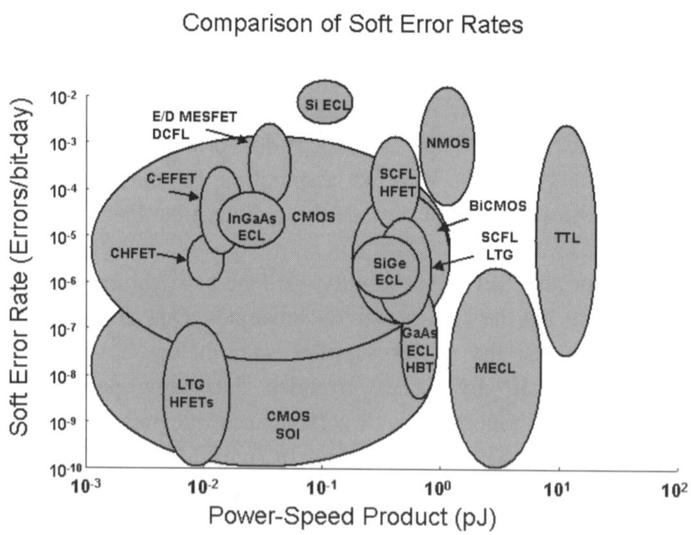

Fig. 1 - Geosynchronous orbit soft error comparisons for various IC technologies.

like a depleted intrinsic detector due to the undoped or low-doped substrate. Carriers ionized in the depleted substrate and/or within a diffusion length of critical depletion regions accumulate carriers due to an electric field in the device. In CMOS/SOS or CMOS/SOI technologies that showed excellent SEU immunity, the sapphire or buried oxide region prevented ionized carriers in the substrate to reach the drain/body junctions in CMOS devices.[40] In GaAs no wide band-gap or insulator barrier existed, and the high mobility of the GaAs substrate with the nanosecond carrier lifetime in undoped material provided diffusion lengths on the order of 5 to 10 microns. CMOS technologies on p-doped silicon epitaxy provided more recombination than undoped silicon to carriers migrating from the Si substrate.

Also attributable to the semi-insulating substrate is the floating body for FETs. Most of the GaAs ICs utilized FET structures that were susceptible to backchannel and/or bipolar effects.[25] A similar phenomenon is present in partially depleted SOI transistors today. Holes collected in the floating body of a CMOS/SOI n-MOSFET back gate the transistor adding source electrons to the initial carrier generation.[41] The difference in the GaAs substrate was that the EL2 defect and the floating body were present. Various experiments and device simulation studies examined backgating and bipolar mechanisms that contributed to the SEE susceptibility.[24,25] The non-insulated gate of GaAs FETs also contributed by providing holes to be collected on the gate terminal directly turning on the FET.[40]

3.3. *How can GaAs ICs be made immune to SEE?*

The use of buried oxides in III-Vs was not available in the 1990's to mimic the CMOS/SOI approach to harden SEE. The only insulator available to utilize was superlattices.[37] The superlattice approach was in the right direction; to limit carriers reaching the channel from the substrate but the band structure of the superlattice trapped carriers that assisted in backgating.[23] Additionally DX centers were present in AlGaAs-GaAs superlattices that contributed to trapping carriers. Other than using a potential barrier to block free carriers in the substrate, the only other approach was to kill the lifetime in the substrate. Proton and neutron irradiation had been observed to reduce upset cross sections of GaAs ICs.[27,42] In 1995 the use of defect rich non-stoichiometric low-temperature grown GaAs buffer layers to reduce lifetime was proposed.[43]. Earlier work had shown that the LTG GaAs buffers eliminated backgating in MESFETs.[44] Implementation in several GaAs processes demonstrated excellent SEU immunity in GaAs memories.[27,28,45] In 1995 digital GaAs ICs hardened to the complete coverage of radiation effects were available. Even though the radiation technical issues were solved no demand existed for rad-hard GaAs ICs. The communications market absorbed the digital GaAs IC foundries for consumer wireless and internet communications. Additionally other technologies were replacing the GaAs FET in digital applications, SiGe and CMOS devices were just under the performance curve of GaAs FETs, and InP-based Heterojunction Bipolar Transistor (HBT) technologies were above the curve. High-speed ECL Si ASICs were known to be SEE sensitive in static tests.[16] SEE studies of III-V ECL HBT technologies at high clock rates were a new issue to be studied for Bit Error Rates (BER) in SEE environments.[27]

The soft errors observed in dynamic switching circuits is referred as Single Event Transient (SET) where the induced photocurrent produce transient signals in the logic producing errors in the bit stream. Various III-V circuits have been investigated for SET in fiber optic link applications under heavy ion tests.[46]

Heavy ion broad beam experiments (as opposed to micro beam experiments) could not easily provide information on which nodes or devices in a VLSI circuit were susceptible. New tools became available in the radiation effects community to investigate individual transistors in a complex circuit. Micro-beam ion beams and lasers were used to locate susceptible devices. Further the laser pulses used to generate carriers in micron-sized locations could be synchronized with the circuit clock to study soft errors in relation to clock phase. Work has progressed to utilize photoconductive sampling probes to measure the SEE induced voltage transients at individual transistors in a design.[47] These experimental tools provided data to provide verification of circuit and device simulations.

3.4. *What was learnt from dynamic testing and analysis?*

Susceptibility for soft errors in clocked III-V ICs was first recognized when incorporating redundancy techniques by performing majority voting.[17] In slower Si-based circuits redundancy was used often, either at the system, or circuit level. However at high speeds, voting becomes a significant portion of the clock period. A compromise between two rates, the scrub rate, i.e. the rate to write the majority vote result back into three locations, and the rate that the probability that 2 of 3 data registers can accumulate soft errors. The overhead of increase voting logic adds to the number of transistors susceptible to SEE. Further knowledge of the photocurrent persistence is important to determine the scrub rate.

To study soft errors in high-speed logic a circuit is tested in a broad beam ion experiment and a bit-error rate is calculated for particular ion fluence.[48] Running the experiment at various clock frequencies give different susceptibilities due to various race conditions in the integrated circuit.[49] Micron-focused laser beams can be used to determine the most susceptible devices; the laser pulse can also be synchronized with the IC's clock to determine which phase of the clock is most susceptible. Photoconductive sampling probes have been used to provide in-situ measurements of voltage transients internal to the IC. Such an approach was used on GaAs MESFET and InP-based HBT technologies.[47] A surprising result to system designers of HBT ECL circuits was the ability at GHz clock frequencies to lose several clock cycles to an SEE, preventing accurate error correction to be applied to streaming data.[50] System or circuit designs to mitigate such long streams of errors require data link resets. The source of the InP-based SEE susceptibility was charge collection from the semi-insulating substrate similar to early GaAs FET issues. A solution was proposed to using lifetime killing buffer layers to eliminate the SEE susceptibility.[51]

3.5. *What does the future hold for III-V ICs for radiation effects?*

The radiation effects of space and weapons environments will continually require radiation-hardened microelectronics. However the flood of microelectronics into every know application and in such large quantities of low power-speed product transistors overlap many radiation sources in our terrestrial surroundings, alphas from solder or packaging, Borophosphosilicate glass (BPSG),[52] cosmic radiation at sea level, and byproducts of neutrons at avionic altitudes. Baumann provide a good review of these terrestrial issues.[53] The neutron environment at aircraft cruising altitudes has required fly-by-wire control electronics to be SEE hardened.

The introduction of low density III-V compounds such as GaN and SiC-based ICs should be promising to ionizing radiation applications. The capability to engineer the substrate by bandgap, introduce insulators (Gallium-on-Insulator)[54] or utilize non-stoichiometric materials to control soft error susceptibility will further provide additional solutions for III-V ICs in soft error applications.[55] The recent capability to manufacture GaAs on Si may not only provide a more competitive product to Si-based ICs, but may allow non-stoichiometric techniques to provide radiation hardness for space or terrestrial needs. The future looks promising with increasing capabilities to engineer III-V materials and devices to improve radiation hardening of ICs for future applications.

References

1. M. Simons, "Radiation effects in GaAs integrated circuits: A comparison with silicon", *IEEE GaAs IC Symp*, (1983) pp.124-128.
2. D. Binder, E.C. Smith and A.B. Holman, "Satellite anomalies from galactic cosmic rays", *IEEE Trans. on Nucl. Sci.* **NS-22, 6**, (1975) pp. 2675-2680.
3. T.C. May and M.H. Woods, "Alpha-particle-induced soft errors in dynamic memories," *Proceedings of the 1978 IEEE International Reliability Physics Symposium*, (1978).
4. J.T. Wallmark, and S.M.Marcus, "Minimum size and maximum packing density of non-redundent semiconductor devices", *Proc.of the IRE*, **50, 3,** (1962) pp. 286-298.
5. E. Normand, "Single-event effects in avionics", *IEEE Trans. on Nucl. Sci.,***43, 2**, (1996) pg. 461-474.
6. M.W. Roberson, "Soft error rates in solder bumped packaging", *Proceedings on the 4th International Symposium on Advanced Packaging Materials*, 15-18 March 1998, Braselton, GA, USA.
7. P. Shapiro, A.B. Campbell, J.C. Ritter, R. Zuleeg, and J.K. Notthoff, *1983 IEEE Natural Space Radiation Effects Conference*, Gatlinburg, TN, July 1983.
8. B.A. Kulp et al, "Electron radiation damage in unipolar transistor devices", *Proc. IRE*, 49, (1961) pp.1437-1438.
9. J.L. McNichols and W.S. Ginell, "Predicted effects of neutron irradiation on GaAs junction field effect transistors", *IEEE Trans. on Nucl. Sci.* **NS-17**, (1970) pp. 52-54.

10. J.M. Borrego et al., "Radiation effects on GaAs MESFETs", *IEEE Trans. on Nucl. Sci.* **NS-25** (1978), pp 1436-1443.
11. M. Simons et al, "Transient radiation study of GaAs MESFETs implanted in Cr-doped and undoped substrates", *J. Appl. Phys.*, **52**, (1981), pp.6630-6638.
12. R. Zuleeg and J.K. Notthoff, "Transient response of GaAs ICs to ionizing radiation", *IEEE Trans. on Nucl. Sci.,* **NS-26**, (1979) pp. 4744-4749.
13. J.J.Wiczer, L.R. Dawson and C.E. Barnes, "Transient effects of ionizing radiation in photodiodes", *IEEE Trans. on Nucl. Sci.* **NS-28**, (1981), pp.4397-4402.
14. R. Zuleeg et al, "Latch-up in GaAs ICs during ionizing radiation exposure", *1982 GaAs IC Symposium Digest* (1982), pp.123-126.
15. R. Zuleeg, J.K. Notthoff and D.K. Nichols, "SEU of Complementary GaAs static RAMs due to heavy ions", *IEEE Trans. on Nucl. Sci.* **NS-30**, (1984), pp. 4533-4539.
16. T.R. Weatherford, A.B.Campbell, D. McMorrow, J.B.Langworthy, W.J. Stapor, A.R.Knudson, E.L.Petersen, D. Wilson, and L. Tran, "Can Digital GaAs Be Used in a Space Environment?: A Look at Single Event Upset in GaAs"", *IEEE 1991 GaAs IC Symposium, Technical Digest,* Monterey, California, October 20-23, 1991.
17. R. Schneiderwind, D. Krening, S. Buchner,K. Kang, T.R. Weatherford, "Laser confirmation of SEU experiments in GaAs MESFET combinational logic for space applications", IEEE Trans. on Nucl. Sci. , **NS-39** (1992), pp. 1665 –1670.
18. R.J. Chaffin, "Microwave Semiconductor Fundamentals and Radiation Effects, John Wiley and Sons, New York (1973).
19. R. Zuleeg, "Radiation Effects in GaAs ICs", Chapter 8, VLSI electronics: microstructure science / edited by Norman G. Einspruch, 1985.
20. J.R.Srour and J.M. McGarrity, "Radiation Effects on Microelectronics in Space", *Proceedings of the IEEE* , **76-11**, (1988), pp. 1443-1469.
21. R. Zuleeg, "Radiation effects in GaAs FET devices", *Proceedings of the IEEE,* **77-3**, (1989), pp. 389 –407.
22. R. Zuleeg, J.K. Notthoff, "SEU Hardening of GaAs E-JFET SRAMs", *Journal of Radiation Effects, Research and Engineering*, 6-2, (1988), pp. 136.
23. T.R. Weatherford, D. McMorrow, W.R. Curtice, A.R. Knudson and A.B. Campbell, "Single event induced charge transport modeling in GaAs MESFETs", *IEEE Trans. on Nuclear Science*, **NS-40**, (1993), pp. 1867-1871.
24. D. McMorrow, A.B. Campbell, A.R. Knudson, S. Buchner, T.R. Weatherford," Transient charge collection from ion tracks in semiconductors", *Nuclear Instruments and Methods in Physics Research Section B*, **67** n 1-4 (1992), pp 384.
25. B.W. Hughlock, G.S. LaRue, A.H. Johnston, "Single-event upset in GaAs E/D MESFET logic", *IEEE Trans. on Nucl. Sci.*, **37-6**, (1990), pp. 1894-1901.
26. D. McMorrow, T.R. Weatherford, S. Buchner, A.R. Knudson, J.S. Melinger, L.H. Tran, and A.B. Campbell, "Single-event phenomena in GaAs devices and circuits", *IEEE Trans. on Nucl. Sci.*, **NS-43**, **2**, (1996), pp. 628 –644.
27. P.W. Marshall, C.J. Dale, T.R. Weatherford, M.Carts, D. McMorrow, A.Peczalski, S.Baier, J. Nohava and J. Skogen, " Heavy ion SEU immunity of a GaAs complementary HIGFET circuit fabricated on a low temperature grown buffer layer", *IEEE Trans. on Nuclear Science,* **NS-42**, (1995), pp. 1850-1855.
28. M. LaMacchia, J. Abrokwah, B.Bernhardt, D. Foster, B. Crawforth, B. Mathes, T. McGuire, and T. Weatherford, "Radiation hardened complementary GaAs (CGaAs^{TM})", *IEEE GaAs IC Symposium Technical Digest,* October 1997.
29. A. Meulenberg, H.-L.A. Hung, K.E. Peterson, W.T. Anderson, "Total dose and transient radiation effects on GaAs MMICs", IEEE Trans. on Elec. Dev., **35-12**, (1988) , pp. 2125-2132.
30. R. Zuleeg, K. Lehovec and J.K. Notthoff, "Effects of total dose ionizing radiation on GaAs JFETs and ICs", *IEEE GaAs IC Symposium Digest*, (1981) pp. 14.

31. J.M. Borrego et al., "Radiation Effects on signal and noise characteristics of GaAs MESFET microwave amplifiers", *IEEE Trans. on Nucl. Sci.* **NS-26** (1979), pp 5092-5099.
32. M. Citterio, S. Rescia, and V. Radeka, "Radiation effects at cryogenic temperatures in Si-JFET, GaAs MESFET, and MOSFET devices", *IEEE Tran. On Nucl Sci.* **NS-42**, (1995) pp. 2266-2270.
33. T.R. Weatherford, "Single event upset in GaAs integrated circuits", PhD thesis, North Carolina State University, 1993.
34. J.L.Andrews, J.E. Schroeder, B.L. Gingerich, W.A. Kolasinski, R. Koga and S.E. Diehl, "Single event error immune CMOS SRAM", *IEEE Trans. on Nucl. Sci.* **NS-29** (1982), pp. 2040-2043.
35. J.R. Hauser, "SEU-hardened silicon bipolar and GaAs MESFET SRAM cells using local redundancy techniques", *IEEE Trans. on Nucl. Sci.* **NS-39**, (1992) , pp. 2-6.
36. R.L. Remke et al, "SEU measurements on HFETs and HFET SRAMs", *IEEE Trans. on Nucl. Sci.* **NS-36**, (1989) pp.2362-2366.
37. C.A.Klein, "Bandgap dependance and related features of radiation ionization energies in Semiconductors", *Journal of Applied Physics*, vol. 39, no. 4, pp. 2029-2038 , March 1968.
38. T.R. Weatherford, "From carriers to contacts, A review of SEE charge collection processes", *IEEE 2002 Natural Space Radiation Effect Conference Short Course notes*, July 2002.
39. T.R. Weatherford, J.R. Hauser, and S.E. Diehl, "Comparisons of single event vulnerability of GaAs SRAMs", *IEEE Trans. on Nucl. Sci.*, **NS-33**, (1986), pg. 1121-1127.
40. A.B. Campbell et al, "Charge collection studies in CMOS/SOS structures", *IEEE Trans. on Nucl. Sci.* **NS-32**, (1985) pp 4128-4132.
41. S.E. Kerns, L.W. Massengill, D.V. Kerns Jr., M.L. Alles, T.W. Houston, H. Lu and L.R. Hite, "Model for CMOS/SOI single-event vulnerability", *IEEE Trans. on Nucl.Sci.*, **NS-36**, (1989) pp. 2305-2310.
42. T.R. Weatherford, L. Tran, W.J. Stapor, E.L. Petersen, J.B. Langworthy, D. McMorrow, W.G. Abdel-Kader, and P.J. McNulty, "Proton and heavy ion upsets in GaAs MESFET devices", *IEEE Trans. on Nucl. Sci.*, **NS-38**, (1991), pp. 1450-1456.
43. T.R. Weatherford, D. McMorrow, W.R. Curtice and A.B. Campbell, "Significant reduction in the soft error susceptibility of GaAs field-effect transistors with a low-temperature grown GaAs buffer layer", *Applied Physics Letters , 67,5,* 31 July 1995, pg. 703-705.
44. F. W. Smith, A. R. Calawa, and C. L. Chen, "New MBE buffer used to eliminate backgating in GaAs MESFETs," *IEEE Elec. Dev. Let.*, **9-2**, (1988) pp. 77–80.
45. T.R. Weatherford, P.W. Marshall, C.J. Marshall, D.J. Fouts, B. Mathes and M. LaMacchia, "Effects of low-temperature buffer-layer thickness and growth temperature on the SEE sensitivity of GaAs HIGFET Circuits", *IEEE Trans. on Nucl. Sci.,* **NS-45,** (1997), pp. 2298-2305.
46. C.J. Marshall, P.W. Marshall, M.A. Carts, R. Reed, S.Baier, and K. Label , "Characterization of transient error cross sections in high speed commercial fiber optic data links", *IEEE 2001 Radiation Effects Data Workshop*, pg. 142-145.
47. G. David, T. Yun, M. Crites, J.F. Whitaker, T.R. Weatherford, K. Jobe, S. Meyer, M. Bustamante, W. Goyette, S. Thomas III, and K. Elliott, " Absolute potential measurements inside microwave digital ICs using a micromachined photoconductive sampling probe", *IEEE Trans. on Microwave Theory and Techniques,* **46-12**, (1998), pp. 2330 –2337.
48. M. Shoga, K. Jobe, M. Glasgow, M. Bustamante, E. Smith and R. Koga, "Single event upset at gigahertz frequencies", *IEEE Trans. on Nucl. Sci.*, **NS-41**, (1994), pp. 2252 –2258.
49. R.A. Reed, M.A.Carts, P.W.Marshall, C.J. Marshall, S. Buchner, M. La Macchia, B. Mathes, D. McMorrow, "Single event upset cross sections at various data rates", *IEEE Trans. on Nucl. Sci.* **NS-43**, (1996), pp. 2862-2867.
50. T.R.Weatherford, J. Whitaker, K. Jobe, S. Meyer, M. Bustamante, S.Thomas III, and K. Elliott, "Single event induced voltage transients within InP HBT circuits", *Journal of Radiation Effects*, September 2001.

51. T.R.Weatherford and P.K. Schiefelbein, "SEE Analysis of digital InP-Based HBT circuits at gigahertz frequencies", *IEEE Trans. on Nucl. Sci.*, **NS-48**, (2001), pp. 1980-1986.
52. R.C. Baumann, T. Hossain, S. Murata, H. Kitagawa, "Boron compounds as a dominant source of alpha particles in semiconductor devices", *Proceedings of the 33rd Annual IEEE International Reliability Physics Symposium,* (1995), pp. 297-302.
53. R.C. Baumann, "Soft errors in advanced semiconductor devices-part I: the three radiation sources", *IEEE Transactions on Device and Materials Reliability*, **1-1**, (2001) pp. 17-22.
54. P. Parikh, P. Chavarkar, U. Mishra, " First demonstration of GaAs on insulator (GOI) technology", *54th Annual Device Research Conference Digest,* (1996), pp. 134 –135.
55. P.Specht, M.J.Cich, H. Sohn and E.R. Weber, "P-Conductive LT-GaAs: The alternative for device applications?" *2nd Symposium On Non-Stoichiometric III-V Compounds, Physik Mikrostrukturierter Halbleiter 23*, Friedrich-Alexander-Universitat 2000, pg. 7-12.

International Journal of High Speed Electronics and Systems
Vol. 13, No. 1 (2003) 293–326

RADIATION EFFECTS IN III-V SEMICONDUCTOR ELECTRONICS

B.D. WEAVER and DALE McMORROW
Naval Research Laboratory, 4555 Overlook Avenue, SW, Washington, DC 20375

L.M. COHN
DTRA, Alexandria, VA, 22310

Particle irradiation effects in III-V semiconductor devices and selected circuits are reviewed. Radiation effects concerns in III-V devices are associated primarily with displacement damage and single-event upset. In conventional transistors, displacement damage decreases the gain, increases leakage and shifts the collector-emitter offset voltage. In reduced dimensional devices, such as high electron mobility transistors and resonant tunneling diodes, the main displacement damage effect is to reduce current by increasing scattering out of the two-dimensional transport state. The current understanding of single-event effects in III-V circuits and devices, and approaches for mitigating their impact, are also discussed here. Single-event effects are a serious concern for high-speed III-V semiconductor devices operating in radiation-intense environments. GaAs integrated circuits (ICs) based on field effect transistor technology exhibit single-event upset sensitivity to protons and very low linear energy transfer (LET) particles; this sensitivity becomes more significant as clock rates and operating speeds increase.

Keywords: III-V semiconductors; semiconductor electronics; radiation effects.

1. Introduction

Devices fabricated from III-V semiconductor compounds offer significant advantages over silicon-based technology in terms of speed, range of power, and versatility of application. III-V semiconductors can be used to make conventional-style diodes, transistors, solar cells and lasers, as well as emitters and detectors of light from the ultraviolet (UV) to the infrared (IR). III-V semiconductors are used to develop devices such as heterojunction bipolar transistors (HBTs), high electron mobility transistors (HEMTs), tunnel diodes, and multi-quantum-well (MQW) and superlattice devices. The advantages of low power and high speed make many III-V devices ideal for application in satellite environments where ambient radiation is high. In order to be useful in space, devices must be able to withstand the harmful effects of radiation.

Currently, few textbooks or review articles on semiconductor devices discuss radiation effects; and few texts on radiation effects focus on III-V devices.[1-3] Those that do go too quickly out of date due to the rapid rate of development of III-V technology. In this article, an attempt is made to bring the subject matter up to date.

The organization of this article is as follows. In Section 2, progress in III-V materials and devices and integrated circuits (ICs) is reviewed and updated since Simon's review in 1983.[3] Various radiation effects are discussed in Section 3 where terms and units are defined, and the variety of damage that can occur is outlined. In Section 4, results on radiation effects in specific III-V technologies are reviewed. Finally, results are summarized and a comparison is made between radiation effects in III-V devices and in other materials systems.

2. III-V Materials and Devices

Twenty years ago, GaAs field effect transistors (FETs) in the form of metal-semiconductor FETs (MESFETs) and junction FETs (JFETs) grown on semi-insulating GaAs substrates comprised the majority of III-V devices. Since then, the number and variety of III-V devices has exploded. Material systems now being exploited include AlAs, AlSb, $Al_xGa_{1-x}N$, $Al_xGa_{1-x}As$, GaN, GaAs, GaSb, InAs, $In_xAl_{1-x}As$, $In_xGa_{1-x}As$, $In_xGa_{1-x}P$, and InP.[4-12]

Two main reasons exist for the rapid increase in III-V semiconductor applications. First, there have been continued advances in fabrication and processing technologies such as molecular beam epitaxy and metal-oxide chemical vapor deposition, and wet- and dry-etch techniques. Second, a wide variety of band gaps and bandstructures can now be engineered. As a result, applications have been extended to higher and lower power levels. Carriers can be forced into undoped intrinsic regions where, lacking scattering centers, they achieve high mobilities and speeds. Using MQW and superlattice structures, meV energy levels and mini-gaps can be made, allowing devices to detect, emit and absorb IR and UV photons. Due to their versatility III-V devices can be found in conventional electronics, optoelectronics, nanoelectronics, photovoltaics, photonics, high power, and now, even in complementary metal-oxide-semiconductor (CMOS)-like applications.[13]

Two aspects of device function are strongly related to radiation tolerance, especially in regard to displacement damage effects. These are (1) minority versus majority carrier status; and (2) conventional versus reduced-dimensional or quantum electronic (RD/QE) carrier transport. In minority carrier devices, radiation damage affects performance mainly by creating traps that act as recombination centers. These centers alter the minority carrier lifetime and carrier diffusion length. In transistors the result is predominantly a decrease in gain, collector current and transconductance. In majority carrier devices, radiation damage affects performance mainly by altering the mean free path and the carrier mobility. Generally, minority carrier devices are significantly more susceptible to radiation damage than are majority carrier devices because the minority carrier concentration depends strongly on the energy levels and concentrations of intrinsic and radiation-induced donors and acceptors.[14] For example, the collector current in minority carrier HBTs is about ten times more sensitive to radiation-induced displacements than is the drain current in majority carrier HEMTs.[4-7]

In a conventional electronic device, carrier transport occurs by the diffusion of thermally-equilibrated carriers from source to drain. However, when device size falls below the carrier mean free path, or when dimensionality is constrained or quantum mechanical transport mechanisms are employed, the physics of device operation is altered. In small InGaAs/InAlAs HEMTs, for example, electrons traverse a two-dimensional electron gas region (2DEG) via ballistic transport (i.e., without collisions). As is discussed below, the novel function of some modern III-V devices gives rise to radiation effects that were unknown as recently as a few years ago.

3. Radiation Effects

Radiation environments vary significantly in their range of particle types and energies. In nuclear environments, for instance, the primary damaging particles are fission neutrons.

In the Earth's Van Allen belts, damage arises mainly from electrons and protons having energies up to around 100 MeV. In other space environments, damage is caused mainly by cosmic rays and solar protons. The primary quantities of radiation exposure are dose and dose rate,[15] both of which can be calculated if the radiation environment, the device characteristics and the quality of shielding are specified.

Like radiation environments, radiation effects take numerous forms. By some conventions, there are four categories of effects: neutron, total dose, dose rate, and single events. We adopt here an alternative but equally sufficient convention of using only three categories: (1) ionizing radiation damage effects that result in the build-up of static charge, (2) non-ionizing radiation damage effects (also called displacement damage effects), and (3) single-event effects. Single-event effects are also caused by ionizing radiation but, because they arise from interactions with single energetic particles rather than the accumulated ionizing dose, a separate classification is warranted.

Non-ionizing radiation causes atoms to be displaced from their original positions. For the radiation environments mentioned above, displacement damage creates primarily small defects such as vacancies, interstitials, and Frenkel pairs – although a small percentage of defect clusters are also formed. The most basic parameter of displacement damage is the non-ionizing energy loss (*NIEL*), which describes the rate at which an incident particle loses energy to displacive events.[16-18] The most common units of *NIEL* are MeVcm^2/g. The displacement damage dose is given by *NIEL* times the particle fluence, Φ. The dose rate is given by *NIEL* times the particle flux, ϕ. Generally, dose rates do not affect displacement damage effects in III-V devices unless the dose rate is so high that the crystal temperature is driven above the annealing temperature. For this reason, dose rate effects are not included in the discussion on displacement damage below. One parameter of interest related to displacement damage dose is the induced defect concentration c. For relatively small particle fluences[19]

$$c = \Phi BA(NIEL), \tag{1}$$

where B depends on the target material density and the displacement energy threshold of target atoms, and A represents the fraction of atoms that do not recombine immediately following irradiation. Units for c are typically “displacements per target-material atom”, or *dpa*.

The effects of ionizing radiation vary with material and device structure. The primary effect of ionizing radiation is the generation of electron-hole pairs. In insulating layers, under the influence of an electric field, these carriers can move to surfaces or interfaces causing the buildup of space charge. Space-charge effects can alter local electric fields and change threshold voltages, effective gate voltages, and leakage currents. In silicon-based devices, the oxides of metal-oxide-semiconductor (MOS) technology are particularly susceptible to ionization effects.[3]

The rate of energy deposition is usually expressed in terms of the mass stopping power for a material, or linear energy transfer (*LET*), which depends on the ion, its energy, and the target material. *LET* is obtained by dividing the energy loss per unit length, *dE/dx*, by the material density, ρ:

$$LET = \frac{dE}{dx}\frac{1}{\rho}. \tag{2}$$

Units of *LET* are MeV·cm^2/mg of material, and the units of *dE/dx* are MeV/cm. The determination of energy loss and range in materials is typically performed by computer codes such as SRIM2000.[20,21] Sometimes, however, the density factor is absorbed into *dE/dx*, such that the units of *dE/dx* are equilavent to those of *LET* noted above. Doses of ionizing radiation are typically expressed in *Rad*(material), where one *Rad* is equivalent to 100 ergs/g. A dose *d* as expressed in *Rads* can be converted to equivalent fluence by

$$d(Rads) = 1.6 \times 10^{-5} \, LET \cdot \Phi, \tag{3}$$

where Φ is in particles/cm^2.

Although MOS devices do currently exist for GaAs-based devices, no radiation damage experiments have yet been performed on them. One advantage that III-V technology has over Si-based technology is that few, if any, III-V devices employ insulating layers as susceptible to charge build-up as is SiO_2. Thus, ionization effects in III-V devices are expected to occur mainly at interfaces. Since no studies on radiation effects in GaAs CMOS have been published, ionization effects are excluded from further discussion here.

Single-event effects (SEEs) refer to phenomena that arise from the interaction of a single energetic particle with a microelectronic device. SEEs can have a significant effect on the performance of III-V technologies, ranging from loss of information to total device failure.[22] SEEs can be nondestructive or destructive; transient or permanent. Single-event upset (SEU) is the most common and the most well characterized nondestructive effect. SEUs are caused by ionization tracks through or near junctions of a semiconductor device, in which the logic state of the circuit is changed. Such effects are also called soft errors because they can be corrected and cause no permanent damage. The impact of an SEE is highly system-dependent, and depends on its type, location, timing, and the specific nature of the application. As discussed below, single-event upset is a significant concern for applications of III-V semiconductors in the space radiation environment, particularly as clock rates increase and critical charges decrease.

4. Radiation Effects in III-V Electronic Devices

For the sake of convenience, the following discussion is divided into two main categories: displacement damage effects and single-event effects. The section on displacement damage is further divided into three subsections: conventional electronic devices, RD/QE devices, and circuits. The single-event effects discussion is broken down into five sections: SEE in GaAs devices and circuits; SEEs in other III-Vs, SEE in dynamic circuits, mitigation approaches, and on-orbit SEE data. Radiation effects in III-V photonic and optoelectronic devices, though of significant technological importance, are not included here because the focus is on purely electronic applications. (We do touch, however, on multi-quantum-well and superlattice IR detectors in the section on RD/QE.) It should also be noted that numerous III-V electronic devices are not discussed here because no relevant radiation effects data exist.

4.1. Displacement damage effects

4.1.1. Conventional electronic devices

Field effect transistors. Conventional GaAs microelectronics such as JFETS, MESFETs and modulation-doped FETs introduced in the 1960s display significant advantages over silicon-based semiconductor devices for both linear and digital applications. GaAs-based analog devices dominate microwave applications above 3 GHz and, for digital applications, provide improvements in speed and power of 2.5 to 8 times over equivalent n-channel metal-oxide-semiconductor FET (MOSFET) devices.[23] However, the low level of integration realizable with GaAs has kept digital circuits from achieving wider utilization.[24]

Radiation-induced reductions in free carrier concentration, minority carrier lifetime and mobility affect the four most significant operating parameters of a JFET or MESFET:[25] pinch-off voltage, maximum channel current, maximum transconductance, and cutoff frequency. In general, the maximum channel current is the most sensitive to displacement damage. The sensitivity of various parameters as a function of neutron flux is shown in Fig. 1,[26] where experimental and simulation results are compared for MESFET and JFET technology. Significant changes in the parameters do not occur until the 10^{14}-10^{15} neutrons/cm^2 range. This level of radiation tolerance is among the best for minority carrier FET devices. Based on these results it can be inferred that GaAs MESFET or JFET integrated circuits should be able to operate at levels up to 10^{15} neutrons/cm^2 with little performance degradation. Additionally, optimized devices with higher channel doping on the order of 10^{18} cm^{-3} might be capable of supporting the fabrication of circuits that can operate at levels up to 10^{15} to 10^{16} neutrons/cm^2.[27]

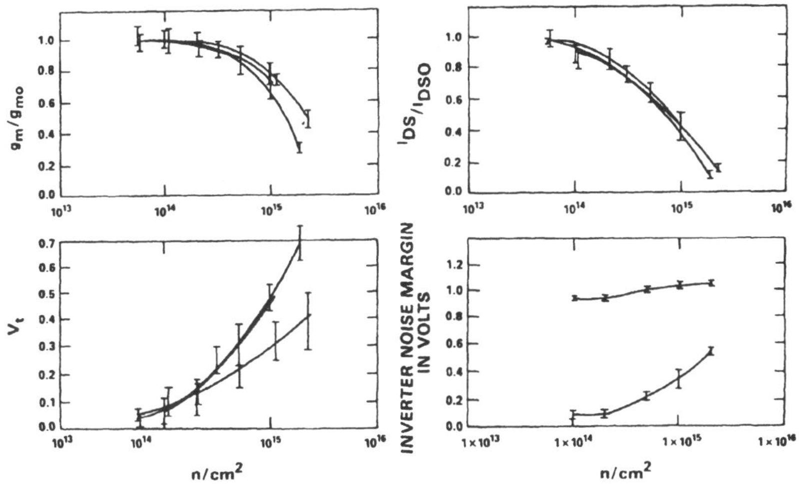

Fig. 1. Neutron effects on pilot line devices. The normalized transconductance g_m/g_{mo} and the normalized drain (collector) current I_{DS}/I_{DSO} decrease with increasing damage while the threshold voltage V_t and noise increase. Solid lines are the result of simulations (after ref. 23).

Subsequent investigations[28] comparing the radiation performance of Si-JFET, GaAs MESFET and MOSFET devices at low temperatures show no significant variations in performance in either noise or dc parameters for a total fluence of 10^{14} neutrons/cm^2 at 90 K and 120 K.

Hetrojunction bipolar transistors. HBTs offer several advantages over Si-based transistors and GaAs FETs. The use of a lightly doped wide band-gap material for the source (emitter) and drain (collector) combined with a thin but highly doped gate (base) and a vertical device structure contribute to decrease power consumption, increase speed, reduce noise and amplify the gain-bandwidth product. Numerous studies on radiation effects in HBTs have been published.[4,5,29-31] A number of simple HBT-based ICs have also been tested, including divide-by-two circuits.[5] Neutron irradiation effects on the common-emitter characteristics of an InP/InGaAs HBT are shown in Fig. 2.

Despite differences in materials systems (GaAs/AlGaAs, InP/InGaAs and SiGe) and device configuration, irradiated HBTs show some general trends in response to irradiation. The gain decreases and the collector-emitter offset voltage offset $V_{ce,off}$ increases. For example in InP/InGaAs HBTs, a dose of 6×10^{14} neutrons/cm^2 nearly doubles $V_{ce,off}$ from 0.19 V to 0.35 V, and in GaAs/AlGaAs HBTs, a dose of 1.3×10^{14} neutrons/cm^2 decreases the collector current by nine percent.[5,31] The gain degradation occurs due to an increased gate current, which in turn is thought to be due to enhanced recombination in the source-gate depletion region. The cause of the offset voltage increase is less certain. One possible explanation is that the increase is due to radiation-induced defects near the gate-collector junction. Another is that the increase is due to defects in the collector itself. Since small concentrations of defects generally have a bigger effect in undoped and lightly doped materials than in highly doped materials, an argument could be made that the latter interpretation is more likely to be correct. In either case, the fact that radiation-induced changes in $V_{ce,off}$ correlate with *NIEL* demonstrates that the shift is a displacement-damage rather than an ionization-damage effect.[31]

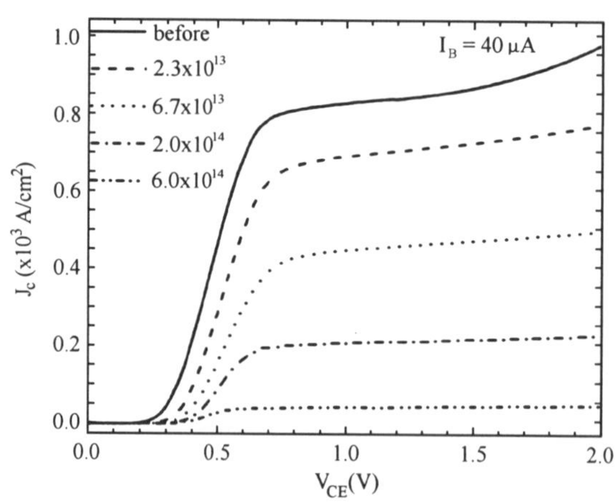

Fig. 2. Common-emitter (I_C-V_{CE}) characteristics of an InP/InGaAs HBT irradiated with neutrons (after ref. 4).

In comparison to other conventional electronic devices, HBTs are near the middle in their radiation-hardness – falling somewhere between highly sensitive Si-based photovoltaic and insensitive majority carrier devices. At least one study suggests that

GaAs/AlGaAs HBTs are significantly more rad-tolerant than Si-based HBTs.[30] It is encouraging that when III-V HBTs are fashioned into simple circuits, the circuits exhibit no additional sensitivity to radiation damage beyond that inherent in the HBTs themselves.[5] Such is not always the case for silicon-based circuits.

4.1.2. Reduced dimensional and quantum electronics

Resonant tunneling diodes. Schematically, RTDs consist of a source, an energy barrier, a quantum well, a second barrier and a drain. Discrete energy levels inside the well allow majority carriers having specific energy and momentum to undergo resonant tunneling across the device, while the barriers restrict other carriers from crossing. In k-space, carriers incident on the first barrier can be characterized as comprising a Fermi sphere. Only those carriers lying on a narrow disk within that sphere can undergo resonant tunneling.[9] In RTDs, resonant tunneling results in n-shaped current-voltage (I-V) curves, as can be seen in the inset of Fig. 3. The parameters of interest are the peak current I_p, the valley current I_v and the peak-to-valley current ratio, I_p/I_v.

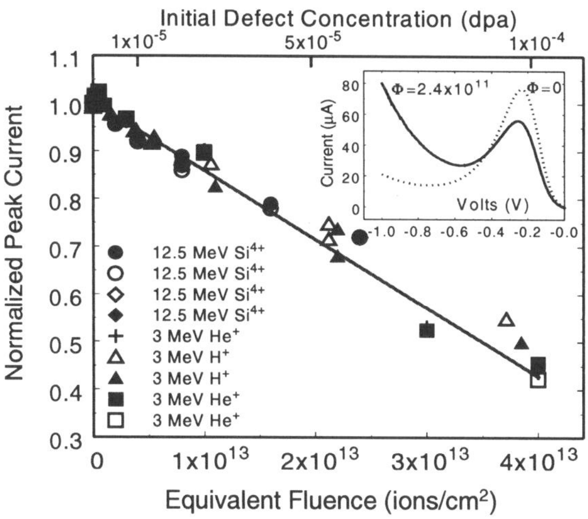

Fig. 3. Normalized I_p vs. equivalent fluence of 3 MeV He^+ for RTDs. I_p decreases linearly with Φ for all arrays. Inset: IV curves before and after irradiation with 12.5-MeV Si^{4+} ions (after ref. 34).

To date, only AlAs/InGaAs/InAs/InGaAs/AlAs RTDs have been tested for radiation effects.[32,33] For testing purposes, devices were wired into arrays containing 100 or 1000 RTDs in parallel, then irradiated with either 3-MeV He^+ ions, 3-MeV H^+, or 12.5-MeV Si^{4+}. I-V curves for a typical RTD are shown in the inset of Fig. 3 for an irradiated and unirradiated device. The peak in the curve is due to resonant tunneling at about –0.2 V. In general, particle irradiation causes I_p and I_p/I_v to decrease while I_v increases due to increased nonresonance (i.e., leakage) current.[32]

In the main body of Fig. 3, the normalized peak current $I_p(\Phi)/I_p(0)$ is plotted vs. equivalent 3-Mev He^+ ion fluence for all arrays tested, and for the different ions used. (To convert 3-MeV H^+ and 12.5-MeV Si^{4+} fluences to equivalent He^+ fluences, scaling factors of 0.052 and 87 were used to compensate for differences in nonionizing energy loss.[34]) The striking result is that I_p decreases linearly with Φ, and thus with defect concentration, as per Eq. (1). This decrease occurs regardless of the cross-sectional area of the device, the incident particle mass or energy, or the number of diodes in an array. With *c* expressed in *dpa*,

$$\frac{I_p(\Phi)}{I_p(0)} \approx 1 - 550c. \tag{4}$$

When defects are introduced into an RTD by irradiation, scattering increases throughout the device. But because RTDs are majority carrier devices, scattering that occurs outside the quantum wells alters the transport current only slightly, by decreasing mobility. Inside a well, however, where the electron density of states is 2-D, scattering from defects easily removes carriers from the transport current by ejecting them from the k-space resonance disk. It has been shown that for narrow resonances and reasonably small values of c in irradiated RTDs, a linear relationship between I_p and the defect concentration follows from an expression for the scattering probability.[34] If instead of a resonance *disk* there were a resonance *sphere*, then the requirements on carrier momentum would not be dimensionally constrained, and scattering from defects would not affect the peak current so strongly. In other words, the main response of RTDs arises from the fact that resonant transport is 2-D. The only previous observation of dimensionally-determined radiation-tolerance was with high temperature cuprate superconductors.[35]

Resonant interband tunneling diodes. RITDs are similar to RTDs in that they contain two barriers and a quantum well across which majority carriers tunnel. In RITDs, however, tunneling requires carrier transitions between conduction and valence bands.[10] In order for this to occur, carriers must interact with phonons or other lattice excitations. The interactions then broaden the resonance disk in *k*-space, extending it from two into three dimensions. Results have been published on irradiated RITDs containing a pair of

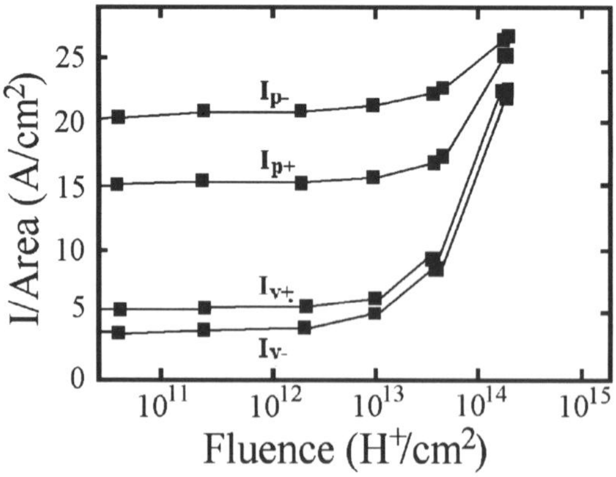

Fig. 4. Fluence dependence for positive and negative bias peak and valley currents for a 13 monolayer AlSb barrier RITD (after ref. 10).

nominally identical AlSb barriers sandwiching a GaSb well.[10] Results are shown in Fig. 4 for both forward- and reverse-bias devices, and for the peak and valley currents. As with RTDs, irradiation causes the off-resonance valley current to increase by increasing leakage channels across the RITD. Surprisingly, irradiation also causes the peak current to increase. This increase is also attributed to increased leakage current through nonresonance channels, and to a less efficient out-scattering of carriers due to the broader (i.e., non-2-D) resonance in k-space as compared to RTDs.[10] The implication is that the sensitivity of I_p to displacement damage is reduced when dimensional constraints are relaxed.

High electron mobility transistors. HEMTs are known by several other names, including but not limited to two-dimensional electron gas field effect transistors and heterostructure field effect transistors (HFETs). In a schematic way they are structurally similar to MESFETs, but operationally they are quite different. The HEMT band structure is designed so that majority carriers from a doped barrier (or donor) layer are injected into an undoped channel layer where they form a nearly-two-dimensional electron gas in which the mean free path and mobility are enhanced.[6,11] Carriers in MESFETs are also constrained to a narrow layer sometimes called a 2DEG even though for currently available devices it is not because transport occurs by diffusion. For example in GaAs/AlGaAs HEMTs, the higher bandgap doped AlGaAs layer injects carriers into a 2DEG in the upper region of a layer of undoped GaAs, where the carriers then travel ballistically. The current at the drain, I_d, depends in part on the mobilities and lifetimes of carriers in the 2DEG.

To date, radiation effects studies have been done on four different HEMT systems: GaAs/AlGaAs, InGaAs/AlGaAs, InGaAs/InGaP and InGaAs/InAlAs.[6,7,11,36-38] The general effect of displacement damage on HEMTs is to decrease the transconductance, shift threshold voltages, increase leakage currents, and decrease drain currents under normal operating conditions. Here, we focus on radiation-induced changes in I_d.

Values of I_d for different Φ were collected for a variety of HEMTs, incident ions and ion energies, then normalized by dividing out the drain current at zero fluence, I_{do}. The rate of change of I_d/I_{do} with fluence, $\Delta(I_d/I_{do})/\Delta\Phi$, was then calculated and plotted versus nonionizing energy loss. Results are shown in Fig. 5, where it can be seen that the data for GaAs/AlGaAs and InGaAs/AlGaAs HEMTs fall on the line

$$\frac{\Delta\left({}^{I_d}\!/_{I_{do}} \right)}{\Delta\Phi} = -d(NIEL) \tag{5}$$

where $d = 5\text{x}10^{-18}$ g/eV. Variations in $\Delta(I_d/I_{do})/\Delta\Phi$ due to different gate lengths, driving voltages etc. are visible as small deviations of the data from the line. The implication of Eq. (5) is that I_d in irradiated (In)GaAs/AlGaAs HEMTs decreases in direct proportion to the induced defect concentration. The same conclusion can be drawn for InGaAs/InGaP HEMTs, except that $d = 3\text{x}10^{-19}$ g/eV. In other words, InGaAs/InGaP HEMTs are about seventeen times more tolerant of displacement damage than (In)GaAs/AlGaAs HEMTs. Furthermore, InGaAs/InAlAs HEMTs are about thirty times more tolerant of disorder than than (In)GaAs/AlGaAs HEMTs. From the data of Fig. 5, it can be concluded that the slope of the linear decrease of I_d with Φ (and c) is determined by the barrier material rather than in the 2DEG.

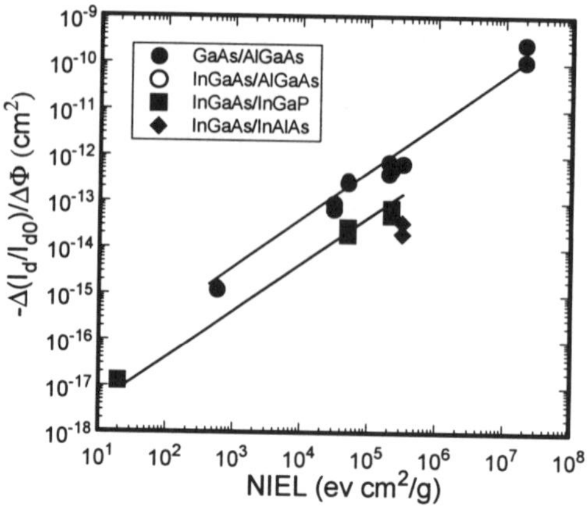

Fig. 5. Radiation-induced change in drain current vs. non-ionizing energy loss in various irradiated HEMTs. Lines are from Eq. (5) in text (after ref. 39).

Displacement damage has a two-part effect on the electronic function of HEMTs. First, scattering from defects in the 2DEG removes carriers from the transport current with high efficiency. Taken alone, out-scattering would cause I_d to decrease linearly in Φ with the same slope for all HEMTs. But unlike in RTDs, electrons scattered out of the 2DEG in HEMTs do not necessarily remain out. Due to a higher conduction band edge in the barrier, it is energetically favorable for scattered carriers to be reinjected. The probability of reinjection depends on the barrier material via the energy difference between barrier and channel. Thus the linear decrease of I_d arises from high-efficiency scattering out of the 2DEG, but the slope is determined by the reinjection rate, which depends on the specific material system.[39]

Multi-quantum-well and superlattice devices. Though currently used only in photonic and optoelectronic applications, MQW and superlattice devices are excellent examples of the state of the art in bandgap engineering and fabrication. A brief discussion on radiation effects in quantum well infrared photodetectors (QWIPs) and antimonide-based superlattice diodes (ASLs) is included here because future electronic devices are expected to be based on structures as intricate on the nano-scale as are these.

The basic components of QWIPs are narrow, low band-gap quantum wells (10 nm of GaAs, say) separated by higher band-gap barriers (AlGaAs). Repeated many times using the proper widths and periods, bound states in the quantum wells are formed which trap electrons in a 2DEG.[40] From the electron gas, carriers can absorb incident photons and undergo transitions into the conduction band, where they are collected. ASLs also contain periodic structures, but no quantum wells or barriers are used. Instead, alternating layers of InAs and GaSb form a two-dimensional superlattice, which develops its own band structure, or "miniband", in which the valence and conduction bands are separated by around a hundred meV (equivalent in energy to an IR photon).[41] QWIPs are majority carrier devices, while ASLs are minority carrier devices.

In irradiated GaAs/AlGaAs QWIPs, a dose of $1\text{x}10^{13}$ 3-MeV He^+/cm^2 reduces the absorbance by about 25%.[42] In InGaAs/InAlAs QWIPs, a dose equivalent to $5.6\text{x}10^{13}$ 3-MeV He^+/cm^2 reduces the absorbance by only 16%.[43] In the ASLs, a dose equivalent to $1.9\text{x}10^{12}$ 3-MeV He^+/cm^2 reduces the dynamic impedance by an order of magnitude.[44]

The reason that GaAs/AlGaAs QWIPs are about eight times more sensitive to displacement damage than InGaAs/InAlAs QWIPS is unknown, but the reason that ASLs are about 32 times more sensitive is most likely due to their minority carrier status.

4.1.3. Circuits

Integrated circuits. The radiation tolerance of various GaAs ICs has been investigated for circuits such as 4K static random access memory (SRAM), ring oscillators, flip-flops, buffers and frequency dividers. These circuits show no serious degradation in electrical performance at total ionizing dose levels up to 10^8 Rad(GaAs).[45,46] Additionally, GaAs MESFET millimeter monolithic integrated circuit amplifiers have demonstrated functionality, but with noticeable degradation in noise figures and RF gain at levels above 10^7 Rad(GaAs).[47-49]

Hybrid nano/microelectronic circuits. Simple circuits containing HEMTs, RTDs and RITDs already exist,[50] as do more advanced hybrid circuits containing RD/QE components (RTDs and HEMTs), conventional electronic components (HBTs) and components more difficult to classify (Schottky, or "interface" devices). Though no irradiation experiments have been performed on hybrid or nanoelectronic circuits, the expectation is that the circuits containing majority carrier devices (RTDs, HEMTs) and HBTs will be more radiation tolerant than current microelectronic III-V circuits.

4.2. Single-Event Effects

General considerations. Single-event effects (SEEs) are associated with localized disturbances in electronic charge caused by the passage of energetic ions through the semiconducting or insulating materials that comprise integrated circuits. A single-event effect is said to occur in a microelectronic device when a single, ionizing particle disrupts the normal operation of that device. As the size and charge storage capacity of microelectronic devices are reduced, their normal operation becomes more vulnerable to the effects of energetic particles, such as those that occur naturally in the radiation environment of space. General discussions of single-event phenomena are given in numerous places.[51,52]

Single-event effects can be either transient disruptions or permanent failures, and can have serious consequences for spacecraft, including loss of information, functional failure or total loss of control. Unlike displacement damage effects, SEEs are not related to the cumulative effects of radiation. Rather, they occur via a stochastic process arising from the random interactions of a device with ions of various energies, species, and angles of incidence. With the passage of an energetic (positively charged) ion through a material, carriers are generated as the ion strips electrons from atoms, leaving behind a track of unbound electrons and holes (a dense electron-hole plasma). This phenomenon is illustrated schematically in Fig. 6 for the passage of an ion through an n-channel HFET (a HEMT). Several effects arise as a result. Single-event breakdown and single-event gate rupture are destructive events caused by charge tracks through gate oxides in silicon power MOSFET transistors, and are not relevant for III-V devices. Single-event upset (SEU), single-event latchup (SEL), single-event snapback (SESB), and single-event transients (SETs) are caused by the passage of ions through or near junctions of an electronic device. Of these, only SEU and SET are important for III-V circuits, and will be the topic of the following discussion.

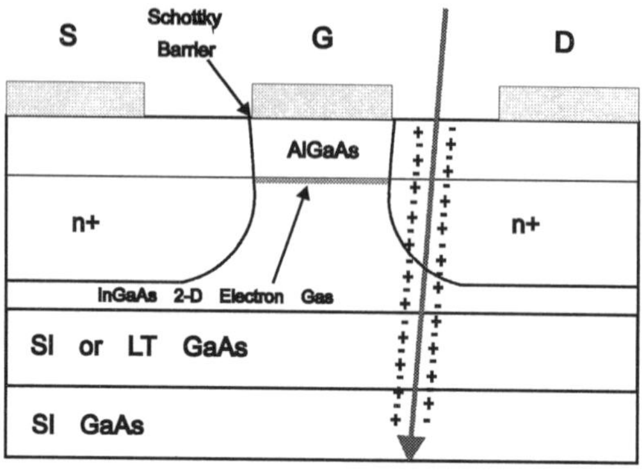

Fig. 6. Schematic diagram of a GaAs n-channel HFET illustrating the passage of a heavy ion with the conventional and low-temperature grown (LT) GaAs configurations indicated.

Since their discovery,[53,54] single-event effects have been investigated and characterized using terrestrial accelerators in an attempt to reproduce the effects of galactic cosmic rays and protons trapped in the Earth's radiation belts. The ionizing energy loss by the ion to the material is characterized by the ion's LET, which is a measure of energy transfer per unit length of material (*cf.*, eq. 2 and accompanying discussion). The metric used to measure the sensitivity of an IC to SEU is expressed in terms of the error cross section, σ, which is the ratio of number of upsets to the total particle fluence, with units of upsets/particles/cm^2. Examples of typical heavy ion cross-section verses LET curves for GaAs MESFET logic are shown in Fig. 7.[55] The two key parameters determined from cross section curves are the threshold LET, LET_{th}, and the saturation cross-section, σ_{sat}. LET_{th} is a measure of the minimum charge required to upset the most sensitive region of the chip; below LET_{th}, no upsets are observed. As the ion LET increases, more regions of the chip become susceptible to upsets, and the cross-section increases until the curve reaches saturation. The saturation cross-section is often interpreted as approximating the total area of the sensitive regions of the chip. This physically appealing interpretation is often complicated by other factors, such as ionization-induced gain and the dependence of the measured cross section on clock rate, for example (*see below*). Different definitions of LET_{th} are found in the literature. In addition to the definition noted above (which can be referred to as the *absolute* threshold), a common definition is the LET for which the cross section is given by $0.1\sigma_{sat}$. This latter definition is of practical utility, since the statistics required for determination of the absolute threshold can lead to prohibitive (expensive) data collection times, and are unwarranted for many applications.

Single-event upsets in semiconductor integrated circuits are caused by energetic protons, neutrons, and heavy ions. Except for the most sensitive devices (photodiodes and charge-coupled devices, for example),[56-58] the stopping power of primary protons or neutrons does not generate sufficient charge to cause an upset. The interaction of protons and neutrons with atoms of the semiconductor lattice can produce nuclear reactions, and the spallation products of such reactions can deposit sufficient charge in the sensitive volumes of devices to cause upsets[59-61]. While the probability of such events is quite low, this is compensated for by the high flux of protons trapped in the Van Allen radiation belts, such that proton events typically dominate the upset rates of sensitive devices that pass through the belts.

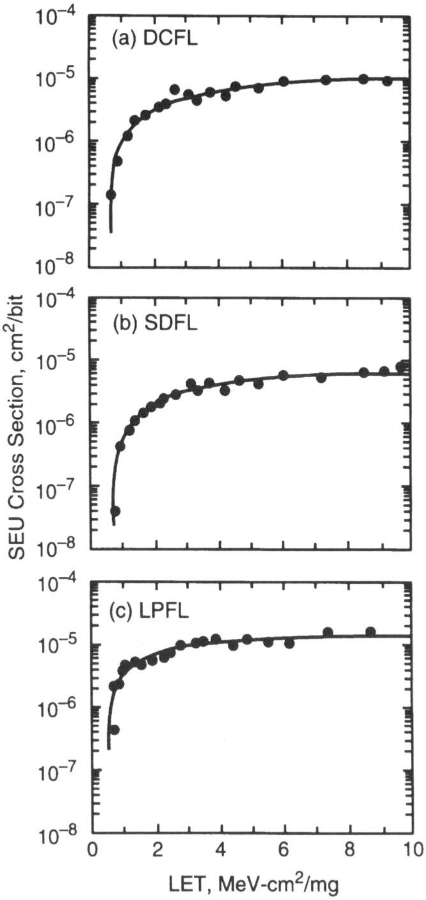

Fig. 7. Typical cross section vs. LET curves for some GaAs E/D MESFET logic families illustrating the low threshold and large saturated cross section: (a) DCFL, direct-coupled FET logic; (b) SDFL, Schottky-diode FET logic; and (c) LPFL, low-power FET logic (after ref. 55).

The saturation cross section and threshold LET determined from terrestrial measurements, such as those of Fig. 7, are used to predict the error rate for a component in a given radiation environment. For such calculations one uses the ion flux (ions/unit area/unit time) as a function of LET for the environment of interest which, for a spacecraft, depends on its orbit. Discussions of the space environment and error rate calculations are beyond the scope of this review and are discussed in detail elsewhere.[62-64]

In what follows we discuss single-event effects in III-V electronic devices and circuits. A significant amount of data has been accumulated on single-event effects in GaAs microelectronic devices since the initial measurement in 1983.[65] That body of work was reviewed several years ago [22], and is updated below. In contrast, very little data on

SEE in other III-V devices is available in the public domain; that available is discussed below.

4.2.1. Single-event effects in GaAs devices and circuits

By the early 1980s, the advantages of GaAs technology in terms of its total ionizing dose and latch-up immunity were well known.[3,66] The sensitivity of GaAs FET-based technologies to SEU, however, rapidly became apparent.[65,67-77] It is now widely recognized that GaAs ICs exhibit SEU sensitivity to protons as well as very low LET particles, making their SEU performance a serious concern for applications in radiation environments.[22] Since those early experiments, significant effort has been expended in an effort to understand and control the mechanisms of error generation and propagation in GaAs devices and circuits.

Early SEU measurements on GaAs parts were performed on complementary enhancement-junction FET (C-EJFET) SRAMs fabricated on semi-insulating substrates.[65,67] Those experiments established a heavy-ion SEU threshold LET of less than 2.65 MeVcm^2/mg, with a predicted geosynchronous SEU error rate on the order of 3x10^{-6} errors/bit-day. This result, which is comparable to those for many unhardened silicon technologies, failed to exhibit the anticipated SEU tolerance. Only a limited data set was obtained in those initial experiments. More recent data on the same technology[76] suggest that the SEU saturated cross section determined in ref. 67 was low by a factor of 40. Application of this correction to the 1983 data gives rise to a predicted error rate of 1.2x10^{-4} errors/bit-day.

SEU testing on GaAs MESFET ICs revealed a higher susceptibility to SEU than C-EJFET devices, with soft error rates on the order of 10^{-3} errors/bit-day predicted from accelerator data.[55,71-75,77] Two factors contribute to this very high error rate. First, unhardened GaAs FETs often exhibit heavy ion LET thresholds below 1 MeVcm^2/mg (LET thresholds for Si ICs are typically above 3 MeVcm^2/mg). Second, the sensitive areas (saturation cross sections) were found to be much larger than expected.[55,73] Normally, SEU upset cross sections measured at LETs well above threshold (e.g., saturated cross sections) for Si MOS circuits approximate the drain area of the sensitive MOSFET. Upset cross sections for GaAs MESFET logic can be an order of magnitude larger than the transistor's active area. The cross section vs. LET curves for three different families of GaAs enhanced-diffusion MESFET logic[55] shown in Fig. 7 illustrate the low LET threshold (<1 MeV cm^2/mg) and large saturated cross section that are typical of GaAs MESFET-based circuitry.

One GaAs logic family, complementary heterostructure field effect transistor technology (C-HIGFET or C-HFET),[78] appears promising for space-based applications. C-HFET technology exhibits SEU cross sections that are one to two orders of magnitude lower than either Si emitter coupled logic or GaAs MESFET alternatives,[79] while maintaining the high speed and total ionizing dose immunity that are characteristic of GaAs. Representative heavy ion upset cross section versus LET curves for C-HFET operating in static and dynamic modes are shown in Fig. 8.[80] The data of Fig. 8 can be compared to those of Fig. 7. In addition to its superior SEU behavior, C-HFET technology exhibits the power characteristics of a complementary cell type, dissipating six times less power than Si CMOS operating at the same speed, making it an attractive choice for operation in satellite environments.[78]

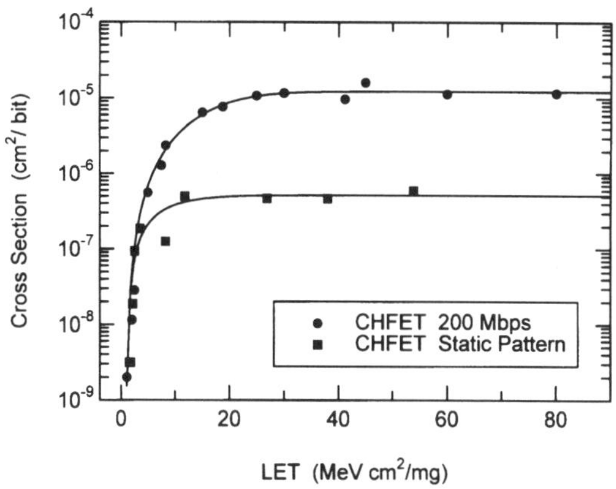

Fig. 8. Heavy ion upset cross section for 32 bit C-HIGFET serial shift register operated in static and dynamic modes (after ref. 80).

Charge-collection processes of GaAs FETs. Charge-collection measurements, when performed as a function of experimentally controlled parameters, provide insight into the physics that occurs when an energetic particle interacts with a semiconductor device. The goal of such investigations is first to understand the charge-collection processes that occur at the device level, and then to use that knowledge to improve the SEU performance of complex ICs through changes in the device and/or circuit design. The evolution in understanding of the charge-collection mechanisms in GaAs FETs leading up to the development of GaAs FETs fabricated with low-temperature-grown buffer layers provides a unique example of how such basic-mechanisms research has resulted in a real solution to the severe SEU problem associated with GaAs technology.

Early charge-collection measurements on GaAs MESFETs revealed that the charge collected at the drain of the transistor could exceed the total charge created by the ion.[81,82] Enhancement factors of two to five are typical, [71,72,81-83] but factors of up to 60 have been reported for α-particle irradiation of GaAs HFETs under certain bias conditions (*cf.*, Fig. 12, *below*).[84] Analogous charge enhancement previously had been observed for electron-beam and optical excitation of MESFETs.[83,85-94] To date, the mechanisms leading to this charge enhancement have been investigated with a variety of techniques, including ion- and laser-induced time-integrated charge-collection studies as a function of various device structural and bias conditions, ion-, laser-, and electron-beam fast transient measurements, and two-dimensional device simulations.[55,71,72,81-87,95-101] Those investigations lead to the significant conclusion that charge-enhancement effects in GaAs FETs are, to a large extent, a consequence of charge deposited in the substrate below the active region of the device.

A typical charge-collection transient measured for picosecond laser excitation of an n-channel HFET is shown in Fig. 9, illustrating the three primary contributions to the charge-collection process. Similar behavior is observed for MESFETs fabricated with buried p-layers or on semi-insulating substrates.[85,86,95,99] The data of Fig. 9 are for above-band-gap pulsed laser excitation; transients measured for α-particle irradiation exhibit similar characteristics.[85,86] The pulsed laser has become a valuable tool for interrogating the charge-collection and SEU mechanisms of III-V semiconductor devices and circuits.

In Fig. 9, the data are partitioned into three different temporal regions, each of which exhibits a distinct time dependence. These three regions behave differently under changes in the various experimental and device parameters, and have been attributed to different charge-collection mechanisms. (The partitions in Fig. 9 are only qualitative and intended primarily to aid the reader in following discussions.) The general characteristics of the transient may be described as follows. At short times there is a narrow, sharp feature labeled (a) that exhibits instrument-limited dynamics. It sometimes appears as a shoulder, or is absent entirely. Device simulations[96,102] suggest that this contribution is very fast, with a width of <10 ps. Next is a broad feature (b) that decays on a timescale of a few hundred picoseconds with a peak that is delayed relative to the peak of (a). At longer times there exists a persistent "background" component (c) that decays on a timescale of one to five nanoseconds.

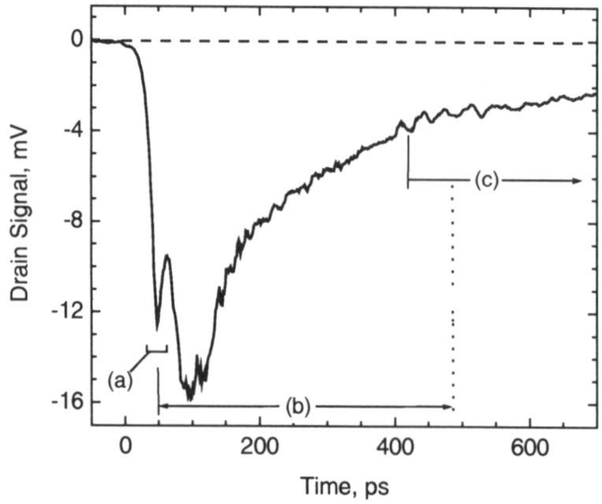

Fig. 9. Typical drain charge-collection transient of an n-channel HFET induced by 60 fC, 610 nm picosecond pulsed laser excitation indicating the different components of the transient (after ref. 86).

Insight into the origin of the different signal contributions illustrated in Fig. 9 has been obtained through systematic studies as a function of various experimental parameters. The data of Fig. 10 illustrate the results of two such studies. In Fig. 10a, drain charge-collection transients are plotted as a function of gate bias for a GaAs n-channel HFET for excitation by 1 ps 610 nm optical pulses focused between the gate and drain electrodes.[86] Similar results are observed for several different types of GaAs FETs,[85,95,97,99] and for alpha particle irradiation[86] These data illustrate the dependence on gate bias of the three components noted above. In particular, when the device is biased strongly "off" (pinch-off for this device is at a gate bias of approximately 0.15 V) contributions from components (b) and (c) are suppressed.

Two dimensional computer simulation studies have been performed to elucidate the mechanisms of charge collection.[96,98,100-103] Analysis of such results indicate that the slow, nanosecond time-scale contribution to the charge-collection transient (contribution (c)) can be identified with a channel-modulation effect in which excess holes present in the substrate beneath the active region of the device serve to backgate the channel. This creates a source-drain current pathway that persists as long as the excess hole density is present.[96,100] When the device is biased beyond pinch-off, as is the case for gate biases below 0.1 V in the data of Fig. 10a, the excess charge density has little effect on the source-drain current. This phenomenon is especially prevalent in buried p-layer MESFET devices because holes tend to accumulate in the potential minimum formed by the p-doped region. The electron and hole density contour plots presented in ref. 96 are particularly illuminating in this regard.

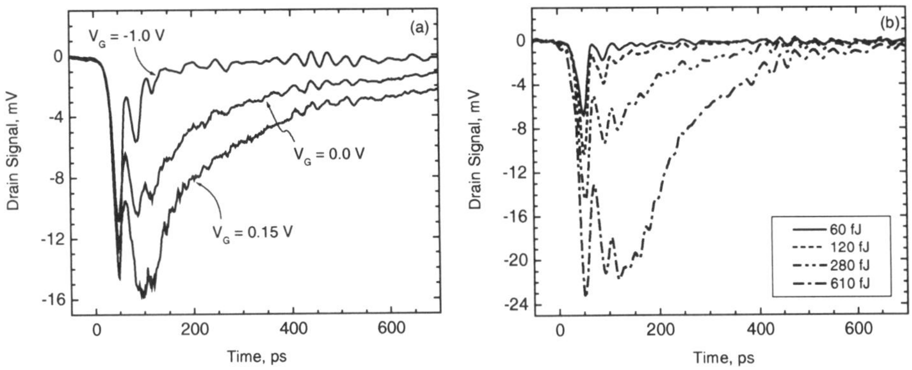

Fig. 10. Drain charge-collection transients of an n-channel HFET excited by above-band-gap laser pulses: (a) gate-bias dependence, 60 fC deposited charge, V_D = 2V; (b) drain charge-collection transients measured as a function of the laser pulse energy illustrating the dependence of component (b) on the excess carrier density; V_D = 2V, V_G = – 0.5V gap (after ref. 86).

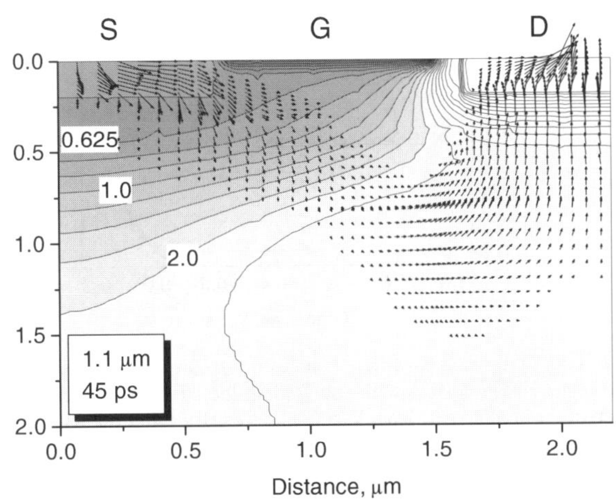

Fig. 11. Constant potential contours for GaAs MESFET indicating the source/substrate/drain current pathway and the evolution of the device potential following perturbation by a deep ion track segment 1.1 μm beneath the surface 45 ps after the ionizing event (after ref. 103). The arrows indicate the direction and magnitude of electron flow.

Variations in the energy of an applied laser are a powerful tool for investigating the mechanisms of charge collection in GaAs FETs and HBTs.[85,86,99,102] Fig. 10b shows charge-collection transients for a n-channel GaAs HFET measured as a function of the injected carrier density.[86] These results provide insight into the physical origin of component (b), which typically exhibits a delayed maximum and decays on a time scale of a few hundred picoseconds. As is evident, component (b) becomes quite prominent at the higher pulse energies, while its contribution is negligible at the lowest pulse energy. This observation reveals that for this particular charge-enhancement process to be efficient, a certain threshold charge density must be achieved in the device. This and related results, when taken together with computer simulation results on MESFETs,[96,102,103] suggest a parasitic bipolar transistor mechanism[82] in which the electron-hole plasma distorts the fields in the device, resulting in a reduction of the source/substrate barrier. Electrons emitted from the source propagate through the substrate and are collected at the drain electrode. A striking illustration of this bipolar-like process in a buried p-layer MESFET device is illustrated by the computer simulation results of Fig. 11. This figure illustrates graphically, with the use of current vectors, the source/substrate/drain current pathway that is characteristic of the parasitic bipolar process in these devices.[103]

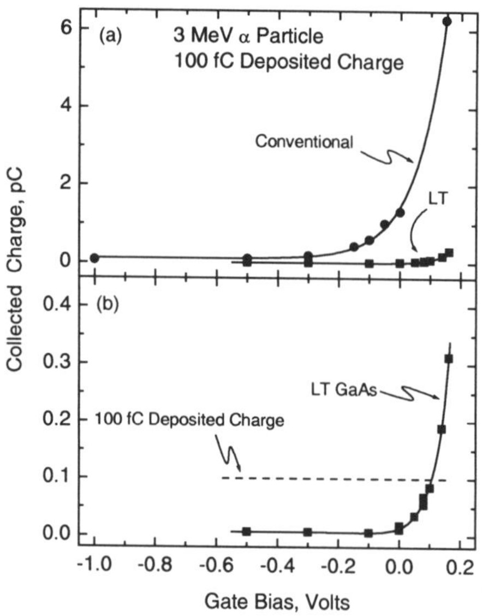

Fig. 12. Time-integrated α-particle induced drain current transients measured for n-channel HFET devices with and without the LT GaAs buffer layer as a function of the applied gate bias; V_D= 2.0 V; V_S = 0 V (after ref. 84).

Charge-enhancement processes play a significant role in the data of Fig. 10. In fact, only component (a) can be identified with direct charge collection in the conventional sense; components (b) and (c), which contribute the bulk of the collected charge, are associated with enhanced source-drain current pathways that are initiated by the ionizing perturbation. These data reveal the sensitivity of the charge-collection process to small changes in gate bias when the FET is biased near its normal operating point of $V_G = 0$. The time-integrated charge-collection data of Fig. 12a (conventional device) illustrate that the collected charge can become quite large when the device is biased near threshold.[84] For this data set, a charge enhancement factor (CEF) of 10 is observed for the normal operating bias of 0 V, with CEF of greater than 60 at the most positive bias tested. This factor of 60 is, we believe, the largest charge enhancement factor reported for a

GaAs FET. Generally smaller CEFs are observed for buried p-layer MESFETs for the same bias range.[104] We note that charge-enhancement factors reported in the literature[71,72,81,82] are typically below five. In many of those studies the bias conditions were not fully specified, and our data suggest that the gate biases were not made sufficiently positive in those studies to observe the very large enhancement factors that are evident in the data of Fig. 12 and ref. 86.

The results discussed in this section lead to a single, significant conclusion: The charge-collection processes of GaAs FETs are dominated by charge-enhancement effects that are a consequence of charge deposited below the active regions of the device. The direct collection of deposited charge, in the conventional sense, contributes only a small fraction to the total collected charge, and may be insignificant in terms of the SEU performance of GaAs field effect transistors. Concepts such as collection depth, collection volume, and sensitive area, which do not consider gain phenomena, are generally inappropriate in discussions of the charge collection and SEU properties of GaAs FETs. This revelation led to a powerful device-level hardening scheme for GaAs FETs, as is discussed below.

4.2.2. SEE in other III-V devices

Experimental results available on the charge collection and SEU characteristics of non-GaAs III-V devices and circuits are limited at the present time. To date, charge-collection measurements have been performed on InGaAs/InAlAs HEMTs and HBTs, AlSb/InAs HEMTs, and on resonant interband tunneling diodes in the AlSb/InAs/GaSb material system. These measurements provide the first insights into the single-event effects performance of these technologies, and provide a basis for future investigations. No SEU measurements are available on these technologies at the present time.

Fig. 13 illustrates the general characteristics of the pulsed laser induced charge-collection transients measured for InGaAs/InAlAs HEMTs lattice-matched to InP substrates.[105] These data reveal the presence of processes that occur on two distinct time scales. The InP devices exhibit significant charge enhancement factors; the temporal signature and behavior under variations in the gate bias are similar to those observed previously for GaAs FETs. Preliminary considerations suggest charge-enhancement mechanisms analogous to those of GaAs FETs. Although these results are preliminary, the very significant charge enhancement measured for these devices suggests that SEU will be a significant concern for digital circuits involving InP-based HEMTs.

Single-event transients induced in InP-based HBT circuits have been measured on-chip using a novel, high-frequency photoconductive sampling probe.[106] These measurements, which are discussed in greater detail below, are unique in that the transients are measured at internal nodes of a circuit operating at 10 GHz.

Highly functional circuits composed of RITDs and HEMTs in the 6.1 Å lattice constant AlSb/InAs/GaSb material system are an attractive approach for obtaining high-speed low-power circuits.[107] RTDs and HEMTs made with InGaAs and InAlAs lattice matched to InP substrates have been combined to develop low-power static random access memory and logic gates operating at 35 GHz.[50,108,109] RITDs and HEMTs composed of InAs, GaSb, and AlSb and their ternary and quaternary alloys offer the potential for circuits with lower power dissipation than obtainable in the InP material system.

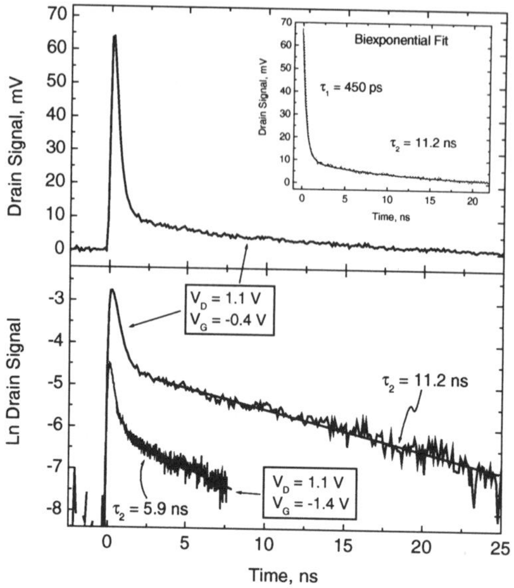

Fig. 13. Typical pulsed laser induced drain charge-collection transients measured for InGaAs/InAlAs HEMT plotted in linear and logarithmic formats for the bias conditions indicated. The inset shows a biexponential fit to the –0.4 V curve: 100 fC deposited charge; 590 nm (after ref. 105).

Gate and drain charge-collection transients have been measured on AlSb/InAs HEMTs for pulsed laser excitation as a function of a range of parameters including gate and drain bias conditions, laser pulse energy, and the laser spot position.[110] The general characteristics of the drain charge-collection transients for these devices are similar to those of the InP HEMTs of Fig. 13. Like the InP data, the InAs HEMT data illustrate the presence of processes that occur on two distinct time scales, with the dependence on gate bias consistent with that observed previously for GaAs FETs (*cf.*, Fig. 10). Although the quantitative interpretation of these laser-based measurements is complicated by the presence of indirect-gap AlSb layers, the clear presence of charge enhancement suggests that SEE mitigation methodologies, analogous to those implemented for GaAs FET-based technologies, may be both appropriate and effective in minimizing the charge collection (*cf.*, Section 4.2.4, *below*).

Fig. 14 shows the first charge-collection measurements performed on a resonant tunneling diode. This figure shows the laser-induced charge-collection transients measured for an AlSb/InAs/GaSb resonant interband tunneling diode (RITD) at two different operating points, one near the peak of the IV curve, and one near the valley.[111] These data reveal the complex and unique character of the charge-collection processes of the RITD. A striking aspect of these data is the distinct difference in the dynamical response at the two operating points. The "peak" transient is of smaller amplitude, and exhibits a distinctly bipolar signature; whereas the "valley" transient is unipolar with a larger amplitude. Both transients, however, decay to the baseline on the same timescale. Conventional charge-collection principles alone are insufficient to explain the charge-collection measurements of the AlSb/InAs/GaSb RITDs.[111] A model that considers transient band-bending effects combined with conventional charge-collection considerations provides a consistent interpretation of the results over a range of bias conditions and laser pulse energy.[111] In terms of SEE vulnerability, these measurements suggest that RITDs may be effectively immune to single-event phenomena when compared to their HEMT and HBT counterparts, but additional measurements are required to verify this conclusion.

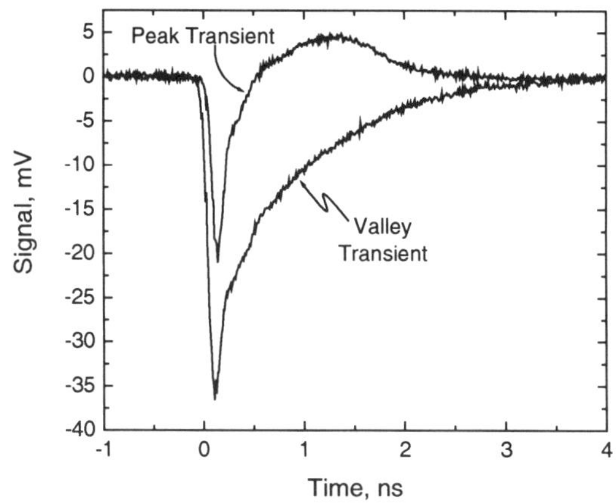

Fig. 14. Pulsed laser-induced charge-collection transients of an AlSb/InAs/GaSb RITD measured near the "peak" and "valley" regions of the I-V curve; 9.1 pC deposited charge; 590 nm (after ref. 111).

4.2.3. SEE in dynamic circuits

A primary motivation for using III-V circuits in space systems is their speed advantage when compared to silicon-based alternatives. The threat to system reliability from single-event effects in dynamic circuits is expected to increase with increases in clock speed. It is important, therefore, to understand the effects of both clock speed and timing effects on the SEU sensitivity of clocked circuits. Dynamic tests on direct-coupled FET logic, source-coupled FET logic, and C-HFET circuits show a dependence of both heavy ion and proton upset cross sections on clock frequency.[80,112-114] Fig. 8 illustrates the dependence of the heavy ion cross section vs. LET curve on the clock rate for a 32 bit GaAs C-HFET serial shift register.[80] The increased cross section at higher clock rates can be interpreted as an increase in the probability of a disruptive ion strike on the circuit. The results of Fig. 8 are typical, and indicate an order-of-magnitude increase in the upset cross section for the dynamic test when compared to the static measurement. These results reveal the importance of testing at the speed at which the circuit will be utilized in space.[80]

Many dynamic SEE effects are independent of the specific technology, and much of the work to date has been performed on silicon parts.[52,115] The important distinction between dynamic SEU effects in combinational and sequential logic is made explicit in ref. 116. Fig. 15 illustrates dependence of the error rate of an inverter circuit on clock frequency and the quantity of deposited charge. These experiments were performed using a pulsed picosecond laser to inject charge into a specific node of the inverter in a well-defined and reproducible fashion. Heavy ion tests show similar results.[114] The data illustrate that, for a given laser pulse energy, the error rate is proportional to the clock frequency, with the proportionality constant (slope) depending on the quantity of charge deposited. Recognizing that the pulse energy deposited in a laser experiment can be related to the LET of a heavy ion,[117] the three curves Fig. 15 can be interpreted as being representative of ions of different LET.

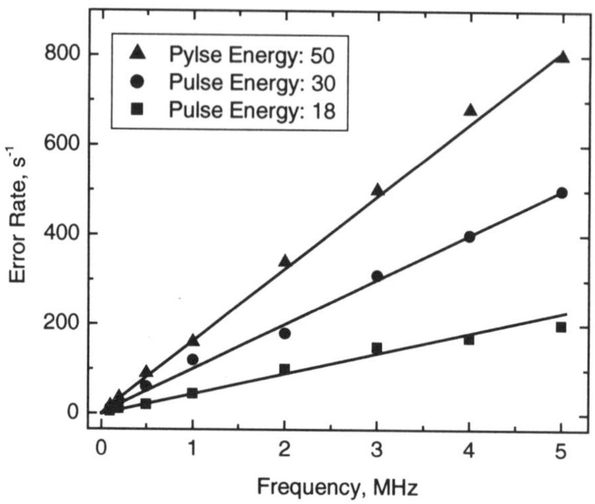

Fig. 15. Pulsed laser induced SEU rates measured as a function of clock frequency for an inverter circuit; the laser pulse energy is given in arbitrary units (after ref. 116).

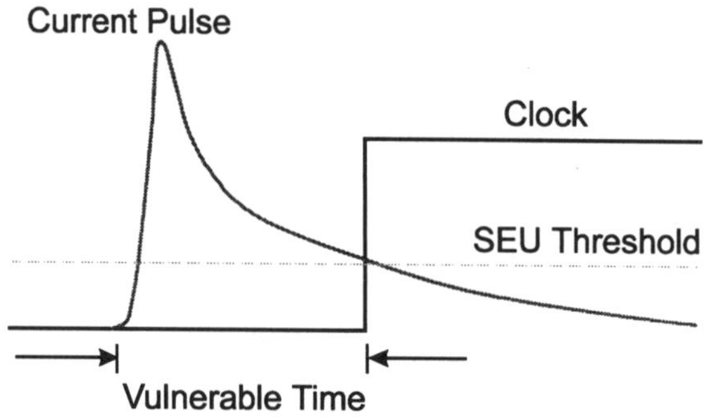

Fig. 16. Schematic diagram illustrating the "window of vulnerability" for clocked circuits (after ref. 113).

The linear dependence of error rate on clock frequency arises because, in combinational logic, transients are only latched into registers at clock edges (in this case, when the clock transitions from low to high). The linear dependence arises simply because the number of clock edges (the probability of a perturbing event) increases linearly with frequency.[113] The dependence of the error rate on deposited charge is best described in terms of the "window of vulnerability", which is illustrated schematically in Fig. 16.[113] An error will be latched into the register if the amplitude of the single-event transient is above some threshold value when the clock transitions from low to high. The window of vulnerability for a given ion LET is defined as the time before a clock edge that the circuit is vulnerable. As may be deduced from Fig. 16, higher LET ions (or higher energy laser pulses) create larger amplitude transients with correspondingly larger windows of vulnerability.

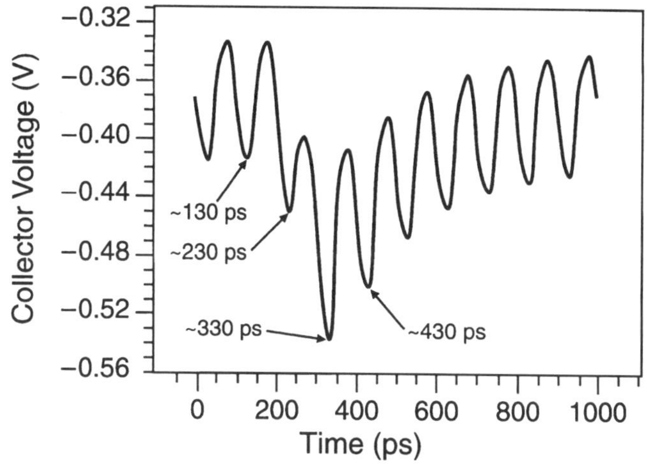

Fig. 17. Device simulation illustrating a laser induced single-event transient measured at an internal node of an InP HBT emitter-coupled logic test circuit operating at 10 GHz (after ref. 118).

An extreme example of this effect that is particularly significant for high-frequency III-V circuits has been demonstrated recently for an InP HBT emitter-coupled logic circuit.[106,118] In this experiment a photoconductive probe is used to map the internal voltage transients induced by a pulsed laser in the HBT logic circuit operating at 10 GHz. Fig. 17 shows the results of a device simulation that illustrates a single-event-induced transient for which the perturbation spans several clock cycles;[117] the corresponding experimental measurements exhibit similar features.[106] In this case, the single-event transient amplitude is such that the window of vulnerability is larger than the inverse clock frequency. Such results illustrate the need for understanding both the charge-collection processes that occur at the device level, and the manner in which single-event-induced transients propagate and are manifested in the output of present-generation high-speed logic circuits. Effects such as those described above are difficult to detect and characterize using the broadbeam ion sources produced by most terrestrial accelerators. The pulsed laser has evolved into a powerful tool for unraveling the complex spatial and temporal aspects of single-event effects in high-speed circuits.[86,101,113,119]

4.2.4. SEE mitigation approaches

Various approaches to SEE mitigation have been applied to GaAs circuits.[68,69,80,84,98,100,112,120-128] Early attempts to harden GaAs systems at the device[123] or circuit[68,69,112,124] level were largely unsuccessful, with the only apparent alternatives involving redundancy at the system or circuit level, and error detection and correction (EDAC) in the system architecture.[122,128] Such methods suffer from reduced performance, increased design complexity, decreased circuit density, and increased power consumption. One circuit-hardening approach, referred to as current shared hardening (CSH), has demonstrated success for GaAs C-HFET circuits.[129] This approach utilizes majority current summing in spatially distributed transistors having three or more nodes and has minimal power overhead.[119] With proper layout, ion strikes to individual nodes will not supply enough current to overrule the other two. One attempt to apply this approach to SiGe HBT logic was unsuccessful, however.[119]

Hardening at the device level through a reduction in the collected charge represents the most fundamental approach to circuit hardening, and suffers from none of the shortcomings noted above. Earlier attempts to reduce charge collection at the device level, utilizing potential barriers between the substrate and channel regions, have met with limited success. Those schemes failed to recognize that the primary source of charge collection in GaAs FETs is associated with enhanced source-drain current, and not simply with the efficient collection of charge deposited by the ion.

Device-level mitigation of SEU in III-V FET-based technologies entails minimization of the charge-enhancement processes that dominate charge collection. One efficient approach for GaAs FETs is based on the incorporation of a "low-temperature-grown" GaAs (LT GaAs) buffer layer beneath the active region of the FET (*cf.*, Fig. 6).[84,98,100,103,120,121,125-127] The unique characteristic of low-temperature grown GaAs relevant to this study is its very short carrier lifetime. The LT GaAs material rapidly lowers the concentration of carriers produced by the ionizing event, effectively minimizing the charge-enhancement processes, resulting in a significant (~100x) reduction in the total collected charge. The initial SEU experiments on GaAs HFET ICs with a low-temperature buffer layer revealed that the circuits do not upset for LETs of up to 90 MeVcm^2/mg.[84,120,121] More recent results have been less spectacular, but still exhibit significant SEU tolerance.[126,127]

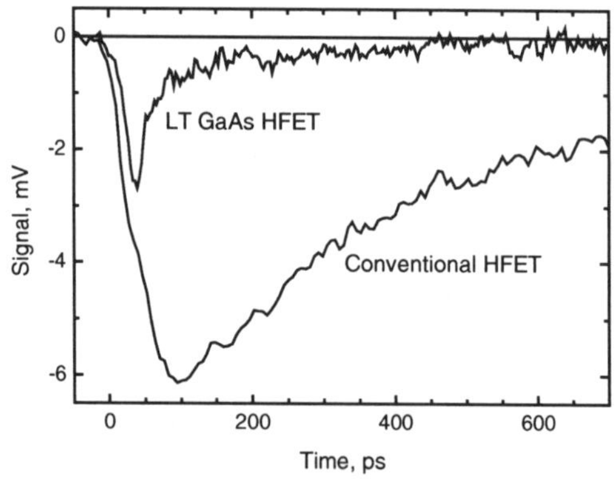

Fig. 18. Charge-collection transients measured for n-channel HFET devices with and without the LT GaAs buffer layer for 3 MeV α-particle irradiation. $V_G = 0$ V, $V_D = 2$ V, and $V_S = 0$ V (after ref. 84).

Fig. 18 illustrates the most fundamental effect of the LT GaAs buffer layer on the SEE response of GaAs devices. This figure shows the charge-collection transients measured for 3 MeV-particle excitation of n-channel HFETs fabricated with and without an LT GaAs buffer layer.[84,120] The structure of the devices investigated is illustrated schematically in Fig. 6. Both devices were biased near pinchoff with $V_G = 0$ V and $V_D = 2$ V. The transient measured for the conventional (non-LT) HFET is consistent with the results of ref. 86, with clear evidence for the charge-enhancement processes discussed above. The LT GaAs device, in contrast, exhibits a rapid recovery from the perturbation, with the charge-collection transient approaching the baseline within 100 ps of the ionizing event. The LT GaAs HFET transient of Fig. 18 exhibits only a prompt response, with little or no evidence for the charge-enhancement mechanisms that dominate the dynamics of the conventional device. More recent experiments and device simulations provide additional insight in to the charge-collection mechanisms of LT GaAs FETs, and demonstrate that the charge-enhancement processes are not entirely eliminated by the LT GaAs buffer.[96,100,103]

Integration of the curves of Fig. 18 reveals that for 3 MeV-particle excitation the LT GaAs buffer layer results in an approximately 100-fold reduction in the collected charge. This 100-fold reduction in the collected charge corresponds to factor of 10^4 reduction in the error rate. Time-integrated charge-collection measurements confirm this result.[84] Fig. 12 compares charge-sensitive preamplifier measurements (time-integrated charge collection) for α-particle irradiation as a function of the gate bias for LT and non-LT GaAs HFET devices. The dashed line corresponds to a charge-collection efficiency of unity (deposited charge = collected charge). As is evident, the charge-collection efficiency of the LT GaAs HFET is reduced significantly when compared to that of the conventional HFET.

The charge-collection data presented above suggest that the SEU behavior of other III-V HEMT based technologies will be similar to that observed for GaAs. For these technologies it is likely that SEU mitigation will be required for critical operations in high-radiation environments. A device-level mitigation methodology with characteristics analogous to LT GaAs is desirable, especially for very high clock rate circuits for which circuit-level hardening schemes may not be inappropriate and EDAC may be ineffective. To date, experimental results incorporating low-temperature grown buffer layers have been reported only for GaAs technologies, but other relevant materials, such as InAlAs and InGaAs lattice matched to InP, exhibit properties similar to those of LT GaAs.[130] Recent charge-collection simulations on an InP HBT circuit, for example, suggest that an LT grown InAlAs buffer layer will be effective for SEU mitigation purposes.[118]

Another related approach may be applicable in some cases.[80] This approach uses particle-induced damage in a post-fabrication step to reduce the carrier lifetime in the material below the active region of the device. The magnitude of the mitigation effect is not as large as for LT GaAs, but is still significant, and may be sufficient for many applications. The post-fabrication aspect of this approach is attractive.

4.2.5. On-orbit SEE results

To date, two space-based SEU experiments involving III-V electronic parts have been flown. The first, the Combined Release and Radiation Effects Satellite (CRRES) contained two GaAs part types: McDonnell-Douglass GaAs C-EJFET SRAMs and Rockwell depletion MESFET/resistor SRAMs.[76] Representative on-orbit data for the GaAs parts flown on CRRES are presented in Fig. 19, together with some data for the 93L422 silicon bipolar SRAM. The 93L422 is known to be particularly susceptible to single-event effects.

The CRRES satellite was flown in a highly elliptical orbit that resulted in two traversals of the proton belts per day, and an extended time at high altitudes near geosynchronous orbit. The data of Fig. 19 are for orbits 638-765, and correspond to the 52 day period following the March 23, 1991 solar flare. The pre-flare data are similar.[76] The error-rates are plotted as a function of the satellite location relative to the Earth's surface (as characterized by the L-shell), and show a clear distinction between the lower and higher altitude radiation environments. The large number of upsets evident for the lower altitudes (L<3) is associated with passage of the satellite through the Earth's proton belts, and reveals the sensitivity of the two GaAs device types to proton upsets. The higher altitudes (L>3) correspond to a heavy ion environment; the enhanced-diffusion MESFET/resistor exhibits a heavy ion error rate between 10^{-4} and 10^{-3} errors/bit-day, consistent with the predictions of ground-based measurements. The C-EJFET exhibited a heavy ion error rate (for L>4) of 4×10^{-5} errors/bit-day, approximately one order of magnitude smaller than predicted from ground-based measurements. Comparison of the error rates for the two different GaAs parts evident in Fig. 19 illustrates the SEU advantage of the C-EJFET technology that is noted above. Unfortunately, due to the short life of the CRRES experiment, the heavy ion SEU statistics for the C-EJFET parts were poor, and the error bars associated with this error rate are necessarily quite large.

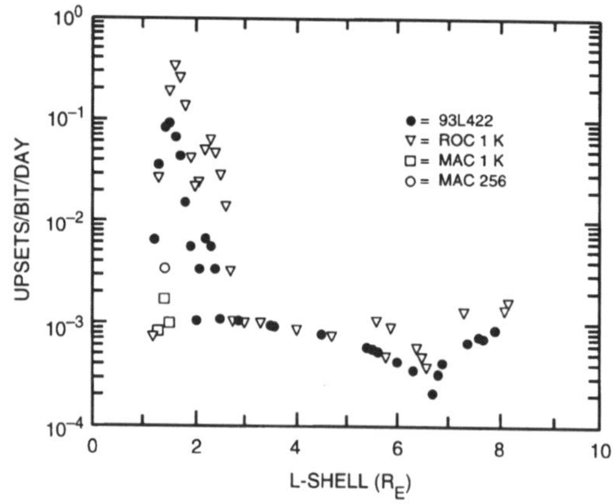

Fig. 19. Soft error rate (number of upsets/bit/day) plotted verses L-shell measured on the CRRES SEU experiment for GaAs C-EJFET (McDonnell-Douglas, MAC 256 and 1K) , GaAs MESFET/resistor (Rockwell , ROC 1K) and a Si 93L422 SRAM (after ref. 76).

The second, and more recent space experiment is the Microelectronics and Photonics Testbed (MPTB).[132,133] Like CRRES, the MPTB orbit is highly elliptical, extending from below the Earth's radiation belts out to geo-stationary. The radiation environment encountered on MPTB is continuously monitored with dosimeters and particle detectors. Two of the sixteen experiments on MPTB contained III-V parts: an InP HBT divide by 16 logic circuit and four GaAs enhancement/depletion MESFET 1K SRAMs in a 4 x 256K configuration. The results of these studies have not yet been fully analyzed, but some preliminary results can be stated here. For a limited data set, the GaAs SRAM exhibited an error rate of approximately 10^{-4} errors/bit-day, in reasonable agreement with the predictions of ground-based analyses for similar technologies.[73,76] The upsets were observed to be fairly equally divided between protons (in the proton belts) and heavy ions. The InP HBT circuit was operational for approximately 1 year with no errors recorded.[134] Because of the low bit count for this device, however, the statistics were quite poor. Details of these experiments will be presented elsewhere.

5. Summary and Conclusions

Displacement damage affects III-V devices in different ways, depending on whether they are minority or majority carrier devices, or are conventional or reduced-dimensional/quantum mechanical in nature. Minority carrier devices are typically more sensitive to displacement damage due to their strong dependence on doping levels and the concentration of donors and acceptors. In reduced-dimensional devices, the main effect of displacement damage is to scatter carriers from the 2-D transport state with high efficiency. In quantum mechanical devices that are not also reduced-dimensional, the main predictor of disorder-tolerance seems to revert back to minority/majority carrier status.

In order to compare the sensitivity to displacement damage of various device types, we have chosen a number of different electronic systems from superconductors to ordinary resistors to InP-based solar cells. For each device, a parameter of merit was chosen (e.g. transition temperature, room temperature resistance, or maximum power). From existing radiation effects data and *NIEL* calculations, we estimated the fluence of 3-MeV He^+ ions that would be required to cause a 5% change in the parameter of merit. Results are shown in Fig. 20. On the far right side of the figure, low-temperature superconductors (LTS) are the hardest materials because they possess a quantum-mechanical immunity to disorder. High temperature superconductors share some of the same immunity but are three orders of magnitude more sensitive due to their reduced dimension. The hardest semiconductor device, a specially-designed SiC diode (f), falls five orders of magnitude below LTS on the rad-hardness scale of Fig. 20. Below the SiC diode come the majority carrier devices: QWIPs, RTDs, RITDs and HEMTs. All of these are III-V devices. Minority carrier devices occupy the remaining three-and-a-half orders of magnitude, with III-V devices interspersed among Si and Ge devices.

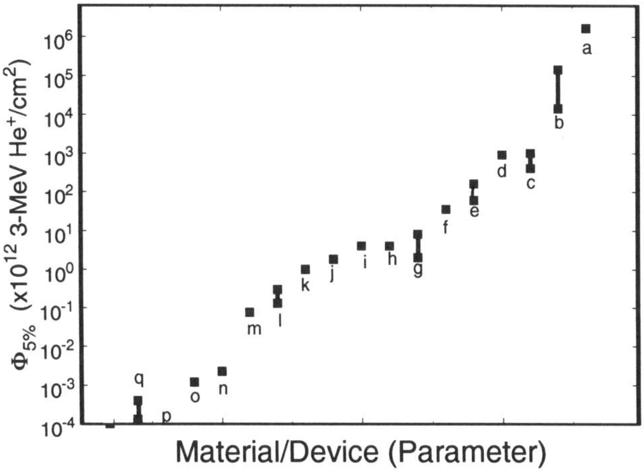

Fig. 20. Fluence of 3-MeV He^+ ions required to change parameters-of-merit by 5%. (a) Transistion temperature T_c of LTS Nb; (b) T_c of LTS A-15 compounds; (c) Resistance R of molded resistors; (d) T_c of HTS $YBa_2Cu_3O_7$; (e) R of ceramic resistors; (f) Forward voltage V_f of hardened SiC diode; (g) Drain current I_d of HEMTs; (h) Peak current I_p of RTD; (i) Valley current I_v of RITDs; (j) Absorbance of GaAs/AlGaAs QWIP; (k) V_f in hardened Si diode; (l) I_d in GaAs/AlGaAs HBT; (m) Threshold voltage in GaAs MESFET and dynamic impedence in ASL detector; (n) Si PNP transistor; (o) Ge PNP transistor; (p) Si OTS diode; (q) Si, Ge, GaAs and InP solar cells (after ref. 44).

Among devices having similar architectures but fabricated with different materials, conventional wisdom holds that InP is more disorder-tolerant than GaAs, which is more tolerant than Si. However, the extent to which an electronic device tolerates disorder depends primarily on the physics of its operation (minority vs. majority carrier; conventional vs. RD/QE) and the actual device type; and only secondarily on the material system. This fact will become increasingly important in the next few years as more and more devices shrink into the region of reduced dimension and quantum mechanical operation.

Single-event effects are a serious concern for high-speed III-V semiconductor technologies operating in radiation-intense environments. GaAs ICs based on field effect transistor technology have demonstrated single-event upset sensitivity to protons and very low LET particles; this sensitivity becomes more significant as clock rates and operational speeds increase. Single-event effects data on other III-V systems is limited, but the general characteristics appear similar to those of GaAs. The motivations for utilizing advanced III-V technologies in satellite systems are well established, but it is likely that SEU mitigation will be required for critical operations in high-radiation environments. Of the different SEU mitigation approaches, device-level hardening using short-recombination-time buffer layers appears to be the most powerful and generally applicable, however, some circuit-level approaches exhibit promise as well.

Acknowledgements

The authors would like to thank Mayrant Simons, Scott Messenger, and Steve Buchner for providing timely and effective advice during the preparation of this manuscript. This work was sponsored in part by the Office of Naval Research and the Defense Threat Reduction Agency.

References

1. Modern Semiconductor Device Physics, S.M. Sze, ed. (Wiley, New York, 1998) p. 230.
2. Handbook of Radiation Effects, A. Holmes-Siedle and L. Adams, eds. (Oxford, England, 1993) pp. 206, 212, 271.
3. M. Simons, "Radiation effects in GaAs integrated circuits: A comparison with silicon", IEEE GaAs Integ. Circ. Symp. Tech. Dig., Oct. 1983, p. 124.
4. A. Shatalov, S. Subramanian and A. Klein, "Correlation between nonionizing energy loss and the offset voltage shift in InP-InGaAs heterjunction bipolar transistors", IEEE Trans. Nucl. Sci., **48**, 2262 (2001).
5. Y. Song, M.E. Kim, A.K. Oki, M.E. Hafizi, W.D. Murlin, J.B. Camou and K.W. Kobayashi, "Effects of neutron irradiation on GaAs/AlGaAs heterojunction bipolar transistors", IEEE Trans. Nucl. Sci. **36**, 2155 (1989).
6. M.J. O'Loughlin, "Radiation effects in high electron mobility transistors: Total dose gamma irradiation", IEEE Trans. Nucl. Sci. **34**, 1808 (1987).
7. H. Ohyama, E. Simoen, S. Kuroda, C. Claeys, Y. Takami, T. Hakuta and H. Sunaga, "Impact of high energy particles on InGaP/InGaAs pseudomorphic HEMTs", IEEE Trans. Nucl. Sci. **45**, 2861 (1998).
8. Q. Zhou, M.O. Manasreh, M. Pophristic, S. Guo and I.T. Ferguson, "Observation of nitrogen vacancy in proton-irradiated $Al_xGa_{1-x}N$", Appl. Phys. Lett. **79**, 2901 (2001).

9. A. Seabaugh, B. Brar, T. Broekaert, F. Morris, P. van der Wagt and G. Frazier, "Resonant-tunneling mixed-signal circuit technology", Sol. State. Electron. **43**, 1355 (1999).
10. R. Magno, B.D. Weaver, A.S. Bracker and B.R. Bennett, "Proton irradiation of InAs/AlSb/GaSb resonant interband tunneling diodes", Appl. Phys. Lett. **78**, 2581 (2001).
11. J.B. Boos, W. Kruppa, B.R. Bennett, D. Park and H.B. Dietrich, "AlSb/InAs HEMTs with high speed and low power consumption", Compound Semicond. **5**, 36 (1999).
12. S.M. Khanna, J. Webb, H. Tang, A.J. Houdayer and C. Carlone, "2 MeV proton radiation damage studies of GaN films through low temperature photoluminescence spectroscopy measurements", IEEE Trans. Nucl. Sci. **47**, 2322 (2000).
13. M. Hong, J.N. Baillargeon, J. Kwo, J.P. Mannaert and A.Y. Cho, "First demonstration of GaAs CMOS", 2000 IEEE International Symposium on Compound Semiconductors, (Monterey, CA, 2-5 Oct. 2000) p. 345.
14. A.S. Grove, Physics and Technology of Semiconductor Devices (Wiley, New York, 1967) pp. 100-105.
15. Effect of Radiation on Materials and Components, J.F. Kircher and R.E. Bowman, eds. (Reinhold, New York, 1964) p.51.
16. J.F. Ziegler, J.P Biersack and U. Littmark, "The Stopping and Range of Ions in Solids", Vol. 1 (Pergamon, 1985). (The latest version of SRIM is available on-line at http://www.research.ibm.com/ionbeams).
17. G.P. Summers, E.A. Burke, P. Shapiro, S.R. Messenger and R.J. Walters, "Damage correlations in semiconductors exposed to gamma, electron and proton radiation", IEEE Trans. Nucl. Sci. **40**, 1372 (1993).
18. S.R. Messenger, R.J. Walters, E.A. Burke, G.P. Summers and M.A. Xapsos, "*NIEL* and damage correlations for high-energy protons in GaAs devices", IEEE Trans. Nucl. Sci. **48**, 2121 (2001).
19. J. Lindhard, V. Nielsen, M. Scharff and P. Thomsen, "Integral Equations Governing Radiation Effects", Mat. Fys. Medd. Dan. Vid. Selsk. **33**, 1 (1963).
20. J.F. Ziegler, SRIM2000; see ref. 16.
21. J.F. Ziegler, "Handbook of Stopping Cross-Sections for Energetic Ions in all Elements", from the series: "The Stopping and Ranges of Ions in Matter", Vol. 5, J.F. Ziegler, ed. (Pergamon Press, NY, 1980).
22. D. McMorrow, T.R. Weatherford, S. Buchner, A.R. Knudson, J.S. Melinger, L.H. Tran and A.B. Campbell, "Single event phenomena in GaAs devices and circuits", IEEE Trans. Nucl. Sci. **43**, 628 (1996).
23. S. Roosild and R. Zuleeg, "Radiation Effects on GaAs Technology", IEEE Nuclear and Space Radiation Effects Conference Short Course, 1988.
24. S. Karp and S. Roosild, "DARPA, SDI, and GaAs", Computer **19**, 17 (1986).
25. W. Shockley, "A unipolar field-effect transistor", IRE Proc. **40**, 1365 (1952).
26. B.K. Janousek, W.E. Yamada, R.J. Krantz and W.L. Bloss, "Neutron radiation effects in GaAs ion-implanted metal-semiconductor field-effect transistors", J. Appl. Phys. **63**, 1678 (1988).
27. B. Buchanan, R. Dolan and S. Roosild, "Comparison of the neutron radiation tolerance of bipolar and junction field effect transistors", IEEE Proc. **55**, 2188 (1967).
28. M. Citterio, S. Rescia and V. Radeka, "Radiation Effects at Cryogenic temperature in Si-JFET, GaAs MESFET, and MOSFET devices", IEEE Trans. Nucl. Sci. **42**, 2266 (1995).
29. S. Zhang, G. Niu, J.D. Cressler, S.J. Mathew, U. Gogineni, S.D. Clark, P. Zampardi and R.L. Pierson, "A comparison of the effects of gamma irradiation on SiGe HBT and GaAs HBT technologies", IEEE Trans. Nucl. Sci. **47**, 2521 (2000)
30. G.A. Schrantz, N.W. van Vonno, W.A. Krull, M.A. Rao, S.I. Long and H. Kroemer, "Neutron irradiation effects on AlGaAs/GaAs heterojunction bipolar transistors", IEEE Trans. Nucl. Sci. **35**, 1657 (1988).
31. A. Shatalov and S. Subramanian, "Neutron irradiation induced degradation of the collector-emitter offset voltage in InP/InGaAs single heterojunction bipolar transistors", J. Appl. Phys. **88**, 3765 (2000).

32. E.M. Jackson, B.D. Weaver and A.C. Seabaugh, "Proton-induced disorder in InP-based resonant tunneling diodes", Appl. Phys. Lett. **75**, 280 (1999).
33. R. Wilkins, S. Shojah-Ardalan, W.P. Kirk, G.F. Spencer, R.T. Bate, A.C. Seabaugh, R. Lake, P. Stelmaszyk, A.D. Wieck and T.N. Fogarty, "Ionization and displacement damage irradiation studies of quantum devices, resonant tunneling diodes and two-dimensional electron gas transistors", IEEE Trans. Nucl. Sci. **46**, 1702 (1999).
34. B.D. Weaver, E.M. Jackson, G.P. Summers and A.C. Seabaugh, "Disorder-effects in reduced dimension: InP-based resonant tunneling diodes", J. Appl. Phys. **88**, 6951 (2000).
35. B.D. Weaver, E.M. Jackson, G.P. Summers and E.A. Burke, "Atomic disorder and the transition temperature in cuprate superconductors", Phys. Rev. B **46**, 1134 (1992).
36. M. Papastamatiou, N. Arpatzanis, G.J. Papaiouannou, C. Papastergiou and A. Christou, "Neutron radiation effects in high electron mobility transistors", IEEE Trans. Electr. Dev. **44**, 364 (1997).
37. H. Ohyama, E. Simoen, S. Kuroda, C. Claeys, Y. Takami, T. Hakata and H. Sunaga, "Radiation induced lattice defects in InGaP/InGaAs p-HEMTs and their effect on device performance", Sol. St. Phenom. **69-70**, 563 (1999).
38. H. Ohyama, E. Simoen, S. Kuroda, C. Claeys, Y. Takami, T. Hakata, K. Kobayashi, M. Nakabayashi and H. Sunaga, "Degradation and recovery of AlGaAs/GaAs p-HEMT irradiated by high-energy particle", Microelectron. Reliab. **41**, 79 (2001).
39. B.D. Weaver and E.M. Jackson, "Universal behavior in irradiated high electron mobility transistors", Appl. Phys. Lett. **80**, 2791 (2002).
40. B.F. Levine, "Quantum-well infrared photodetectors", J. Appl. Phys. 74, R1 (1993)
41. E.H. Aifer, B.R. Bennett, B.Z. Nosho, M.E. Twigg, I. Vurgaftman, J.R. Meyer, J.R. Waterman and M. Goldenberg, "Progress in long-wave InAs/(In)GaSb infrared photodiodes", MSS Specialty Groups on Passive Sensors; Camouflage, Concealment and Deception; Detectors; and Materials, ed. IRIA Center, Vienna VA, filename MTCO2EA.pdf, 5-7 March, 2001.
42. Y. Berhane, M.O. Manasreh and B.D. Weaver, "He^+-ion irradiation effect on intersubband transitions in GaAs/AlGaAs multiple quantum wells", J. Appl. Phys. **89**, 3517 (2001).
43. J. Chen, Q. Zhou, M.O. Manasreh, B.D. Weaver and M. Missous, "Intersubband transitions in InGaAs/InAlAs multiple quantum wells grown on InP substrates", Mat. Res. Soc. Symp. Proc. Vol. 692, p. 402 (2002).
44. B.D. Weaver and E.H. Aifer, "Effects of disorder in reduced dimensional and quantum electronics", B.D. Weaver and E.H. Aifer, Computational Nanoscience and Nanotechnology 2002 (San Juan, PR 21-25 April 2002) p. 310.
45. J. Notthoff, R. Zuleeg and G. Troeger, "Logic upset level of GaAs SRAMs for pulsed ionizing radiation", IEEE Trans. Nucl. Sci. **30**, 4173 (1983).
46. H. Derewonko, A. Bosella, G. Patuat, D. Perie, J.L. Pinsard, C. Sentubery, C. Verbeck, P. Bresy and P. Augier, "Evaluation of GaAs low noise and power MMIC technologies to neutron ionizing dose and dose rate effects", IEEE Trans. Nucl. Sci. **43**, 837 (1996).
47. D. Newell, P. Ho, R. Mencik and J. Pelose, "Total dose hardness of microwave GaAs field effect transistors", IEEE Trans. Nucl. Sci. **28**, 4403 (1981).
48. J. Cazoux, "MMICs for space-borne systems: Status and prospectives", 16th GaAs IC Symposium, p. 3 (1994); Y. Kadowaki, Y. Mitsui, T. Takebe, O. Ishihara and M. Nakatani, 4th GaAs IC Symposium, p. 83 (1982).
49. W.T. Anderson, J. Gerdes and J.A. Roussos, "Temperature dependent GaAs MMIC radiation effects", IEE Trans. Nucl. Sci **40**, 1735 (1993).
50. P. Fay, J. Lu, Y. Xu, G.H. Bernstein, A. Gonzalez, P. Mazumder, D.H. Chow and J.N. Schulman, "Digital integrated circuits using InAlAs/InGaAs/InP HEMTs and InAs/AlSb/GaSb RITDs, Electron. Lett. **37**, 758 (2001).
51. F.W. Sexton, "Measurement of Single Event Phenomena in Devices and ICs", 1992 IEEE Nuclear and Space Radiation Effects Short Course, Chapter III, pp. III1-III55, 1992.
52. S. Buchner, "Single event transients in fast electronic circuits", 2001 IEEE Nuclear and Space Radiation Effects Short Course, Chapter V, pp. V1-V105, 2001.

53. J.T. Wallmark and S.M. Marcus, "Minimum size and maximum packing density of nonredundant semiconductor devices", Proceedings of the IRE, 286 (1962).
54. D. Binder, E.C. Smith, and A.B. Holman, "Satellite anomalies from Galactic cosmic rays", IEEE Trans. Nuc. Sci. 22 (1975) 2675.
55. B.W. Hughlock, G.S. LaRue, and A.H. Johnston, "Single event upset in GaAs E/D MESFET logic", IEEE Trans. Nucl. Sci. **37**, 1894 (1990).
56. P. Marshall, C. Dale, and K. LaBel, "Charged particle effects on optoelectronic devices and bit error rate measurements on 400 Mbps fiber based data links", Proceedings RADECS 93, 266-271, (1994).
57. K.A. LaBel, P. Marshall, C. Dale, C.M. Crabtree, E.G. Stassinopoulos, J.T. Miller, and M.M. Gates", SEDS MIL-STD-1773 fiber optic data bus: proton irradiation test results and spaceflight SEU data", IEEE Trans. Nuc. Sci. **40**, 1638-1644, (1993).
58. P.W. Marshall, C.J. Dale, K.A. Label, "Space radiation effects in high performance fiber optic data links for satellite data management", IEEE Trans. Nuc. Sci. **43**, 2259-2266 (1996).
59. C.S. Guenzer, E.A. Wolicki, and R.G. Allas, "Single event upset of dynamic RAM's by neutrons and protons", IEEE Trans. Nuc. Sci. **26**, 5048 (1979).
60. R.C. Wyatt, P.J. McNulty, P. Toumbas, P.L. Rothwell, and R.C. Fritz, IEEE Trans. Nuc. Sci. **26**, 4905 (1979).
61. P.J. McNulty, G.E. Farrell, R.C. Wyatt, P.L. Rothwell, R.C. Fritz, and N.N. Bradford, "Upset phenomena induced by energetic protons and electrons", IEEE Trans. Nuc. Sci. **27**, 1516-1522 (1980).
62. J. Barth, "Modeling space radiation environments", 1997 IEEE Nuclear and Space Radiation Effects Short Course, Chapter 1, pp. I1-I83, 1997.
63. J.C. Pickel, "Single event effects rate prediction", IEEE Trans. Nuc. Sci., **43**, 483 (1996).
64. E.L. Peterson, J.C. Pickel, J.H. Adams, Jr., and E.C. Smith, "Rate prediction for single event effects–a critique", IEEE Trans. Nuc. Sci. **39**, 1577 (1992).
65. R. Zuleeg and K. Lehovec, "Radiation effects in GaAs junction field-effect transistors", IEEE Trans. Nuc. Sci. **27**, 1343 (1980).
66. P. Shapiro, A.B. Campbell, J.C. Ritter, R. Zuleeg, and J.K.Notthoff, "Single event upset measurements of GaAs E-JFET RAMs", IEEE Trans. Nucl. Sci. **30**, 4610 (1983).
67. R. Zuleeg, J.K. Notthoff, and D.K. Nichols, "SEU of complementary GaAs static RAMS due to heavy ions", IEEE Trans. Nucl. Sci. **31**, 1121 (1984).
68. (a) R. Zuleeg, "GaAs JFET memory technology", DARPA Digital GaAs Technology Review, April 29, 1986; (b) R. Zuleeg and J.K. Notthoff, "SEU hardening of GaAs E-JFET SRAMs", J. Rad. Effects, Research and Engineering, **6-2**, 136 (1988).
69. R. Zuleeg, "Radiation effects in GaAs FET devices", Proc. of the IEEE, **77**, 309 (1989).
70. A.Prabhakar, "Digital GaAs upgrades for military systems", 1989 IEEE GaAs IC Symposium Technical Digest, October 22-25, 1989, p. 15.
71. B.W. Hughlock, T. Williams, A.H Johnston, and R.E. Plaag, "Ion induced charge collection in GaAs MESFETs and its effect on SEU vulnerability", IEEE Trans. Nucl. Sci. **38**, 1442 (1991).
72. B. Hughlock, A. Johnston, T. Williams, and J. Harrang, "A comparison of charge collection Effects Between GaAs MESFETs and III-V HFETs", IEEE Trans. Nucl. Sci. **39**, 1642 (1992).
73. T.R. Weatherford, L. Tran, W.J. Stapor, E.L. Peterson, J.B. Langworthy, D. McMorrow, W.G. Abdel-Kader, and P.J. McNulty, "Proton and heavy ion upsets in GaAs MESFET devices", IEEE Trans. Nucl. Sci. **38**, 1450 (1991).
74. T.R. Weatherford, A.B. Campbell, D. McMorrow, J.B.Langworthy, W.J. Stapor, A.R. Knudson, E.L. Petersen, D. Wilson, and L. Tran, "Can digital GaAs be used in a space environment?: a look at single event upset in GaAs", IEEE GaAs IC Symposium, Technical Digest 1991, Monterey, California, October 20-23, 1991.
75. T.R. Weatherford, Ph.D. Dissertation, "Single event upset in GaAs integrated circuits", North Carolina State University, 1992.
76. T.R. Weatherford, P.T. McDonald, A.B. Campbell, and J.B. Langworthy, "SEU rate prediction of GaAs SRAMs onboard the CRRES satellite", IEEE Trans. Nucl. Sci. **40**,1463 (1993).

77. G.L. Hash, J.R. Schwank, M.R. Shaneyfelt, C.E. Sandoval, M.P. Conners, T.J. Sheridan, F.W. Sexton, E.M. Slayton, J.A. Heise, and C.C. Foster, "Proton irradiation effects on advanced digital and microwave III-V components", IEEE Trans. Nuc. Sci. **41**, 2259-2266, (1994).
78. R.B. Brown, B. Bernhardt, M. LaMacchia, J. Abrokwah, P.N. Parakh, T.D. Basso, S.M. Bold, S. Stetson, C.R. Gauther, D. Foster, B. Crawforth, T. McQuire, K. Sakallah, R.J. Lomax, and R.N. Mudge, "Overview of complementary GaAs technology for high-speed VLSI circuits", IEEE Trans. On VLSI Sys. **6**, 47-51 (1998).
79. J. Cutchin, P. Marshall, T.R. Weatherford, J. Langworthy, E.L. Petersen, A.B. Campbell, S. Hanka, and A. Peczalski, "Heavy ion and proton analysis of a C-HIGFET SRAM", IEEE Trans. Nucl Sci., **NS-40**, 1660 (1992).
80. P.W. Marshall, C.J. Dale, T.R. Weatherford, M.LaMacchia, and K.A. LaBel, "Particle-induced mitigation of SEU sensitivity in high data rate GaAs HIGFET technologies", IEEE Trans. Nucl. Sci. **NS-42**, 1844 (1995).
81. A.B. Campbell, A. Knudson, D. McMorrow, W. Anderson, J. Roussos, S. Espy, S. Buchner, K. Kang, D. Kerns, and S. Kerns, "Ion-induced charge collection in GaAs MESFETs", IEEE Trans. Nucl. Sci. **NS-36**, 2292 (1989).
82. Y. Umemoto, N. Matsunaga, and K. Mitsusada, "A bipolar mechanism for alpha particle-induced soft errors in GaAs integrated circuits", IEEE Trans. Elect. Dev. ED-16, 864 (1989).
83. A.R. Knudson, A.B. Campbell, D. McMorrow, S. Buchner, K.Kang, T. Weatherford, V. Srinivas, G.A. Swartzlander, Jr., and Y.J. Chen, "Pulsed laser-induced charge collection in GaAs MESFETs", IEEE Trans. Nucl. Sci., **NS-37**, 1909 (1992).
84. D. McMorrow, T.R. Weatherford, A.R. Knudson, S. Buchner, J.S. Melinger, L.H. Tran, A.B. Campbell, P.W. Marshall, C.J. Dale, A. Peczalski, and S. Baier, "Charge collection characteristics of GaAs heterostructure FETs fabricated with a low-temperature grown GaAs buffer layer", IEEE Trans. Nuc. Sci., **43**, 918 (1996).
85. D. McMorrow, T. Weatherford, A.R. Knudson, L.H. Tran, J.S. Melinger, and A.B. Campbell, "Single event dynamics of high-performance HBTs and GaAs MESFETs", IEEE Trans. Nucl. Sci. **NS-40**, 1858 (1993).
86. D. McMorrow, J.S. Melinger, N. Thantu, A.B. Campbell, T.R. Weatherford, A.R. Knudson, L.H. Tran, and A. Peczalski, "Charge collection mechanisms of heterostructure FETs", IEEE Trans. Nucl. Sci., **NS-41** (1994).
87. L.D. Flesner, N.M. Davis, and H.H. Wieder, "High-speed response of GaAs metal-semiconductor field effect transistor to electron-beam excitation", J. Appl. Phys. **53**, 3873 (1982); L. Flesner, "Gate charge collection and induced drain current in GaAs FETs", IEEE Trans. Nucl. Sci. **NS-32**, 4110 (1985).
88. W.D. Edwards, "Two and three terminal gallium arsenide FET optical detectors", IEEE Electron Device Lett., EDL-1, 149 (1980).
89. T.F. Carruthers, W.T. Anderson, and J.F. Weller, "Optically induced backgating transients in GaAs FETs", IEEE Electron Device Lett., **EDL-6**, 580 (1985).
90. D.E. Cooper and S.C. Moss, "Picosecond optoelectronic sampling of electrical waveforms produced by an optically excited field effect transistor", in Ultrafast Phenomena V (Springer-Verlag, Berlin, 1986) pp. 117-119.
91. T. Sugeta and Mizushima, "High-speed photoresponse mechanism of a GaAs MESFET", Jap. J. Appl. Phys. **19**, L27 (1980).
92. R.B. Darling and J.P.Uyemura, "Optical gain and large-signal characteristics of illuminated MESFETs", IEEE J. Quant. Elect. **QE-23**, 1160 (1987).
93. J.C. Gammel and J.M. Ballantyne, "The OPFET: a new high speed optical detector", IEDM Technical Digest, 120 (1978).
94. W.T. Anderson, Jr., M. Simons, E.E. King, H.B. Dietrich, and R.J. Lambert, "Reduction of long-term transient radiation response in ion implanted GaAs FETs", IEEE Trans. Nuc. Sci. **NS-29**, 1533 (1982).
95. D. McMorrow, J.S. Melinger, A.R. Knudson, T.R. Weatherford, L.H. Tran, and W.R. Curtice, "Picosecond charge collection dynamics in GaAs MESFETs", IEEE Trans. Nucl. Sci., **NS-39** 1657 (1992).

96. (a) T.R. Weatherford, D. McMorrow, W.R. Curtice, A.R. Knudson, and A.B. Campbell, "Single event induced charge-transport modeling of GaAs MESFETs", IEEE Trans. Nucl. Sci. **NS-40**, 1867 (1993); (b) T.R. Weatherford, D. McMorrow, W.R. Curtice, A.B. Campbell, and A.R. Knudson, "Examination of the SEU sensitivity of GaAs MESFETs via 2-D computer simulation and picosecond charge collection experiments", Proceedings of the Realiability Physics Symposium (IEEE, Piscataway, NJ, 1993) pp. 357-363.
97. D. McMorrow, A.R. Knudson, and A.B. Campbell, "Fast charge collection in GaAs MESFETs", IEEE Trans. Nucl. Sci., **NS-37**, 1902 (1990).
98. T.R. Weatherford, D. McMorrow, A.B. Campbell, and W.R. Curtice, "Significant Reduction in the soft error susceptibility of GaAs field effect transistors with a low-temperature grown GaAs buffer layer", Appl. Phys. Lett. **67**, 703 (1995).
99. J.R. Schwank, F.W. Sexton, T.R. Weatherford, D. McMorrow, A.R. Knudson, and J.S. Melinger, "Charge collection in GaAs MESFETs fabricated in semi-insulating substrates", IEEE Trans. Nucl. Sci. **NS-42**,1585 (1995).
100. D. McMorrow, W.R. Curtice, S. Buchner, A.R. Knudson, J.S. Melinger, and A.B. Campbell, "Charge collection characteristics of GaAs MESFETs fabricated with a low-temperature grown GaAs buffer layer: computer simulation", IEEE Trans. Nuc. Sci., **43**, 2904 (1996).
101. J.S. Melinger, D. McMorrow, A.B. Campbell, S. Buchner, L.H. Tran, A.R. Knudson, and W.R. Curtice, "Pulsed laser induced single event upset and charge collection measurements as a function of optical penetration depth", J. App. Phys., **84**, 690 (1998).
102. D. McMorrow, J.S. Melinger, A.R. Knudson, S. Buchner, L.H. Tran, A.B. Campbell, and W.R. Curtice, "Charge-enhancement mechanisms of GaAs field effect transistors: experiment and simulation", IEEE Trans. Nuc. Sci., **45**, 1494 (1998).
103. D. McMorrow, A.R. Knudson, J.S. Melinger, and S. Buchner, "Charge collection efficiency of GaAs field effect transistors fabricated with a low-temperature grown buffer layer: dependence on charge deposition profile", IEEE Trans. Nuc. Sci., **47**, 498 (2000).
104. D. McMorrow and J.S. Melinger, unpublished work.
105. D. McMorrow, J.B. Boos, D. Park, S. Buchner, A.R. Knudson, and J.S. Melinger, "Charge collection dynamics of InP-based high electron mobility transistors (HEMTs)", Proceedings, RADECS 2001; IEEE Trans. Nuc. Sci., **49**, June, 2001, *in press*.
106. T.R. Weatherford, J. Whitaker, K. Jobe, S. Meyer, M. Bustamante, S. Thomas III, and K. Elliott, "Single event induced voltage transients within InP HBT circuits", Government Microcircuit Applications Conference (GOMAC 2000), Digest of Papers, 479-482 (2000).
107. P. Mazumder, S. Kulkarni, M. Bhattacharya, J.P. Sun, and G.I. Haddad, "Digital circuit applications of resonant tunneling devices", Proc. of the IEEE, **86**, 664 (1998).
108. J.P.A. van der Wagt, A.C. Seabaugh, and E.A. Beam, III, "RTD/HFET low standby power SRAM gain cell", IEEE Elect. Dev. Lett. **19**, 7 (1998).
109. K. Maezawa, H. Matsuzaki, M. Yamamoto, and T. Otsuji, "High-speed and low-power operation of a resonant tunneling logic gate MOBILE", IEEE Elect. Dev. Lett. **19**, 80 (1998).
110. D. McMorrow, J.B. Boos, A.R. Knudson, S. Buchner, M.J. Yang, B.R. Bennett, and J.S. Melinger, "Charge collection characteristics of low-power, ultrahigh speed, metamorphic AlSb/InAs high-electron mobility transistors (HEMTs)", IEEE Trans. Nuc. Sci., **47**, 2662 (2000).
111. D. McMorrow, R. Magno, A.S. Bracker, B.R. Bennett, S. Buchner, and J.S. Melinger, "Charge collection dynamics of AlSb-InAs-GaSb resonant interband tunneling diodes (RITDs)", IEEE Trans. Nuc. Sci., **48**, 1973 (2001).
112. R. Schneiderwind, D. Krening, S. Buchner, K. Kang, and T.R. Weatherford, "Confirmation of SEU ion experiments by laser experiments in radiation-hardened GaAs MESFET combinational logic", IEEE Trans. Nucl. Sci. **NS-39**, 1665 (1992).
113. S. Buchner, K.Kang, D. Krening, G.Lannan, and R. Schneiderwind, "Dependence of the SEU window of vulnerability of a logic circuit on magnitude of deposited charge", IEEE Trans. Nucl Sci. **NS-41**, 1853, (1993).
114. R.A. Reed, M.A. Carts, P.W. .Marshall, C.J. Marshall, S. Buchner, M. LaMacchia, B. Mathes, and D. McMorrow, "Single event upset cross sections at various data rates", IEEE Trans. Nuc. Sci., **43**, 2862 (1996).

115. M. Shoga, K. Jobe, M. Glasgow, M. Bustamante, E. Smith, and R. Koga, "Single event upset at gigahertz frequencies", IEEE Trans. Nuc. Sci. **41**, 2252-2258 (1994).
116. S. Buchner, M. Baze, D. McMorrow, and J. Melinger, "Comparison of error rates in combinational and sequential logic", IEEE Trans. Nuc. Sci. **44**, 2209-2216 (1997).
117. D. McMorrow, J.S. Melinger, S. Buchner, T. Scott, R.D. Brown, N.F. Haddad, "Application of a pulsed laser for evaluation and optimization of SEU-hard designs", IEEE Trans. Nuc. Sci. **47**, 559 (2000).
118. T.R. Weatherford and P.K Schienfelbein, "SEE Analysis of digital InP-based circuits at gigahertz frequencies", IEEE Trans. Nuc. Sci., **48**, 1980 (2001).
119. P.W. Marshall, M.A. Carts, A.B. Campbell, D. McMorrow, S. Buchner, R. Stewart, B. Randall, B. Gilbert, and R.A. Reed, "Single event effects in circuit-hardened SiGe HBT logic at gigabit per second data rates", IEEE Trans. Nuc. Sci., **47**, 2669 (2000).
120. D. McMorrow, T.R. Weatherford, W.R. Curtice, A.R. Knudson, S. Buchner, J.S. Melinger, L.H. Tran, and A.B. Campbell, "Elimination of charge-enhancement effects in GaAs FETs with an LT GaAs buffer layer", IEEE Trans. Nucl. Sci. **NS-42**, 1837 (1995).
121. P.W. Marshall, C.J. Dale, T.R. Weatherford, M. Carts, D. McMorrow, A. Peczalski, S. Baier, J. Nohava, and J. Skogen, "Heavy ion SEU immunity of a GaAs Complementary HIGFET circuit fabricated on a low temperature grown buffer layer", IEEE Trans. Nucl. Sci. **NS-42**, 1850 (1995).
122. J.R. Hauser, "SEU hardened silicon bipolar and GaAs MESFET SRAM cells using local redundancy techniques", IEEE Trans. Nucl. Sci. **NS-39**, 2 (1992).
123. Y. Umemoto, N. Nasuda and K. Mitsusada, "Effects of a buried p-layer on alpha particle immunity of MESFET's fabricated on semi-insulating GaAs substrates", IEEE Electron Dev. Lett. **EDL-7**, 396 (1986).
124. F. Barber, "GaAs radiation hardened design techniques", IEEE Proceedings, **76**, 1501 (1988).
125. T.R. Weatherford, R. Radice, D. Eskins, J. Devers, D.J. Fouts, P.W. Marshall, C.J. Marshall, H. Dietrich, M. Twigg, and R. Milano, "SEU design considerations for MESFETs on LT GaAs", IEEE Trans. Nuc. Sci., **44**, 2282 (1997).
126. T.R. Weatherford, P.W. Marshall, C.J. Marshall, D.J. Fouts, B. Mathes, and M. LaMacchia, "Effects of low-temperature buffer layer thickness and growth temperature on the SEE sensitivity of GaAs HIGFET circuits", IEEE Trans. Nuc. Sci., **44**, 2298 (1997).
127. D.J. Fouts, T.R. Weatherford, C.J. Dale, P.W. Marshall, H.B. Dietrich, D. McMorrow, J. Abrokwah, M. Lamacchia, and R. Milano, "Soft-error immune gallium arsenide ICs using COTS designs and foundries", Government Microcircuit Applications Conference (GOMAC 1997), Digest of Papers, 93 (1997).
128. H.T. Nagle, T.K. Miller, and D.F. McAllister, "Fault tolerance for single events", J. Rad. Effects, Research and Engineering, **6-2**, 155 (1988).
129. M. LaMacchia, *et al.*, "SEU hardening approach for high speed logic", presented at the 1997 Nuclear and Space Radiation Effects Conference; U.S. Patent number 5,600,260.
130. P.W. Juodawlkis, Z. Zhou, C.M. Verber, D.T. McInturff, M.C. Hargis, and S.E. Ralph, "Subpicosecond nonlinear absorption recovery dynamics of low-temperature grown, $In_{0.53}Ga_{0.47}As/In_{0.52}Al_{0.48}As$ multiple quantum well p-i-n Diodes", Ultrafast Electronics and Optoelectronics, Vol. 13, Optical Society of America, pp. 284-289, 1997.
131. M.S. Gussenhoven, E.G. Mullen, and R.C. Sagalyn, "CRRES/SPACERAD experiment descriptions", AFGL-TR-85-0017, Environmental Research Papers, No. 906, Air Force Geophysics Laboratory, Hanscom AFB, MA 01731 (24 January 1985).
132. J.C. Ritter, "Microelectronics and photonics test bed", Advances in the Astronautical Sciences, published for the American Astronautical Society by Univelt, Vol. 94, pp 139-150, The Proceedings of the 20th annual AAS Guidance and Control Conference, AAS97-024, 1997.
133. S. Buchner, T. Meehan, A. Campbell, K. Clark, and D. McMorrow, "Characterization of single event upsets in a flash analog-to-digital converter (AD9058)", IEEE Trans. Nucl Sci. **47**, 2358 (2000).
134. Paul Marshall, *private communication.*

International Journal of High Speed Electronics and Systems
Vol. 13, No. 1 (2003) 327–349

RELIABILITY AND RADIATION HARDNESS OF COMPOUND SEMICONDUCTORS*

SAMMY A. KAYALI
Jet Propulsion Laboratory, California Institute of Technology
Pasadena, California 91109,1 USA
sammy.a.kayali@jpl.nasa.gov

ALLAN H. JOHNSTON
Jet Propulsion Laboratory, California Institute of Technology
Pasadena, California 91109, USA
allan.h.johnston@jpl.nasa.gov

A thorough understanding of reliability and radiation hardness is required in order to use compound semiconductors in space, or in other environments involving radiation and/or extended temperature operation. This paper discusses those issues for several types of compound semiconductors that are of interest for high-performance applications.

Keywords: Reliability, space radiation, compound semiconductors.

1. Introduction

This paper discusses reliability and radiation hardness of mainstream compound semiconductor devices. The material discussed in the paper draws on recent material in the literature as well as experience of the authors in evaluating compound semiconductor devices for space applications. Conventional concepts of reliability and radiation hardness are reviewed and compared with trends in more advanced devices.

Three basic types of compound semiconductor devices are considered in this work. The first category is that of discrete transistors, fabricated with heterostructures, primarily intended for applications in very high frequency or microwave applications. The second category is integrated and hybrid circuits, including MMIC devices. The third category is optoelectronics. We will limit the discussion to reliability and radiation effects in optical emitters – light emitting diodes and laser diodes – in this paper.

One of the key points in the paper is that reliability and radiation effects in compound semiconductors involve different effects and mechanisms compared to silicon technology semiconductors, which have been the focus of most work on these topics. Different methodologies are required to deal with compound semiconductors that may

*The research in this paper was carried out at the Jet Propulsion Laboratory, California Institute of Technology, under contract with the National Aeronautics and Space Administration (NASA).

not be a straightforward extension of the knowledge base that exists for silicon technology. The main issues are dealing with emerging types of semiconductors, where there is limited experience in manufacturing and reliability; and device technologies with extremely small dimension.

2. Background

2.1. *Reliability issues*

The recent growth of the compound semiconductor industry has resulted in substantial improvements in processing methods, fabrication yield, and overall quality of commercially viable compound semiconductor devices. This, coupled with large volume production and the utilization of statistical process control, has greatly reduced the infant mortality population without having to impose traditional high reliability part specifications. However, reproducibility of a product does not guarantee reliability in the intended application. For critical space applications where the success or failure of a mission hinges on the lifetime and performance of a single device; it is critical that all aspects of the reliability and the various known failure modes and mechanisms be addressed prior to the insertion of the component in the application [1].

The selection and application of microelectronic components in high reliability space systems requires knowledge of the component design, fabrication process, and applicable tests. In addition, reliability analysis and detailed knowledge of the application environment are necessary to determine the suitability of the selected component for the application. These issues are of particular importance for the application of compound semiconductor devices in high reliability systems due to the need for the utilization of large numbers of these devices at the upper limit of their performance and stress capabilities.

The user of compound semiconductor devices must gain an understanding of not only the technology performance capabilities but also of the limitations of the technology and must employ methods to utilize it in a reliable fashion. The user must also understand that many of the failure mechanisms associated with silicon devices do not apply to GaAs and other compound semiconductors, and that new device structures bring new failure mechanisms. In addition, many of the traditional assumptions for mean-time failure rate predictions do not hold for those new devices. Thus, today's high reliability user must be more aware of measurement-based predictions of long term failure rate over calculation based predictions.

This article provides a brief overview of reliability issues relating to compound semiconductor devices and some common practices for determining suitability of these devices for application in high reliability space systems.

2.2. *Radiation environments*

2.2.1. Fundamental Interactions

Five types of particles are usually considered in radiation environments: gamma rays, electrons, protons, neutrons, and energetic nuclei from space or from nuclear fission.

Gamma rays and electrons are often lumped together because the primary way in which they lose energy is through ionization. Ionization produces electron-hole pairs within semiconductors and insulators. This can cause charge to be trapped at interfaces between semiconductors and insulators. Although that mechanism is extremely important for silicon devices, it is much less important for compound semiconductors because insulating materials do not exist in compound devices with the high quality and low interface state density that is present in silicon dioxide. Consequently, most compound semiconductors are relatively immune to ionization radiation damage at levels below 1 Mrad(GaAs) [2].

Protons also produce ionization, but they also interact with lattice atoms, where they can impart sufficient energy to move atoms from their normal lattice position to a resting point that can be up to 1000 angstroms away from the starting point. These displacement effects introduce a great deal of damage in the lattice. Neutrons produce similar displacement effects. Displacement damage has three important effects. Minority carrier lifetime is reduced; carrier mobility is reduced; and the effective doping level can be altered because of carrier removal. Displacement damage is generally far more important than ionization damage for compound semiconductors, and that will be the main focus of Section V.

High-energy nuclear particles also produce displacement damage and ionization. However, the most important consideration is usually single-event upset, not permanent damage effects. Single-event upset is a circuit effect that occurs because the interaction of a *single* charged particle produces a small but significant amount of charge that can be collected at sensitive circuit nodes, causing the stored information in a memory or flip-flop to be altered. Integrated circuits manufactured with compound semiconductors can be very sensitive to single-event upset effects.

2.2.2. Space Environments

Space environments can generally be divided into (a) permanent damage effects from high-energy electrons and protons that produce uniform damage within each individual region of a semiconductor device, and (b) highly localized ionization or displacement effects from the single interaction of a cosmic ray or trapped proton.

Radiation levels for typical spacecraft are summarized in Table 4, assuming that a 100-mil aluminum shield surrounds the electronics within the spacecraft. The shield is very important because it removes most of the low energy particles from the environment. Planetary space missions that go near Jupiter have very high radiation requirements because of intense trapped radiation belts that extend to very large distances.

Table 1. Total Dose Requirements for Representative Space Missions

Description	Orbit	Operating Time (years)	Total Dose rad(SiO_2)
Space Station	500 km 54 degree	10	5×10^3
High-inclination earth orbiter	705 km 98 degree	5	2×10^4
Geostationary	36,000 km	5	5×10^4
Mars Surface Exploration	NA	3	10^4
Mission near Jupiter	NA	9	1.5×10^5 – 2×10^6

Requirements for single-event effects are much more difficult to define because there is a distribution of ion types and energies in space, as well as solar activity. Proton single-event upset is also influenced by the South Atlantic anomaly in the earth's trapped radiation belts. See References 3 and 4 for more details. The error rate for memory cells or registers is usually used as a benchmark for single-event upset. Commercial silicon-based devices have error rates on the order of 10^{-6} to 10^{-8} errors per bit day in a deep space or geostationary environment. It is often possible to use error-detection-and-correction or other system level approaches to deal with these types of upset effects. The alternative is to use special hardened circuits. As discussed in Section V, logic circuits that use compound semiconductors are somewhat more sensitive to single-event upset than comparable silicon technologies. However, compound semiconductor structures do not exhibit latchup or other catastrophic failure effects, which is an advantage in space applications.

2.2.3. Nuclear Reactor Environment

For nuclear reactors, the primary concern is displacement damage from neutrons and decay products from activated material, and ionization damage. Very high radiation levels – in the multi-Megarad region - are often required, particular for worst-case operational scenarios. Many compound semiconductor devices are highly resistant to ionization and displacement damage, making them good candidates for use at nuclear reactor facilities.

3. Compound Semiconductor Construction and Physics

3.1. Typical materials

Initial work was done by extending the principles of silicon technology, adding dopants to GaAs to produce p-n junctions with a single semiconductor type (homogeneous semiconductor junctions). The much higher electron mobility of GaAs provides a significant advantage in high-frequency devices. Earlier work concentrated on MESFET structures. In 1978 heterostructures were developed that allowed new material combinations to be developed [5], although the concept of heterostructures was first proposed by William Shockley. The initial material combination was AlGaAs on GaAs substrates. This work was extended to other material types, using advanced processing techniques such as liquid-phase growth and molecular beam epitaxy to form precisely controlled layers of different material types and dopants. These processes allow very thin layers to be formed, providing the ability to form extremely thin regions of different semiconductor materials. Quantum-well structures can be deliberately fabricated with these techniques. One of the key properties of heterojunctions is that the barrier between the materials is mainly determined by bandgap difference, not by bias conditions. This allows junctions to be formed in extremely narrow regions.

Several different material combinations can be used for compound semiconductor junctions. The key properties are the bandgap, which can be tailored by varying the material composition; and lattice mismatch, which generally must be below 0.2% in order to minimize defects that reduce mobility. The table below shows a AlGaAs alloy and InGaAs alloy that are lattice-matched to GaAs, as well as an InGaAs alloy that is lattice-matched to InP. Many other material combinations have been developed.

Table 2. Properties of Some Compound Semiconductor Alloys

Material	Electron Mobility (cm^2/V-s)	Lattice Constant	Eg (eV)
GaAs	8500	5.65	1.42
$Al_{0.3}Ga_{0.7}As$	2300	5.66	1.65
$In_{0.15}Ga_{0.85}As$	5800	5.71	1.05
InP	4600	5.87	1.35
$In_{0.53}Ga_{0.47}As$	13,000	5.87	0.78

Other factors that are important for compound semiconductors are materials and growth methods for contacts. Thermal conductivity is another key property. Note for example that the thermal conductivity of GaAs is only about 1/3 that of silicon. The thermal conductivity of SiGe is also much lower than for silicon. These are key parameters in the design of devices that operate at high power densities, such as RF amplifiers, but are less important for high-speed logic circuits.

3.2. *Strained lattices*

Earlier material development for heterojunctions assumed that a close lattice match was required in order to keep the defect density at acceptable levels. However, it is possible to accommodate strain within the lattice if the lattice dimensions are kept below a critical thickness, usually < 100 Å. Lattice mismatch up to 1.5% can be used in strained lattices.

Strained materials not only allow a wider range of options for "bandgap engineering" but can also modify material properties. By introducing selected amounts of strain, it is possible to increase both electron and hole mobility. This can improve the performance of heterojunction bipolar structures [6]. Strained layers are frequently used in silicon-germanium HBTs, as well as in laser diodes (see the discussion in III-C, below). Although it would appear that strained layer devices would be less reliable than conventional semiconductors, extensive work has been done that demonstrates comparable reliability between strained and unstrained materials in a given technology as long as the misfit between the different materials is not too large [7].

3.3. *Transistor technologies*

3.3.1. MESFETs and HFETs

GaAs MESFETs were the first compound semiconductor device to be widely used, particularly for microwave applications. The basic principles are similar to those of silicon MESFETs, with smaller dimensions in order to optimize high-frequency performance. Most GaAs MESFETs are fabricated on semi-insulating substrates, which affects their radiation performance (see Section V).

Heterostructure field-effect transistors (HFETs) use thin layers of different materials to allow quantum-mechanical confinement of carriers within the active region. The earliest HFETs used Schottky gates, but it is also possible to fabricate HFETs with insulating gate structures. Many different material systems can be used, including strained layers. It is possible to produce complementary transistors with HFET technology, providing a major advantage over MESFETs for high-performance logic. The main applications of HFETs are in very high-speed and low-noise applications. Cost and yield prevent direct competition with large-scale silicon devices, but HFETs are selectively used in many applications that require high speed, including fiber optic data busses and RF communication systems.

3.3.2. Heterojunction Bipolar Transistors

Heterojunction bipolar transistors have been developed using the bandgap discontinuity between the emitter and base materials for carrier injection. This allows the base width to be reduced to about 0.1 μm. It also allows transistors to be fabricated with heavily doped base regions [8]. The net effect is a very compact transistor with high gain and very high gain bandwidth product.

Modern HBTs can achieve gain-bandwidth products of 100 GHz or more [9]. SiGe, InP and InGaAs heterostructures have been used in advanced heterojunction bipolar devices. HBTs have many commercial applications, and further development of these technologies is an area of active research.

3.4. *Optical devices*

Additional material properties are important for optical devices. For LEDs and laser diodes, photon containment requires that the index of refraction in the confined (active optical) layer is less than the index of refraction of surrounding layers. Another key property is the band structure. Efficient optical transitions are only possible for materials with direct bandgap. Figure 1 shows how bandgap and wavelength can be tailored for AlGaAs/GaAs heterostructures. When the mole fraction of aluminum exceeds 0.45, the bandgap becomes indirect. Thus, that limits the shortest wavelength to about 630 nm for that material system. However, by varying the composition it is possible to use AlGaAs alloys for optical devices with wavelengths between 630 and 950 nm.

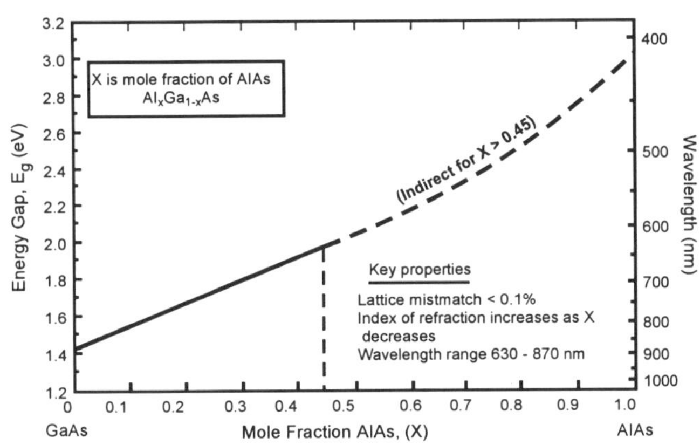

Figure 1. Effect of material composition on wavelength for AlGaAs/GaAs

Other material combinations can also be used, such as the three material systems shown in Table 3. InGaAsP/InP is usually used in strained layer lasers, where the presence of strain increases hole mobility, decreasing the threshold current of the laser by as much as a factor of two [10,11].

Table 3. Material Combinations Used for Optical Emitters in the Visible and Near Infrared Regions of the Spectrum

Material Combination	Wavelength Range (mm)
AlGaAs/GaAs	650 – 850
InGaAsP/InP	1100 – 1600
InGaAs/GaAs	900 – 1100

4. Reliability

4.1. *Conventional reliability concepts*

Device reliability involves probability statistics, time, and a definition of failure. Given a failure criterion, the most direct way to determine reliability is to submit a large number of samples to actual use conditions and monitor their performance against the failure criteria over time. Since most applications require device lifetimes of many years, this approach is not practical. To acquire device reliability data in a reasonable amount of time, an accelerated life test at high temperatures is used. This type of accelerated test is based on the observation that most failure mechanisms are thermally activated. By exposing the devices to elevated temperatures, it is possible to reduce the time to failure of a component, thereby enabling data to be obtained in a shorter time than would otherwise be required. Such a technique is known as "accelerated testing" and is widely used throughout the semiconductor industry. The rate at which many chemical processes take place is governed by the Arrhenius equation:

$$R = A \exp(-E_a/kT) \qquad (1)$$

where

R = rate of the process
A = a proportional multiplier
E_a = activation energy, a constant
k = Boltzmann's constant, $8.6\text{x}10^{-5}$ (eV/K)
T = Absolute temperature in Kelvin

This equation has been adopted by the semiconductor industry as a guideline by which the operation of devices under varying temperature conditions can be monitored. Experimental data obtained from life tests at elevated temperatures are processed via the Arrhenius equation to obtain a model of device behavior at normal operating temperatures. Rearranging the Arrhenius equation allows the temperature dependence of component failure to be modeled as follows:

$$\ln t_2/t_1 = E_a/k\,(1/T_2 - 1/T_1) \qquad (2)$$

where

$t_{1,2}$ = time to failure
E_a = activation energy in electron volts
T = absolute temperature in Kelvin

4.2. *Common failure mechanisms*

Several failure mechanisms are important for compound semiconductors that either have no counterpart in silicon technology, or are not significant issues.

Failures in electronic devices can be classified as either catastrophic failures or degradation failures. The exact mechanism that causes failure is normally dependent on the material structure, processing methods, application, and stress conditions. Device bias, resultant channel temperature, passivation, and material interactions may all cause or contribute to different failure mechanisms. Furthermore, device handling, choice of materials for packaging and the application environment may also cause failures. Some common failure mechanisms affecting the device at die level are discussed below:

4.2.1. Gate-Metal Sinking

The performance of GaAs-based devices relies heavily on the quality of the active channel area of the device. The Schottky gate metal-to-semiconductor interface directly influences device electrical parameters, such as drain saturation current and reverse breakdown. The gate structures are based on the industry standard multi-layer Au/Pt/Ti or Au/Pd/Ti on GaAs. Inter-diffusion of gate metal with GaAs results in a reduction of the active channel depth and a change in the effective channel doping. This effect is termed "gate sinking." This process is affected by the surface conditions of the GaAs material at the time of deposition, the deposition parameters, and the choice of deposited materials [12,13].

4.2.2. Ohmic Contact Degradation

The most common system for ohmic contacts is AuGe/Ni, which is alloyed into the GaAs at temperatures in excess of 400°C to provide the necessary low contact resistance (0.1 to 0.5 Ω/mm). A thick Au layer is then deposited on top of the alloyed contacts to provide conduction. This structure, employed at the drain and source contacts, has been shown to degrade at elevated temperatures (>150 °C). The degradation is the result of Ga out-diffusion into the top Au layer and the diffusion of Au into the GaAs causing an increase in the contact resistance. The Ni layer used in the ohmic contact is intended as a Au- and Ga-diffusion barrier. Some other materials such as Cr, Ag, Pt, Ta, and Ti have been used as barrier materials with varying degrees of success [14]. The activation energy associated with ohmic contact degradation varies between 0.5 eV and 1.8 eV. This activation energy may provide reasonable contact life at low operating temperatures (<100 °C) but it also indicates rapid deterioration at elevated temperatures [15].

4.2.3. Channel Degradation

Degradation observed in device parameters can sometimes be attributed to changes in the quality and purity of the active channel area and a reduction in the carrier concentration beneath the gate Schottky contact area. These changes have been postulated to be a result of diffusion of dopants out of the channel or diffusion of impurities or defects from the substrate to the channel. Deep level traps have also been postulated to cause similar degradation in MESFETs [16].

HEMT devices, being strongly dependent on the properties of the interface of the AlGaAs/GaAs heterostructure, can suffer a related failure mechanism. A decrease in electron concentration in the channel, caused by a de-confinement of the 2-Dimensional Electron Gas (2DEG), was postulated to be the cause of the observed failure mechanism.

HEMT devices can also suffer from metal-diffusion-related mechanisms, which are manifested as channel-related degradation. Lateral diffusion of Al into the gate recess region changes the conduction band discontinuity and consequently the confinement of the channel electrons. Gold diffusion from the ohmic contact into the active channel region under the gate can also cause similar degradation. Lastly, vertical diffusion of Al from the AlGaAs donor layer and Si from the n^+ AlGaAs layer into the channel layer causes an increase in the impurity scattering in the undoped GaAs, thus deteriorating the high electron mobility of the 2DEG [17].

4.2.4. Surface State Effects

The performance of GaAs-based devices depends highly on the quality of the interface between metal and GaAs or the passivation layer (Si_3N_4 or SiO_2) and GaAs. The quality of the interface can depend on the surface cleaning materials and procedures, the deposition method and conditions, and the composition of the passivation layer. As shown in Fig. 2, the main effect of an increase in surface state density is the lowering of the effective electric field at the drain/gate region, which results in an increase in the depletion region and a change in the breakdown voltage.

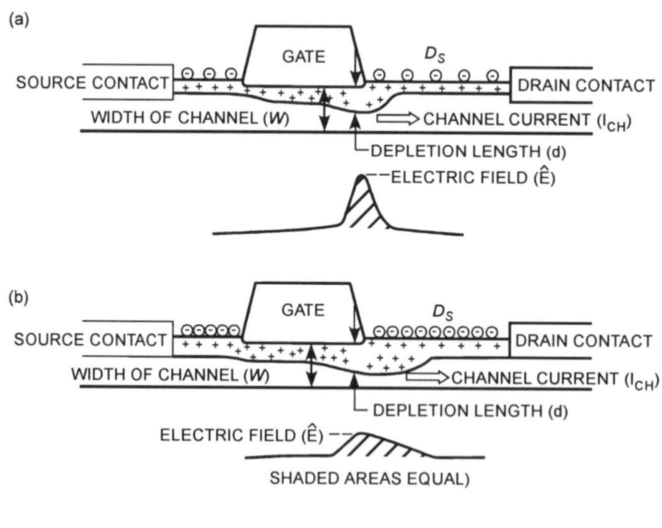

Figure 2. Schematic cross section of a MESFET with different surface charges. (a) with low density of surface states, and (b) with high density of surface states[18].

Unpassivated devices can be susceptible to surface oxidation and loss of arsenic, which may result in an increase in gate leakage current and a reduction of the breakdown voltage. Devices passivated using SiO_2 may experience surface erosion due to the interaction of SiO_2 with GaAs [19].

4.2.5. Electromigration

The movement of metal atoms along a metallic strip due to momentum exchange with electrons is termed electromigration. Since the mechanism is dependent on momentum transfer from electrons, electromigration is dependent on the temperature and number of electrons. Therefore, this failure mechanism is generally seen in narrow gates and in power devices where the current density is greater than $2x10^5$ A/cm^2, which is normally used as a threshold current density for electromigration to occur. As shown in Fig. 3, this effect is observed both perpendicular and along the source and drain contact edges and also at the interconnect of multilevel metallizations.

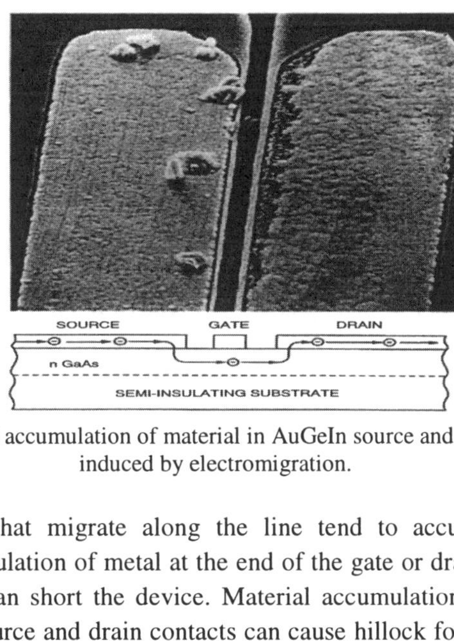

Figure 3. Depletion and accumulation of material in AuGeIn source and drain ohmic contacts induced by electromigration.

The metal atoms that migrate along the line tend to accumulate at the grain boundaries. The accumulation of metal at the end of the gate or drain contact can create fingers of metal that can short the device. Material accumulation and void formation perpendicular to the source and drain contacts can cause hillock formation over the gate structure. This may result in shorting the gate to the source or drain that may result in catastrophic failure.

4.2.6. Hot Electron Trapping

Under RF drive, hot electrons are generated near the drain end of the channel where the electrical field is the highest. A few electrons can accumulate sufficient energy to tunnel into the Si_3N_4 passivation to form permanently charged traps. As shown in Fig. 4, these traps can result in lower open-channel drain current, transconductance, higher knee voltage, increased leakage current, and decreased breakdown voltage. Since the traps are located above the channel, there is usually little change in the dc or small signal parameters near the quiescent point. Further, since the traps are located beside the channel, Schottky-barrier height and the ideality factor often remain constant. This selective change in device characteristics helps distinguish hot-electron effects from thermal or environmental effects [20].

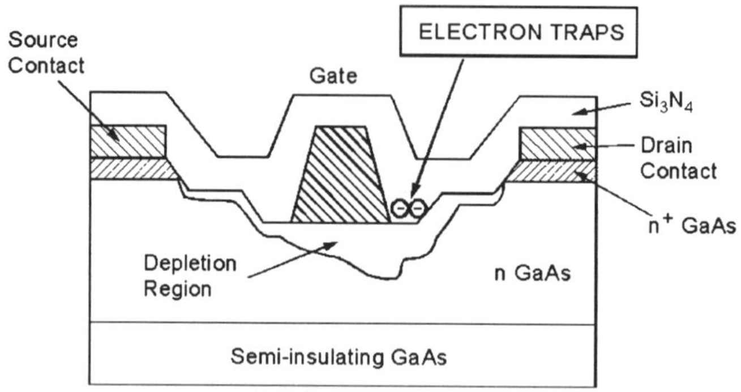

Figure 4. Location of electron traps after tunneling into passivation regions.

4.2.7. Hydrogen Effects

Degradation in I_{DSS}, V_p, g_m, and output power was observed on GaAs and InP devices tested in hermetically sealed packages or under hydrogen atmosphere. The source of the degradation has been attributed to hydrogen gas desorbed from the package metals (Kovar, plating, etc.). The exact mechanism by which hydrogen degrades the device performance and the path by which hydrogen reaches the active area of a device are not known and have been under investigation [21].

Earlier research, [22], on GaAs transistors identified the diffusion of atomic hydrogen directly into the channel area of the device where it neutralizes the silicon donors as the possible mechanism. It is believed that atomic hydrogen diffuses into the GaAs channel and forms Si-H, thereby neutralizing the donors. Experiments have shown that exposure of Si-doped GaAs to RF hydrogen plasma results in neutralization of the Si donors. Infrared spectroscopy data have also given evidence of $(SiAs_3)As$-H complexes[23].

The neutralization of donors can decrease the carrier concentration in the channel, which, in turn, can decrease the drain current, transconductance, and gain of the device. Hydrogen effects in FETs with either Pt or Pd gate metals have been observed. Recent research has concluded that the diffusion of hydrogen may occur at the Pt side-walls and not at the Au surface of the Au/Pt/Ti gate metal [24].

Other research, an example of which is shown in Fig. 5, on GaAs PHEMT and InP HEMT in a hydrogen atmosphere has shown that the drain current may increase in some cases. This observation has led to the conclusion that the hydrogen diffuses into the semiconductor surface where it is thought to change the metal-semiconductor built-in potential.

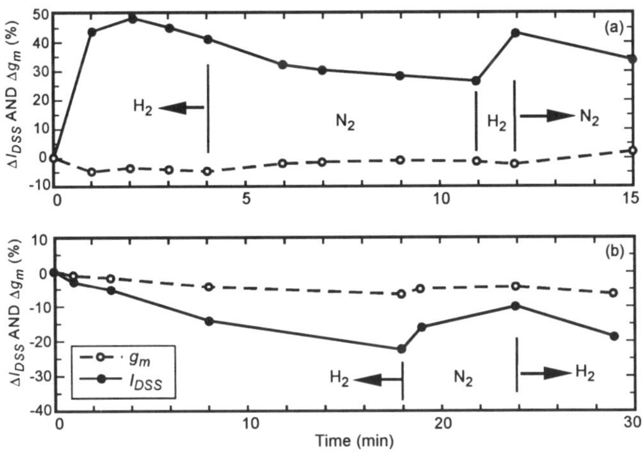

Figure 5. Changes in peak transconductance, g_m, and drain current at zero bias, I_{dss}, of (a) InP HEMT and (b) GaAs PHEMT under nitrogen and 4% hydrogen treatment at 270°C[18].

Manufacturers and users of GaAs devices used in hermetically sealed packages are currently pursuing an acceptable solution to this problem. Some of the possible solutions include thermal treatment of the packaging materials to reduce the amount of desorbed hydrogen after the seal, the use of hydrogen getter materials in hermetically sealed packages, and the use of barrier materials that do not contain the Pt/Ti or Pd/Ti structure. These solutions have limitations and possible instability problems that must be fully understood prior to implementation in high reliability systems.

4.2.8. Packaging Effects

The package serves to integrate all the components required for a system application in a manner that minimizes size, cost, mass and complexity. In doing so, the package must provide for mechanical support, protection from the environment, a stable thermal dissipation path, and electrical connection to other system components. For compound semiconductors, the package must satisfy all these characteristics and allow for reliable device performance over a wide range of conditions.

Understanding the packaging effects on the reliability of compound semiconductors is essential to attaining a reliable space system. In most applications, packaging of compound semiconductor devices is similar to that developed for silicon based technologies. However, the choice of packaging materials plays more of a critical role due to differences in the coefficient of thermal expansion. In addition, compound semiconductors are more fragile and may exhibit mechanical stresses causing device degradation and failure.

The stability and reliability of the die attach is largely determined by the ability of the structure to withstand the thermo-mechanical stress created by the difference in the Coefficient of Thermal Expansion (CTE) between the die and the packaging material. These stresses are concentrated at the interface between the die and the die-attach material and the interface between the die-attach material and the package [25]. The

Coffin-Manson relation relates the number of thermal cycles a die attachment can withstand before failure:

$$Nf \propto \gamma^m \{2*t/L*\Delta CTE*\Delta T\} \qquad (3)$$

where

γ = shear strain for failure
m = constant dependent on the material
L = diagonal length of the die
t = die-attach material thickness

The number of thermal cycles before failure can be significantly reduced by the presence of voids in the die attach material, since voids cause areas of concentrated localized stress which can lead to premature die delamination. In addition, voids cause localized heating which in turn causes an increase in the thermal resistance of the die attach material leading to device degradation and possible catastrophic failure.

Infrared imaging techniques can provide for a qualitative and sometimes a quantitative measure of the adequacy of the thermal path and a visual representation and mapping of possible void locations. Figure 6 shows a comparison of an optical and an Infrared image of the same die.

Figure 6. Optical (left) and IR image (right) of the same die. The IR image shows thermal gradient and location of hot spots and possible void locations.

4.2.9. Light-Emitting and Laser Diodes

Power output from optical emitters can change during operation. The high internal optical power density in those devices causes different wearout mechanisms to occur compared to conventional electronic components. In some cases the degradation is gradual, while in others sudden, catastrophic damage occurs. Facet damage in laser diodes is an example of catastrophic damage [26]. It is caused by localized heating, due to very high localized optical power levels. It is a more severe problem for AlGaAs than for other laser materials. Another degradation mechanism is associated with internal crystal dislocation defects, which grow during operation (dark-line defects, or DLDs) [27]. DLDs can produce abrupt changes in the threshold characteristics of laser diodes over operating times of several hundred hours or longer, or may develop more gradually over extended time periods. A third mechanism is attributed to gradual increase of point defects, increasing non-radiative recombination losses. That mechanism can occur for both laser diodes and LEDs, with a typical activation energy of 0.5 eV.

Although earlier laser diode technologies had very limited operating life, improvements in laser diode technology have decreased the internal power densities by many orders of magnitude, increasing reliability to the point where operation over periods of 10,000 hours or more can be achieved [28]. The reliability of strained-layer lasers has comparable reliability [29]. Note, however that achieving high operating lifetime requires derating below the maximum operating power levels.

LEDs operate at lower power densities than laser diodes, but they also degrade during operation. Non-radiative defects gradually increase with time, changing the slope of the I-V characteristics as well as decreasing light output. The rate of the increase depends on operating conditions and temperature, and varies for different samples. In some cases LEDs undergo rapid initial decrease in light output, followed by a plateau region. The initial rapid decrease is due to dark-line defects, which cannot be described by the Arrhenius model. The gradual degradation region does follow the Arrhenius relationship, with an activation energy of about 0.6 eV [30].

Although a great deal of information is available about LED reliability in conventional environments, one of the key issues for space applications is whether wearout effects can be considered separately from the degradation that occurs from space radiation. Wearout data for light-emitting diodes is shown in Figure 7. Three different LEDs were subjected to an extended test, using the maximum recommended operating current. The devices were operated with a heat sink to keep the case temperature at 25 °C. Note the gradual deterioration in output power.

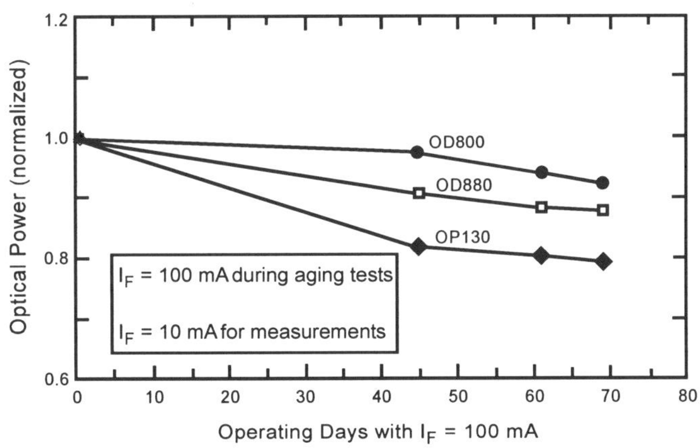

Figure 7. Degradation of light-emitting diodes with operating time. The devices were kept at room temperature during the aging tests.

Radiation tests on aged samples showed that aging and radiation damage could be considered separately, with no synergistic effects even though both environmental stress conditions decrease power output and increase the forward voltage required at a given light output [31]. However, it should be noted that much less degradation occurs when the LEDs are operated at low power compared to the maximum rated value. For example, tests of a different sample of the devices in Figure 7 at 50 mA instead of 100 mA decreased the degradation by about a factor of five at the longest operating times.

5. Radiation Hardness

5.1. *Ionizing radiation damage*

Although ionizing radiation is often one of the dominant radiation problems for devices made with silicon technology, ionization is usually of secondary importance for compound semiconductors. The basic reason is that it is not possible to make extremely high-quality insulators (such as SiO_2) in compound semiconductor systems.

Figure 8 shows how electron damage affects InP/InGaAs heterojunction bipolar transistors [32]. Very little damage occurs until the total dose is above 50 Mrad, even at low current densities. For comparison, silicon transistors can exhibit large decreases in gain at total dose levels between 10 and 50 krad(SiO_2). Although not all compound semiconductors are as robust as the InP device in Figure 8, they generally show little degradation until radiation levels above 10 Mrad(GaAs).

Older work on MMIC devices has shown similar radiation hardness for 30 GHz microwave integrated circuits that used GaAs MESFETs [33]. Changes in power gain first started to occur at 10 Mrad(GaAs), and the devices were useable to levels of about 50 Mrad(GaAs). Thus, both bipolar and MESFET structures are only affected by ionizing radiation after exposure to very high radiation levels.

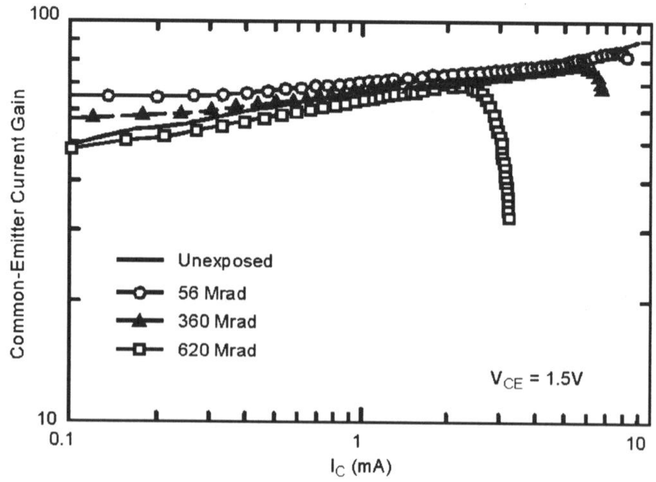

Figure 8. Total dose degradation of an InP/InGAAs heterojunction bipolar transistor.

5.2. *Displacement damage*

5.2.1. Heterostructure-Based Transistors and MESFETs

Displacement damage is more critical for compound semiconductors than ionization damage. However, different mechanisms may be involved compared to displacement damage in silicon devices. Most compound semiconductors are only slightly affected by minority carrier lifetime damage either because they are majority carrier devices, or they are minority carrier devices with extremely narrow carrier transport dimensions (such as transistor base width). However, at high fluences carrier removal will alter the effective doping concentration. Typical carrier removal rates for high-energy protons are on the order of 40 cm^{-1}; this means that a 4% change in carrier concentration will occur for material doped to 10^{15} at a proton fluence of 10^{12} p/cm^2. Most high-frequency compound semiconductor devices have doping concentration above 10^{16} cm^{-3}, and consequently they do not exhibit significant damage until they are exposed to radiation levels above 10^{13} p/cm^2.

Silicon-germanium bipolar transistors are beginning to dominant applications at very high frequencies. Figure 9 shows proton radiation test results for silicon-germanium heterojunction bipolar transistors that are intended for RF applications [34]. There is significant degradation at low currents, although it should be emphasized that the figure shows results at a relatively high proton fluence. However, typical RF applications use the device at high current densities where the degradation is extremely small (note the callout in the figure regarding normal biasing for RF applications). Thus, this device is useable at fluences above 10^{14} p/cm^2, far higher than the radiation levels encountered by most space systems.

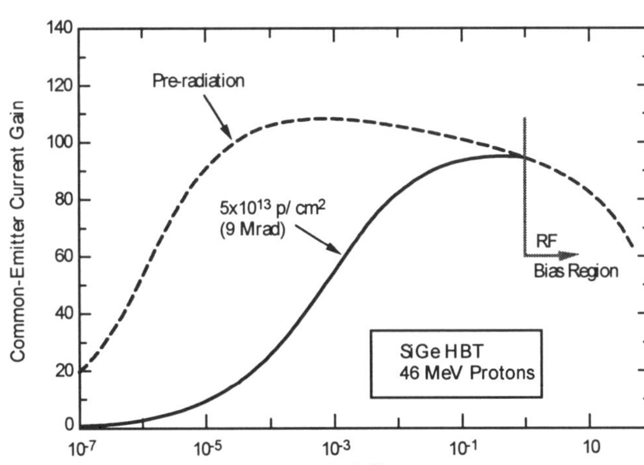

Figure 9. Proton degradation of a silicon-germanium heterojunction bipolar transistor.

5.2.2. Optoelectronics

Displacement damage is a much more severe issue for optoelectronic devices. There is a tremendous range in sensitivities due to differences in fabrication techniques and principles of operation.

5.2.2.1. LED displacement damage

Light emitting diodes can be fabricated in many different ways. One older method, still widely used, relies on *amphoteric* doping. This process uses a single type of dopant (silicon) in GaAs, which is n-type when grown at high temperature, but p-type when it is grown at low temperature [35]. The junction is formed by changing the temperature during the growth process. This process produces LEDs with very high efficiency because the material is closely compensated, reducing free carrier absorption; and the energy of the emitted radiation is below the bandgap energy (due to the compensated doping), eliminating band-to-band absorption. These LEDs have graded junctions that extend over relatively long distances. Consequently, they require very long minority carrier lifetime in order to operate efficiently.

Modern LEDs are usually made with shallow, double heterojunctions that do not require long lifetime. These types of LEDs have much shorter turn-on times, but are less efficient. Consequently there are still many applications of the older amphoterically doped LEDs. Figure 10 shows how proton damage affects various types of LEDs [36]. Note that there is a difference of about two orders of magnitude in damage sensitivity. Amphoterically doped LEDs are so sensitive to proton damage that the proton fluence from a single solar flare can significantly degrade the LED.

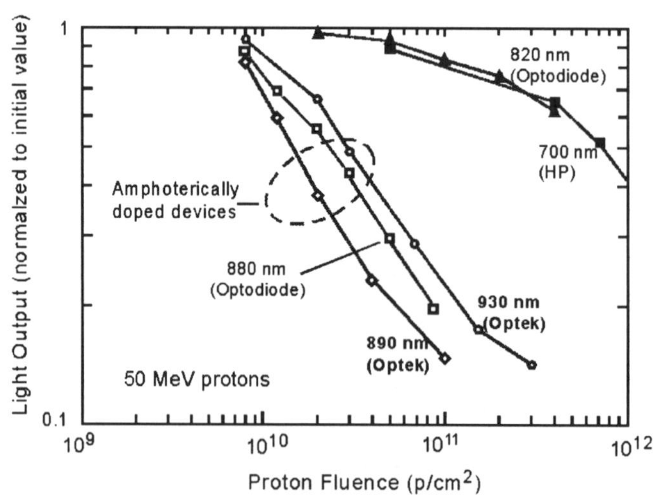

Figure 10. Proton degradation of various types of LEDs. Note the extreme sensitivity of amphoterically doped devices.

One important application of LEDs is in optocouplers. However, the type of LED within an optocoupler usually is not specified, only the overall electrical performance.

Amphoterically doped LEDs are frequently used in optocouplers, resulting in extremely poor radiation performance [37]. Unless the manufacturer explicitly controls the type of LED, this can cause extreme variability in the radiation hardness of the optocoupler.

5.2.2.2. Laser displacement damage

Semiconductor lasers are far less affected by proton displacement damage than light-emitting diodes. The reason is that they use heterojunctions with thin layers and high carrier densities, and do not require long lifetimes for efficient operation. Displacement damage increases internal (non-radiative) losses, increasing the threshold current as well as the slope efficiency [38,39]. However, the internal optical power level of a semiconductor laser is quite high, much higher than that of LEDs. Thus, there is less margin for change in operating characteristics. Figure 11 shows an example of the effect of proton damage on the output power characteristics of a semiconductor laser. Although the change in threshold current is relatively small, the laser diode must be driven at much higher internal power levels after irradiation in order to maintain light output, which will reduce reliability. Most space applications require that laser diodes operate well below maximum ratings to maintain reliable operation.

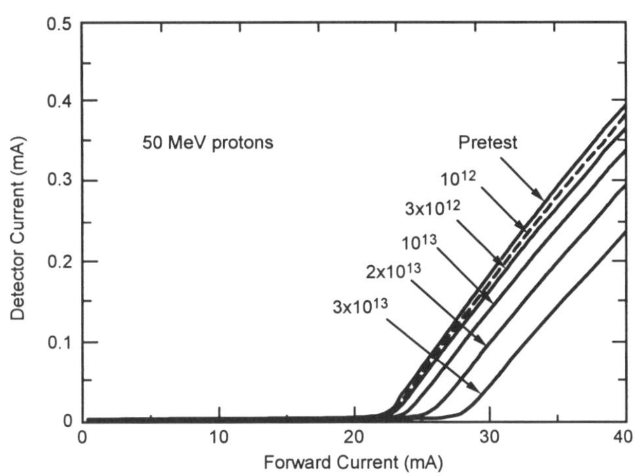

Figure 11. Degradation of the output power characteristics of a semiconductor laser after irradiation with high-energy protons.

In addition to degradation of the laser, most laser diodes include an internal monitor diode that is used to measure the light output. The output of the monitor diode is typically used in a control circuit that establishes a fixed light output condition from the laser. However, the monitor diode may also degrade from radiation, which can cause the laser to be driven into an operational region that is beyond the normal range, affecting reliability. In some cases the monitor diodes degrade more rapidly than the lasers [39]. Controlling operating temperature and maintaining light output over a restricted range are essential in order to use laser diodes in space.

5.3. *Single-particle effects*

Because of their fast response time, logic circuits made with compound semiconductors are highly sensitive to upset from cosmic rays and protons. Fabrication details play a large role. (The upset is caused by the generation of a short-duration charge when the cosmic ray or proton interacts with the device). The most sensitive devices are those made on a semi-insulating (SI) substrate. Excess carriers generated within the SI substrate can be collected by a MESFET or HFET structure, allowing the device to be upset with a particle that has lower specific charge generation (linear energy transfer) compared to a structure where charge collection is limited to the active region.

Another important factor is the parasitic bipolar transistor that is present in MESFET and HFET structures. Figure 12, after Hughlock, et al. [40], compares the single-event upset threshold LET and cross section for three different logic devices, all manufactured with 1-μm feature size. The reason for the high sensitivity of the GaAs MESFET is the high parasitic bipolar gain. The other two technologies have lower gain, which reduces single-event sensitivity. Hughlock, et al. observed the same response in special GaAs MESFETs that were fabricated without the gate region, demonstrating that the bipolar effect was the underlying mechanism for the high sensitivity of those devices.

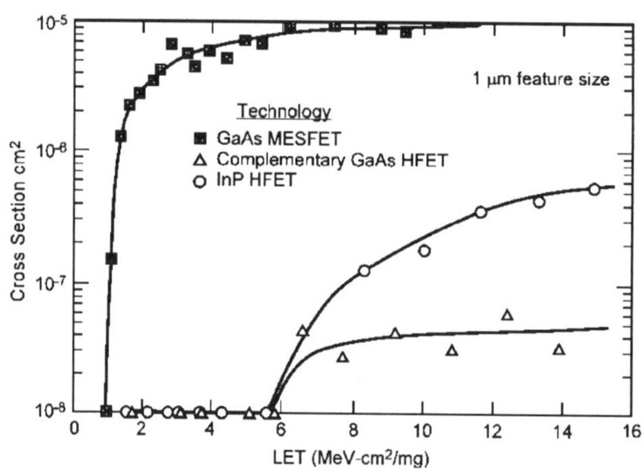

Figure 12. Single-event upset sensitivity of GaAs MESFETs (on a semi-insulating substrate) compared with other compound semiconductor logic devices.

Recent work by McMorrow, et al. showed that the excess charge collection process observed for GaAs MESFETs also occurs in HEMTs fabricated with AlSb/InAs [41]. They performed charge-collection experiments on devices with gate lengths of only 0.1 μm. Their results suggest that single-event upset sensitivity of advanced compound semiconductor devices may still be a problem for space use. Threshold LET is typically about 1 MeV-cm^2/mg, about a factor of three lower than that of advanced silicon devices (such as memories and microprocessors), and so low that even alpha particles from packaging can cause upsets. The cross section of compound semiconductors has typically been about two orders of magnitude higher than that of silicon devices with equivalent feature size because of the excess charge from parasitic bipolar devices.

Although most small-area compound semiconductor devices are highly sensitive to single-event upset, they are usually not used in high-density circuits (memories and processors, which are dominated by CMOS technology). Thus, the practical effect of the extreme sensitivity to transient pulses and logic errors is usually slight compared to CMOS technology in most system applications simply because there are relatively small numbers of flip-flops or other storage devices within the types of circuits that are designed with compound semiconductors. However, there are cases where upset effects can be an issue, including high-speed fiber-optic data buses. The effect of the upsets is to increase the bit-error rate in space compared to terrestrial applications [42]. This can be overcome by increasing optical power, making the subsystem less sensitive to the short-duration "noise" pulses that are produced by ions in space.

6. Conclusions and Future Predictions

This paper has discussed reliability and radiation hardness of compound semiconductors, pointing out how differences in material technology and properties as well as fabrication technology affect the overall reliability of these devices. It is important to understand that one cannot simply extend the knowledge of silicon technology to compound semiconductors. The key issues for reliability stem from differences in fabrication, along with the requirement that some types of compound semiconductors have to operate at very high power densities in order to achieve maximum benefit. This, along with the lower thermal conductivity of GaAs (and some other compound semiconductor materials) increases the importance of packaging technology and thermal management to overall device reliability.

Another important issue is the design of transistors with extremely small dimensions. This places more demands on contacts and metallization, which can potentially affect reliability. Some advanced compound semiconductor devices use physical dimensions of 0.1 μm or less, and fabrication methods are still evolving for reliable manufacture of devices with such small dimensions.

From the standpoint of radiation hardness, most compound semiconductors are much less affected by ionization or proton displacement damage than silicon-based technologies. However, light-emitting diodes are an important exception. Older, highly efficient LEDs are among the most sensitive components to radiation damage from protons. This not only affects LED response, but also causes certain types of optocouplers to be highly sensitive to radiation damage in space applications.

Logic devices fabricated with compound semiconductors are highly sensitive to upset effects from cosmic rays and protons, and this can be an important effect for some applications. However, this limitation is relatively unimportant unless compound semiconductor circuits are being used with large numbers of storage elements, which is usually not the case. Thus, most compound semiconductor devices are excellent choices for space applications because of high radiation damage tolerance.

References

1. S. Kayali, "Reliability of compound semiconductor devices for space applications," *Microelectronics Reliability* **39** (1723), 1999.
2. J. R. Srour and J. M. McGarrity, "Radiation effects on microelectronics in space," *Proc. of the IEEE*, **76**(11), pp. 1443-1469 (1988).
3. J. C. Pickel, "Single-event effects rate prediction," *IEEE Trans. Nucl. Sci.*, **43**(2), pp. 483-495 (1996).
4. E. L. Petersen, "Approaches to proton single-event rate calculations," *IEEE Trans. Nucl. Sci.*, **43**(2), pp. 496-504 (1996).
5. R. Dingle, et al., "Electron mobilities in modulation-doped GaAs-AlGaAs heterojunction superlattices," *Appl. Phys. Lett.*, **33**, 665 (1978).
6. G. C. Osbourn, "InGaAs/InGaAsa strained superlattices: Proposal for Useful, New Electronic Materials," *Phys. Rev. B* **27**, pp. 5126-5128 (1983),
7. R. People and J. C. Bean, "Calculation of critical layer thickness vs. mismatch for GeSi/Si strained layer heterostructures," *Appl. Phys. Lett.*, **47**, pp. 322-324 (1985).
8. H. Kroemer, "Heterostructure bipolar transistors and integrated circuits," *Proc. of the IEEE*, **70**, pp. 13-25 (1982).
9. K. Shinohara, et al., "Ultrahigh-speed s pseudomorphic InGaAs/InAlAs HEMTs with 400-GHz cutoff frequency," *IEEE Elect. Dev. Lett.*, **22**(11), 507 (2001).
10. C. H. Lin and Y. H. Lo, "Empirical formulas for design and operation of 1.55 μm InGaAs/InGaAsP strained-quantum-well lasers," *IEEE Phot. Tech. Lett.*, **5**(3) , pp. 288-290 (1993).
11. P. J. A. Thijs, L. F. Tiemeijer, J. J. M. Bisma and T. van Dongen, "Progress in long wavelength strained layer InGaAs(P) quantum-well semiconductor lasers and amplifiers," *IEEE J. Quant. Elect.*, **30**(2), pp. 477-499 (1994).
12. C. Canali, F. Castaldo, F. Fantini, D. Ogliari, L. Umena and E. Zanoni, "Gate metallization sinking into the active channel in Ti/W/Au metallized power MESFETs," *IEEE Elect. Dev. Lett.*, **7**, pp. 185-187 (1986).
13. J. Wurfl and H. L. Hartnagel, "Thermal stability and degradation mechanisms of Ti-Pt-Au, Ti-W-Au and Wsi_2-Ti-Pt-Au schottky contacts on GaAs," *Proc. 4th Int. Conf. on Quality in Electronic Components*, Bordeaux, pp. 133-138 (1989).
14. J. F. Bresse, "Reliability of the structure Au/Cr/Au-Ge/Ni/GaAs in low noise dual gate GaAs FET," *Microelect. Reliability*, **25**, pp. 411-425 (1985).
15. J. C. Irvin, "Reliability of GaAs FETs," Chapter 6 in *GaAs FET Principles and Technology*, J. DiLorenzo and D. Kahandelwal, eds., Mass.: Artech House, 1985, pp. 348-400.
16. M. Omori, J. Wholey and J. Gibbons, "Accelerated active life test of GaAs FET and new failure modes," *Proc. Int. Reliability Physics Symposium*, pp. 134-139 (1980).
17. A. Christou, "Reliability of high-speed HEMT integrated circuits and multi-2DEG structures," in *Semiconductor Device Reliability*, A. Christou and B. A. Unger, eds., Kluwer Academic, 1990, pp. 454-555.
18. P. C. Chao, M. Kao, K. Nordheden and A. Swanson, "HEMT degradation in hydrogen gas," IEEE Elect. Dev. Lett., **15**(5), 151 (1994).
19. P. H. Ladbrooke and S. R. Blight, "Low field low mobility dispersion of transconductance in GaAs MESFETs with implication for other rate-dependent anomalies," *IEEE Trans. Elect. Dev.*, **35**, pp. 257-67 (1988).
20. Y. A. Tkachenko, Y. Lan, D. Whitefield, C. J. Wei, J. C. M. Huang, L. Aucoin and S. Shanfield, "Hot-electron-induced degradation of metal-semiconductor field effect transistors," *Proc. of IEEE GaAs IC Symp.*, pp. 259-262 (1994).
21. S. Kayali, "Hydrogen effects on GaAs device reliability," *Int. Conf. on GaAs Manufacturing Technology*, San Diego, CA, 1996, pp. 80-82.

22. J. Chevallier, W. Dautremont-Smith, C. Tu and S. Pearton, "Donor neutralization in GaAs(Si) by atomic hydrogen," *Appl. Phys. Lett.*, **47**, p. 108 (1985).
23. A. Jahil, J. Chevallier, J. Pesant and R. Mostefaoui, "Infrared spectroscopic evidence of silicon related hydrogen complexes in hydrogenated n-type GaAs doped with silicon," *Appl. Phys. Lett.*, **50**(8), p. 439 (1987).
24. D. Eng, R. Culbertson and K. MacWilliams, "The effects of hydrogen and deuterium incorporated on the electrical performance of GaAs MESFETs," *Proc. GaAs IC Symposium* (1995).
25. T. Tuhus and A. Bjorneklett, "Thermal cycling reliability of die bonding adhesives," *IEEE Annual Reliability Physics Symposium Digest*, pp. 208 (1993).
26. M. Ettenberg, H. S. Sommers, Jr., H. Kressel and H. F. Lockwood, "Control of facet damage in GaAs laser diodes," *Appl. Phys. Lett.*, **18**, pp. 571-573 (1971).
27. M. P. Panish, "Heterostructure injection lasers," *Proc. of the IEEE*, **64**(10), pp. 1512-1540 (1976).
28. S. L. Yellen, et al., "Reliability of GaAs-based semiconductor diode lasers: 0.6 – 1.1 μm," *IEEE J. Quant. Elect.*, **29**(6), 2058 (1993).
29. J. J. Coleman, "Strained-layer InGaAs quantum-well heterostructure lasers," *IEEE J. on Selected Topics in Quant. Elect.*, **6**(6), pp. 1008-1013 (2000).
30. M. Fukuda, *Reliability and Degradation of Semiconductor Lasers and LEDs*, Artech House: Boston, London, 1991.
31. A. H. Johnston and T. F. Miyahira, "Characterization of proton damage in LEDs," *IEEE Trans. Nucl. Sci.*, **47**(6), pp. 2500-2507 (2000).
32. A. Shataov, S. Subramanian, S. Chandrasekhar, A. Denati and S. M. Goodnick, "Electron irradiation effects in polyimide passivated InP/InGaAs single heterojunction bipolar transistors," *IEEE Trans. Nucl. Sci.*, **46**(6), 1707 (1999).
33. W. T. Anderson, et al., "GaAs MMIC technology radiation effects," *IEEE Trans. Nucl. Sci.*, **32**(6), 4040 (1985).
34. S. Zhang, et al., "The effects of proton irradiation on the RF performance of SiGe HBTs," *IEEE Trans. Nucl. Sci.*, **46**(6), 1716 (1999).
35. I. Ladany, "Electroluminescence characteristics and efficiency of GaAs:Si diodes," *J. Appl. Phys.*, **42**(2), pp. 654-656 (1974).
36. A. H. Johnston and T. F. Miyahira, "Characterization of proton damage in light-emitting diodes," *IEEE Trans. Nucl. Sci.*, **47**(6), pp. 2500-2507 (2000).
37. B. G. Rax, C. I. Lee, A. H. Johnston and C. E. Barnes, "Total dose and proton damage in optocouplers," *IEEE Trans. Nucl. Sci.*, **43**(6), pp. 3167-3174 (1996).
38. S. C. Lee, Y. F. Zao, R D. Schrimpf, M. A. Neifield and K. F. Galloway, "Comparison of lifetime and threshold current damage factors for multi-quantum-well GaAs/AlGaAs laser diodes," *IEEE Trans. Nucl. Sci.*, **46**(6), pp. 1797-1804 (1999).
39. A. H. Johnston T. F. Miyahira and B. G. Rax, "Proton damage in advanced laser diodes," *IEEE Trans. Nucl. Sci.*, **48**(6), pp. 1764-1772 (2001).
40. B. Hughlock, A. Johnston, T. Williams and J. Harrang, "A comparison of charge collection between GaAs MESFETs and III-V HFETs," *IEEE Trans. Nucl. Sci.*, **39**(6), 1642 (1992).
41. D. McMorrow, J. B. Boos, A. R. Knudsen, S. Buchner, M.-J. Yang, B. Bennett and J. S. Melinger, "Charge-collection characteristics of low-power ultrahigh speed, metamorphic AlSb/InAs high-electron mobility transistors," *IEEE Trans. Nucl. Sci.*, **47**(6), pp. 2662-2668 (2001).
42. M. A. Carts, et al., "Single-event test methodology and test results of commercial gigabit per second fiber channel hardware," *IEEE Trans. Nucl. Sci.*, **44**(6), pp. 1878-1884 (1997).